AF775422

MODELING AND SIMULATION OF CATALYTIC REACTORS FOR PETROLEUM REFINING

MODELING AND SIMULATION OF CATALYTIC REACTORS FOR PETROLEUM REFINING

JORGE ANCHEYTA

WILEY

A JOHN WILEY & SONS, INC., PUBLICATION

Copyright © 2011 by John Wiley & Sons, Inc. All rights reserved

Published by John Wiley & Sons, Inc., Hoboken, New Jersey
Published simultaneously in Canada

No part of this publication may be reproduced, stored in a retrieval system, or transmitted in any form or by any means, electronic, mechanical, photocopying, recording, scanning, or otherwise, except as permitted under Section 107 or 108 of the 1976 United States Copyright Act, without either the prior written permission of the Publisher, or authorization through payment of the appropriate per-copy fee to the Copyright Clearance Center, Inc., 222 Rosewood Drive, Danvers, MA 01923, (978) 750-8400, fax (978) 750-4470, or on the web at www.copyright.com. Requests to the Publisher for permission should be addressed to the Permissions Department, John Wiley & Sons, Inc., 111 River Street, Hoboken, NJ 07030, (201) 748-6011, fax (201) 748-6008, or online at http://www.wiley.com/go/permission.

Limit of Liability/Disclaimer of Warranty: While the publisher and author have used their best efforts in preparing this book, they make no representations or warranties with respect to the accuracy or completeness of the contents of this book and specifically disclaim any implied warranties of merchantability or fitness for a particular purpose. No warranty may be created or extended by sales representatives or written sales materials. The advice and strategies contained herein may not be suitable for your situation. You should consult with a professional where appropriate. Neither the publisher nor author shall be liable for any loss of profit or any other commercial damages, including but not limited to special, incidental, consequential, or other damages.

For general information on our other products and services or for technical support, please contact our Customer Care Department within the United States at (800) 762-2974, outside the United States at (317) 572-3993 or fax (317) 572-4002.

Wiley also publishes its books in a variety of electronic formats. Some content that appears in print may not be available in electronic formats. For more information about Wiley products, visit our web site at www.wiley.com.

Library of Congress Cataloging-in-Publication Data:

Ancheyta, Jorge.
Modeling and simulation of catalytic reactors for petroleum refining / Jorge Ancheyta.
p. cm.
Includes bibliographical references and index.
ISBN 978-0-470-18530-8 (cloth)
1. Catalytic reforming–Simulation methods. I. Title.
TP690.45.A534 2011
665.5′3–dc22

2010030993

oBook ISBN: 9780470933565
ePDF ISBN: 9780470933558
ePub ISBN: 9781118002162

MIX
Paper from responsible sources
FSC® C013604

CONTENTS

PREFACE

The reactor is the heart of a chemical process, and a thorough understanding of the phenomena occurring during the transformation of reactants into the desired products is of vital importance for the development and optimization of the process. Particularly in the petroleum refining industry, in which apart from the reactors, other operations (separations, heating, cooling, pumping, etc.) are carried out in series or in parallel and each plant is connected with others, improper design and operation of reactors can cause shutdown of a plant or, even worse, of the entire refinery, with the consequent loss in production and income. It is thus essential to have a thorough knowledge of the fundamental equations critical to chemical reactor design, such as reactor sizing and optimal operating conditions.

The reactors used during petroleum refining are among the most complex and difficult to model and design. The composition and properties of the various petroleum fractions that are converted in reactors is such that the reaction system can involve various phases, catalysts, reactor configuration, continuous catalyst addition, and so on, making the development of a model a challenging task. In addition, the presence of hundreds of components undergoing different reaction pathways and competing for the active sites of catalysts, contributes to increasing the complexity of the formulation of the kinetics and reactor models.

Over the years, many excellent textbooks have been published dealing with various aspects of reactors: chemical reactor design, modeling of chemical reaction kinetics, reaction mechanisms, chemical reaction engineering, scale-up, and so on. The level of sophistication in each book varies from academic reactions (e.g., $A \rightarrow B$), represented by simple kinetic models (e.g., the power-law model, $-r_A = kC_A^n$) and using integrated equations for the design of ideal reactors (e.g., PFR, CSTR), to complex catalytic reaction systems employing a set of differential equations to solve for mass and energy balances. However, detailed descriptions of the various reactor models, reaction kinetics, and real examples of the application of these models for the simulation of experimental reaction units and commercial plants have not previously been treated in detail. Moreover, most books do not discuss the modeling of the reactors that are typically used during the conversion of oil distillates in the petroleum refining industry, and do not describe reactor models in an uncluttered or thorough manner.

Modeling and Simulation of Catalytic Reactors for Petroleum Refining is designed to give an up-to-date treatment of all the important aspects of reactor modeling, with particular emphasis on reactors employed in the petroleum refining industry. We explain and analyze approaches to modeling catalytic reactors for steady-state and dynamic simulations and discuss such aspects as thermodynamics, reaction kinetics, process variables, process schemes, and reactor design. To validate the models developed, experimental data obtained directly from laboratory and commercial plants are used. Our goal is that the book will become an essential reference for chemical and process engineers, computational chemists and modelers, catalysis researchers, and professionals in the petroleum industry, as well for use as a textbook either for full courses in chemical reaction engineering or as a supplement to related courses.

The book is organized in five chapters, each with individual reference and nomenclature sections. About 500 references are cited and discussed, covering most of the published literature regarding the modeling of reactors used in the petroleum refinery industry. Chapter 1 provides an in-depth introduction to topics related to petroleum refining, such as petroleum properties, separation processes, upgrading of distillates, and upgrading of heavy feeds. A brief description of all the conversion and separation processes is given in this chapter. Detailed experimental data on light, medium, and heavy crude oil assays are also provided.

General aspects of reactor modeling in the petroleum refining industry are treated in Chapter 2. The emphasis is on reactors, deviations from ideal flow patterns, kinetic modeling approaches, estimation of model parameters, and classification and description of reactor models. The fundamental equations are given for each reactor model, together with their advantages and disadvantages. A generalized reactor model is proposed from which each previously reported reactor model can easily be derived.

Chapter 3 is devoted to the modeling of catalytic hydrotreating reactors. The most important features of this type of reactor are highlighted in the first sections, such as the characteristics and classification of hydrotreating reactors, process variables, other process aspects (quench systems, reactor internals), and fundamentals of hydrotreating (chemistry, thermodynamics, kinetics, and catalysts). The final section covers hydrotreating reactor modeling, with examples of the modeling and simulation of reactors operating with catalysts of different particle shapes, steady-state operation, hydrotreating reactors with quenching, dynamic simulation, and co-current and countercurrent operations for both laboratory and commercial reactors.

The modeling of catalytic reforming reactors is the subject of Chapter 4. The description and types of processes, process variables, and fundamentals of catalytic reforming are described at the beginning of the chapter, followed by a section on reactor modeling in which the development of a kinetic reforming model is reported. Validation of the model developed, with bench-scale isothermal reactor experiments and simulation of commercial semiregenerative reforming reactors, is discussed. The effect of benzene precursors in the feed in both laboratory and commercial reactors is also simulated, and use of the reactor model to predict other process parameters is highlighted.

In Chapter 5, Dr. Maya-Yescas describes the modeling and simulation of the fluid catalytic cracking reactor. Descriptions of the process, reaction mechanism, transport phenomena, thermodynamics, and kinetics are provided in the initial sections. Simulations used to estimate kinetic parameters from laboratory and commercial reactors, to determine the controlling reaction steps, of steady-state operation, of scale-up kinetic factors, of the regenerator reactor, of burning nonheterogeneous coke, of side reactions during the burning of heterogeneous coke, and of the energy balance in the regenerator are discussed in detail. Other sections deal with modeling a catalyst stripper, simulation of the controlled unit, pilot-plant emulation, and industrial plant emulation.

Detailed experimental data and comparisons with reactor model predictions are provided in each chapter. Also, all data and parameters required to build up each reactor and kinetic model are detailed, so that readers can adapt their own computer programs for use in reactor simulation, optimization, and design purposes.

It is our intention that *Modeling and Simulation of Catalytic Reactors for Petroleum Refining* will quickly become a leading book in this field through its emphasis on detailed descriptions of catalytic reactor modeling used in the petroleum refining industry, its use of laboratory and commercial data for model validations, the details provided of results of simulations in steady-state and dynamic operations, and in general its focus on more practical issues regarding reactor modeling than have been available in previous textbooks on chemical reactor engineering.

ACKNOWLEDGMENTS

I would like especially to acknowledge Dr. Rafael Maya-Yescas, Professor of Chemical Reaction Engineering. Universidad Michoacana de Nicolás de Hidalgo, Morelia, Michoacán, México, who kindly agreed to write Chapter 5. I also thank all the M.Sc., Ph.D., and postdoctoral students who over a period of many years have contributed enormously to the preparation of this book.

Jorge Ancheyta

ABOUT THE AUTHOR

Jorge Ancheyta, holds a bachelor's degree in petrochemical engineering (1989), a master's degree in chemical engineering (1993), and a master's degree in administration, planning, and economics of hydrocarbons (1997) from the National Polytechnic Institute of Mexico. He split his Ph.D. between the Metropolitan Autonomous University of Mexico and the Imperial College London (1998), and was awarded a postdoctoral fellowship in the Laboratory of Catalytic Process Engineering of the CPE-CNRS in Lyon, France (1999). He has also been a visiting professor at the Laboratoire de Catalyse et Spectrochimie, Université de Caen, France (2008, 2009, 2010), and Imperial College London (2009).

Dr. Ancheyta has worked for the Mexican Institute of Petroleum (IMP) since 1989, where his present position is project leader of research and development. He has also worked as a professor on the undergraduate and postgraduate levels at the School of Chemical Engineering and Extractive Industries at the National Polytechnic Institute of Mexico since 1992 and for the IMP postgraduate program since 2003. He has supervised about 100 B.Sc., M.Sc., and Ph.D. theses as well as a number of postdoctoral and sabbatical-year professors.

Dr. Ancheyta has worked on the development and application of petroleum refining catalysts, kinetic and reactor models, and process technologies, primarily in catalytic cracking, catalytic reforming, middle distillate hydrotreating, and heavy oils upgrading. He is the author or co-author of a number of patents, books, and about 200 scientific papers, and has been awarded the highest distinction (level III) as a national researcher by the Mexican government and is a member of the Mexican Academy of Science. He has also been guest editor of various international journals: *Catalysis Today*, *Petroleum Science and Technology*, *Industrial Engineering Chemistry Research*, *Energy and Fuels*, *Chemical Engineering Communications*, and *Fuel*. Dr. Ancheyta has also chaired numerous international conferences and is a member of the scientific boards of various prestigious journals.

1

PETROLEUM REFINING

1.1 PROPERTIES OF PETROLEUM

Petroleum is the most important substance consumed in modern society. It provides not only fuel and energy for transportation but is also used in plastics, paint, fertilizer, insecticide, medicine, and elsewhere. The exact composition of petroleum varies widely from source to source, but the percentage of chemical elements changes over fairly narrow limits. Hydrogen and carbon are the major components, and sulfur, nitrogen, oxygen, and metals are present in relatively lower quantities (Table 1.1). Usually, petroleum or crude oil comes from deep underground, where the vestiges of plants and animals from millions of years ago have been heated and pressurized over time. It is blackish in color and has a characteristic odor that comes from the presence of small amounts of chemical compounds containing sulfur, nitrogen, and metals.

The change in crude oil quality around the world (e.g., heavy petroleum production has been increased in recent years) has obliged crude oil refiners to reconfigure current refineries and to design new refineries specifically to process heavier feedstocks (i.e., blends of various crude oils with elevated amount of heavy petroleum). These new feeds are characterized by high amounts of impurities (sulfur, metals, nitrogen, asphaltenes) and low distillate yields, which make them more difficult than light crude oils to process.

Comparisons of some properties of various crude oils are presented in Tables 1.2 and 1.3. It is clear that light and heavy crude oils have remarkable

Modeling and Simulation of Catalytic Reactors for Petroleum Refining, First Edition.
Jorge Ancheyta.
© 2011 John Wiley & Sons, Inc. Published 2011 by John Wiley & Sons, Inc.

TABLE 1.1. Typical Elemental Composition of Petroleum

Element	Weight Percentage
C	84–87
H	11–14
O	0.1–0.5
N	0.1–2
S	0.5–6
Metals	0–0.1

TABLE 1.2. Range of Properties of Various Types of Petroleum

	Extra-light Crude Oil	Light Crude Oil	Heavy Crude Oil	Extra-Heavy Crude Oil
API gravity	>50	22–32	10–22	<10
Hydrocarbons (wt%)				
Asphaltenes	0–<2	<0.1–12	11–25	15–40
Resins	0.05–3	3–22	14–39	
Oils	—	67–97	24–64	
Impurities (wt%)				
Total sulfur	0.02–0.2	0.05–4.0	0.1–5.0	0.8–6.0
Total nitrogen	0.0–0.01	0.02–0.5	0.2–0.8	0.1–1.3
Ni + V (wppm)	<10	10–200	50–500	200–600

TABLE 1.3. Properties of Various Crude Oils

Crude Oil	Lagrave	Isthmus	Maya	Lloyminster	Athabasca
Country	France	Mexico	Mexico	Canada	Canada
API gravity	43	33.34	21.31	15.0	8.0
Sulfur (wt%)	—	1.46	3.57	—	1.25
Nitrogen (wt%)	—	0.1467	0.32	4.30	7.95
Insolubles in nC_7 (wt%)	4	1.65	11.32	12.9	15.0

differences. Heavy petroleum is characterized by low API gravity, large amounts of impurities, and low distillates yields; light petroleum is of much better quality. In general, the lower the API gravity (i.e., the heavier the crude oil), the higher the impurities content and the lower the distillates yield. Such properties make processing of heavy petroleum different from that used for light crude oil refining. In other words, a refinery capable of processing light petroleum cannot, without changes in some units or even complete reconfiguration, be employed to process 100% heavy petroleum.

TABLE 1.4. SARA Analysis and Physical Properties of Petroleum

		Physical Properties		
		Non-polar	Low density	Low aromaticity
Saturates	Maltenes	↓	↓	↓
Aromatics	Maltenes	↓	↓	↓
Resins	Maltenes	↓	↓	↓
Asphaltenes		↓	↓	↓
		Most polar	High density	High aromaticity

In general, light crude oil is rich in light distillates, and heavy crude oil, in residuum. However, the petroleum composition may vary with its API gravity and origin. Physical properties and exact chemical composition of crude oil also vary from one source to another. As a guide to chemical composition, Table 1.4 provides qualitative data on saturate, aromatic, resin and asphaltene (SARA) contents in the heavy fractions present in various crude oils. The most complex impurity of petroleum is asphaltene, which consists of condensed polynuclear aromatics containing small amounts of heteroatoms (S, N, O) and traces of nickel and vanadium. Asphaltenes are typically defined as brown and black powdery material produced by the treatment of petroleum, petroleum residua, or bituminous materials with a low-boiling liquid hydrocarbon (e.g., pentane or heptane); and soluble in benzene (and other aromatic solvents), carbon disulfide, and chloroform (or other chlorinated hydrocarbon solvents). Asphaltene molecules are grouped together in systems of up to five or six sheets, which are surrounded by the maltenes (all those structures different from asphaltenes that are soluble in *n*-heptane) and resin.

The properties of petroleum, such as viscosity, density, boiling point, and color, may vary widely, and the ultimate or elemental analysis varies over a narrow range for a large number of samples. Metals have a tendency to concentrate more in the heavier fraction (asphaltene) than in the saturated and aromatic fractions. The higher the asphaltene content in crude oil, the higher the metal content; however, the increase in vanadium concentration is not proportional to that of nickel. Nitrogen and sulfur can be present in traces in light petroleum, but with heavier or extra heavy crude oil, the sulfur and nitrogen contents also increase.

1.2 ASSAY OF CRUDE OILS

It is important to determine the physical and chemical characterizations of crude oil through a crude oil assay, since they are used in different areas in the petroleum refining industry. The most common applications of petroleum assays are:

- To provide extensive detailed experimental data for refiners to establish the compatibility of a crude oil for a particular petroleum refinery
- To anticipate if the crude oil will fulfill the required product yield, quality, and production
- To determine if during refining the crude oil will meet environmental and other standards
- To help refiners to make decisions about changes in plant operation, development of product schedules, and examination of future processing ventures
- To supply engineering companies with detailed crude oil analyses for their process design of petroleum refining plants
- To facilitate companies' crude oil pricing and to negotiate possible penalties due to impurities and other nondesired properties

A *crude oil assay* is a compilation of laboratory (physical and chemical properties) and pilot-plant (distillation and product fractionation) data that characterize a specific crude oil. Assay analyses of whole crude oils are carried out by combining atmospheric and vacuum distillation units, which when combined will provide a true boiling-point (TBP) distillation. These batch distillation methods, although taking between 3 and 5 days, allow the collection of a sufficient amount of distillation fractions for use in further testing. The values of the distillation ranges of the distilled fractions are usually defined

TABLE 1.5. Typical Distillation Range of Fractions in Petroleum Assays

TBP Distillation Range (°C)	Distillate
IBP–71	Light straight-run naphtha
71–177	Medium straight-run naphtha
177–204	Heavy straight-run naphtha
204–274	Jet fuel
274–316	Kerosene
316–343	Straight-run gasoil
343–454	Light vacuum gasoil
454–538	Heavy vacuum gasoil
R 538°C+	Vacuum residue

on the basis of their refinery product classifications. The most common distillation ranges used in international assays of crude oils are reported in Table 1.5.

There are various types of assays, which vary considerably in the amount of experimental information determined. Some include yields and properties of the streams used as feed for catalytic reforming (naphtha) and catalytic cracking (gas oils). Others give additional details for the potential production of lubricant oil and/or asphalt. At a minimum, the assay should contain a distillation curve (typically, TBP distillation) for the crude oil and a specific gravity curve.

The most complete assay includes experimental characterization of the entire crude oil fraction and various boiling-range fractions. Curves of TBP, specific gravity, and sulfur content are normal data contained in a well-produced assay. As an example, assays of various Mexican crude oils are presented in Table 1.6. The API gravity of these crude oils ranges from 10 to 33°API. *API gravity* is a measure of the relative density of a petroleum liquid and the density of water (i.e., how heavy or light a petroleum liquid is compared to water). Although, mathematically, API gravity has no units, it is always referred to as being in "degrees." The correlation between specific gravity (sg) and degrees API is as follows (the specific gravity and the API gravity are both at 60°F):

$$\text{API gravity} = \frac{141.5}{\text{sg}_{60^\circ\text{F}}^{60^\circ\text{F}}} - 131.5 \tag{1.1}$$

Viscosity must be provided at a minimum of three temperatures so that one can calculate the sample viscosity at other temperatures. The most common temperatures used to determine viscosity are 15.5, 21.1, and 25°C. If viscosities of the sample cannot be measured at those temperatures, the sample needs to be heated and higher temperatures are used, such as in the case of the 10 and 13°API crude oils reported in Table 1.6. Once viscosities at three temperatures are available, a plot of a double logarithm ($\log_{10}$) of viscosity against the temperature can be constructed, and viscosities at other temperatures can be obtained easily, as shown in Figure 1.1.

The characterization factor (K_{UOP} or K_{Watson}) of the Mexican crude oils reported in Table 1.6 ranges from 11.5 to 12.0. The K factor is not determined experimentally; rather, it is calculated using the following equation (for petroleum fractions):

$$K = \frac{\sqrt[3]{\text{MeABP}}}{\text{sg}_{60^\circ\text{F}}^{60^\circ\text{F}}} \tag{1.2}$$

where MeABP (in degrees Rankine) is the mean average boiling point of the sample calculated with distillation curve data.

In general, if $K > 12.5$, the sample is predominantly paraffinic in nature, while $K < 10.0$ is indicative of highly aromatic material. The characterization

TABLE 1.6. Assay of Various Mexican Crude Oils

	ASTM Method	Crude Oil				
		10°API	13°API	Maya	Isthmus	Olmeca
Specific gravity, 60°F/60°F	D-1298	1.0008	0.9801	0.9260	0.8584	0.8315
API gravity	D-287	9.89	12.87	21.31	33.34	38.67
Kinematic viscosity (cSt)	D-445					
At 15.5°C		—	—	299.2	16.0	5.4
At 21.1°C		—	—	221.6	12.5	4.6
At 25.0°C		—	19,646	181.4	10.3	4.1
At 37.8°C		—	5,102	—	—	
At 54.4°C		7,081	1,235	—	—	
At 60.0°C		4,426	—	—	—	
At 70.0°C		2,068	—	—	—	
Characterization factor, K_{UOP}	UOP-375	11.50	11.60	11.71	11.95	12.00
Pour point (°C)	D-97	+12	0	—	−33	−39
Ramsbottom carbon (wt%)	D-524	20.67	16.06	10.87	4.02	2.10
Conradson carbon (wt%)	D-189	20.42	17.94	11.42	4.85	2.76
Water and sediments (vol%)	D-4007	1.40	0.10	0.20	<0.05	<0.05
Total sulfur (wt%)	D-4294	5.72	5.35	3.57	1.46	0.99
Salt content (PTB)	D-3230	744.0	17.7	15.0	4.1	3.9
Hydrogen sulfide (mg/kg)	UOP-163	—	—	—	44	59
Mercaptans (mg/kg)	UOP-163	—	—	—	65	75
Total acid number (mg KOH/g)	D-664	0.48	0.34	0.30	0.61	0.46
Total nitrogen (wppm)	D4629	5650	4761	3200	1467	737
Basic nitrogen (wppm)	UOP-313	1275	1779	748	389	150
nC_7 insolubles (wt%)	D-3279	25.06	18.03	11.32	1.65	0.68
Toluene insolubles (wt%)	D-4055	0.41	0.20	0.11	0.09	0.11
Metals (wppm)	Atomic absorption					
Nickel		94.2	83.4	53.4	8.9	1.6
Vanadium		494.0	445.0	298.1	37.1	8.0
Total		588.2	528.4	351.5	46.0	9.6
Chloride content (wppm)	D-808	86	10	4	10	9

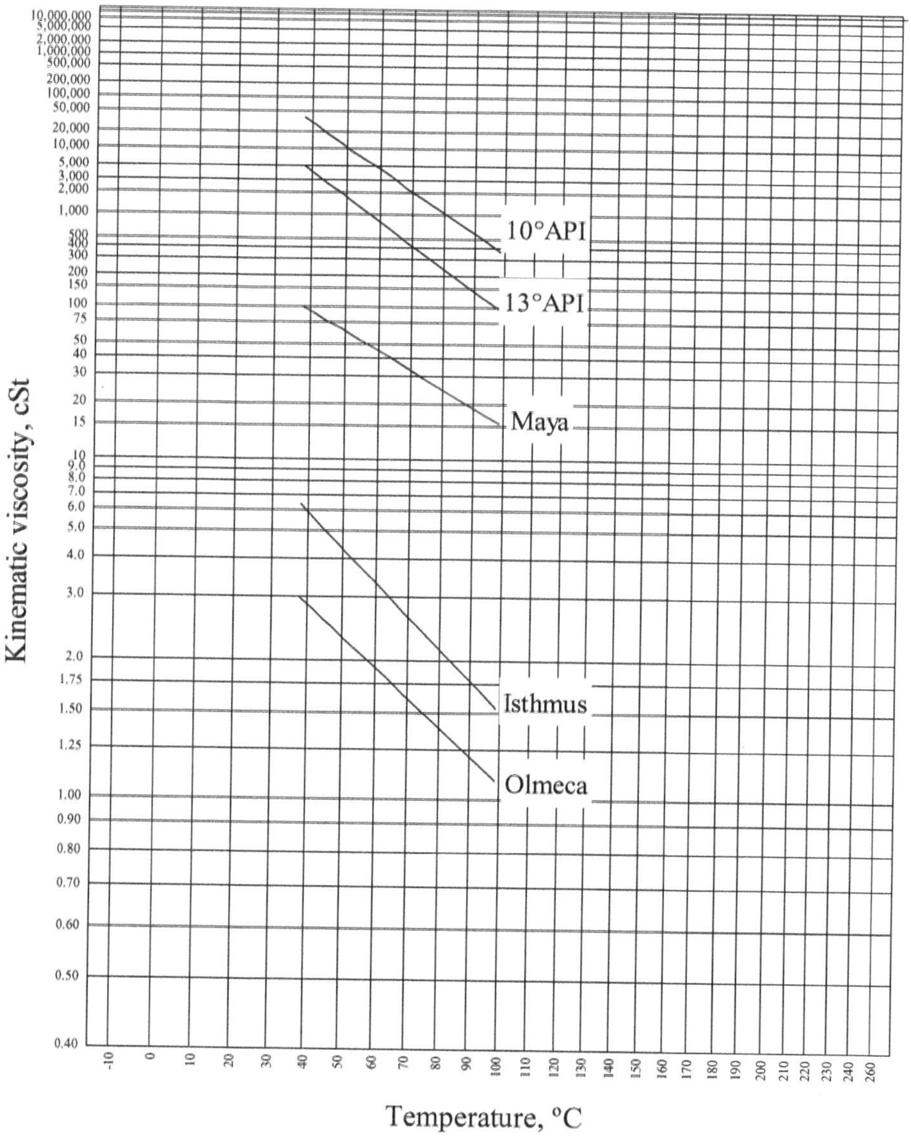

Figure 1.1. Kinematic viscosities of several Mexican crude oils.

factor thus provides a means for roughly identifying the general origin and nature of petroleum solely on the basis of two observable physical parameters, sg and MeABP. More detailed relationships of the *K* factor to the nature of the sample are given in Table 1.7. The characterization factor has also been related to other properties (e.g., viscosity, aniline point, molecular weight, critical temperature, percentage of hydrocarbons), so it can be estimated using a number of petroleum properties.

TABLE 1.7. Relationship of Type of Hydrocarbon to the Characterization Factor

K Factor	Type of Hydrocarbon
12.15–12.90	Paraffinic
11.50–12.10	Naphthenic–paraffinic
11.00–11.45	Naphthenic
10.50–10.90	Aromatic–naphthenic
10.00–10.45	Aromatic

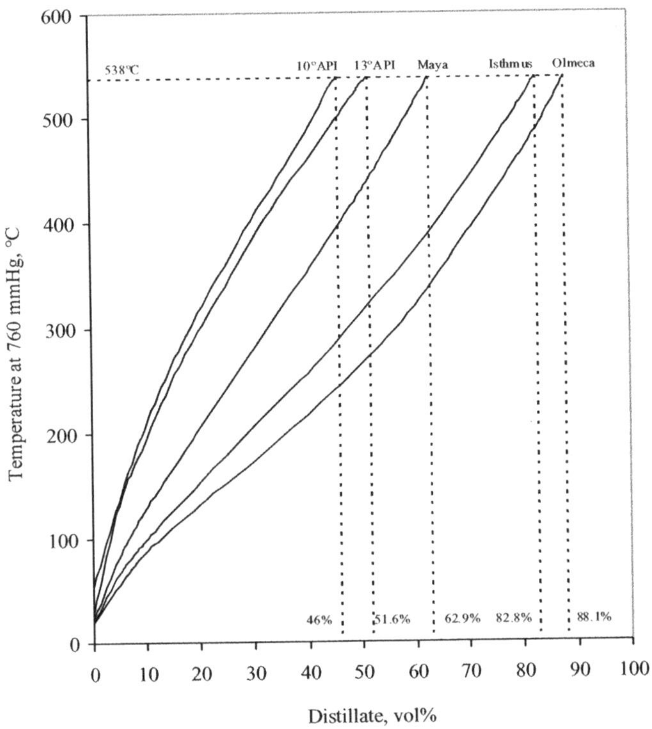

Figure 1.2. True boiling-point curve of various Mexican crude oils.

Asphaltenes, which are generally reported as *n*-heptane insolubles, are, strictly speaking, defined as the weight percentage of *n*-heptane insolubles (HIs) minus the weight percentage of toluene insolubles (TIs) in the sample (wt% of asphaltenes = wt% of HI − wt% of TI). For the crude oils given in Table 1.6, their asphaltene contents are 24.65, 17.83, 11.21, 1.56, and 0.57 wt% for the 10°API, 13°API, Maya, Isthmus, and Olmeca crude oils, respectively.

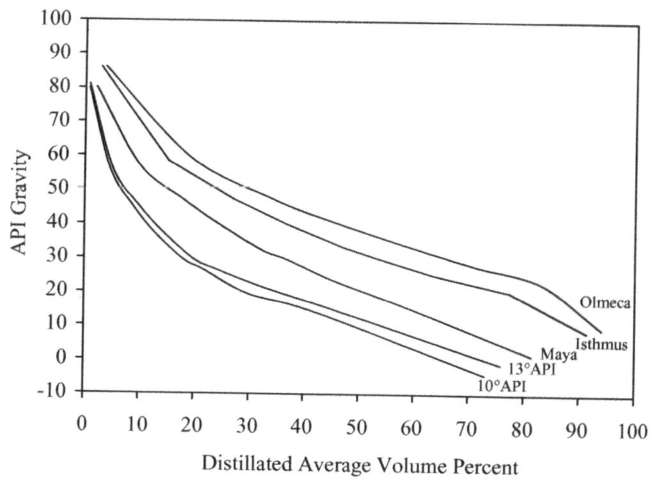

Figure 1.3. API gravity of distillates versus average volume percentage.

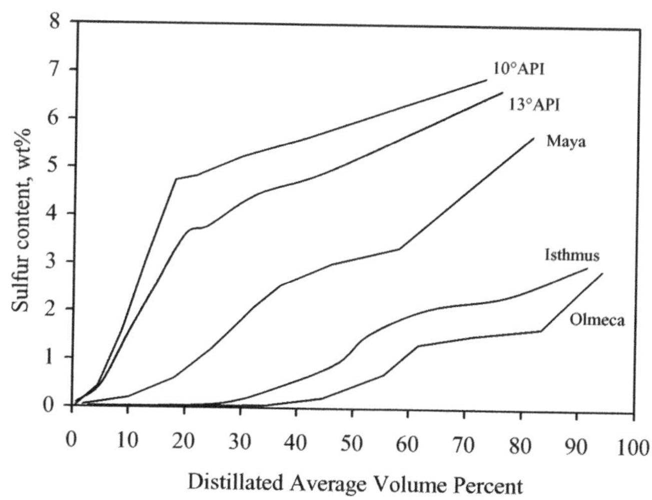

Figure 1.4. Sulfur content of distillates versus average volume percentage.

TBP distillations for Mexican crude oils are presented in Figure 1.2. It is clear that light crude oils that have high API gravity values present also the highest amounts of distillates [e.g., Olmeca crude oil (38.67°API) has 88.1 vol% distillates, whereas the 10°API has only 46 vol% distillates]. Figures 1.3 and 1.4 illustrate plots of API gravity and the sulfur content of distillates against the average volume percentage of distillates of the various crude oils. Distillates of heavier crude oils have lower API gravity and a higher sulfur content than those obtained from light crude oils.

1.3 SEPARATION PROCESSES

1.3.1 Crude Oil Pretreatment: Desalting

Desalting is the first separation process that takes place at the front end of a petroleum refinery (i.e., prior to atmospheric distillation; Figure 1.5). Its primary objective is to prevent corrosion and fouling of downstream lines and equipment by reducing the oil's salt content significantly. Desalting is normally considered a part of the crude distillation unit since heat from some of the streams in the atmospheric distillation is used to heat the crude in the desalting process. The most common salts in crude oil are sodium, calcium and magnesium chlorides ($NaCl \sim 70$ to 80 wt%, $CaCl_2 \sim 10$ wt%, and $MgCl_2 \sim 10$ to 20 wt%), which are in the form of crystals or ionized in the water present in the crude. If salt is not removed, the high temperatures present during crude oil refining could cause water hydrolysis, which in turn allows the formation of hydrochloric acid (HCl), provoking serious corrosion problems in the equipment. Part of the salt that has not been removed can also cause fouling problems in pipes, heat transfer equipment, and furnaces. Deactivation of catalysts (e.g., the zeolite-type catalysts used in fluid catalytic cracking) may be enhanced by the metals in salts, particularly sodium. Typically, the maximum salt content allowed in the feed to crude distillation units is 50 PTB (pounds of salt per thousand barrels of crude oil).

Desalting consists of washing the crude oil with water and caustic (NaOH) so that the salts can be diluted in water and washed from the organic phase.

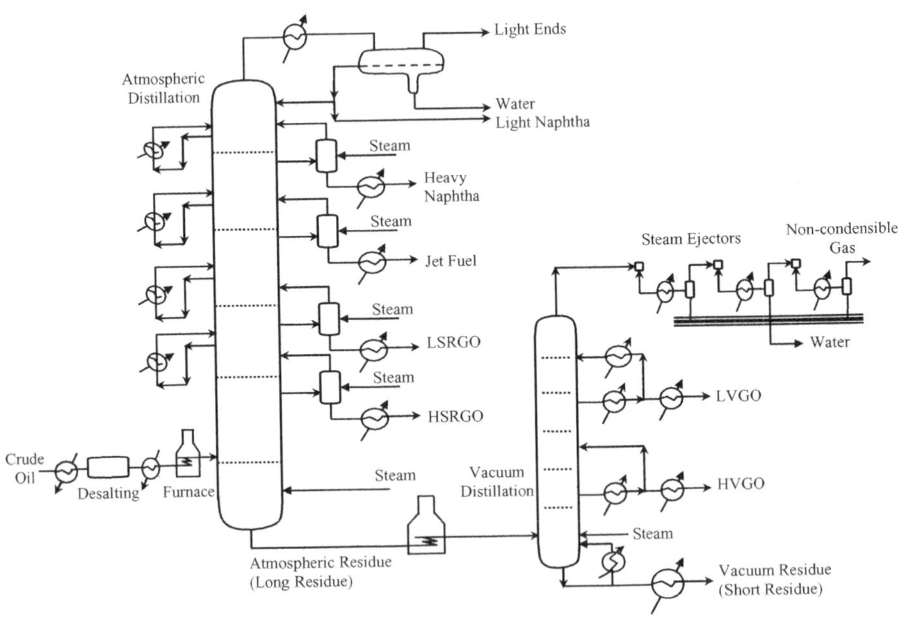

Figure 1.5. Desalting and atmospheric and vacuum distillations of crude oil.

Some of the mixed water forms an emulsion that must be demulsified to separate water from oil. Emulsifiers are present in the form of clay, metallic salts, and asphaltenes, whose contents are higher in heavy crude oils. By this means, dissolved salts are removed and acid chlorides ($MgCl_2$ and $CaCl_2$) are converted to a neutral chloride (NaCl), which prevents the formation of hydrochloric acid when residual chlorides enter the refinery. Some naphthenic acids are also converted to their respective carboxylate salts and removed as part of the aqueous effluent.

The reactions occurring during desalting are

$$MgCl_2(aq) + 2NaOH(aq) \rightarrow Mg(OH)_2(aq) + 2NaCl(aq)$$

$$RCOOH + NaOH(aq) \rightarrow RCOONa(aq) + H_2O$$

The carboxylate salts produced during the conversion of naphthenic acids are surface active and can form stable solutions. This process is controlled by coalescing and decanting the suspended water droplets, which possess an electric charge, under the influence of an electric field (~700 to 1000 V/cm). This electric field destabilizes the electric array in the droplets.

Desalting can be carried out in a single stage (dehydration efficiency of ~95%) or in two stages (dehydration efficiency of ~99%). The dehydration efficiency can be compared with the desalting efficiency, as most of the salt passes from the organic phase into the water phase if mixing is good. The decision as to whether to use a single or a double stage depends on the requirements of the refinery. Typical desalters have two electrodes which generate an electric field within the emulsion, causing the droplets to vibrate, migrate, and collide with each other and coalesce. Voltage (16,000 to 30,000 V ac) is what makes coalescence possible, so that the larger drops settle under the effect of gravity. Electric current does not participate in this process.

The principal steps during desalting are:

- Preheating of water and oil and mixing in a 1:20 ratio.
- Addition of a demulsifier substance (~0.005 to 0.01 lb/bbl).
- Mixing in a valve (5 to 20 psi pressure drop). The better the mixing, the higher the salt removal, so that the salt content in oil is washed with the water and a water–oil emulsion is formed.
- Entrance of the emulsion into the desalter, where an intense electric field is present. The desalter operates at temperatures between 95 and 150°C. The oil leaves the desalter.

Apart from removing salt, electrostatic desalting also eliminates water and suspended solids in crude oil. Water removal is important to reduce pumping costs and to avoid vaporization when the water is passing through the preheater train (i.e., the water heat of vaporization reduces the crude preheater capacity). Otherwise, due to the high pressure, it causes disturbances and vibrations and eventually plant shutdown. Elimination of suspended solids is necessary to avoid their going all the way through the plant to be expelled with the flue gas. This causes flue gas opacity that does not meet environmental requirements, resulting in mandatory additional treatment prior to being expelled.

1.3.2 Atmospheric Distillation

The main separation step in any crude oil refinery is atmospheric or primary distillation. Atmospheric distillation fractionates the crude oil into various distillates, fractions, or cuts of hydrocarbon compounds based on molecular size and boiling-point range [e.g., light ends, propane, butanes, straight-run naphthas (light and heavy), kerosene, straight-run gas oils (light and heavy), and atmospheric residue] (Figure 1.5). The term *atmospheric distillation* is used because the unit operates slightly above atmospheric pressure. Separation is carried out in a large tower, which contains a number of trays where hydrocarbon gases and liquids interact. The heated desalted crude enters the fractionation tower in a lower section called the *flash zone*. The unvaporized portion of the crude oil leaves the bottom of the tower via a steam stripper section, while the distillate vapors move up the tower countercurrent to a cooler liquid reflux stream. The cooling and condensing of the distillation tower overhead is provided partially by exchanging heat with the incoming crude oil and partially by either an air- or a water-cooled condenser. Additional heat is removed from the distillation column by a pump-around system, which is simply an internal condenser that ensures a continued reflux stream flow. The overhead distillate fraction from the distillation column is naphtha, which is allowed to leave the top of the tower to be condensed and collected in the overhead drum. A portion of this stream is returned as reflux, while the rest is delivered to the light-end processes for stabilizing and further distillation. The other fractions removed from the side of the distillation column [i.e., from selected trays (draw-off trays)] at various points between the column top and bottom are jet fuel, kerosene, light gas oil, and heavy gas oil, which are steam stripped, cooled by exchanging heat with the incoming crude oil, and sent to other treatment areas and/or to storage. The heavier material (i.e., atmospheric residue oil) is withdrawn from the bottom of the tower.

Each stream is converted further by changing the size and structure of the molecules through cracking, reforming, and other conversion processes. The converted products are then subjected to various treatment and separation processes to remove undesirable constituents or impurities (e.g., sulfur, nitrogen) and to improve product quality (e.g., octane number, cetane number). Atmospheric distillation is a crucial step, since it routes the molecules to the appropriate conversion units in the refinery. The cut point of the atmospheric residue depends on the prevailing fuel specifications and crude slate used. The atmospheric residue leaves the bottom of the unit and is processed further in the vacuum distillation unit.

It is important not to subject crude oil to temperatures above 370 to 380°C because the high-molecular-weight components will undergo thermal cracking and form coke. The coke, by operating the distillation units at a high temperature, would result in plugging the tubes in the furnace that heats the crude oil fed to the distillation column. Plugging would also occur in the piping from the furnace to the distillation column as well as in the column itself.

1.3.3 Vacuum Distillation

The main objective of a *vacuum* or *secondary distillation unit* is to recover additional distillates from atmospheric residue (long residue). The atmospheric residue is distilled to provide the heavy distillate streams used to produce lube oil or as feed to conversion units. The primary advantage of vacuum distillation is that it allows for distilling heavier materials at lower temperatures than those that would be required at atmospheric pressure, thus avoiding thermal cracking of the components. Vacuum distillation is often integrated with the atmospheric distillation as far as heat transfer is concerned. This unit's integration is called *combined distillation*. Generally, the atmospheric residue is received hot from the atmospheric distillation and is sent to the fired heater of the vacuum unit. The vacuum distillation unit is operated at a slight vacuum, which is most often achieved by using multiple stages of steam jet ejectors (absolute pressures as low as 10 to 40 mmHg). This allows the hydrocarbons to be separated at lower temperatures and prevents undesirable chemical reactions.

Atmospheric residue is separated into light vacuum gas oil, heavy vacuum gas oil, and vacuum residue (Figure 1.5). The vacuum gas oils are sent to the catalytic cracking unit for further processing, while the vacuum residue (short residue) can be used as feedstock for further upgrading (i.e., coking, hydrocracking, etc.) or as a fuel component.

Vacuum distillation follows very much the same pattern as that of atmospheric distillation. One difference is that neither the vacuum residue that leaves the bottom of the tower nor the sidestreams are steam stripped. The technology of vacuum distillation has developed considerably in recent decades. The main objectives have been to maximize the recovery of valuable distillates and to reduce the energy consumption of the units. The vacuum distillation column internals must provide good vapor–liquid contact while maintaining a very low pressure increase from the top of the column to the bottom. Therefore, the vacuum column uses distillation trays only where withdrawing products from the side of the column. Most of the column uses packing material for the vapor–liquid contact because such a packing has a lower pressure drop than that of distillation trays. This packing material can be either structured sheet metal or randomly dumped packing such as Raschig rings.

1.3.4 Solvent Extraction and Dewaxing

Since distillation separates petroleum products into groups only by their boiling-point ranges, impurities such as sulfur and nitrogen may remain. Solvent refining processes, including solvent extraction and solvent dewaxing, usually remove these undesirables at intermediate refining stages or just before sending the product to storage.

Solvent extraction processes are employed primarily for the removal by dissolution or precipitation of constituents that would have an adverse effect on the performance of the product in use. An important application is the removal of heavy aromatic compounds from lubricating oils. Removal improves the viscosity–temperature relationship of the product, extending the temperature range over which satisfactory lubrication is obtained. The usual solvents for the extraction of lubricating oil are phenol, furfural, and cresylic acid. Solvents used less frequently are liquid sulfur dioxide, nitrobenzene, and 2,2′-dichloroethyl ether.

Solvent dewaxing is used to remove wax from either distillate or residua at any stage in the refining process. The general steps of solvent dewaxing processes are (1) mixing the feedstock with a solvent, (2) precipitating the wax from the mixture by chilling, and (3) recovering the solvent from the wax and dewaxed oil for recycling by distillation and steam stripping. Usually, two solvents are used: toluene to dissolve the oil and maintain fluidity at low temperatures, and methyl ethyl ketone (MEK) to dissolve a little wax at low temperatures and act as a wax-precipitating agent. Other solvents that are sometimes used are benzene, methyl isobutyl ketone, propane, petroleum naphtha, ethylene dichloride, methylene chloride, and sulfur dioxide. In addition, a catalytic process is used as an alternative to solvent dewaxing.

1.3.5 Deasphalting

The separation of vacuum residue into fractions by distillation without decomposition is not practiced commercially since it is very difficult and expensive. *Solvent deasphalting* (SDA), a nondestructive liquid–liquid extraction process, is preferred to achieve this goal, whereby the last of the molecules that can be refined to valuable products are extracted from the vacuum residue. SDA is a molecular-weight-based separation process member of the family of carbon rejection technologies, which has been used for more than 50 years to separate heavy fractions of crude oil beyond the range of economical commercial distillation. Use of SDA has been reported for production of lube oil feedstocks from vacuum residue using propane as a solvent, for preparation of feedstocks for catalytic cracking, hydrocracking, and hydrodesulfurization units, as well as for the production of specialty asphalts. In most of these conversion units the performance of the catalyst is greatly affected by the presence of heavy metals and the high Conradson carbon content of the residue feed, which are concentrated in the asphaltene molecules, so that removing asphaltenes also eliminates these impurities.

Deasphalting is an extraction process that separates the residue into several fractions on the basis of relative solubility in a solvent (normally, a light hydrocarbon such as propane, butane, pentane, or hexane). The yield of deasphalted oil increased with increases in the molecular weight of the solvent, but its quality decreases. SDA produces a low-contaminant deasphalted oil (DAO) rich in paraffinic-type molecules and a pitch product rich in aromatic compounds and asphaltenes containing, of course, the majority of the feed impurities. The DAO produced has a lower carbon residue and metals content than that of the untreated oil, but SDA is not as effective in lowering the sulfur or nitrogen content in DAO.

1.3.6 Other Separation Processes

Gas and Liquid Sweetening *Gas sweetening* is a process used to remove hydrogen sulfide and carbon dioxide (acid gases) from refinery gas streams. The acid gases are highly concentrated in H_2S, which comes mainly from hydrotreating processes within the refinery. Acid gases are required to be removed:

- For environmental reasons. If H_2S and CO_2 are not removed, they combine with the atmosphere to form very dilute sulfuric acid, and carbonic acid, respectively, which are considered injurious to personal health.
- To purify gas streams for further use in a process. Acid gases cause excessive corrosion to metals.

Gas sweetening is commonly carried out using an amine gas-treating process which uses aqueous solutions of various alkanolamines: MEA, monoethanolamine; DEA, diethanolamine; MDEA, methyldiethanolamine; DIPA, diisopropylamine; DGA, aminoethoxyethanol or diglycolamine—MEA, DEA, and MDEA being the most commonly used amines. Among them, MEA has become the preferred amine commercially, due to its high acid gas absorbency. Apart from amine gas treating, hot potassium carbonate (Benfield) is another process that can be used for acid gas sweetening. There are also other alternatives, based on physical solvent processes (e.g., Sulfinol, Selexol, Propylene Carbonate, Rectisol) and dry adsorbent processes (e.g., molecular sieve, activated charcoal, iron sponge, zinc oxide).

A typical amine gas-treating process consists of the following steps:

- Passing the acid gas stream through an absorber unit (contactor), in which the downflowing amine solution absorbs H_2S and CO_2 from the upflowing gas to produce an H_2S-free gas called *sweetened gas* and an amine solution rich in absorbed acid gases.
- Sending the rich amine to a regenerator, which consists of a stripper with a reboiler, to produce regenerated or lean amine.
- Cooling and recycling the regenerated amine for reuse in the absorber.
- Sending the H_2S-rich stripped gas stream to a Claus process to convert it into elemental sulfur, which is produced by burning H_2S with a controlled airstream. This gas stream can also be sent to a WSA process to recover sulfur as concentrated sulfuric acid.
- Washing the sweetened gas with water to remove any entrained amine before leaving the top of the contactor.

In the case of *liquid sweetening*, there are different treating processes, aiming at the elimination of unwanted sulfur compounds (hydrogen sulfide, thiophene, and mercaptans). The crude oil liquid fractions that require sweetening either at an intermediate stage in the refining process or just before sending them to storage are gasoline, jet fuel, and sometimes kerosene, to improve color, odor, and oxidation stability. Acids, solvents, alkalis, and oxidizing and adsorption agents are the most common materials used for liquid sweetening. Selection of the treatment method depends on:

- The properties of the liquid distillate and the origin of the crude
- The amounts and types of impurities in the liquid distillate
- The degree of impurities removal achieved by the treating method
- The specification of the final product

LPG, naphthas, jet fuel, and kerosene have a sulfur content, predominately in the form of mercaptans, that can be removed by converting them to liquid hydrocarbon disulfides. The most common process used to achieve this target is Merox (mercaptan oxidation), licensed by the UOP. This process requires an alkaline environment provided by either a strong base (commonly aqueous solution of sodium hydroxide) or a weak base (ammonia). Although the Merox process is more economical than catalytic hydrodesulfurization, some refiners still select it to remove sulfur compounds from debutanized naphtha.

Sour Water Treatment In general, the term *sour water* is applied to any water that contains hydrogen sulfide, although it may also contains ammonia, phenol, and cyanide. It is also important to eliminate selenium since it causes mutagenic effects in wildlife. Prior to disposal, sour water must be treated to remove these contaminants. The various sources of sour water in a refinery are:

- Effluent water from the crude unit overhead condenser
- Water phase from the desalter
- Condensed water from the vacuum unit's hot well
- Water condensate from the hydrotreater product steam strippers

Sour water is typically treated by a stripping unit with steam by means of which H_2S and NH_3 are released at the top of the stripping tower. The H_2S-free water is treated in a biological wastewater treatment plant where the remaining ammonia is nitrified and then denitrified. Due to the physics and chemistry of H_2S treatment systems, removal amounts of ammonia, selenium, phenol, salts, and other constituents are lower than that of hydrogen sulfide. In a typical stripping unit, the sour water is fed on to the top tray of the tower while steam is introduced below the bottom tray, which lends itself to tray-by-tray mass and heat transfer. The sour water stripping unit is almost always located in the process area of the refinery and can be a single tower with no reflux or a single trayed tower with an overhead reflux stream.

Other processes for treatment of sour water are: caustic/acid neutralization, caustic oxidization, and oil removal by settling.

1.4 UPGRADING OF DISTILLATES

The main objective of a petroleum refinery is the production of fuels (e.g., gasoline, diesel). Straight-run distillates cannot be used directly as fuels since they possess high amounts of impurities and octane and cetane numbers that are not appropriate for gasoline and diesel engines. These straight-run distillates need treatment to make them suitable for fuel production, which is carried out in various refining processes, as illustrated in Figure 1.6. A brief description of the fundamentals of the various processes used for fuels production is presented in this section. More details on the most important refining processes are given in subsequent chapters.

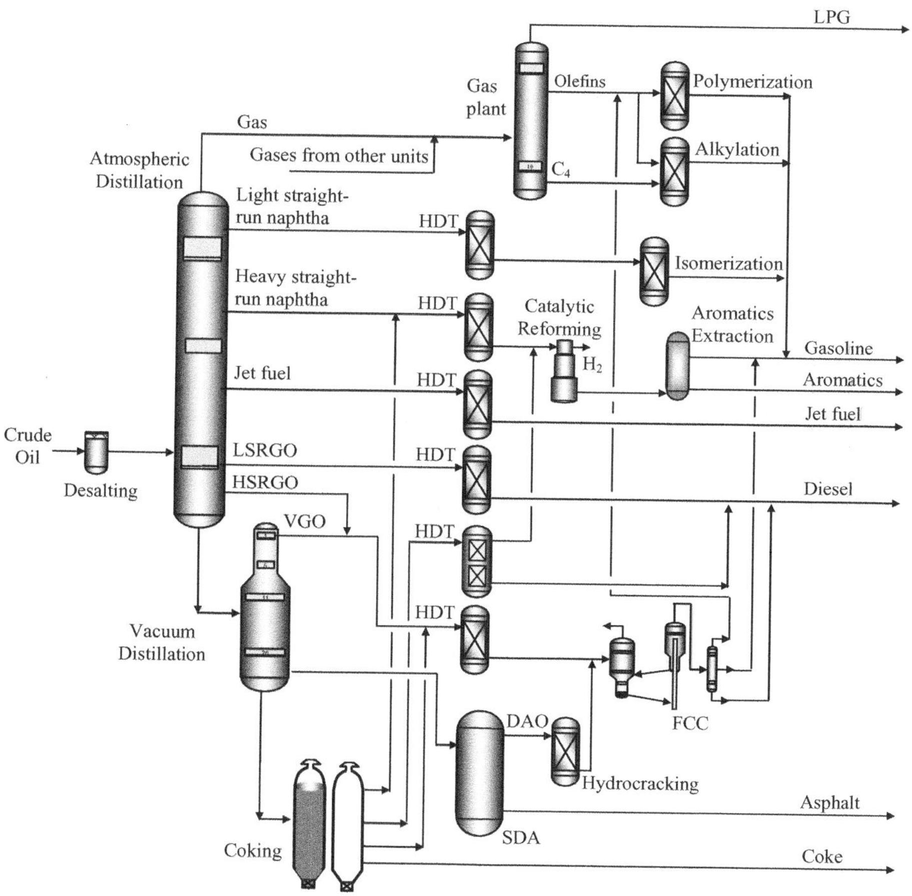

Figure 1.6. Typical process scheme of a petroleum refinery.

1.4.1 Catalytic Reforming

Catalytic reforming is used to convert low-octane straight-run naphtha into high-octane gasoline, called *reformate*, and to provide aromatics (BTX: benzene, toluene, and xylene) for petrochemical plants. The reformate has higher aromatic and cyclic hydrocarbon contents. The main reactions occurring in catalytic reforming are:

- Dehydrogenation of naphthenes to aromatics
- Isomerization of paraffins to branched-chain structures
- Isomerization of naphthenes
- Dehydrocyclization of paraffins and olefins to aromatics
- Hydrocracking of high-boiling hydrocarbons to low-molecular-weight paraffins (hydrocracking of paraffins is undesirable due to increased light ends made)

The objective of these reactions is to restructure and crack some of the molecules present in the feed to produce a product with hydrocarbons that have more complex molecular shapes, whose overall effect is the production of a reformate with a higher octane number than that of the feed. Apart from producing high-octane gasoline, catalytic reforming also produces very significant amounts of hydrogen gas as a by-product, which is released during catalyst reaction and is used in other processes within the refinery (e.g., catalytic hydrotreating and hydrocracking).

A typical catalytic reforming process includes the following steps (Figure 1.7):

- Mixing the feed (naphtha) with recycle hydrogen, heating, and passing through a series of catalytic reactors. The feed must be almost free of sulfur, since even in extremely low concentrations, it poisons the noble metal catalysts (platinum and rhenium) used in the catalytic reforming units.
- Since most of the reactions are highly endothermic, each reactor effluent is reheated before entering the following reactor.
- The effluent from the final reactor is separated into hydrogen-rich gas and reformate, and the hydrogen is recycled or purged for using in other processes. Hydrogen recycle reduces the formation of carbon.
- Reformate product is sent to gasoline blending.

1.4.2 Isomerization

Isomerization is an ideal choice to produce a gasoline blending component from light paraffins. The objective of isomerization is to convert low-octane *n*-paraffins to high-octane *i*-paraffins by using a chloride-promoted fixed-bed reactor. The main steps of a typical isomerization process are (Figure 1.8):

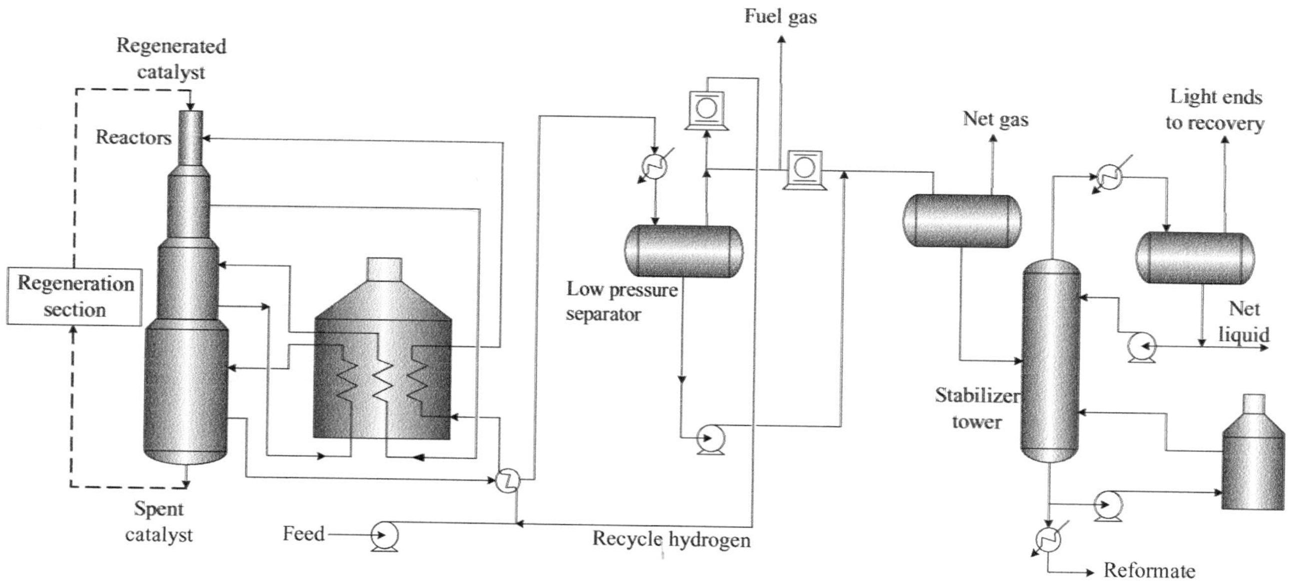

Figure 1.7. Typical process scheme of a catalytic reforming unit.

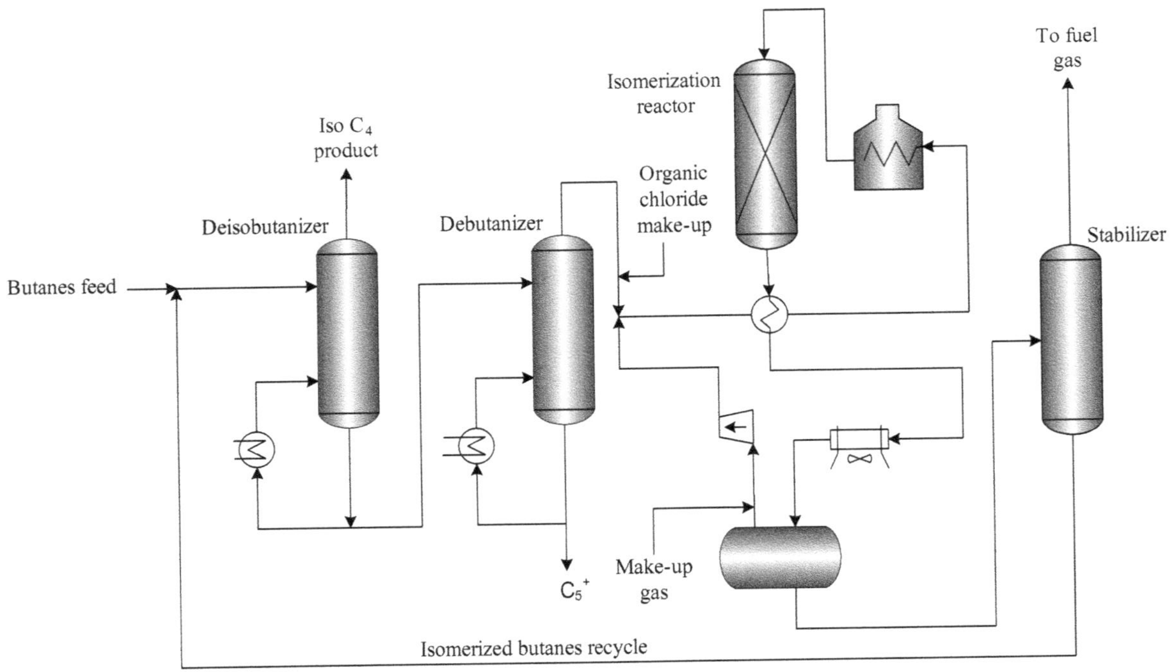

Figure 1.8. Typical process scheme of an isomerization unit.

- Drying the previously desulfurized feed and hydrogen in fixed beds of solid desiccant prior to mixing together
- Heating the mixed feed and passing it through a hydrogenation reactor to saturate olefins to paraffins and to saturate benzene
- Cooling the hydrogenation effluent and passing it through an isomerization reactor, where the isomerization reaction takes place in the catalyst bed
- Cooling the final effluent first by heat exchange with the incoming feed and then by water or air cooling
- Separating the cooled effluent into hydrogen and a liquid stream
- Sending the liquid stream to a reboiled stripper column, where a debutanized isomerate liquid leaves as the bottom product, and the butanes and lighter components leave at the top
- Partially condensing to the gas stream provide reflux to the column and a liquid product rich in butanes and propane (LPG)
- When it leaves the stripper condenser drum, sending the uncondensed overhead to the fuel gas.
- Sending the debutanized isomerate as a product for gasoline blending

As result of the isomerization reactions, highly branched, high-octane paraffinic blending components are obtained, which by themselves can satisfy the strictest gasoline environmental requirements. However, production of this isomerate is low, and other streams for gasoline blending are still necessary. Isomerization of *n*-butane is also one source for the isobutane required in alkylation.

1.4.3 Alkylation

The objective of the *alkylation* process is to combine light olefins (primarily a mixture of propylene and butylene) with isobutane to form a high-octane gasoline (highly branched C_5–C_{12} *i*-paraffins), called *alkylate*. The major constituents of alkylate are isopentane and isooctane (2,2,4-trimethyl pentane), the latter possessing an octane number of 100. Among all refinery processes, alkylation is a very important process that enhances the yield of high-octane gasoline. The reaction occurs in the presence of a highly acidic liquid catalyst (HF: hydrofluoric acid or H_2SO_4: sulfuric acid). As a consequence of the environmental problems associated with the use of these liquid catalysts, solid acid catalysts have also been proposed, having as a major problem rapid deactivation due to coke formation.

The main steps of a typical hydrofluoric alkylation unit are (Figure 1.9):

- Mixing the olefins coming from fluid catalytic cracking process with isobutane and feeding the mixture to the reactor where the alkylation

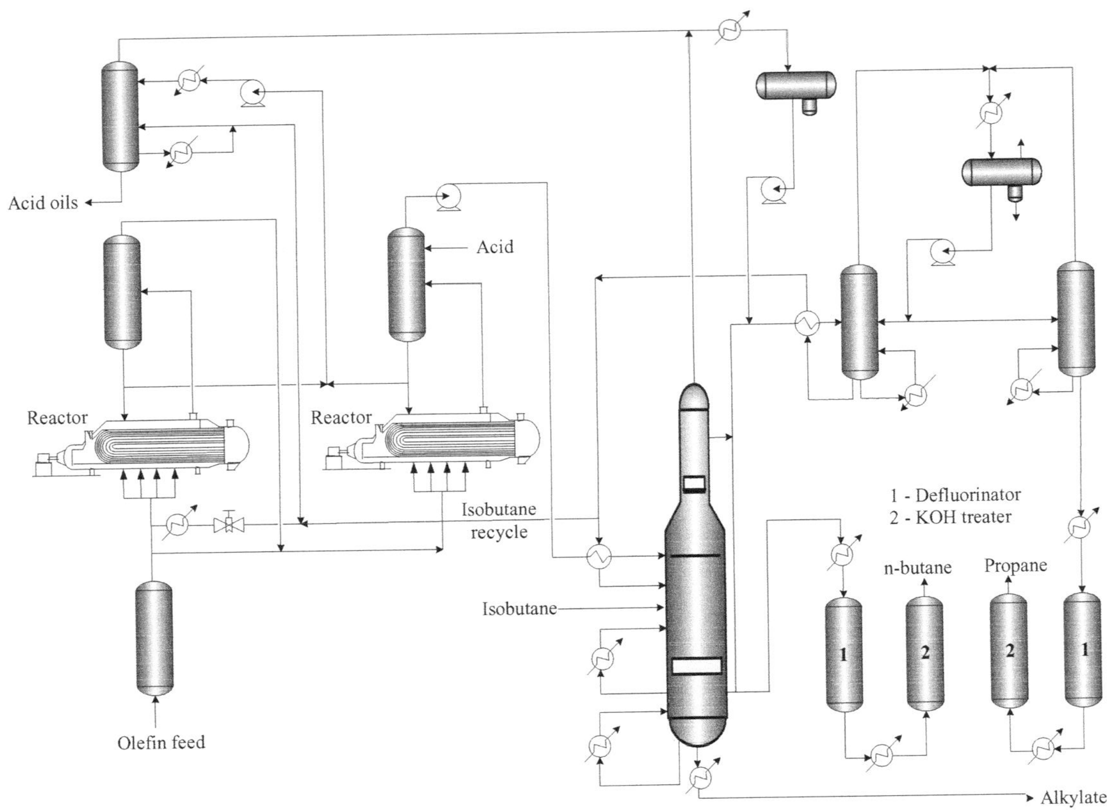

Figure 1.9. Typical process scheme of an alkylation unit.

reaction occurs. Prior to mixing, the olefin feed needs pretreatment to remove H_2S and mercaptans.

- Separation of the free HF from the hydrocarbons in an acid settler and recycling the acid back to the reactor.
- Regeneration of part of the HF to remove acid oils formed by feed contaminants or hydrocarbon polymerization.
- Sending the hydrocarbons from the acid settler to the de-isobutanizer, where propane and isobutane are separated from *n*-butane and alkylate.
- Fractionation of propane from isobutane. Isobutane in then recycled to the reactor.
- *n*-Butane and alkylate are defluorinated in a bed of solid adsorbent and fractionated as separate products. Propane and *n*-butane are nonreactive hydrocarbons.

The function of the acid catalyst is to protonate the olefin feed to produce reactive carbocations, which alkylate isobutane. Alkylation reaction is very fast with 100% olefin conversion. It is important to keep a high isobutene-to-olefin ratio to prevent side reactions, which can produce a lower-octane product. This is the reason that alkylation units have a high recycle of isobutane.

1.4.4 Polymerization

The objective of a *polymerization* unit is to combine or polymerize the light olefins propylene and butylene into molecules two or three times their original molecular weight. The feed to this process consists of light gaseous hydrocarbons (C_3 and C_4) produced by catalytic cracking, which are highly unsaturated. The polymer gasoline produced has octane numbers above 90. Although the amount of polymer gasoline is very small, it is an important part of a refinery since the polymerization process increases the yield of gasoline possible from gas oil. For example, the numbers of barrels of polymer gasoline per barrel of olefin feed is about half those of alkylate, but capital and operating costs are much lower in polymerization because it operates at low pressures compared with alkylation. The polymerization reaction consists of passing the C_3–C_4 hydrocarbon stream with a high proportion of olefins through a reactor containing a phosphoric acid–supported catalyst, where the carbon–carbon bond formation occurs.

Polymerization comprises the following main steps (Figure 1.10):

- Contacting the feed with an amine solution to remove H_2S and washing with caustic to remove mercaptans
- Scrubbing with water to remove any caustic or amines
- Drying by passing through a silica gel or molecular sieve bed

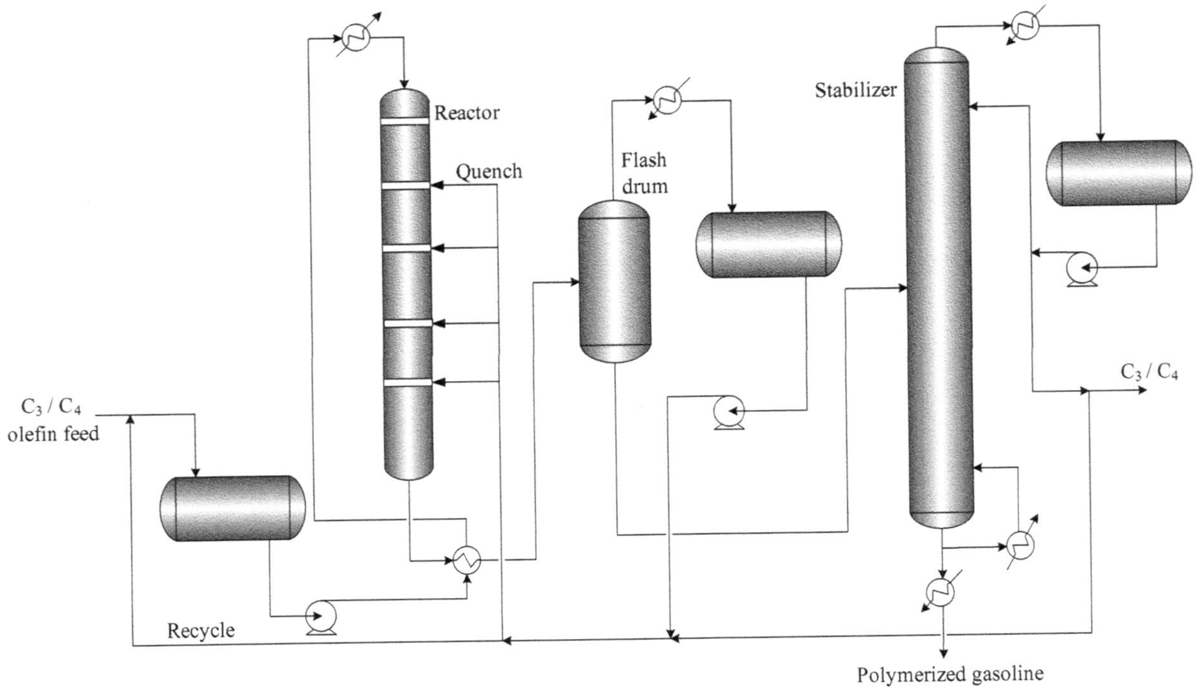

Figure 1.10. Typical process scheme of a polymerization unit.

- Adding a small amount of water to promote ionization of the acid before heating the olefin feedstream and passing over the catalyst bed
- Injecting a cold propane quench or by generating steam to control the reaction temperature since the polymerization reaction is highly exothermic
- Fractionating the product after leaving the reactor to separate the butane and lighter hydrocarbons from the polymer gasoline

1.4.5 Catalytic Hydrotreating

Catalytic hydrotreating (HDT) is one of the most important processes in the petroleum refining industry. The HDT process is applied to treat a great variety of refinery streams, such as straight-run distillates, vacuum gas oils [fluid catalytic cracking (FCC) feed], atmospheric and vacuum residua, light cycle oil, FCC naphtha, and lube oils. The main differences in the hydrotreating processes of each feed are the operating conditions, type of catalyst, reactor configuration, and reaction system. Depending on the feed and the main objective of the treatment, the process can be called *hydrodesulfurization* (HDS), as in the case of the HDS of straight-run naphtha, which is used as reforming feed where sulfur is the main undesirable heteroatom. For straight-run gas oil, the process is called *hydrotreating* because, in addition to sulfur removal, aromatic saturation and nitrogen removal are also desired for diesel fuel production. A hydrodemetallization process is used for the removal of vanadium and nickel from heavy oils. When a change in the molecular weight of the feed is required, a hydrocracking process is used.

Sulfur is removed primarily to reduce the sulfur dioxide (SO_2) emissions caused during fuel combustion. Removal of sulfur is also desired to have better feed for subsequent processes (e.g., catalytic reforming, fluid catalytic cracking). For naphtha HDS it is necessary to remove the total sulfur from the feed down to a few parts per million to prevent poisoning the noble metal catalysts in the catalytic reforming. For gas oil HDS, the production of ultralow-sulfur diesel (ULSD) requires the use of highly selective catalyst together with appropriate reaction conditions.

During hydrotreating a number of reactions are carried out: hydrogenolysis, by which C–S, C–N or C–C bonds are cleaved, and hydrogenation of unsaturated compounds. The reacting conditions of the HDT process vary with the type of feedstock; whereas light oils are easy to desulfurize, the desulfurization of heavy oils is much more difficult. The hydrotreating reactions take place in catalytic reactors at elevated temperatures and pressures, typically in the presence of a catalyst consisting of an alumina base impregnated with cobalt, nickel, and molybdenum. A typical hydrotreating unit involves the following steps (Figure 1.11):

- Mixing the liquid feed with a stream of hydrogen-rich recycle gas.
- Heating the resulting liquid–gas mixture to the desired reaction temperature.

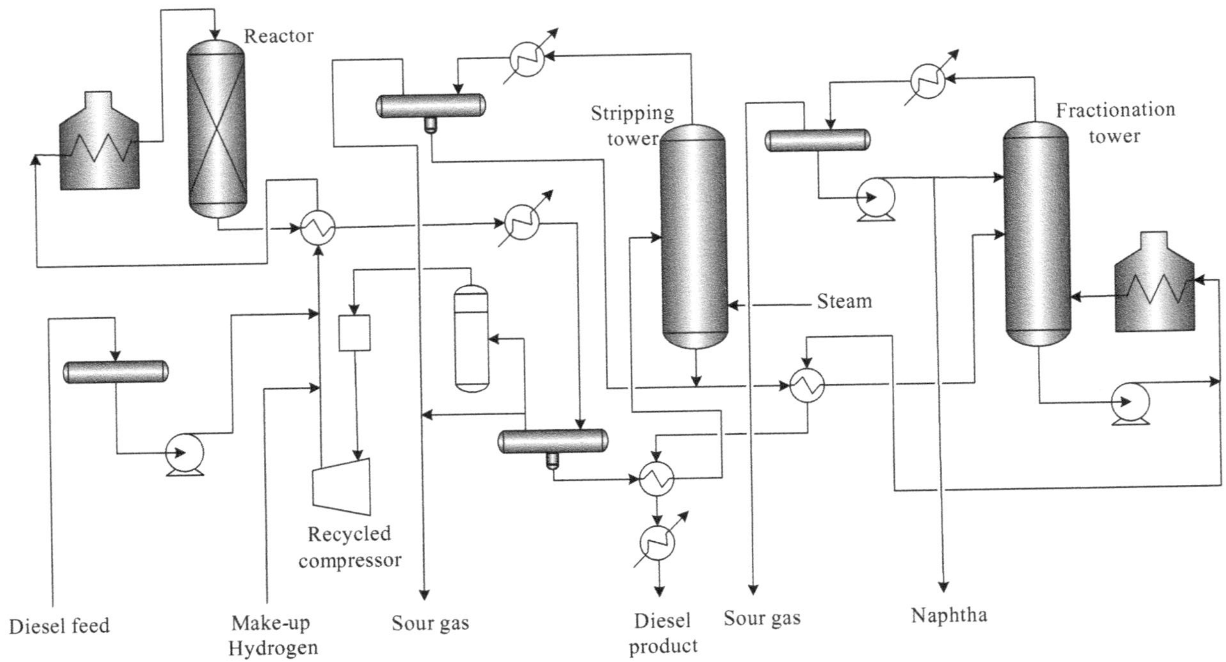

Figure 1.11. Typical process scheme of a hydrotreating unit.

- Feeding the mixture to the catalytic reactor, where the hydrotreating reactions take place.
- Cooling the reaction products and feeding them to a gas separator vessel.
- Sending most of the hydrogen-rich gas separated from this vessel through an amine contactor for removal of H_2S.
- Recycling the H_2S -free hydrogen-rich gas to the reactor.
- Sending the liquid from the gas separator vessel through a stripper distillation tower. The bottoms product from the stripper is the final desulfurized liquid product, while the overhead sour gas (i.e., hydrogen, methane, ethane, H_2S, propane, butane, and some heavier components) is sent to the amine gas treating. Subsequently, the H_2S removed and recovered is converted to elemental sulfur in a Claus process unit.

1.4.6 Fluid Catalytic Cracking

The *fluid catalytic cracking* (FCC) process is the heart of a modern refinery oriented toward maximum gasoline production. Within the entire refinery process, this process offers the greatest potential for increasing profitability; even a small improvement giving higher gasoline yields can result in a substantial economic gain. The FCC process increases the H/C ratio by carbon rejection in a continuous process and is used to convert the high-boiling, high-molecular-weight hydrocarbon fractions (typically, a blend of heavy straight-run gas oil, light vacuum gas oil, and heavy vacuum gas oil) to more valuable gasoline, olefinic gases, and other products.

The process consists of two main vessels: a reactor and a regenerator, which are interconnected to allow for transferring the spent catalyst from the reactor to the regenerator and the regenerated catalysts from the regenerator to the reactor. During catalytic cracking the feed is vaporized and the long-chain molecules are cracked into much shorter molecules by contacting the feed with a fluidized powdered catalyst at high temperature and moderate pressure.

Catalytic cracking reactions are believed to follow the carbonium ion mechanism, involving the following steps:

- *Initiation:* which starts from an early contact of an olefin with an active site of the catalyst at high temperature to produce the active complex corresponding to the formation of a carbocation
- *Propagation:* represented by the transfer of a hydride ion from a reactant molecule to an adsorbed carbenium ion
- *Termination:* corresponding to the desorption of the adsorbed carbenium ion to produce an olefin while the initial active site is restored

According to this mechanism, a catalyst promotes the removal of a negatively charged hydride ion from a paraffin compound or the addition of a positively charged proton (H^+) to an olefin compound, which results in the formation of a carbonium ion. Carbonium ion is a positively charged molecule that has only a very short life as an intermediate compound and transfers the positive charge through the hydrocarbon. This carbonium transfer continues as hydrocarbon compounds come into contact with active sites on the surface of the catalyst that promote the continued addition of protons or the removal of hydride ions. The result is a weakening of carbon–carbon bonds in many of the hydrocarbon molecules and a consequent cracking into smaller compounds. These ions also react with other molecules, isomerize, and react with the catalyst to terminate a chain. Coke formation is unavoidable in the catalytic cracking process, which is probably formed by the dehydrogenation and condensation of polyaromatics and olefins. Fast deactivation by blocking the active pores of the catalyst is a consequence of coke deposition. During these reactions, the catalytic cracked gasoline produced contains large amounts of aromatics and branched compounds, which is beneficial for the gasoline's octane level.

A typical modern FCC unit consists of the following steps (Figure 1.12):

- Preheating the feed and mixing with the recycle slurry oil from the bottom of the distillation column.
- Injecting the combined feed into the catalyst riser, where vaporization occurs.

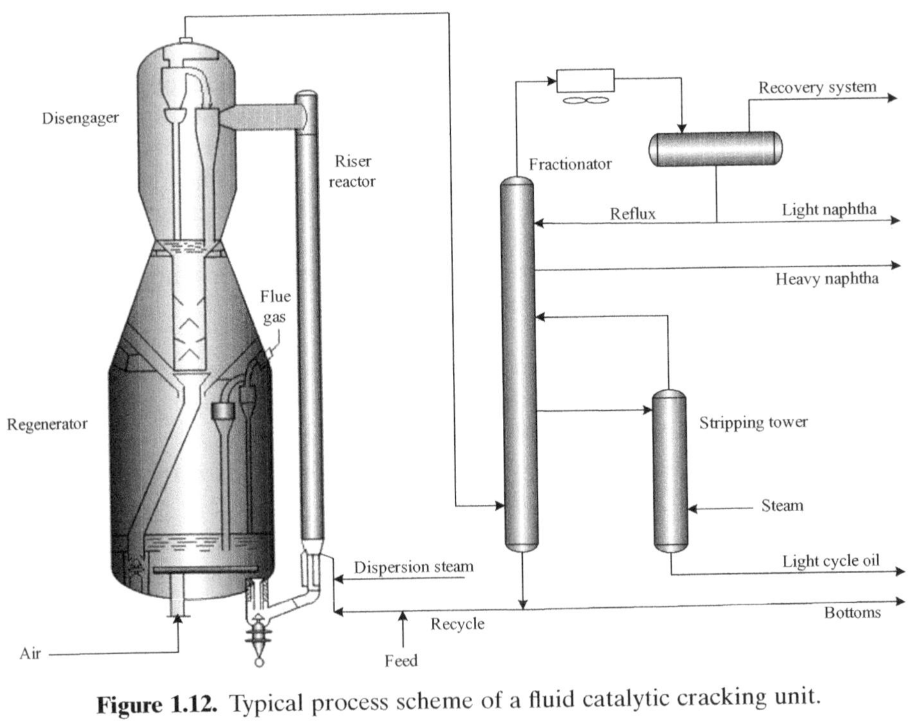

Figure 1.12. Typical process scheme of a fluid catalytic cracking unit.

- Cracking the vaporized feed into smaller molecules by contact with the hot powdered catalyst coming from the regenerator.
- Separation of the cracked product vapors from the spent catalyst by flowing through a set of two-stage cyclones.
- Stripping the spent catalyst with steam to remove any hydrocarbon vapors before the spent catalyst returns to the regenerator.
- Regeneration of the spent catalyst to burn off the deposited coke with blown air. This reaction is exothermic and produces a large amount of heat, which is partially absorbed by the regenerated catalyst and provides the heat required for feed vaporization and the endothermic cracking reactions that take place in the catalyst riser.
- Passing the hot flue gas leaving the regenerator through multiple sets of cyclones that remove entrained catalyst from the flue gas.
- Suitably separating the cracked product vapors from the reactor from entrained catalyst particles by cyclone and sending them to the recovery section of the FCC unit to meet the product stream requirements.

1.5 UPGRADING OF HEAVY FEEDS

Heavy feeds are characterized by low API gravity and high amounts of impurities. In general, it is known that the lower the API gravity, the higher the impurities content. Such properties make the processing of heavy feeds different from that used for light distillates, causing several problems:

- Permanent catalyst deactivation in catalytic cracking and hydrocracking processes, caused by metals deposition
- Temporary deactivation of acid catalysts, due to the presence of basic nitrogen
- Higher coke formation and lower liquid product yield, as a result of high Conradson carbon and asphaltene contents
- Products with high levels of sulfur

To reduce such problems, numerous catalytic and noncatalytic technologies are commercially available to upgrade heavy oils, which are summarized in the following sections.

1.5.1 Properties of Heavy Oils

Heavy oils exhibit a wide range of physical properties. Whereas properties such as viscosity, density, and boiling point may vary widely, the ultimate or elemental analysis varies over a narrow range for a large number of samples. The carbon content is relatively constant, while the hydrogen and heteroatom contents are responsible for the major differences in various heavy oils.

Heavy oils are comprised of heavy hydrocarbons and several metals, predominantly in the form of porphyrines. Heavy feeds also contain aggregates of resins and asphaltenes dissolved in the oil fraction, held together by weak physical interactions. With resins being less polar than asphaltenes but more polar than oil, equilibrium between the micelles and the surrounding oil leads to homogeneity and the stability of the colloidal system. If the amount of resin decreases, the asphaltenes coagulate, forming sediments. Asphaltenes are complex polar structures with polyaromatic character containing metals (mostly Ni and V) that cannot be defined properly according to their chemical properties, but they are usually defined according to their solubility. Thus, asphaltenes are the hydrocarbon compounds that precipitate by addition of light paraffin in the heavy oil. Asphaltenes precipitated with *n*-heptane have a lower H/C ratio than those precipitated with *n*-pentane, whereas asphaltenes obtained with *n*-heptane are more polar, have a greater molecular weight, and display higher N/C, O/C, and S/C ratios than those obtained with *n*-pentane.

Asphaltenes are constituted by condensed aromatic nuclei carrying alkyl groups, alicyclic systems, and heteroelements. Asphaltene molecules are grouped together in systems of up to five or six sheets, which are surrounded by the maltenes (all those structures different from asphaltenes that are soluble in *n*-heptane). The exact structure of asphaltenes is difficult to obtain, and several structures have been proposed for the asphaltenes present in various crudes. An asphaltene molecule may be 4 to 5 nm in diameter, which is too large to pass through micropores or even some mesopores in a catalyst. Metals in the asphaltene aggregates are believed to be present as organometallic compounds (porphyrine structure) associated with the asphaltene sheets, making the asphaltene molecule heavier than its original structure (Figure 1.13).

The complex nature of heavy oil fractions is the reason that refining of these feeds becomes so difficult. Therefore, an evaluation of the overall chemical and physical characteristics of petroleum feeds is mandatory to determine the processing strategy. Apart from having low API gravity (high density), high viscosity, and a high initial boiling point, heavy oils exhibit higher contents of sulfur, nitrogen, metals (Ni and V), and high-molecular-weight material (asphaltenes).

Generally, the majority of the sulfur and nitrogen species present in a crude oil is found in the heaviest fractions. These heteroatoms are removed from hydrocarbon streams in downstream refining units to produce ecologically acceptable fuels and/or to provide better quality feeds to subsequent processes: for example, feed with a low concentration of basic nitrogen is required to avoid the temporary poisoning effect on acid catalysts typically used in fluid catalytic cracking (FCC) and hydrocracking (HCR). Metals are found in most heavy oils in the form of metalloporphyrins and are concentrated exclusively in the residual fraction. The problem with metal-containing feeds is the permanent catalyst deactivation experienced in FCC, residue fluid catalytic cracking (RFCC), and HCR units. Asphaltenes are the most complex structures and

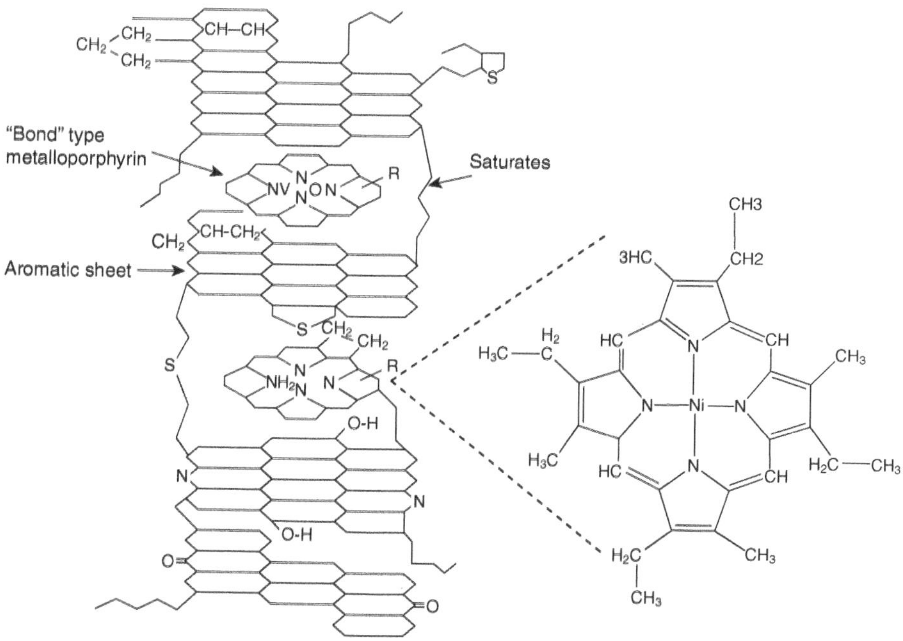

Figure 1.13. Hypothetical structure of an asphaltene molecule.

cause many problems in refining operations. Known as *coke precursors*, they reduce catalyst cycle life and liquid yield and are the main contributors of solids formation, producing fouling in all types of equipment.

The properties of petroleum residue vary widely, depending on the crude of origin, as shown in Table 1.8. Crude oils and their respective residua have a similar composition (e.g., sulfur, metals, and asphaltene contents), and the latter represents a significant portion of a barrel of crude oil. In the case of heavy petroleum, the yield of residue may be as high as 85%. For this reason, in the near future the material at the bottom of the barrel will be the main raw material for obtaining valuable liquid products, to keep up with fuel demand.

1.5.2 Process Options for Upgrading Heavy Feeds

General Classification One way to establish the quality of heavy oils is by the hydrogen-to-carbon (H/C) ratio. Values of about 1.5 indicate high-quality feed, while poor-quality oils may have an H/C ratio as low as 0.8. Therefore, to improve the quality of heavy oil, its H/C ratio needs to be increased either by increasing the hydrogen content or by decreasing the carbon content. Based on this consideration, processes for upgrading of heavy oils can be classified into two groups:

TABLE 1.8. Properties of Various Atmospheric Residua (AR), 343°C+

Crude Oil	Origin	API Gravity	Sulfur (wt%)	Ni + V (wppm)	Carbon Residue (wt%)	Yield of AR (vol%)
Ekofisk	North Sea	20.9	0.4	6	4.3	25.2
Arabian Light	Arabia	17.2	3.1	50	7.2	44.6
West Texas Sour	United States	15.5	3.4	29	9.0	41.6
Isthmus	Mexico	15.5	2.9	82	8.1	40.4
Export	Kuwait	15.0	4.1	75	—	45.9
North Slope	Alaska	14.9	1.8	71	9.2	51.5
Arabian Heavy	Arabia	13.0	4.3	125	12.8	53.8
Bachaquero	Venezuela	9.4	3.0	509	14.1	70.2
Maya	Mexico	7.9	4.7	620	15.3	56.4
Hondo	United States	7.5	5.8	489	12.0	67.2
Cold Lake	Canada	6.8	5.0	333	15.1	83.7
Athabasca	Canada	5.8	5.4	374	—	85.3
Ku-Maloob-Zaap	Mexico	3.7	5.8	640	20.4	73.7

1. *Hydrogen addition:* hydroprocesses such as hydrotreating and hydrocracking, hydrovisbreaking, and donor-solvent processes
2. *Carbon rejection:* coking, visbreaking, and other processes, such as solvent deasphalting

Both hydrogen addition and carbon rejection processes have disadvantages when applied to upgrading heavy oils. For example, removal of nitrogen, sulfur, and metals by exhaustive hydrodenitrogenation (HDN), hydrodesulfurization (HDS), and hydrodemetallization (HDM) is very expensive (excessive catalyst utilization), due to metal and carbon deposition. Noncatalytic processes yield uneconomically large amounts of coke and low liquid yield.

Processes for upgrading heavy oils are evaluated on the basis of liquid yield (i.e., naphtha, distillate, and gas oil), heteroatom removal efficiency (HDS, HDN, HDM), feedstock or residue conversion (RC), carbon mobilization (CM) and hydrogen utilization (HU), along with other process characteristics. Heteroatom removals and feedstock conversion are calculated from their corresponding amounts in feed and product:

$$\text{HDS, HDN, or HDM} = \frac{I_{\text{feed}} - I_{\text{product}}}{I_{\text{feed}}} \times 100 \qquad (1.3)$$

$$\text{conversion (RC)} = \frac{538°\text{C}^+_{\text{feed}} - 538°\text{C}^+_{\text{product}}}{538°\text{C}^+_{\text{feed}}} \times 100 \qquad (1.4)$$

where I_{feed} and I_{product} represent the amount of impurity (sulfur, nitrogen, or metals) in the feed and product, respectively. $538°\text{C}^+_{\text{feed}}$ and $538°\text{C}^+_{\text{product}}$ are the petroleum fractions in the feed and product, respectively, with a boiling point higher than 538°C (i.e., vacuum residue).

Carbon mobilization and hydrogen utilization are defined as follows:

$$CM = \frac{\text{carbon}_{\text{liquids}}}{\text{carbon}_{\text{feedstock}}} \times 100 \tag{1.5}$$

$$HU = \frac{\text{hydrogen}_{\text{liquids}}}{\text{hydrogen}_{\text{feedstock}}} \times 100 \tag{1.6}$$

High values of CM and HU correspond to high feedstock conversion processes such as hydrocracking (hydrogen addition). Since hydrogen is added, HU can be greater than 100%. On the contrary, low CM and HU correspond to low feedstock conversion, such as coking (carbon rejection).

The focus on the downstream and upstream petroleum sectors for each country may vary depending on the quality of crude oil. Significant advances have been made in these sectors over the last few decades. The downstream sector has traditionally been in charge of petroleum refining. However, with the increasing production of heavy petroleum, the upstream sector has entered into the upgrading area to increase the value of the oil produced. Thus, nowadays, both sectors are looking for better alternatives to upgrade and refine heavy petroleum.

Heavy oil upgrading is usually carried out directly by using the residue as feed after crude distillation. There is a wide range of catalytic and noncatalytic conversion processes that can be classified into the carbon rejection and hydrogen addition processes, as presented in Table 1.9. These processes use a variety of reactor designs and configurations, such as multi-fixed-bed systems, ebullated-bed reactors, fluidized reactors, and moving-bed reactors. Examples of some of these upgrading technologies are presented in Figure 1.14. The process technologies differ principally on the basis of the feedstock and process conditions (reactor) and catalyst used by the various licensers.

Carbon Rejection Processes The carbon rejection route is based on the removal of carbon in the form of coke with a low atomic hydrogen/carbon

TABLE 1.9. General Classification of Technologies for Upgrading of Heavy Petroleum Feeds

	Carbon Rejection	Hydrogen Addition
Noncatalytic	Solvent deasphalting Coking Visbreaking	Hydrovisbreaking
Catalytic	Catalytic cracking of residue	Hydrotreating Hydrocracking

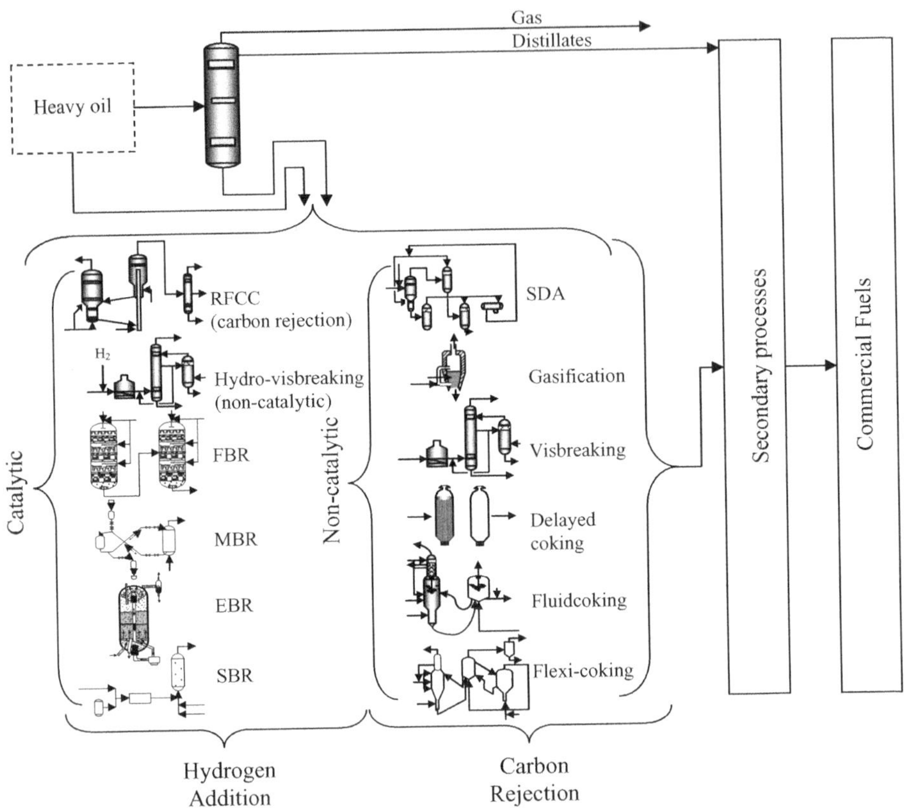

Figure 1.14. Process alternatives for upgrading of heavy oils.

ratio or in the form of asphalt (in the case of deasphalting), producing a moderate yield of liquid products. The following processes belong to this category: solvent deasphalting, thermal cracking processes such as coking and visbreaking, and catalytic cracking of residue.

Carbon rejection is an important process for residue conversion and is the most common method used commercially. In general, thermal cracking of residue is carried out at relatively moderate pressure and is often called the *coking process*. It is conducted at temperatures between 480 and 550°C and vapor-phase residence times of 20 or more, providing a significant degree of cracking and dehydrogenation of the feed, which makes subsequent processing more cumbersome and produces low-value by-products such as gas and coke. The coking process transfers hydrogen from the heavy molecules to the lighter molecules, resulting in the production of coke or carbon. The residue is hydrogen donors at high temperature.

The thermal conversion of heavy oil has attracted great interest in recent years, due to the decrease in middle distillate or increase in low-quality crude oil. Thermal processes produce a relatively high amount of gas, such as methane, ethene, propene, butane, and secondary products such as LPG and dry gas. Coke is a significant by-product whose formation mechanism is different from that of other products. Some of the thermal processes are coupled with catalytic processes. The catalytic pyrolysis of heavy oil may be a good option for a petrochemical refinery but not for the transportation of fuel oil.

Solvent deasphalting (SDA), described earlier, is a separation process in which the asphaltenic fraction is precipitated from the residue using a light paraffinic solvent (i.e., propane, butane, pentane, or *n*-heptane). The product is a low-sulfur/metal deasphalted oil (DAO) rich in paraffins that is normally used as feed for FCC and hydrocracking. The advantages of this method are the relatively low cost, the flexibility to adjust the DAO quality in a wide range, and the elimination of fouling problems in subsequent units. However, disposal of the SDA pitch (asphaltenic fraction) is still a matter of concern.

Thermal cracking processes are the most mature technologies for converting heavy feeds. They are carried out at moderate pressure in the absence of a catalyst. Coking processes (i.e., delayed coking, fluid coking, and flexicoking) are capable of eliminating the heaviest fractions from crude oils, producing coke that contains the majority of sulfur, nitrogen, and metals of the original oil. Delayed coking has been the upgrading process of choice, due to its flexibility to handle any type of feed and its ability to remove carbon and metals completely, along with partial conversion to liquids. Fluid coking and flexicoking are advanced processes that employ fluidized-bed technology, derived from FCC technology. Technically, fluid coking is only marginally better than delayed coking, as it offers a slightly higher liquid yield, less coke formation, and lower operating costs. *Visbreaking*, on the other hand, is a mild thermal decomposition process to improve the viscosity of heavy oils and residue, without significant conversion to distillates. In general, thermal processes appear to be attractive, due to low investment and operating costs; however, they suffer from the disadvantage of producing uneconomically large amounts of coke and having a low liquid yield. Additionally, liquid products require extensive posttreatment to meet the specifications of commercial fuels.

Catalytic cracking of residue (RFCC) is the only catalytic process found in this class of upgrading technologies. It is an extension of conventional FCC, which is employed for converting heavy feedstocks into high-octane gasoline blending components. RFCC exhibits better selectivity to gasoline and a lower gas yield than thermal cracking and hydroprocessing. However, the main drawback of RFCC is the need for good-quality feed (low metals content and H/C ratio) to avoid high coke production and excessive catalyst use; therefore, the application of RFCC directly to residues derived from heavy oil is not likely.

Additional details regarding these processes are given in the following sections.

Solvent Deasphalting Since asphaltenes cause many problems during various steps of petroleum refining, it is more convenient to remove them from heavy oil and make it a trouble-free feedstock. For example, if asphaltene separation is carried out before hydroprocessing, the following main problems encountered when handling heavy feeds can be avoided:

- Pipeline deposition and its plugging
- Efficiency decrease in refinery plants
- Precipitation of asphaltene due to blending of light hydrocarbon streams
- Sludge and sediment formation during storage as well as processing
- Catalyst deactivation in downstream processes

The most common method used for asphaltene precipitation is *solvent deasphalting* (SDA). This process uses a solvent (light paraffin such as C_3, C_4, C_5, C_6, and C_7) to separate a residue into a deasphalted oil (DAO) and a pitch (asphaltene), the latter containing most of the impurities of the feedstock. The insoluble pitch will precipitate out of the mixed feedstock as asphaltene. Separation of the DAO phase and the pitch phase occurs in an extractor. The extractor is designed to separate the two phases efficiently and to minimize contaminant entrainment in the DAO phase. At a constant solvent composition and pressure, a lower extractor temperature increases the DAO yield and decreases the quality. With an increase in solvent ratio the DAO yield remains constant, improves the degree of separation of individual components, and results in the recovery of a better quality DAO. The solvent recovered under low pressure from the pitch and DAO strippers is condensed and combined with the solvent recovered under high pressure from the DAO separator, which is then recycled to the initial stage. DAO is normally used as fluid catalytic cracking or hydrocracker feed.

Solvent deasphalting is used in refineries to upgrade heavy bottoms streams to deasphalted oil that may be processed to produce transportation fuels. The process may also be used in the oil field to enhance the value of heavy crude oil before it gets to the refinery. Thus, SDA is an economically attractive and environmentally friendly process to upgrade heavy petroleum.

Gasification *Gasification* involves complete cracking of residue, including asphaltenes, into gaseous products. The gasification of residue is carried out at a high temperature (>1000°C) having synthesis gas (consisting primarily of hydrogen, carbon monoxide, carbon dioxide, and water), carbon black, and ash as major products. The syngas can be converted to hydrogen or used by cogeneration facilities to provide low-cost power and steam to refineries. An integrated SDA-gasification facility is an attractive alternative for upgrading of heavy petroleum. The following are some of the benefits obtained in integrating deasphalting and gasification:

- Heavy oils can be upgraded economically.
- Capital and operating costs of both processes can be reduced.

- Higher yields of DAO are possible.
- Lower emissions are possible.
- Profit margins of a refinery can be increased.

Coking Depending on feedstock properties, coker unit design, and operating conditions, the solid product (petroleum coke or "petcoke") can be:

- *Fuel-grade coke:* the most common type of coker is the fuel grade, whose main objective is to maximize liquid yields and reduce low-value coke formation. This coke is used as fuel in process heaters and power generation facilities.
- *Anode-grade coke:* which is produced from low-sulfur and metals feeds, and is used for anodes in the aluminum industry.
- *Needle-grade coke:* which is produced from highly aromatic feedstocks with low asphaltenes, sulfur, and ash contents. This coke, with high strength and a low coefficient of thermal expansion, is used to manufacture large electrodes for the steel industry and the production of synthetic graphite.

The physical and chemical properties of fuel coke, anode coke, and needle coke vary substantially.

Three main coking processes are in use:

1. *Delayed or retarded coking:* which can produce shot coke (a type of fuel coke), sponge coke (used to produce anode coke or as a fuel coke), or needle coke. This process accounts for the majority of the coke produced in the world today.
2. *Fluid coking:* which produces fluid coke typically used as fuel coke.
3. *Flexicoking:* which produces a type of fluid coke that is gasified to generate a low-Btu synthesis gas.

1. DELAYED COKING *Delayed coking* is a semicontinuous thermal cracking process used in petroleum refineries to upgrade and convert bottoms from atmospheric and vacuum distillation of crude oil into liquid and gas product streams, leaving behind a solid concentrated carbon material, petroleum coke, whose value will depend on its properties, such as sulfur or metals. The products of a delayed coker are wet gas, naphtha, light and heavy gas oils, and coke. The coke produced in the delayed coker is almost pure carbon and is utilized as fuel or, depending on its quality, in the manufacture of anodes and electrodes.

In a delayed coker the feed enters the bottom of the fractionator, where it mixes with recycle liquid condensed from the coke drum effluent. It is then pumped through the coking heater, then to one of two coke drums through a switch valve. The total number of coke drums required for a particular application depends on the quality and quantity of the feed and the coking cycle desired. A minimum of two drums is required for operation, with one drum receiving the heater effluent while the other is being decoked.

A delayed coking unit is frequently designed with the objective of maximizing the yield of liquid product and minimizing the yields of wet gas and coke. The conversion is accomplished by heating the feed material to a high temperature and introducing it into a large drum to provide soaking or residence time for the three major reactions to take place:

- Partial vaporization and mild cracking (visbreaking) of the feed as it passes through the coker's furnace.
- Thermal cracking, the mechanism through which high-molecular-weight molecules are decomposed into smaller, lighter molecules that are fractionated into the products. The reaction is highly endothermic. The coker heaters supply the heat necessary to initiate the cracking reaction. Heater temperature and residence time are strictly controlled, so that coking in the heaters is minimized.
- Polymerization, the reaction through which small hydrocarbon molecules are combined to form a single large molecule of high molecular weight. The result of this reaction is the formation of coke. Polymerization reactions require a long reaction time and the coke drums provide the necessary residence time for these reactions to proceed to completion.

Delayed coking has been selected by many refiners as their preferred choice for upgrading the bottom of the barrel, because of the process's inherent flexibility to handle any type of residua. The process provides essentially complete rejection of metals and carbon while providing partial conversion to liquid products (naphtha and diesel). The product selectivity of the process is based on the operating conditions, mainly pressure and temperature. This process is more expensive than SDA, although still less expensive than other thermal processes. The disadvantages of this process are the very high coke formation and low yield of liquid products. Despite these disadvantages, delayed coking is the favorite process of all refiners for residue processing. Advances in delayed coking have increased light products while decreasing coke production, lowering pressure and oil recirculation.

2. FLUID COKING AND FLEXICOKING *Fluid coking* is a continuous process that uses the fluidized-solids technique to convert residue feedstock to more valuable products. The heated coker feeds (petroleum residua) are sprayed into a fluidized bed of hot, fine coke particles which are maintained at 20 to 40 psi and 500°C. The use of a fluid bed permits the coking reactions to be conducted at higher temperatures and with shorter contact times than in delayed coking. These conditions result in lower yields of coke and higher yields of liquid products. Fluid coking uses two vessels, a reactor and a burner. Coke particles are circulated between them to transfer heat to the reactor.

This heat is generated by burning a portion of the coke. The reactor contains a fluidized bed of the coke particles, which is agitated by the introduction of steam below. The residue feed is injected directly into the reactor and is distributed uniformly over the surface of the coke particles, where it cracks and vaporizes. The feed vapors are cracked while forming a liquid film on the coke particles. The particles grow by layers until they are removed and new seed coke particles are added. Coke is a product and a heat carrier. *Flexicoking* is an extension of fluid coking which includes the gasification of the coke produced in the fluid coking operation and produces syngas, but the temperature (1000°C) used is insufficient to burn all coke.

Both fluid coking and flexicoking are fluid-bed processes developed from fluid catalytic cracking technology. In both processes, the circulating coke carries heat from the burner back to the reactor, where the coke serves as reaction sites for the cracking of the residua into lighter products. Fluid coking can have liquid yield credits over delayed coking. The shorter residence time can yield higher quantities of liquids and less coke, but the products are lower in quality. Fluid coking is a slightly better process than delayed coking because of the advantage of a slightly improved liquid yield, and because delayed coking has a higher utilities cost and higher fuel consumption.

Visbreaking *Visbreaking* (viscosity reduction or breaking), a mature process that may be applied to both atmospheric residua (AR) and vacuum residua (VR) and even solvent deasphalted pitch, improves viscosity by means of its mild thermal decomposition. The thermal conversion of the residue is accomplished by heating at high temperatures in a specially designed furnace. A common operation is to visbreak residue in combination with a thermal cracker to minimize fuel oil while producing additional light distillates.

Visbreaking is a process in which a residue stream is heated in a furnace (450 to 500°C) and then cracked during a low specific residence time, to avoid coking reactions within a soaking zone under certain pressure and moderate temperature conditions. The cracked product leaves the soaking zone after the desired conversion is reached, and is then quenched with gas oil to stop the reaction and prevent coking, although increased conversion during visbreaking will turn to more sediment deposition. The residence time, temperature, and pressure of the furnace's soaking zone are controlled to optimize the thermal free-radical cracking to produce the desired products. In general, visbreaking is used to increase refinery net distillate yield. The main objectives of visbreaking are to reduce the viscosity of the feed stream and the amount of residual fuel oil produced by a refinery and to increase the proportion of middle distillates in the refinery output.

Carbon rejection processes are characterized by having lower investment and operating costs than those of hydroprocessing, but the yield of light products tends to be lower, which is not favored by refiners. Moreover, liquid products obtained from thermal processes contain S, N, and metals (e.g., V, Ni) that need further purification by hydrotreating processes such as HDS, HDN, and HDM, respectively. Thus, thermal processes and coking-based technologies suffer from the disadvantages of producing a large amount of low-value by-products and require further extensive processing of the liquid products. Therefore, the importance of thermal processes remains lower than that of catalytic processes, but due to their lower investment, these processes remain most common for residue upgrading.

Some of the advantages and disadvantages of carbon rejection processes are as follows:

- Visbreaking is the least expensive process but provides only a modest degree of residue conversion. Its applicability is constrained further by oil-quality considerations involving stability and compatibility.
- Delayed coking is relatively easy to implement, requires a moderate investment, provides a high degree of conversion, but may produce a large amount of low-value coke.
- Fluid coking is similar to delayed coking in many aspects but produces higher yields. However, the coke produced usually has a lower value, and the gas oils are somewhat more difficult to refine.

Residue Fluid Catalytic Cracking Fluid catalytic cracking is a well-established process for converting a significant portion of the heavy fractions (typically, heavy straight-run gas oil and light and heavy vacuum gas oils) of the crude barrel into a high-octane gasoline blending component. Residue fluid catalytic cracking (RFCC) is an extension of conventional FCC technology developed during the early 1980s which offers better selectivity to high gasoline and lower gas yield than that of hydroprocessing and thermal processes. The RFCC process uses reactor technology similar to that of the FCC process, in which the catalyst is in a fluidized bed at 480 to 540°C and is targeted for residual feeds greater than 4 wt% Conradson carbon. Because RFCC requires better feed quality (e.g., a high H/C ratio, a low metal and asphaltene content), it makes this process less likely than hydroprocessing. The need for good feedstock quality is to avoid unreasonable high coke yield, high catalyst consumption, and unit operability. However, such feeds are high in price and limited in refineries.

To control heat balance and to recover part of the heat for steam production, RFCC process design includes two-stage regeneration: mix temperature control and catalyst cooler. The catalyst properties also play an important role in resisting metal content and carbon deposition. In this respect catalyst pore structure limits the diffusion of residue on the catalytic sites. The catalyst used for RFCC is an acidic matrix such as crystalline aluminosilicate zeolite in an inorganic matrix, which fulfills the required physical–chemical properties.

Hydrogen Addition Processes The hydrogen addition route, better known as *hydroprocessing*, reduces coke formation in favor of liquid products by means of a hydrocracking or hydrogenolysis mechanism. Hydroprocessing is a hydrogen addition process which increases the H/C ratio in products. It is the most attractive route for upgrading heavy crudes and residua. In general, hydroprocessing requires hydrogen to hydrogenate the oil at high pressures and temperatures in the liquid phase because such oils have a very high concentration of carbon. Asphaltene conversion is more complicated for heavy oils, since a wide range of molecular changes occur with temperature. Except for hydrovisbreaking, a mild process based on visbreaking but in a hydrogen atmosphere, hydroprocessing is carried out in the presence of a catalyst. Catalytic hydroprocessing is extremely relevant in petroleum refining for upgrading a variety of streams; ranging from straight-run naphtha to vacuum residues or even heavy and extra-heavy crude oils. When handling heavy feeds, hydroprocessing has the virtue of reducing the contents sulfur, nitrogen, metals (Ni and V), and asphaltenes, and contributing simultaneously to the production of liquid fuels by HCR. Nevertheless, processing of this type of feed presents many difficulties caused principally by enhanced metal and carbon deposition on the catalyst.

There are numerous hydroprocessing technologies for converting heavy feeds, differing mainly in catalyst type, reactor technology, and operating conditions. The catalyst system is chosen based on activity, selectivity, and cycle life and is generally composed of CoMo/NiMo alumina-supported catalysts designed for specific objectives, such as hydrodesulfurization (HDS), hydrodemetallization (HDM), hydrodenitrogenation (HDN), hydrodeasphaltenization (HDA), HCR, and Conradson carbon (CCR) removal. Commercial reactor technologies for hydroprocessing of heavy feeds can be classified according to the type of catalytic bed: fixed, moving, ebullated, and slurry. Selection of the reactor is generally a function of the quality and composition of the feedstock and desired level of conversion and impurities removal. Typically, dirty feeds can be processed effectively in ebullated-bed reactors, since the major disadvantage of fixed-bed reactors is the catalyst deactivation with time onstream. The major selection criterion between each type of reactor is based on the catalyst deactivation rate, which depends on the contents of metals and asphaltenes in the feed, as the products formed during their removal are known as *catalyst deactivating species*. Reaction severity [i.e., pressure, reaction temperature, hydrogen-to-oil ratio, and liquid hourly space velocity (LHSV)] also depends on the properties of the feed and the product quality desired; in general, for higher-boiling-point feeds, more severe conditions are required.

Traditionally, fixed-bed reactors were employed for processing light feeds, but they were gradually adapted for tougher feeds, such as vacuum gas oil and residues. The main disadvantage of using fixed-bed reactors for upgrading heavy feeds is the loss of catalyst activity during time-on-stream. This reduces the length of run drastically, due to the frequent shutdowns required for replacing the catalyst. However, recent advances in the field have led to the development of layered catalyst systems that extend significantly the length of run. Typically, these systems comprise a front-end HDM catalyst, a midsection catalyst with balanced HDM/HDS activity, and a tail-end highly active HDS/HCR catalyst. The front-end catalyst exhibits a high-metal-uptake capacity, and its main function is to disaggregate asphaltene molecules for metals removal, so that the downstream catalysts can operate with hydrocarbons of low metal and coke precursor content. In the midsection, there is additional metals elimination and partial HDS, whereas the tail-end catalyst provides hydroconversion. The combination of catalysts is selected according to the objectives of each situation.

Among all the possibilities available for treatment of heavy oils, hydrogen addition processes lead to high hydrogen consumption but higher liquid yields. These processes, which provide the feedstock for subsequent processes, require the use of well-designed catalysts capable of dealing with the high concentration of metals and asphaltenes present in the feedstock. Moreover, the multifunctional catalysts used for hydrocracking processes become poisoned by coke deposition and the heavy metals present in the feed, creating a hazardous waste which has to be disposed off properly and with safety. A high catalyst demetallization function is necessary because vanadium destroys the zeolitic catalyst used in the subsequent FCC process. Moreover, the concentration of nitrogen compounds must be reduced to a minimum to avoid poisoning the catalyst acid sites in this and subsequent FCC process. Although in the hydrocracking process the amount of metals is not as critical as in FCC, the elimination of nitrogen compounds is determinant to avoid poisoning of the catalyst acid sites.

Catalyst cycle life does not represent a problem in moving- and ebullated-bed reactors. Such technologies allow for replacing spent catalyst without interrupting operation; therefore, they are adequate for handling the most problematic feeds (high contents of metals and asphaltenes). Moving-bed reactors combine fixed-bed operation in plug-flow mode with the possibility of replacement during time-on-stream portions of spent catalyst. Catalyst replacement is a batch operation, typically carried out once or twice a week. The application of these reactors is specifically in front-end residue demetallization to protect subsequent fixed-bed reactors for HDS and HCR. Ebullated-bed reactors represent the most advanced hydroprocessing technology, suited specifically for upgrading extra-heavy feeds, directly without any type of pretreatment. The continuous catalyst replacement feature in these reactors allows using conventional high-activity HDT/HCR catalysts. Operation of these reactors is very flexible, hydroconversion is very efficient (up to 90 vol%), and products have low levels of sulfur, metals, and nitrogen. Nevertheless, ebullated-bed technologies suffer from considerable sediment formation and high catalyst consumption. Also, scale-up and design of such reactors is more difficult, due to the complex hydrodynamics.

Residue desulfurization processes (RDS/VRDS) are of common use to meet a variety of objectives, such as preparing feed for FCC, RFCC, coker, and HCR, and are available from major licensors (Chevron, Unocal, UOP, Shell, and Exxon). IFP's Hyvahl-F is another process of this nature, characterized by a system of fixed-bed reactors in series in combination with a graded catalyst system employed for atmospheric and vacuum residue hydrotreating and conversion to liquid fuels. Aiming to increase run length, this process was modified by introducing a swing reactor system (SRS) in front of the conventional fixed-bed reactor train. The SRS, under the brand name Hyvahl-S, has two reactors that can alternate in operation when the catalyst in one reactor is deactivated. Other advances in this field include systems for shortening the time required to replace spent catalyst, such as Shell's quick catalyst replacement system (QCR), which avoids opening and closing the reactor to load and unload catalyst. Examples of commercial developments using the moving-bed approach are the Hycon process developed by Shell and Chevron's onstream catalyst replacement reactor (OCR).

Ebullated-bed reactors are employed for residue hydrocracking as well as for desulfurization and demetallization. The two major commercialized technologies that use this type of reactor are H-Oil, licensed by Axens (IFP), and LC-Fining, licensed by Chevron Lummus. The two have very similar characteristics in terms of process parameters and reactor design, but differ in some mechanical details.

In summary, all of these processes have serious drawbacks when applied individually to the conversion of heavy feeds; thermal processes alone yield large amounts of coke, while catalytic processes suffer from excessive catalyst consumption due to rapid deactivation and high-hydrogen inputs in the case of hydroprocessing. Therefore, a careful inspection of feed characteristics, desired goals, and available technologies is required to define the best refining strategy. Either way, hydroprocessing will play an essential role in this matter, as it is favorable for primary upgrading and certainly offers much better selectivity to liquid yield and substantially cleaner products than thermal processes. Optimal hydroprocessing can be achieved by proper matching of reactor technology, catalyst, and reaction severity with the properties of heavy feeds.

Hydrovisbreaking Processing

1. HYCAR This is one type of noncatalytic process, based on visbreaking and involving treatment with hydrogen at mild conditions. This process is completed in three reactors:

- Visbreaking. This reactor carries out a moderate thermal cracking process in the presence of hydrogen. Hydrogen leads to more stable products than those obtained with straight visbreaking, which means that higher conversions can be achieved, producing a lower-viscosity product while no coke formation is induced.
- Hydrodemetallization. This reactor is to remove contaminants, particularly metals, prior to HCR. The product coming from the visbreaker is fed to the demetallization reactor in the presence of catalyst, which provides pore of sufficient size for diffusion and adsorption of high-molecular-weight constituents.

- Hydrocracking. In this reactor desulfurization and denitrogenation take place along with hydrocracking. Hydrocracking and hydrodemetallization reactors may employ inexpensive catalyst (CoMo) to remove metals and for cracking of complex molecules, respectively.

2. AQUACONVERSION Another type of hydrovisbreaking technology is aquaconversion, which is a catalytic process that uses catalyst-activated transfer of hydrogen from water added to the feedstock in slurry mode. The homogeneous catalyst is added in the presence of steam, which allows the hydrogen from the water to be transferred to the heavy oil when contacted in a coil–soaker system, normally used for the visbreaking process. Reactions that lead to coke formation are suppressed and there is no separation of asphaltene-type material.

The main characteristics of aquaconversion are:

- Hydrogen incorporation is much lower than that obtained when using a deep hydroconversion process under high hydrogen partial pressure.
- Hydrogen saturates the free radicals, formed within the thermal process, which would normally lead to coke formation.
- A higher conversion level can be reached, and thus higher API and viscosity improvements, while maintaining syncrude stability.
- It does not produce coke.
- It does not require a hydrogen source or high-pressure equipment.
- It can be implanted in the production area, thus eliminating the need for external diluent and its transport over large distances. Light distillates from the raw crude can be used as a diluent for both the production and desalting processes.

The presence of an oil-soluble catalyst and water prevents the coke formation and deposition of sediment that often occurs during visbreaking. In this process catalyst may be used as a support or mixed directly with the feedstock. The metals (metal salts) used for hydrovisbreaking are alkali metals such as potassium or sodium. The role of the catalyst is to enhance the dissociation of H_2O to release hydrogen (H^+) ions, which is subsequently consumed in hydroprocessing.

Fixed-Bed Hydroprocessing Hydroprocessing of residue in a fixed-bed reactor is well established and reported in the literature. The general characteristic of hydroprocessing is the simultaneous or sequential hydrogenation of hydrocarbon feed in the presence of sulfided catalyst by reacting with hydrogen. The main problem with fixed-bed catalyst is deactivation over time, which can be minimized by a guard-bed reactor in order to reduce metal deposition on the downstream reactors. Several combinations using two or three processing steps can be implemented in the refining. The catalyst in the guard-bed reactor is typically an HDM catalyst or large-pore catalyst with a high metal retention capacity.

Various improvements have been reported in the last decade to increase the efficiency of fixed-bed hydroprocessing, such as run length, conversion, and product quality. Some of these improvements have been focused on mechanical design, such as the use of a bunker, swing reactors, guard-bed reactors, feed distribution, and a coke and metal deactivation-resistant catalyst, including pore and particle grading and onstream catalyst replacement. Despite all disadvantages, mainly short catalyst life, up until now most residue hydroprocessing units have used fixed-bed reactors.

1. RDS/VRDS The RDS process is used for atmospheric residuum hydrotreating and the VRDS process for vacuum residuum desulfurization to remove sulfur and metallic constituents while part of the feedstock is converted to lower-boiling products. In both processes, AR or VR feedstocks contact with catalyst and hydrogen at moderate temperatures and pressures, consuming about 700 to 1300 standard cubic feet (SCF) H_2/bbl of feed. The conversion increases with temperature, but due to the high coke deposition, the process is not appropriate for use at high temperature. The RDS and VRDS processes do not convert directly to transportation fuel, but this process is able to produce acceptable feedstock for RFCC or delayed coking units to achieve minimal production of residual products in a refinery. The basic process flow and catalyst are the same for RDS and VRDS.

A combination of RDS/VRDS and RFCC has gained wide acceptance due to the selective conversion of residue and smaller amount of by-products. The limitation for RFCC is deposited metals, since Ni deposition increases olefin yields through dehydrogenation, and as a result more coke formation is obtained, while deposition of V metal destroys the zeolite structure. Also, the combination of a desulfurization step and VRDS is often seen as an attractive alternative to the atmospheric residuum desulfurizer. In addition, either RDS or VRDS can be coupled with other processes (such as delayed coking and solvent deasphalting) to achieve the optimum refining performance.

2. HYVAHL-F AND HYVAHL-S These processes are used to hydrotreat AR and VR feedstocks to convert them into more valuable products (naphtha and middle distillates). Hyvahl processes are designed primarily for feedstock containing high concentrations of asphaltene, maltenes, and metals, which strongly limit catalyst performance. The reactors can be used in the classical fixed-bed (Hyvahl F) or swing-mode system (Hyvahl S). Hyvahl-F uses fixed-bed reactors in which liquid and gas flow downstream co-currently in a trickle-flow regime. The first catalyst is resistant to fouling, coking, and plugging by asphaltene constituents, has a high metals retention capacity, and is used for both hydrodemetallization and most of the conversion. So the highly active second catalyst is protected from metal poisons and deposition of cokelike products and can carry out its deep hydrodesulfurization and refining functions. Its main

application is for processing distillate fractions and some atmospheric residua. For vacuum residua and heavy oils, it may require frequent catalyst replacements, which may be complex and uneconomical. Optimization of catalyst properties, space velocity, and maximum reaction temperature may extend the life of the catalyst, depending mainly on the metals content in the feed. Hyvahl-S, also called the Hyvahl-F-swing reactor, uses two guard reactors in a swing arrangement and switchable operation, which are fixed-bed reactors with simple internals. This configuration allows fast switching of the guard reactor in operation with a deactivated catalyst to the other guard reactor with fresh catalyst, without shutting down the plant. The major feature of this process is a fixed bed using the swing-mode reactor concept at high temperature, high hydrogen pressure, and low contact time. The switching of guard reactor and adjusting of conditions are fast and controlled by a conditioning package.

3. RESIDUE HYDROCRACKING The growing demand for middle distillates has increased the need for HCR in terms of process flexibility as well as configuration and product composition. The catalysts used for HCR should have dual functionality [i.e., cracking and hydrogenation (HYD) functions]. The process scheme of a typical HCR fixed-bed system contains two reactors. The first reactor (first-stage HCR) contains an HDT catalyst of high activity for the removal of heteroatoms or metal, while the second reactor (second-stage HCR) contains the actual HCR catalyst. In general, the first-stage reactor contains NiMo catalyst, which removes S, N, metals, and hydrogenate aromatics, while the second reactor possesses an acidic support (zeolite, mixed oxides)–based catalyst that promotes hydrogenation as well as hydrocracking reaction.

In contrast to HDT, the support plays an active role in the conversion of the feed in HCR catalyst. The performance of HCR catalyst is determined by the ratio or balance between the hydrogenation metal (sulfide) site and the acid sites of support. When the number of hydrogenating sites is low compared with the number of acid sites, secondary cracking processes can take place, resulting in light products. Additionally, the hydrogenation function also prevents the oligomerization and coking over the acid sites. A deficient hydrogenation function will lead to enhanced deactivation of the catalyst. On the other hand, when a very strong hydrogenation function is used, cracking is suppressed in favor of isomerization. In the ideal HCR the catalyst requires a balance between metal and acid functions.

Several fixed-bed hydrocracking processes are used by refiners on the basis of their product selectivity:

- *IFP hydrocracking.* This process features a dual-catalyst system. The first catalyst is a promoted NiMo amorphous catalyst which acts to remove sulfur and nitrogen and hydrogenate aromatic rings, while the second catalyst is a zeolite which finishes the hydrogenation and promotes the hydrocracking reaction. There are two versions of this process:

- Single-stage process. The first reactor effluent is sent directly to the second reactor, followed by separation and fractionation steps. The fractionator bottoms are recycled to the second reactor or sold.
- Two-stage process. The feedstock and hydrogen are heated and sent to the first reaction stage, where conversion to products occurs. The reactor effluent phases are cooled and separated and the hydrogen-rich gas is compressed and recycled. The liquid leaving the separator is fractionated, the middle distillates and lower-boiling streams are sent to storage, and the high-boiling stream is transferred to the second reactor section and then recycled to the separator section.

- *Isocracking.* Depending on the feedstock properties, there are various process flow schemes: single-stage once-through liquid; single-stage partial recycle of heavy oil; single-stage extinction recycle of oil (100% conversion); and two-stage extinction recycle of oil. The isocracking process uses multibed reactors and a number of catalysts. The catalysts are dual function, being a mixture of hydrous oxides (for cracking) and heavy metal sulfides (for hydrogenation). The catalysts are used in a layered system to optimize the processing of the feedstock, which undergoes changes in its properties along the reaction pathway.
- *Mild hydrocracking.* This process uses operating conditions (and a flow scheme) similar to those of a vacuum gas oil desulfurizer to convert the feed into significant yields of lighter products. The conditions for mild hydrocracking are typical of many low-pressure desulfurization units, and the process is a simple form of hydrocracking.
- *MRH.* MRH is a hydrocracking process designed to upgrade heavy feedstocks containing large amounts of metals and asphaltene, such as VR and bitumen, and to produce mainly middle distillates. The reactor is designed to maintain a mixed three-phase slurry of feedstock, fine powder catalyst, and hydrogen, and to promote effective contact.
- *Unicracking.* There are various versions of this process:
  - *Basic unicracking.* This is a fixed-bed catalytic process designed as a single-stage or two-stage system with provisions to recycle to extinction. The process operates satisfactorily for a variety of feedstocks. The catalysts, which induce desulfurization, denitrogenation, and hydrocracking, are based on both amorphous and molecular-sieve-containing supports.
  - *Advanced partial conversion unicracking* (APCU). This is a recent advancement in the area of ultralow-sulfur diesel (ULSD) production and feedstock pretreatment for catalytic cracking units.
  - *HyCycle unicracking.* This is designed to maximize diesel production for full-conversion applications.

Moving-Bed Hydroprocessing There are a few types of hydroprocessing reactors with moving catalyst beds in which the catalyst goes in downflow through the reactor by gravitational forces. In general, catalyst replacement is commonly a batch operation, which is done typically once or twice a week. The fresh catalyst enters at the top of the reactor and the deactivated catalyst leaves the reactor at the bottom, while the hydrocarbon goes either in counter- or co-current flow through the reactor. With this moving-bed system, the catalyst can be replaced either continuously or in batch operation. Catalyst transfer is the most critical section. The countercurrent mode of operation seems to be the best configuration since the spent catalyst contacts the fresh feed at the bottom of the moving-bed reactor while the fresh catalyst reacts with an almost already hydrodemetallized feed at the top of the moving-bed reactor, resulting in lower catalyst consumption.

1. HYCON PROCESS The Hycon process is used to improve the quality of residual oils by removing sulfur, metals, and asphaltene constituents and is typically operated in fixed-bed mode, but with increasing metal content in the feedstock, one or more moving-bed "bunker" reactors are added as the leading reactors for HDM. The process is suitable for a wide range of the heavy feedstocks, particularly high in metals and asphaltene constituents. This process enables easy catalyst replacement (to remove or add portions of catalyst) without interrupting operation by means of valves of lock hoppers. The catalyst and heavy oil are fed in co-current flow; the fresh catalyst enters at the top of the reactor and deactivated catalyst is removed from the bottom. The catalyst is replaced at a rate that will ensure a total plant run time of at least a year, which depends on the metal contaminants in the feed. In this way, the bunker reactor technology combines the advantages of plug-flow fixed-bed reactor operation with easy catalyst replacement and provides extra process flexibility if it is used upstream from the desulfurization reactor, especially with reference to the processing of feedstocks with a high metal content. For feeds containing a large amount of metal, metal sulfide can be better accommodated on the catalyst in a bunker-flow reactor than on other reactor systems. Operating conditions and the catalyst addition and withdrawal rates can also be adjusted to ensure that the catalyst taken out is completely spent, while retaining an acceptable average activity in the reactor.

2. OCR PROCESS OCR (onstream catalyst replacement) is a moving-bed reactor for hydroprocessing of heavy oils and residua with a significant amount of metals operating in a countercurrent mode at high temperature and pressure. Fresh catalyst is added at the top of the reactor and the feed into the bottom, and both move through the reactor in a countercurrent flow, causing the feed with the highest content of impurities to contact the oldest catalyst first. The fresh catalyst can be added at the top of the reactor and the spent catalyst removed from the OCR reactor while the unit is onstream. An OCR moving-bed reactor can be incorporated in the processing scheme either before or after fixed-bed reactors, so that heavier feeds with higher levels of

contaminants can be processed while maintaining constant the product quality and economical fixed-bed reactor run lengths.

3. HYVAHL-M This process employs countercurrent moving-bed reactors and is recommended for feeds containing large amounts of metals and asphaltenes. It requires special equipment and procedures for safe and effective catalyst transfer into and out of the high-pressure unit, similar to the OCR reactor. Catalyst is taken at atmospheric pressure and transferred to a reactor operating under hydrogen pressure, and then the catalyst is taken from the reactor at high operating conditions and discharged to the atmosphere.

Ebullated-Bed Hydroprocessing In ebullated-bed hydroprocessing, the catalyst within the reactor is not fixed. In such a process, the hydrocarbon feed stream enters the bottom of the reactor and flows upward through the catalyst, which is kept in suspension by the pressure of the fluid feed. The hydrocarbon feed and hydrogen are fed upflow through the catalyst bed, expanding and backmixing the bed, and minimizing bed plugging and ΔP. The oil is separated from the catalyst at the top of the reactor and recirculated to the bottom of the bed to mix with the new feed. Alternatively, fresh catalyst is added to the top of the reactor and spent catalyst is withdrawn from the bottom of the reactor.

Ebullating-bed reactors are capable of converting the most problematic feeds, such as AR, VR, and all other heavy oil feedstocks, which have high contents of asphaltenes, metals, and sulfur. Ebullating-bed reactors can perform both HDT and HCR functions; thus, these reactors are referred as dual-purpose reactors. Ebullating-bed catalysts are made of pellets or grains that are less than 1-mm in size to facilitate suspension by the liquid phase in the reactor.

There are three main ebullated-bed processes, which are similar in concept but different in mechanical aspects.

1. *H-Oil.* The H-Oil ebullated-bed process uses a single-stage, two-stage, or three-stage ebullated-bed reactors and can operate over a wide range of conversion levels. It is particularly adapted to process heavy vacuum residues with high metals and Conradson carbon to convert them into distillate products as well as to desulfurize and demetallize feeds to coking units or residue fluid catalytic cracking units, for production of low-sulfur fuel oil or for production in asphalt blending. An H-Oil process maintains constant product properties during cycle length. Since an H-Oil reactor has the unique characteristic of stirred-reactor-type operation with a fluidized catalyst, it has the ability to handle exothermic reactions, solid-containing feedstock, and flexible operation while changing feedstocks or operating objectives.
2. *T-Star.* T-Star is an extension of the H-Oil process which can maintain global conversions in the range 20 to 60% and specifically, an HDS of 93 to 99%. This process can act as an FCC pretreater or vacuum gas oil (VGO) hydrocracker. H-Oil catalyst can be used in the T-Star process. A T-Star reactor

can also be placed in-line with an H-Oil reactor to improve the quality of H-Oil distillate products. In mild hydrocracking mode, the T-Star process can reach conversions up to the 60%, with a catalyst not sensitive to sulfur and nitrogen levels in the feed and will provide constant conversion, product yields, and product quality. This consistency in output is due to the catalyst being replaced while the unit remains online.

3. *LC-Fining.* The LC-Fining ebullated-bed process is a hydrogenation process that can be operated for HDS, HDM, and HCR of atmospheric and vacuum residues. LC-Fining is well suited for extraheavy residue, bitumen, and vacuum residue feedstock HDT and has demonstrated long cycle lengths. The general advantages of LC-Fining are low investment, more light-end recovery, lower operating costs, and lower hydrogen losses. This process yields a full range of high-quality distillates; heavy residue can be used as fuel oil, synthetic crude, or feedstock for RFCC, coker, visbreaker, or SDA. The LC-Fining process can achieve conversions for HDS of 60 to 90%, HDM of 50 to 98%, and CCR reduction of 35 to 80%. The process parameters and reactor design are marginally different from the H-Oil process. The reaction section uses a commercially proven low-pressure hydrogen recovery system. An internal liquid recycle is provided with a pump to expand the catalyst bed continuously. As a result of an expanded bed operating mode, small pressure drops and isothermal operating conditions are accomplished. Small-diameter extruded catalyst particles as small as 0.8 mm ($\frac{1}{32}$ in.) can be used in this reactor. Separating the reactor effluent and purifying the recycled hydrogen at low pressure results in lower capital cost and allows design at lower gas rates.

Slurry-Bed Hydroprocessing Slurry-bed reactor can also be used for hydroprocessing of feeds with very high metals content to obtain lower-boiling products using a single reactor. SBR-based technologies combine the advantages of the carbon rejection technologies in terms of flexibility with the high performances peculiar to the hydrogen addition processes. SBR achieves a similar intimate contacting of oil and catalyst and may operate with a lower degree of backmixing than EBR. In contrast to FBR and EBR, in SBR a small amount of finely divided powder is used, which can be an additive or a catalyst (or catalyst precursors). The catalyst is mixed with the feed (heavy oil), and both are fed upward with hydrogen through an empty reactor vessel. Since the oil and catalyst flow co-currently, the mixture approaches plug-flow behavior. In an SBR the fresh catalyst is slurried with the heavy oil prior to entering the reactor, and when the reaction finishes, the spent catalyst leaves the SBR together with the heavy fraction and remains in the unconverted residue in a benign form.

1. CANMET Canmet is a hydrocracking process for heavy oils, atmospheric residua, and vacuum residua which was developed to upgrade heavy oil and

tar sand bitumen as well as residua. The process uses an additive to inhibit coke formation, thus allowing high conversion to lower-boiling products using a single reactor. The vertical reactor vessel is free of internal equipment and operates in a three-phase mode. The solid additive particles are suspended in the primary liquid hydrocarbon phase through which the hydrogen and product gases flow rapidly in bubble form. The spent additive leaves with the heavy fraction and remains in the unconverted vacuum residue. Typical operating conditions are a reactor temperature of 440 to 460°C and a pressure of 10 to 15 MPa.

2. MICROCAT-RC The Microcat-RC or M-Coke process is a catalytic ebullated-bed hydroconversion process which operates at relatively moderate pressures and temperatures. The catalyst particles are dispersed uniformly throughout the feed, which results in less distance between particles and less time for a reactant molecule or intermediate to find an active catalyst site. The hydrocarbon feed, microcatalyst, and hydrogen are fed to the reactor. The effluent is sent to a flash separation zone to recover hydrogen, gases, and liquid products. The residuum from the flash step is then fed to a vacuum distillation tower to obtain a 565°C– product oil and a 565°C+ bottoms fraction that contains unconverted feed, microcatalyst, and essentially all of the feed metals.

3. MRH The MRH process is a hydrocracking process designed to upgrade heavy feedstocks containing large amounts of metals and asphaltene, such as VR and bitumen, and to produce mainly middle distillates. The reactor is designed to maintain a mixed three-phase slurry of feedstock, fine powder catalyst, and hydrogen, and to promote effective contact. In the process, a slurry consisting of heavy feedstock and fine powder catalyst is preheated in a furnace and fed into the reactor vessel. From the lower section of the reactor, bottom slurry oil containing the catalyst, uncracked residuum, and a small amount of vacuum gas oil fraction are withdrawn. Vacuum gas oil is recovered in the slurry separation section, and the remaining catalyst and coke are fed to the regenerator.

4. VCC AND HDH PLUS The Veba Combi Cracking (VCC) process is a hydrocracking and hydrogenation process for converting residua and other heavy feedstocks. The VCC technology was transferred to HDH (hydrocracking distillation hydrotreating), which has been developed in parallel by Intevep since 1984. Only the HDH process survives. Recently, Intevep announced the implementation of HDH Plus technology in two refineries in Venezuela: Puerto La Cruz and El Palito. In the process, the heavy feedstock, slurried with a small amount of finely powdered additive and mixed with hydrogen and recycle gas, is hydrogenated (hydrocracked) using a commercial catalyst and liquid-phase hydrogenation reactor operating at 440 to 485°C and 2175 to 4350 psi pressure.

5. ENI SLURRY TECHNOLOGY (EST) The EST process is based on the slurry hydrotreatment of heavy feedstock at relatively low temperature in the presence of hydrogen and a dispersed catalyst, which is recycled to the slurry reactor via solvent deasphalting together with the asphaltene recycle. EST has demonstrated residue conversion of 98 to 99%, HDS > 80%, HDM > 99% and CCR removal > 96% on a small pilot-plant scale. Part of the feedstock is converted directly to light and medium distillates, while the other products represent suitable feedstocks for FCC or hydrocracking. Some of the reported advantages of EST process are feedstock flexibility, optimal hydrogen utilization/consumption, and product slate flexibility.

2

REACTOR MODELING IN THE PETROLEUM REFINING INDUSTRY

2.1 DESCRIPTION OF REACTORS

Multiphase catalytic packed-bed reactors (PBRs) operate in two modes: (1) trickle operation, with a continuous gas phase and a distributed liquid phase, and the main mass transfer resistance located in the gas, and (2) bubble operation, with a distributed gas and a continuous liquid phase, and the main mass transfer resistance located in the liquid phase. For three-phase reactions (gas and liquid phases in contact with a solid catalyst), the common modes of operation are trickle- or packed-bed reactors, in which the catalyst is stationary, and slurry reactors, in which the catalyst is suspended in the liquid phase (Figure 2.1). In these reactors, gas and liquid move co-currently downflow or gas is fed countercurrently upflow. Commercially, the former is the most used reactor, in which the liquid phase flows mainly through the catalyst particles in the form of films, rivulets, and droplets (Figure 2.2).

Based on the direction of the fluid flow, PBRs can then be classified as trickle-bed reactors (TBRs) with co-current gas–liquid downflow, trickle-bed reactors with countercurrent gas–liquid flow, and packed-bubble reactors, where gas and liquid are contacted in co-current upflow. To carry out the catalyst and reactor selection and process design properly, knowledge of what each reactor type can and cannot do is very important. When a fixed-bed reactor is chosen, the question frequently asked is whether to use an upflow or downflow mode of operation.

Modeling and Simulation of Catalytic Reactors for Petroleum Refining, First Edition.
Jorge Ancheyta.
© 2011 John Wiley & Sons, Inc. Published 2011 by John Wiley & Sons, Inc.

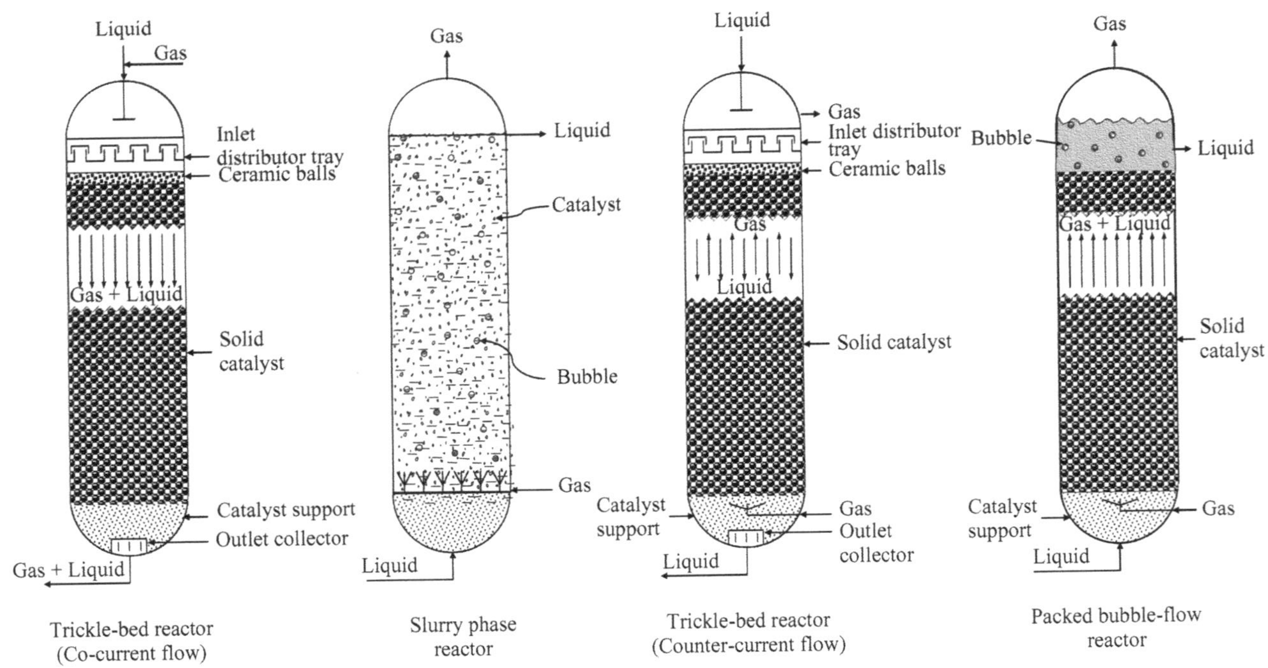

Figure 2.1. Various types of multiphase catalytic reactors.

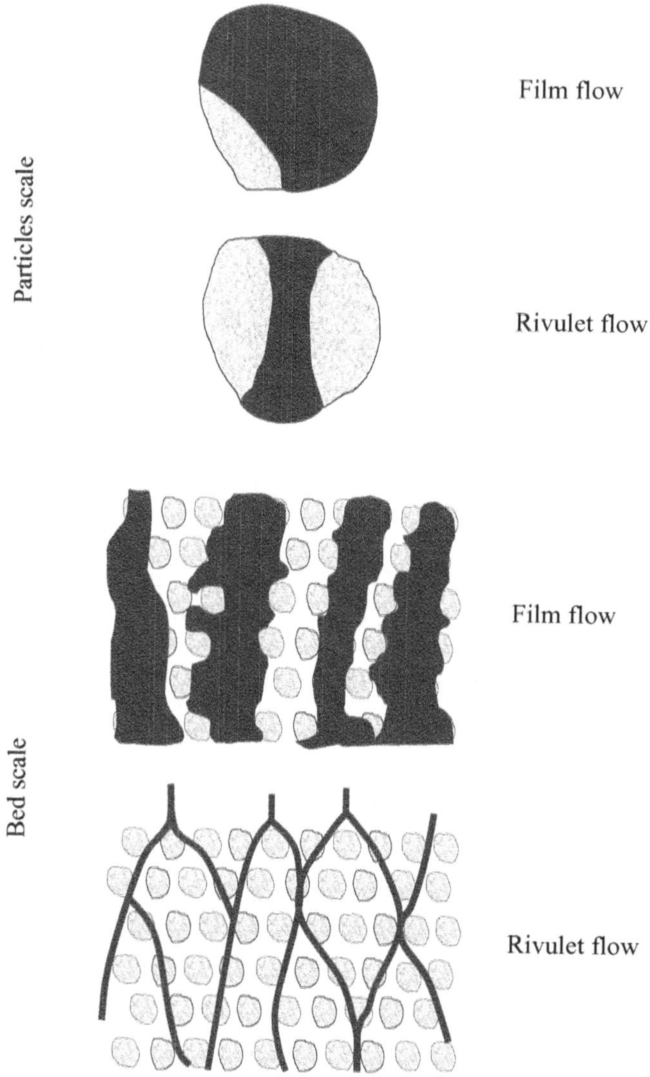

Figure 2.2. Liquid flow texture found during the trickle-flow regime in a TBR.

In the case of catalytic packed beds with two-phase flow, such as those used for straight-run naphtha hydrodesulfurization, from a reaction engineering perspective, a large catalyst-to-liquid volume ratio and plug flow of both phases are preferred, and catalyst deactivation is very slow or negligible, which facilitates reactor modeling and design. However, for three-phase catalytic reactors such as those employed for hydrotreating of middle distillates and heavy petroleum fractions, the reaction occurs between the dissolved gas and the liquid-phase reactant at the surface of the catalyst, and the choice of upflow versus downflow operation can be based on rational considerations regarding the limiting reactant at the operating conditions of interest (Duduković et al., 2002).

2.1.1 Fixed-Bed Reactors

In a TBR the catalyst bed is fixed (Figure 2.1), the flow pattern is much closer to plug flow, and the ratio of liquid to solid catalyst is small. If heat effects are substantial [i.e., highly exothermic reactions such as those occurring in hydrotreating of unsaturated feeds (light cycle oil from fluid catalytic cracking units)], they can be controlled by recycling of the liquid product stream, although this may not be practical if the product is not relatively stable under reaction conditions or if very high conversion is desired, as in HDS, since recycling causes the system to approach the behavior of a continuous-stirred-tank reactor (CSTR). For such high-temperature increases, the preferred solution is quenching with hydrogen, although the use of other streams has also been reported (Muñoz et al., 2005; Alvarez and Ancheyta, 2008). Even when a completely vapor-phase reaction in a fixed catalyst bed may be technically feasible, a TBR may be preferred to save energy costs due to reactant vaporization. The limiting reactant may be essentially all in the liquid phase or in both the liquid and gas phases, and the distribution of reactant and products between the gas and liquid phases may vary with conversion.

TBR with Co-current Gas–Liquid Downflow A TBR consists of a column that may be very high (above 10 to 30 m), equipped with one or various fixed beds of solid catalysts, throughout which gas and liquid move in co-current downflow. Figure 2.3 shows the typical film flow texture found during a trickle-flow regime (Gianetto and Specchia, 1992). In this mode, gas is the continuous phase and liquid holdup is lower. This operation is the one most used in practice, since there are less severe limitations in throughput than in countercurrent operation.

For gas-limited reactions (high liquid reactant flux to the catalyst particle, low gas reactant flux to the particle), especially at partially wetted conditions, a downflow reactor is preferred, as it facilitates transport of the gaseous reactant to the catalyst (Duduković et al., 2002). In contrast to commercial TBR, in the case of bench-scale TBR operating at equivalent space velocity, the liquid velocity and the catalyst bed length have important effects on the performance of the reactor. The principal advantages and disadvantages of TBR with downflow co-current operation are given below.

Advantages

- Recommended for gas-limited reactions
- Liquid flow approaches plug-flow behavior, which leads to high conversions

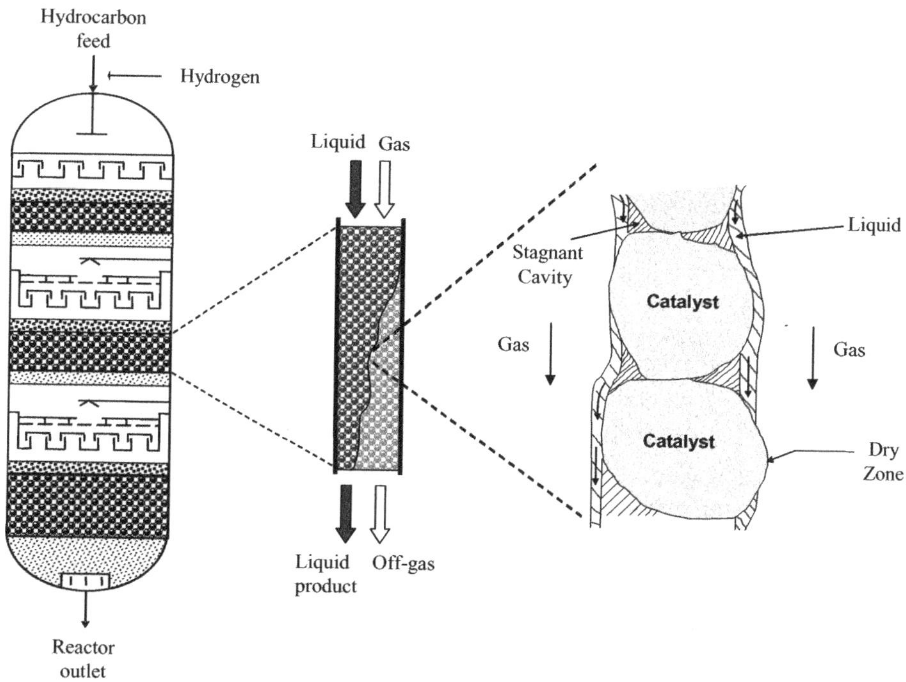

Figure 2.3. Nonideal TBR suffering from liquid maldistribution.

- Low liquid–solid volume ratio: fewer occurrences of homogeneous side reactions
- Possibility of varying the liquid rate according to catalyst wetting and heat and mass transfer resistances
- A variety of flow regimes allowed; most flexible with respect to varying throughput demands
- The downflow mode also helps keep the bed in place, although with catalysts that are soft or deformable, this might hasten undesired cementation
- Compared with countercurrent flow operation, for co-current flow of the two phases, no limitation on the throughput arises from the phenomenon of flooding, and the quantities of the phase that can be passed depend only on the upstream pressure available because of vaporization effects
- At higher gas loadings, the texture of the liquid is modified by gas-phase friction, the liquid distribution is improved (lower liquid wall flow), and the pressure drop rises (less rapidly in co-current than in countercurrent flow)
- Easy operation with fixed adiabatic beds; for exothermic reaction systems, gas or liquid streams as quench, and the liquid and/or gas recycle limit temperature rises

- Possibility of operating at higher pressure and temperature
- Pressure drop through the bed is relatively low, thus reducing pumping costs
- Larger reactor size, and generally of simple construction, as there are no moving parts
- Lower investment and operating costs, and low catalyst loss, which is important when costly catalysts are used

Disadvantages

- Limitations on the use of viscous or foaming liquids
- Limited to reasonably fast reactions
- Lower catalyst effectiveness, due to the use of large catalyst particle size
- Particle size cannot be smaller than 1 mm because of pressure drop; risk of increasing pressure drop or obstructing catalyst pores when side reactions lead to fouling products
- Reactor-scale maldistribution, channeling, and incomplete and/or ineffective external catalyst wetting (poor contacting effectiveness) can occur with low liquid flow rates and reactor diameter/particle size ratios (<25)
- Sensitivity to thermal effects, although this drawback can be limited by recycling part of the outlet liquid or injecting cooled gas (quenching)
- Difficulties in the recovery of reaction heat
- Lower liquid holdup compared with co-current gas–liquid upflow
- Deactivation of the catalyst by deposits
- Dismantling of the reactor during catalyst replacement
- In hydrotreating (HDT) reactors, most of the bed is under the H_2S and NH_3 reach regime and its inhibiting effect is strongest in the region where the refractory sulfur compounds have to be converted. NH_3, particularly, strongly suppresses the activity of the acidic function of the hydrocracking catalyst
- H_2 partial pressure will be lowest at the HDT reactor outlet due to the combined effect of pressure drop, hydrogen consumption, and reduction of hydrogen purity as gaseous by-product yields (H_2S, NH_3, and H_2O) increase along the reactor
- Used in downward mode in the refining industry with less conversion; the inhibition effect of H_2S and NH_3 on the catalyst results in a poorer performance

TBR with Countercurrent Gas–Liquid Flow TBRs operating in countercurrent gas–liquid flow (Figure 2.1) provide an opportunity for selective removal of by-products that may act as inhibitors (e.g., in hydrodesulfurization, where hydrogen sulfide may have an inhibitory effect). The introduction of FBRs with countercurrent flow in a number of refining operations is probably either via redesign of existing reactors or by introduction of new technology. As mentioned earlier, the goal is not an improvement in reactant (hydrogen) mass transfer, which is not rate limiting, but enhanced removal of inhibitory by-products or in situ product separation. That is why countercurrent flow will become more prominent in the future for processes that suffer from by-product catalyst inhibition (Duduković et al., 2002).

A catalytic PBR with countercurrent mode is a suitable alternative to TBRs for reactions conducted over catalysts with a very large surface area-to-volume ratio. However, the main problem of the countercurrent reactor for commercial application is due to hardware limitations. There is therefore a need to develop improved hardware configurations that allow countercurrent contacting of gas and liquid in the presence of small catalyst particles (Kundu et al., 2003). The main advantages and disadvantages of TBRs with countercurrent flow are given below.

Advantages

- Countercurrent operation is preferred over co-current when a large heat of reaction is involved
- Countercurrent operation gives a more favorable flat axial temperature profile
- Large surface area for vapor–liquid mass transfer
- High ratio of number of active sites to reactor volume
- Easy catalyst handling
- For the HDT process, the major part of the bed is in an H_2S-lean regime, which protects from inhibition by H_2S formed in a large part of the bed.
- H_2 partial pressure is highest at the end of the bed, and temperature in this part can be lowered and more active, less sulfur-tolerant catalysts can be used in the downstream part of the bed, which will favor the chemical equilibrium for reversible reactions [i.e., hydrodearomatization (HDA) reaction]. The effect of equilibrium-limited conversion and product inhibition is reduced
- The major part of the bed is in the NH_3 [a by-product of hydrodenitrogenation (HDN) reaction]-lean regime, which favors the HDT reaction by protection from the inhibition of NH_3 and H_2S. This operation has great advantages through omitting two separate reactor stages
- The concentration of gas impurities formed during reaction is less in most parts of the bed. This favors the conversion of reactions normally limited by chemical equilibrium and enables handling more difficult feedstocks to obtain higher levels of conversion. Figure 2.4 shows typical partial pressure profiles of H_2S along the bed length for co-current and

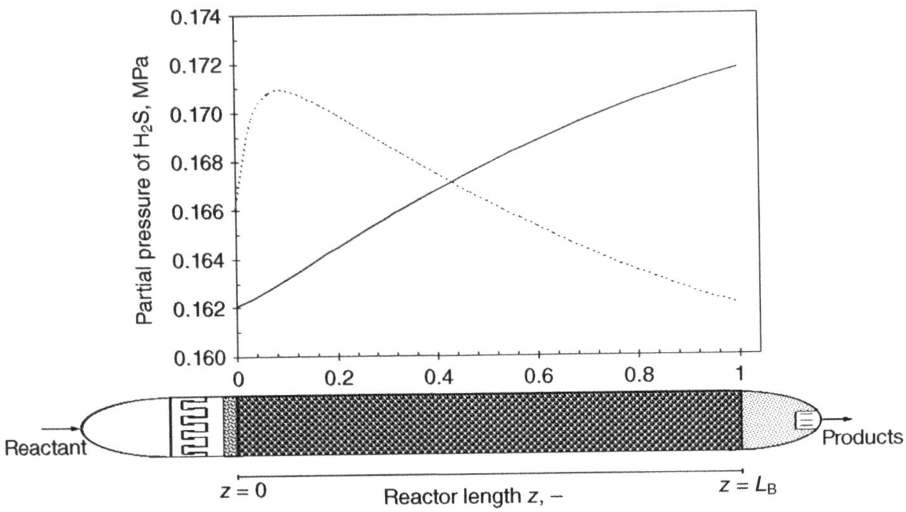

Figure 2.4. Profiles of H_2S partial pressure along the catalytic bed in an HDT reactor (—, co-current; ---, countercurrent).

countercurrent operations during hydroprocessing, in which the aforementioned behavior is clearly observed

- Countercurrent operation provides the highest hydrogen purity in that part of the bed where the least reactive compounds need to be converted

Disadvantages

- Presence of flooding at high liquid throughputs
- Estimation of liquid holdup, pressure drop, and mass transfer coefficients is difficult since correlations employed to calculate these parameters do not include data for the small porous catalyst packing typically used in PBRs with two-phase flow
- Limited to low velocities far below those of industrial interest, due to the occurrence of excessive pressure drop and flooding problems
- It is not possible to use smaller (1 to 5 mm) catalyst particles than those used in co-current downflow TBRs
- High axial dispersion effects in the liquid phase

Packed Bubble-Flow Reactors with Co-current Gas–Liquid Upflow This classification includes upflow reactors, upflow co-current reactors, packed-bubble columns, upflow packed-bubble columns, and flooded fixed-bed reactors. In bubble-flow operation a continuous liquid phase, together with a dispersed gas phase, move upward co-currently through the packed bed (Figure 2.1). Such an operation would be recommended in cases where liquid reactants are treated with a relatively small amount of gas, as in the hydration of nitro compounds and olefins, or where a relatively large liquid residence time is required for the degree of conversion desired. Use of these reactors assures complete external wetting of the catalyst and high liquid holdup. In this mode the liquid is typically the continuous phase.

Bubble operation is also advantageous when the reactor diameter/particle diameter ratio is relatively small, because the liquid catalyst contact is more effective than in trickle operation. Compared with empty bubble columns, the packed bed has the advantage of reducing substantially backmixing in the flowing phases as well as the coalescence of gas bubbles. Under any conditions the wall heat transfer coefficient should also be higher than it is in trickle operation (Hofmann, 1978).

For liquid-limited reactions (low liquid reactant flux to the catalyst particle, high gas reactant flux to the particle), an upflow reactor should be preferred, as it provides complete catalyst wetting and the fastest transport of the liquid reactant to the catalyst. For very shallow catalyst beds, upflow operation gives much better conversions than downflow operation under the same reaction conditions. The gas and liquid flow rates typically used in a bench-scale downflow trickle-bed HDS reactor are such that when they are used in co-current upflow operation, a bubble flow regime will be generated.

The performance of a reactor under this hydrodynamic flow condition should be considerably different from the one obtained under trickle-flow conditions. In an upflow system the low-boiling components, which are generally more reactive, pass into the vapor phase and are swept out more rapidly than the high-boiling material, which progresses relatively slowly through the bed. This superior performance of upflow processing is attributed to the long residence time of the heavy liquid fractions, but a more important factor may be the very low liquid flow used (Satterfield, 1975).

When both gas and liquid flow upward, maldistribution of liquid or incomplete catalyst wetting should not be very important, particularly when the hydrodynamic conditions of bubble flow prevail within the reactor. An upflow (flooded bed) reactor, which should give good solid–liquid contacting, could be used instead of an autoclave to obtain information on the intrinsic kinetics. The main advantages and disadvantages of TBRs with co-current upflow are given below.

Advantages

- Recommended for liquid-limited reactions
- Liquid holdup is higher. The liquid holdup is larger in an upflow operation than in a downflow operation under similar conditions
- Better effective wetting

- Better thermal stability for highly exothermic reactions
- High liquid saturation
- The liquid flow can be more uniformly distributed (better distribution of liquid throughout the catalyst bed)
- The gas–liquid and liquid–solid mass transfer coefficients are larger in an upflow operation than in a downflow operation
- In backmix flow conditions, where variations in gas and liquid flow rates change the conversion, upflow operation gives better results than downflow operation under the same conditions
- Larger effective residence time
- If a catalyst gradually becomes deactivated by the deposit of polymeric or tarry materials, the upflow reactor may maintain its activity longer by washing off these deposits more effectively
- For rapid and highly exothermic reactions, heat transfer between liquid and solid may also be more effective in upflow than in downflow operation

Disadvantages

- For HDT operations, conversions of sulfur, metals, and asphaltenes decrease with an increase in gas and liquid flow rates at constant temperature and pressure. Conversion of sulfur in upflow operation is reduced faster with time than in downflow operation; however, the conversion is always highest
- Higher pump requirements in order to overcome the hydrostatic head of the liquid
- The need of some designs to avoid the fluidization of the catalyst unless the catalyst was held in place by an extra weight or suitable mechanical methods
- If limiting reactant is present in both phases, over a range of operating conditions in which catalyst pellets filled with liquid are diffusion limited, an upflow reactor would be expected to exhibit a lower reaction rate than a partially wetted TBR
- Formation of stagnant zones inside the catalyst bed
- Higher axial dispersion compared with the downflow mode of operation

2.1.2 Slurry-Bed Reactors

The best alternative to the use of a fixed-bed reactor with two-phase flow, either upward or downward, is a slurry-bed or ebullating-bed reactor in which the catalyst particles, which must be substantially smaller, are in motion. These reactors are sometimes termed three-phase fluidized-bed reactors or suspended-bed reactors (Figure 2.1). The main advantages and disadvantages of slurry-bed reactors are given below.

Advantages

- High heat capacity to provide good temperature control
- Potentially high reaction rate per unit volume of reactor if the catalyst is highly active
- Easy heat recovery
- Adaptability to either batch or flow processing
- Much lower pressure drop
- The catalyst may readily be removed and replaced if its working life is relatively short
- Continuous removal of solid material formed in reaction
- Because of high intraparticle diffusion rate, small particles can be used, which may allow for operating at catalyst effectiveness factors approaching unity, of special importance if diffusion limitations cause rapid catalyst degradation or poorer selectivity
- Lower external mass transfer resistance by means of a high stirring speed

Disadvantages

- Residence-time distribution patterns are close to those of a CSTR, which makes it difficult to obtain high degrees of conversion except by staging and/or increasing operation temperature
- Generation of fine particles by abrasion of the catalyst
- Catalyst removal by filtration may provoke problems with possible plugging difficulties on filters, further time of operation, and the costs of filtering systems may be a substantial portion of the capital investment
- Higher catalyst consumption than that of fixed-bed reactors
- Difficult to scale up
- The high liquid-to-solid ratio in a slurry-bed reactor allows homogeneous side reactions to become more important, if any are possible
- Potential hazard of localized overheating in the reactor because of bad fluidization
- Backmixed flow and the volume of the reactor are not fully utilized

2.2 DEVIATION FROM AN IDEAL FLOW PATTERN

2.2.1 Ideal Flow Reactors

Plug-flow and perfectly mixed patterns are the two ideal extremes in which continuous reactors commonly lie (Figure 2.5). The ideality of these two flow patterns allows simple mathematical description of the reactors, and reliable and easy treatment of experimental data since radial and axial mass and heat dispersion terms in the continuity equations are neglected. Macrogradients of

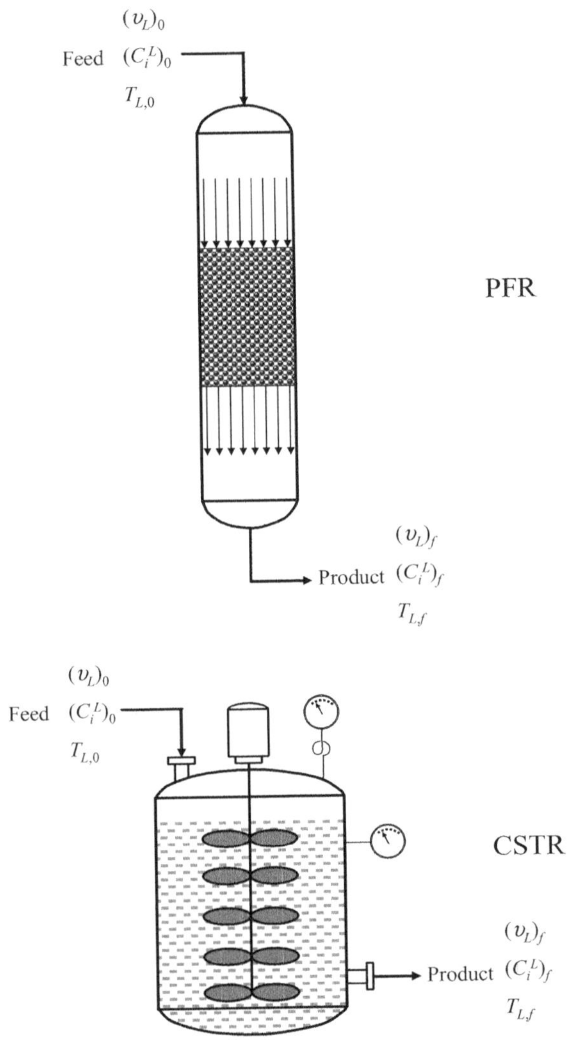

Figure 2.5. Ideal flow patterns.

concentration or temperature in the reactor are the major reasons for deviations from ideal flow patterns, which may result from imperfect mixing in the case of CSTR or from axial dispersion, wall effects, or catalyst bypassing in the case of a plug-flow reactor (PFR). These intrareactor gradients affect the evolution of the chemical reactions occurring inside. Therefore, they have to be taken into account to avoid errors during experimental data interpretation.

Following are some observations about deviation from ideal plug flow in TBRs:

- Axial liquid dispersion in commercial reactors is not significant relative to the bulk flow.
- Reduction of reactor bed length will cause greater deviation from plug flow than will reduction of reactor bed diameter.
- Wall flow leads to axial dispersion because of higher fluxes in the area near the wall, where the packing void fraction is higher.

Once the criteria for elimination of intrareactor transport effects are fulfilled, the substitution of new catalyst and/or feedstocks in existing TBRs can also be studied with considerably scaled-down reactors. The way in which laboratory and bench-scale reactors have to be operated to ensure the ideality of flow pattern for measuring intrinsic reaction rates is described in the following sections.

Plug-Flow Continuous Reactor The mass balance equation for any flow pattern is given by

$$\rho_B \zeta \nu_{i,j} r'_{\text{app},j} = \frac{1}{V}\frac{\partial n_i}{\partial t} + \frac{\partial \upsilon_L C_i^L}{\partial V} - \varepsilon_L D_a^L \frac{\partial^2 C_i^L}{\partial z^2} - \varepsilon_L \frac{D_r^L}{r}\frac{\partial}{\partial r}\left(r\frac{\partial C_i^L}{\partial r}\right) \qquad (2.1)$$

D_a^L and D_r^L are the overall axial and radial dispersions in the liquid phase, respectively, which are made up of contributions from axial molecular diffusion, convective dispersion in the packing, and macroscopic velocity distributions in the reactor. When the reacting fluid moves through the PBR in turbulent flow, the fluid moves like a plug or piston. In the PFR, the condition of turbulent flow generally ensures the absence of concentration, temperature, and velocity gradients in the radial direction, that is, well-mixed radially and zero-mixed axially (Perego and Paratello, 1999). Hence, the space–time (τ) expression from the steady-state mass balance equation for ideal plug flow in PBR becomes

$$\bar{\tau} = \frac{V}{\upsilon_L} = \frac{-(C_i^L)_0}{\rho_B \zeta \nu_{i,j}} \int_0^{X_i} \frac{dX}{r'_{\text{app},j}} \qquad (2.2)$$

A plug-flow regime implies that the residence time of each volume element of the fluid must be the same as that in a large number of mixers in series. Hence, axial molecular diffusion may be the main cause of spread in the residence time in the reactor; meanwhile, the axial convective dispersion within the packing results from statistical variations in the width, length, and direction of individual channels (Sie, 1996).

If a commercial TBR is assumed to be close to an ideal integral FBR, although the hydrodynamics is different, the condition for a kinetically representative laboratory reactor is that it should also be sufficiently approximate to ideality behavior. A well-designed industrial FBR can be considered as a close approximation to an ideal integral reactor if the following criteria are fulfilled (Sie, 1996):

- All volume elements of the feed must last the same time inside the reactor so as to contribute equally to the overall conversion (e.g., PFR behavior, narrow residence time distribution of the liquid).
- All parts of the catalyst bed must contribute maximally to the overall conversion (all catalyst particles in good contact with the reactant stream, i.e., complete contacting of the catalyst with the liquid phase).

Therefore, hydrodynamics in laboratory and pilot-plant reactors is of importance only to the extent that they affect compliance with the following requirements: plug-flow pattern, complete catalyst particles wetting, pressure drops, and heat transport phenomena.

Perfectly Mixed Continuous Reactor Perfectly mixed reactors include an external recycle reactor (recycle-to-feed flow ratio > 25) and a CSTR. On the macroreactor scale, both types of reactors are characterized by the absence of intrareactor concentration and temperature gradients throughout the reaction volume. According to the ideal behavior, the space–time may be obtained directly from a simple algebraic equation expressed in terms of effluent conversion:

$$\bar{\tau} = \frac{V}{\upsilon_L} = \frac{-(C_i^L)_0 X_i}{\rho_B \zeta v_{i,j} r'_{\text{app},j}} \tag{2.3}$$

2.2.2 Intrareactor Temperature Gradients

Temperature differences inside a reactor may originate from mass and temperature intrareactor gradients, which can cause deviations from isothermality. Uniformity of reactor temperature profile is a crucial characteristic for performing appropriate experiments in order to obtain intrinsic reaction rates, since it is very difficult to cope up with intrinsic rate constants from nonisothermal data. Therefore, both axial and radial gradients must be minimized. Due to the exothermic or endothermic nature of reactions, axial temperature gradients in FBRs are always present; however, by increasing the ratio of bed length to catalyst particle diameter (L_B/d_{pe}) the gradients can be minimized. Dilution of either the feed or the catalyst and decreasing the reactor diameter are the most common experimental methods for reducing radial temperature gradients. If the reactor diameter is decreased, the particle diameter should also be reduced, to maintain a constant d_R/d_{pe} ratio and fulfill the PFR condition without introducing flow bypassing problems. However, dilution of catalyst particles with smaller inert particles produces high pressure drops. This problem can be solved by mixing catalyst particles with suitably sized inert particles so that good fluid distribution and low pressure drop are guaranteed.

The mathematical criteria reported in the literature to neglect temperature gradients generally compare two functions, one representing measurable parameters of the actual system and the other calculated from theoretical parameters. If the value of the former is greater (or lower) than that of the latter, it is assumed that the temperature gradient observed may cause only a certain deviation (e.g., a 5 or 10% difference in reaction rate).

Radial Heat Dispersion Radial and axial transport effects within a reactor are not easy to evaluate and control. Radial temperature gradients probably cause unreliable data in TBRs, which are attributed to heat conduction problems due to low effective thermal conductivity of the catalyst. When the reaction rate and heat release are large, severe radial temperature gradients are expected. If these gradients are neglected, they can lead to reaction rates several orders of magnitude greater than those calculated with the temperature at the wall, thus causing unreliable results in PFRs (Mears, 1971).

Typical diameters of experimental reactors (e.g., pilot scale and bench scale) may provoke undesired radial temperature profiles if the reactor is not operated in true adiabatic mode. It has been shown that in these experimental reactors, radial heat transfer effects at the hot-spot location are more important than axial heat transfer effects (Mears, 1976). These intrareactor temperature gradients are nearly always more severe than interphase temperature gradients, which are generally more severe than intraparticle temperature gradients. Therefore, isothermal operation of the reactor is critical to the generation of reliable laboratory data. For this task, mathematical criteria aid in understanding what reactor system features can be manipulated to achieve better isothermal control. Based on this, the following recommendations can help to establish isothermal operation:

- Low conversion levels
- Small catalyst particles
- Dilution of catalyst bed with inert particles to decrease bed voidage
- High thermal conductivity catalyst support
- Feed diluents with high thermal conductivity
- High flow rates
- Decreased reactor diameter

Mears (1971) has pointed out the importance of decreasing the diameters of both the reactor and the catalyst particles to minimize radial interparticle heat transport limitations in experimental FBRs with heat exchange at the wall. However, this may not be appropriate to avoid the wall effect. The approximate criterion derived by Mears (1971) for determining the existence of radial interparticle heat transfer limitations often causes severe deviations from the isothermal plug-flow operation desired. Mears (1971) established that <5% deviation in the reaction rate due to radial temperature gradients

TABLE 2.1. Equations for the Criteria for Intrareactor Temperature Gradients

Criterion	Eq.
$\frac{(d_R/2)^2 \lvert \Delta H_{Rj} \rvert [\rho_B \zeta r'_{app,j}]_z}{\lambda^S_{er} T_W} < 0.4 \frac{RT_W}{E_a}$	(2.4)
$\frac{(d_R/2)^2 \lvert \Delta H_{Rj} \rvert [\rho_B \zeta r'_{app,j}]_z}{\lambda^S_{er} T_W} < \frac{0.4 RT_W/E_a}{1+8(d_{pe}/d_R)\mathrm{Bi}_{hW}}$	(2.5)
$\gamma' \beta' (r'_{app,j})_0 = \frac{\lambda^f_{ea}(-\Delta H_{Rj})\rho_B \zeta}{(G_f Cp_f)^2 (T_{f,0} - T_W)} (r'_{app,j})_0 = \frac{(-\Delta H_{Rj})\rho_B \zeta d_{pe}}{(G_f Cp_f)(T_{f,0} - T_W)\mathrm{Bo}^f_{a,h}} (r'_{app,j})_0 << 1$	(2.6)
$\max_z \left\lvert \frac{1}{\mathrm{Bo}^f_{a,h}} \frac{dT_f}{d(z/d_{pe})} \right\rvert << 1$	(2.7)
$\mathrm{Da}_I \left\lvert \frac{n}{\mathrm{Bo}^f_{a,m}} - \frac{\mathrm{Ar} \cdot \omega}{\mathrm{Bo}^f_{a,h}} \right\rvert < 0.5$	(2.8)

for high values of d_R/d_{pe} (>100), where heat transfer resistance at the wall of the reactor is negligible, is found if Eq. (2.4) (Table 2.1) is fulfilled. However, for small laboratory reactors, heat transfer resistance at the wall cannot be neglected. If it is significant, for such a deviation in the reaction rate in cases when $d_R/d_{pe} < 100$, the interparticle criterion becomes Eq. (2.5). This criterion is conservative when applied at the inlet section of a reactor, but it does apply at any cross section in reactors with linearly decreasing dilution, and at the hot spot in undiluted or uniformly diluted beds.

Doraiswamy and Tajbl (1974) have indicated that a reduction in reactor diameter, and thus in the aspect ratio (d_R/d_{pe}), is helpful in minimizing radial temperature gradients. The practical lower limit of d_R/d_{pe} appears to be about 4, so the contribution of heat transfer at the inside tube wall is not significant. According to Butt and Weekman (1974), the critical design factor to avoid the influence of radial temperature gradients is reactor diameter and hence the d_R/d_{pe} ratio, since if the reactor diameter is reduced beyond a limit, the wall effect will appear. There are not analytical criteria to choose the d_R/d_{pe} ratio. However, most researchers have reported values between 10 and 20. Once this ratio has been specified, the criteria given by Eqs. (2.4) and (2.5) may be used to estimate the probable influence of radial temperature gradients.

Carberry (1976) stated that a reactor diameter of five to six times the particle diameter is the maximum value at which reactor isothermality is certain to be maintained. However, this criterion is in conflict with the general PBR rule that a minimum of 8 to 15 particle diameters is necessary to minimize wall effects. This means that no matter what is done invariably, some error will be made (Tarhan, 1983). Although the catalyst bed will be diluted with a good heat-conducting material [e.g., silicon carbide (SiC), α-Al_2O_3, or quartz], it is recommended that temperatures at the wall and at the center of the bed always be measured if the reactor diameter is greater than 3 cm (Gierman, 1988).

Axial Heat Dispersion The sensible heat term can be neglected in the differential energy balance when the axial temperature gradient is zero. This occurs at any cross section in reactors with linearly decreasing dilution and at hot-spot locations. Axial heat conduction can also be neglected for beds with a sufficient length-to-particle diameter ratio ($L_B/d_{pe} > 30$), so that plug flow is approached.

Young and Finlayson (1973) proposed the criterion given by Eq. (2.6) to neglect axial heat dispersion at the initial reactor section without catalyst particles. If Eq. (2.6) [and Eq. (2.4) for axial mass dispersion] is not properly satisfied, axial mass and heat dispersions will be important at the inlet section of the reactor, although axial dispersion may be of less or no importance at the outlet. Neglecting these axial dispersion effects would lead to errors at the inlet, and consequently, in the entire reactor. Young and Finlayson (1973) also proposed another criterion [Eq. (2.7)] to determine the importance of axial heat dispersion in the reactor section with catalyst particles. Based on the criterion proposed by Young and Finlayson (1973), Mears (1976) derived another criterion [Eq. (2.8)] to consider cases where the inlet reactor temperature is equal to the wall reactor temperature ($T_{f,0} = T_W$) or for endothermic reactions (Figure 2.6).

2.2.3 Intrareactor Mass Gradients

Radial Mass Dispersion Compared with other hydrodynamic parameters, relatively few studies have been reported on the radial liquid distribution in TBRs. The reactor-to-particle size ratio (d_R/d_{pe}) has been demonstrated to have a significant effect on radial distribution (Saroha et al., 1988). In large commercial units, a uniform radial mass can be achieved by employing a suitable design for a multipoint distributor and other reactor internals (Alvarez et al., 2007). In laboratory reactors, a single-point distributor and a certain amount of inert material (e.g., a helly pack) are generally used to achieve a prior uniform radial distribution. An adequate radial flow of reactants and products through the catalyst beds can minimize the pressure drop in FBRs.

Bischoff and Levenspiel (1962) have presented a criterion [Eq. (2.9), Table 2.2] to neglect the radial dispersion effect in packed beds based on the ratio of the catalytic bed length to the reactor diameter. For large ratios of reactor to particle diameter ($d_R/d_{pe} > 25$), D_r^L is approximately constant over the radius in packed beds. By assuming a size of one particle diameter for individual stirred tanks, Levenspiel and Bischoff (1963) found that $\varepsilon_L D_r^L / u_L d_{pe} = 0.21$ for flow in packed beds with a Reynolds number higher than 100. Satterfield (1970) confirmed that liquid and gas radial dispersion is constant for a particle

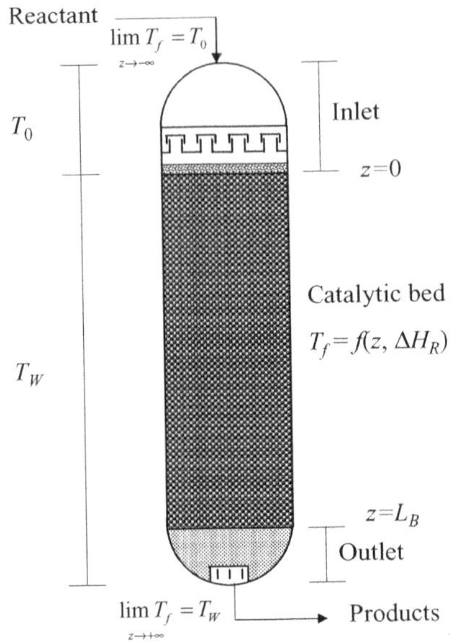

Figure 2.6. Geometry of a nonisothermal reactor to derive the criterion of Young and Finlayson (1973).

Reynolds number (Re) higher than 100, whereas axial dispersion is constant for Re > 10. According to Carberry and White (1969), the product yield calculated for an FBR using a two-dimensional model is quite sensitive to a radial heat transport effect but is virtually insensitive to a radial mass transport effect. This suggests that if a reactor design meets the criteria for eliminating a radial temperature gradient, the radial mass transport effect can be neglected by assuming uniform radial concentration.

Axial Mass Dispersion Contrary to radial mass dispersion, mass transfer in the axial direction is always present, but it can be minimized to prevent significant deviations from plug flow by selecting the appropriate ratio of bed length to particle diameter (L_B/d_{pe}). Table 2.3 summarizes the values of this ratio proposed by different researchers for neglecting axial mass dispersion. The axial dispersion of the reactant in an FBR can also be described in terms of an equivalent number of CSTR in series (N) or by the dimensionless Peclet number (Pe) defined by Eq. (2.10), in which the reactor length (L_B) features as the characteristic dimension. For axial dispersion caused only by packing, the dimensionless Bodenstein number (Bo), defined by Eq. (2.11), can be used to describe it; in this case the equivalent particle diameter (d_{pe}) is chosen as the characteristic dimension. In Eqs. (2.10) and (2.11), D_a^f is the overall axial-dispersion coefficient, which includes contributions from axial molecular diffusion, convective dispersion in the packing, and macroscopic velocity distributions in the reactor (Sie, 1996). By combining these two equations, the relationship given by Eq. (2.12) between the Peclet and Bodenstein numbers can be derived.

As a general criterion, to have ideal plug-flow behavior, N or Pe should be higher than a certain minimum value. This minimum value depends on the reaction order (n) and the degree of conversion (X_i). The minimum Peclet number is about twice the equivalent number of CSTRs in series [Eq. (2.13)]. Based on a variety of published data of axial dispersion in fixed beds for trickle flow, Gierman (1988) developed a correlation between Bodenstein and liquid particle Reynolds numbers which shows that an increased Bo number requires an increased Re number when Re > 10. However, the Bo number reaches a more or less constant value at low Re numbers, which are typically found in small laboratory reactors (0.001 < Re < 0.1). This is valid for single-phase flow as well as for trickle flow with a lower constant value of Bo for trickle flow, indicating greater axial dispersion in a trickle bed at similar velocity and particle size. The mean Bodenstein value for the low Reynolds region of interest in laboratory TBRs is 0.04.

Székely (1961) and Petersen (1965) applied an asymptotic solution approach to isothermal first-order reactions and found that the effect of backmixing is negligible ($L_{B,P}/L_B \approx 1$) if the reaction rate is slow. They proposed the criterion given by Eq. (2.14) for plug flow. Mears (1971) obtained a more conservative design criterion by utilizing perturbation solutions for power-law kinetics, which holds the deviation in the required reactor length to less than 5% [Eq. (2.15)]. This criterion shows that the dispersion effect can be negligible except for cases involving short beds and high conversion.

Levenspiel and Bischoff (1963) developed a criterion to neglect deviations from plug flow in isothermal reactors with first-order reactions. They proposed Eqs. (2.16) and (2.17), which involve conversion and size of the actual reactor and the reactor calculated using the plug-flow model. By combining Eqs. (2.11) and (2.17) and assuming the mean residence time of fluid in the system ($\bar{\tau}$) with a maximum error of 5%, Eqs. (2.18) and (2.19) in terms of concentration and conversion can be obtained. This criterion is best suited for design purposes and might not be valid for large $\varepsilon_L D_a^L/(u_L L_B)$ values (Young and Finlayson, 1973). By taking the size of an individual CSTR equal to one particle diameter, the value $\varepsilon_L D_a^L / u_L d_{pe} = 0.5$ can be reproduced for flow in packed beds with Re > 100 (Satterfield, 1970).

For a 4% deviation of plug flow in adiabatic reactors with single nth-order reactions, Hlavacek and Marek (1966) proposed the criteria given by Eqs. (2.20) and (2.21). One of the most used design criteria based on the minimum bed length required to neglect axial dispersion or backmixing effects on three-phase reactor behavior was developed by Mears (1971), who was the first to use a relationship between Bo number, reaction order n, and conversion X_i. The criterion establishes that deviation in the required actual reactor length with respect to the ideal length where plug flow is achieved (for a given conversion) is less than 5% if Eq. (2.22) is fulfilled.

TABLE 2.2. Equations for the Criteria for Intrareactor Mass Gradients

Criterion	Eq.	Criterion	Eq.
$\frac{L_B}{d_R} > 0.04 \frac{u_L d_R}{\varepsilon_L D_r^L}$	(2.9)	$\frac{V}{V_P} = \frac{L_B}{L_{B,P}} = 1 + \left(\rho_B \zeta k'_{\text{app},j} \bar{\tau}\right) \frac{\varepsilon_L D_a^L}{u_L L_B}$ for $X_i = (X_i)_P$	(2.16)
$\text{Pe}_{a,m}^f = \frac{L_B u_f}{\varepsilon_f D_a^f}$	(2.10)	$\frac{(C_i^L)_f}{(C_i^L)_{f,P}} = 1 + \left(\rho_B \zeta k'_{\text{app},j} \bar{\tau}\right)^2 \frac{\varepsilon_L D_a^L}{u_L L_B}$ for $V = V_P$	(2.17)
$\text{Bo}_{a,m}^f = \frac{d_{pe} u_f}{\varepsilon_f D_a^f}$	(2.11)	$\frac{(C_i^L)_f}{(C_i^L)_{f,P}} = 1 + \left(\frac{\rho_L \rho_B \zeta k'_{\text{app},j}}{G_L}\right)^2 \frac{d_{pe} L_B}{\text{Bo}_{a,m}^L} < 1.05$	(2.18)
$\text{Pe}_{a,m}^f = \text{Bo}_{a,m}^f \frac{L_B}{d_{pe}}$	(2.12)	$(X_i)_P - X_i = \frac{(C_i^L)_f - (C_i^L)_{f,P}}{(C_i^L)_0} = \left(\frac{\rho_L \rho_B \zeta k'_{\text{app},j}}{G_L}\right)^2 \frac{d_{pe} L_B}{\text{Bo}_{a,m}^L} \exp\left(-\frac{\rho_L \rho_B \zeta k'_{\text{app},j} L_B}{G_L}\right) << 1$	(2.19)
$\text{Pe}_{a,m}^f = 2N > f(n, X_i)$	(2.13)	$\text{Bo}_{a,m}^L \frac{L_B}{d_{pe}} > 400$ for high reaction rate	(2.20)
$\alpha = \frac{\sqrt{\rho_B \zeta k'_{\text{app},j} \varepsilon_L D_a^L}}{u_L} < 1$	(2.14)	$\text{Bo}_{a,m}^L \frac{L_B}{d_{pe}} > 100$ for low reaction rate	(2.21)
$\alpha = \frac{\sqrt{\rho_B \zeta k'_{\text{app},j} \varepsilon_L D_a^L}}{u_L} < 0.22$	(2.15)	$\frac{L_B}{d_{pe}} > \frac{20n}{\text{Bo}_{a,m}^f} \ln \frac{(C_i^f)_0}{(C_i^f)_f}$	(2.22)

$$\gamma\beta(r'_{\mathrm{app},j})_0 = \left(\frac{\rho_f}{G_f}\right)^2 \frac{D_a^f \rho_B \zeta}{(C_i^f)_0}(r'_{\mathrm{app},j})_0 = \frac{\rho_f \rho_B \zeta d_{pe}}{G_f (C_i^f)_0 \mathrm{Bo}_{a,m}^f}(r'_{\mathrm{app},j})_0 << 1 \tag{2.23}$$

$$\max_z \left| \frac{1}{\mathrm{Bo}_{a,m}^f} \frac{dX_i}{d(z/d_{pe})} \right| << 1 \tag{2.24}$$

$$\left| \frac{n\mathrm{Da}_I}{\mathrm{Bo}_{a,m}^f} - \frac{\mathrm{Ar} \cdot \mathrm{Da}_{III}}{\mathrm{Bo}_{a,h}^f} \right| < 0.05 \tag{2.25}$$

$$\frac{L_B}{d_{pe}} > 20 \left[\frac{\rho_B \zeta (r'_{\mathrm{app},j})_0 \bar{\tau}}{(C_i^f)_0} \right] \left| \frac{n}{\mathrm{Bo}_{a,m}^f} - \frac{\mathrm{Ar} \cdot \omega}{\mathrm{Bo}_{a,h}^f} \right| \tag{2.26}$$

$$\frac{\mathrm{Da}_I}{\mathrm{Bo}_{a,m}^f} < \frac{0.05}{n} \tag{2.27}$$

$$\frac{L_B}{d_{pe}} > \frac{8n}{\mathrm{Bo}_{a.m}^f} \ln \frac{1}{1 - X_i} \tag{2.28}$$

$$\mathrm{Pe}_{a,m}^L \geq -(21 - 20R_{CM}^*) + \left[(21 - 20R_{CM}^*)^2 - 21 + 20R_{CM}^*\right]^{1/2} \tag{2.29}$$

$$F_D = \frac{(C_i^L)_{f,A}}{(C_i^L)_{f,P}} = \frac{1 + 2\mathrm{Pe}_{a,m}^L}{(1 + \mathrm{Pe}_{a,m}^L)^2} R_{CM} + \frac{(\mathrm{Pe}_{a,m}^L)^2}{(1 + \mathrm{Pe}_{a,m}^L)^2} \leq 1.05 \tag{2.30}$$

$$(\mathrm{Pe}_{a,m}^L)_c \geq -(20R_{CM}^* - 19) + \left[(20R_{CM}^* - 19)^2 - 20R_{CM}^* + 19\right]^{1/2} \tag{2.31}$$

$$R_{CM} = \frac{(C_i^L)_{f,M}}{(C_i^L)_{f,P}} = \frac{g\left((C_i^L)_{f,\exp}\right)}{(C_i^L)_{f,\exp}} \tag{2.32}$$

$$\mathrm{Pe}_{a,m}^L > 100 \tag{2.33}$$

$$\frac{L_B}{d_{pe}} > \frac{\sqrt{20}n}{\mathrm{Bo}_{a,m}^f} \ln \frac{1}{1 - X_i} \tag{2.34}$$

TABLE 2.3. Rule of Thumb for Axial Mass Dispersion

L_B/d_{pe}	Application	Reference
>15–20	From data of Hochman and Effron (1969) on TBR with a liquid mass velocity of 4 $kg/m^2 \cdot s$	Gianetto and Specchia (1992)
>25	To minimize axial dispersion in packed towers	Scott (1935)
>30	Simple rule of minimum length to neglect axial mass dispersion and heat conduction	Mears (1971)
	Axial dispersion and axial heat conduction can be neglected, ensuring that plug flow is closely approached	Doraiswamy and Tajbl (1974)
>50	For long isothermal reactors	Carberry and Wendel (1963)
	For the flow velocities in industrial practice the effect of axial dispersion of heat and mass on conversion is negligible	Froment and Bischoff (1990)
	For two-phase FBRs	Perego and Paratello (1999)
	For neglecting backmixing effects in two-phase FBRs	Ancheyta et al. (2002)
>100	For TBRs	Perego and Paratello (1999)
	To avoid backmixing in TBRs	Sie (1991)
>350	Minimum axial dispersion for bench-scale HDT units processing straight-run gas oil	Mears (1971, 1974)
	Backmixing is minimum	Kumar et al. (2001)

Equation (2.22) was derived from perturbation solutions using the one-parameter piston diffusion (PD) model for power-law kinetics. Its applicability is valid only under the conditions of liquid-limiting reactions (it cannot be applied to complete external wetting), *n*th-order irreversible reaction kinetics, isothermal TBR, and conversion of less than 90%. The minimum length for negligible deviation increases with an increase in conversion or a decrease in Bodenstein number. As Mears (1971) pointed out, no simple rules such as $L_B/d_{pe} > 30$ are adequate for all cases.

Young and Finlayson (1973) examined the importance of both axial and radial dispersion in nonadiabatic reactors. A criterion was derived [Eq. (2.23)] (Figure 2.6) to determine if axial mass dispersion is important at the inlet section of nonisothermal PBRs with cooling or heating at the walls and with inlet and outlet sections free of catalyst particles. Even though criteria for axial heat dispersion [Eq. (2.6)] and axial mass dispersion [Eq. (2.23)] are satisfied, if the maximum axial temperature or conversion gradient occurs at $z \neq 0$, discrepancies in temperature and conversion profiles from those with plug flow can also occur. An indication of the importance of axial dispersion inside the reactor can be obtained by comparing the fluxes with and without axial dispersion. If axial mass dispersion is to be negligible, the absolute difference in the fluxes for these two cases must also be negligible [Eq. (2.24)]. The problem with using the criteria for axial heat dispersion [Eq. (2.7)] and axial mass dispersion [Eq. (2.24)] is that their importance inside the reactor requires knowledge of the maximum temperature and conversion gradients.

Shah and Paraskos (1975) extended Mears' (1971) criterion for isothermal TBRs to predict significant axial dispersion effects in commercial and pilot-scale adiabatic hydroprocessing TBRs. The criteria were validated for irreversible reactions following power-law kinetics ($n = 1$ and 2) and $\mathrm{Bo}_{a,m}^{L} > 3$. It was observed that at high conversions, adiabatic operation produces a larger axial dispersion effect than that produced by isothermal operation. At low conversions, the opposite results were obtained. Plots of $\left(\mathrm{Pe}_{a,m}^{L}\right)_{c}$ in the plane of $\mathrm{Pe}_{a,m}^{L}$ versus R_n' for first- and second-order reactions were used to show the criteria for $L_B = 0.95L_{B,P}$. For vapor-phase reactors, these criteria do not apply in cases where the external film heat transfer is important.

Mears (1976) applied the inlet rate criterion proposed by Young and Finlayson (1973) to evaluate the axial dispersion effect of heat and mass in an nonisothermal reactor cooled or heated at the wall, with a uniform temperature along the wall reactor [Eq. (2.25)]. This criterion predicts that the reaction rate deviation from plug-flow conditions at the reactor inlet will be less than ±5%. It is more precise and less conservative than the criteria given by Eqs. (2.6) and (2.23). The criterion may also be expressed in terms of minimum length for insignificant axial dispersion effect at the inlet section [Eq. (2.26)]. From a criterion expressed in this way, it is possible to see that the minimum length increases with increases in conversion, reaction order, heat of reaction, and activation energy for the specific case of isothermal reactors with first-order reactions throughout the bed and non-first-order reactions at the inlet section before the catalyst bed [Eq. (2.27)].

Mears' (1971) conservative criterion to neglect axial dispersion effects on TBRs was slightly modified by Gierman (1988). The former is attained to estimate concentration deviation values from plug flow of less than 5%; the latter, for deviations less than 10%. Gierman's criterion is based on the more relaxed criterion that the temperature requirement for the same conversion should not be higher than theoretical by about 1°C, which is considered to be the maximum accuracy of temperature definition in practical cases. For the usual magnitude of activation energy of reactions of interest, the criterion proposed is that given by Eq. (2.28) (Sie, 1996).

Both Mears and Gierman criteria apply only for liquid-limiting reactions. According to the authors, much higher L_B/d_{pe} ratios are required in laboratory experiments than those expected from the rule of thumb of $L_B/d_{pe} > 50$. For FBRs, the latter simple criterion is indeed sufficient, provided that the particle

Reynolds number is above 10. However, laboratory reactors are usually operated with a particle $Re_L < 0.1$ being more the rule than the exception.

Cassanello et al. (1994) demonstrated that Mears' model is severe only in certain conditions and formulated another criterion for liquid or gaseous limited reactions, based on a general approximation to the solution of the axial dispersion model, to establish the conditions under which the liquid axial dispersion affects the behavior of three-phase FBRs for both downflow and upflow operations [Eq. (2.29)]. This criterion implies that the deviation with respect to the plug-flow model should not be larger than 5%.

Cassanello et al. (1996) developed a general criterion to analyze the influence of plug-flow deviations on TBRs' behavior. The criterion may be used for reactor design purposes by determining operating conditions and geometric features to ensure negligible axial dispersion effects based on the assumption that deviations with respect to the plug-flow model should not be larger than 5% if Eq. (2.30) is fulfilled. By applying a general approximate solution of the axial dispersion model and considering the ratio of product concentrations, Eq. (2.30) can be expressed as Eq. (2.31). This criterion is valid for any type of kinetics and very useful when the geometrical characteristics and operating conditions of the reactor are already defined. However, the kinetic constant is unknown. Similar to the criterion developed by Cassanello et al. (1992), for this critical Peclet value, deviation of outlet concentration with respect to that from plug flow is close to 0.95. The main disadvantage of this criterion is the need of reactant concentration at the reactor outlet according to perfect mixing ($Pe^L_{a,m} \rightarrow 0$) and plug-flow ($Pe^L_{a,m} \rightarrow \infty$) models. However, the evaluation of these concentrations is much easier than that corresponding to the axial dispersion model.

It is possible to evaluate $Pe^L_{a,m}$ from correlations reported in the literature and to measure the reactant outlet concentration. If the experimental outlet concentration is considered as that obtained by assuming a plug-flow model, the apparent kinetic constant can be estimated in a first approximation. The value obtained in this way is used to calculate the outlet concentration with the perfect mixing model. Hence, the ratio given by Eq. (2.32) is determined, which can be applied in Eq. (2.30) to estimate whether there is an axial dispersion effect in the experimental measurements.

From data of Hochman and Effron (1969) on TBRs with a liquid mass velocity of $4\,kg/(m^2 \cdot s)$, Gianetto and Specchia (1992) established that plug flow should be approximated in beds in which $L_B/d_{pe} > 15$ to 20. For kinetic experiments at high conversion, van Herk et al. (2005) indicated that the Peclet number is a more suitable parameter to be considered and proposed the rule of thumb given by Eq. (2.33) for neglecting axial mass dispersion. Since achieving the criteria proposed by Mears (1971) and Gierman (1988) might still be too strict for some cases [e.g., hydrotreating TBR when producing ultralow-sulfur diesel due to the very high conversions (99.5% or even higher)], Chen et al. (2009) have proposed an even more relaxed criterion with 15% deviation from plug flow [Eq. (2.34)].

2.2.4 Wetting Effects

As mentioned before, the second condition for an ideal reactor is that all catalyst particles must contribute equally to the overall conversion. In TBR, each catalyst particle should therefore be surrounded by a flowing film of liquid. The flux of liquid and gas should be the same in any part of the bed cross section. However, a special situation may prevail, particularly at low liquid velocities, in which liquid flows preferentially through a certain part of the bed while gas passes predominantly through another part in the interstitial spaces that are not occupied by liquid. In the latter part there are catalyst particles that are not totally contacted by liquid reactant and hence do not contribute to the overall conversion (Figure 2.3). This situation leads to incomplete utilization of the bed and is known as *macroscopic maldistribution of liquid* or *incomplete wetting*. Its effect on the reactor scale is measured as *contacting effectiveness* or the *catalyst utilization fraction* [Eq. (2.35), Table 2.4].

Satterfield (1975) has shown that the contacting efficiency (CE) depends on the liquid mass velocity, which can be approximated by a power-law function with an exponent of roughly two-thirds [Eq. (2.36)]. This correlation assumes that full catalyst utilization is achieved at liquid mass velocities equal to or greater than those required by meeting plug-flow criteria. Although the term *incomplete wetting* is appropriate in cases where some catalyst particles indeed remain dry, the term *uneven irrigation* is preferred. This problem aggravates by poor initial maldistribution of liquid in commercial reactors (e.g., inefficient reactor internals), but it can also be developed at low velocities even if the initial distribution is adequate.

On a particle scale, incomplete wetting is not very likely to occur under normal hydrotreating conditions in the sense of interfacial tensions because the hydrocarbon feedstocks used (i.e., viscous oil) generally tend to spread easily over the porous catalyst surface. This complete particle wetting does not mean optimal catalyst utilization. If liquid flow around the catalyst particles is very uneven, part of the liquid film of variable thickness covering them is not refreshed at a sufficiently high rate, and part of the catalyst particle may not be used as it should be, as illustrated in Figure 2.7 for the case of a pore with a diffusion-reaction process where the renewal of liquid is essential. Hence, complete wetting is an insufficient condition for an ideal reactor, whereas even irrigation appears to be a more stringent and descriptive requirement which correctly associates the phenomenon of liquid maldistribution with dynamic (flow-related) forces instead of with static force balance (interfacial tensions). Figure 2.3 shows that at the contact points between adjacent catalyst particles there are pockets of liquid that are stagnant. Mass transfer between these stagnant pockets and the flowing liquid has to occur by diffusion.

Satterfield and Ozel (1973) provided visual evidence of ineffective wetting in laboratory TBRs and showed downward flow of liquid in rivulets which tended to maintain their positions with time. It was also observed that some catalyst pellets were covered with trickling liquid film, whereas others were

TABLE 2.4. Equations for the Criteria for Wetting and Wall Effects

Criterion	Eq.	Criterion	Eq.
$\eta_{CE} = \dfrac{\text{LHSV}_{\text{nom}}}{\text{LHSV}_e}$	(2.35)	$R^*_{\text{exp}} \geq 0.95 \dfrac{\Phi_j^L \tanh(\Phi_j^L)}{1+\Phi_j^L \tanh(\Phi_j^L)/\text{Bi}_{mi}^S}$	(2.43)
$\eta_{\text{CE}} = \left(\dfrac{G_L}{G_{L,P}}\right)^{2/3}$ for $G_L < G_{L,P}$ $\eta_{\text{CE}} = 1.0$ for $G_L \geq G_{L,P}$	(2.36)	$\dfrac{R^*_{\text{exp}}}{\text{Bi}_{mi}^S} \leq \dfrac{\left[-b_f - (b_f^2 - 4a_f c_f)^{1/2}\right]}{2a_f}$, $a_f = 7.8 - 15.8 f_w + 8 f_w^2$ $b_f = -0.19 - 7.61 f_w + 15.8 f_w^2 - 8 f_w^3$, $c_f = 0.19(f_w - f_w^2)$	(2.44)
$f_w = \dfrac{a_w}{a_S} = \dfrac{k'_{\text{app},j}}{k'_{in,j}}$	(2.37)	$\overline{C}_i^L = \dfrac{(C_i^L)_0 + (C_i^L)_f}{2}$ $\bar{r}_{\text{exp}} = \left[(C_i^L)_0 - (C_i^L)_f\right]\dfrac{u_L}{L_B}$, $R^*_{\text{exp}} = \dfrac{(V_p/S_p)^2 \bar{r}_{\text{exp}}}{(1-\varepsilon_B) D_{ei}^L \overline{C}_i^L}$	(2.45)
$\dfrac{(dP/dz)_{\text{flow}}}{(dP/dz)_{\text{gravity}}} = \dfrac{180\mu_L(1-\varepsilon_L)^2 u_L}{\rho_L d_{pe}^2 g \varepsilon_L^4} > 1$	(2.38)		
$W = \dfrac{\mu_L u_L}{\rho_L d_{pe}^2 g} > 4 \times 10^{-6}$	(2.39)	$\dfrac{R^*_{\text{exp}}}{\text{Bi}_{mi}^S} = \dfrac{\left[1-(C_i^L)_f/(C_i^L)_0\right]\text{LHSV}}{\left[1+(C_i^L)_f/(C_i^L)_0\right] 1800 k_i^S (S_p/V_p)(1-\varepsilon_B)}$	(2.46)
$W > 5 \times 10^{-6}$	(2.40)	$\dfrac{L_B}{d_{pe}} > \dfrac{250(1-\zeta)}{\zeta\delta}$	(2.47)
$\dfrac{(r'_{\text{app},j})_0 (V_p/S_p)\rho_S}{k_{\text{H}_2}^S (C_{\text{H}_2}^L)_0} > 1$	(2.41)	$L_B > \dfrac{0.25 d_R^2}{\sqrt{d_{pe}}}$	(2.48)
$\eta_j(\Phi_j^L)^2 = \left(\dfrac{(r'_{\text{app},j})_0 (V_p/S_p)^2 \rho_S}{D_{e\text{H}_2}^L (C_{\text{H}_2}^L)_0}\right) > \dfrac{\Phi_j^L \tanh(\Phi_j^L)}{1+\Phi_j^L \tanh(\Phi_j^L)/\text{Bi}_{m\text{H}_2}^L}$	(2.42)		

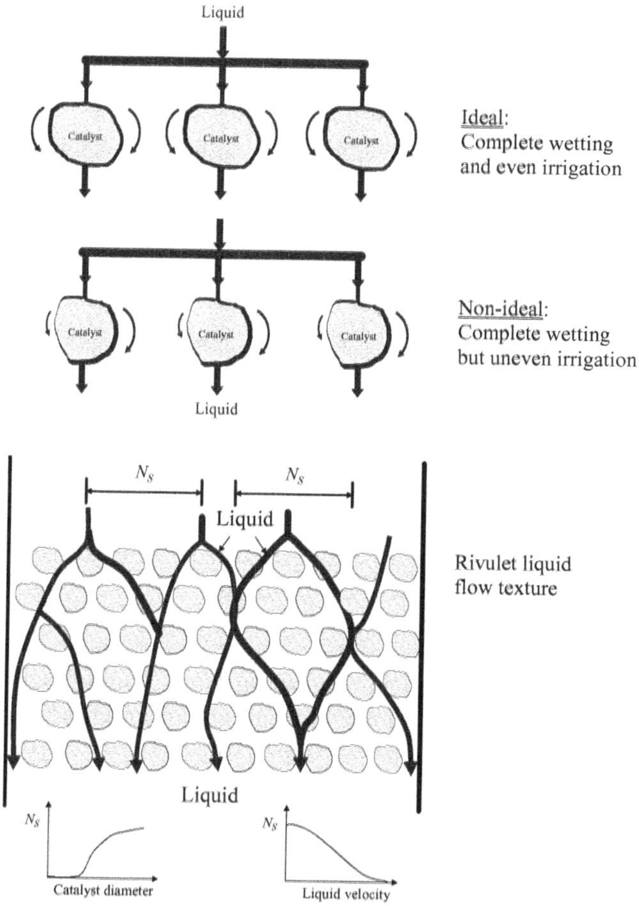

Figure 2.7. Effect of irrigation on catalyst utilization for a pore-diffusion-limited reaction and rivulet liquid flow texture at the bed scale in a TBR.

not, although being wet. It then seems reasonable to assume that the reaction rate is proportional to the fraction of external catalyst surface effectively (freshly) wetted by the flowing liquid, being expressed as effective wetting or catalyst wetting efficiency [Eq. (2.37)]. With respect to the internal wetting efficiency of catalyst pores (f_{iw}), there is general consensus that pores are completely filled with liquid due to capillary forces, even in the presence of partial external wetting, except for heavy oils, for which some researchers have found that $f_{iw} < 1$.

On a bed scale, nonhomogeneous liquid flow can be distinguished. Particularly at low liquid velocities with large catalyst particles, the liquid passes through the bed as small streams following different paths, the rivulets. The characteristic distance between adjacent rivulets (N_S) is a function of catalyst particle diameter and liquid velocity, as indicated in Figure 2.7. This tendency of the liquid to bypass parts of the bed and thus to exclude catalyst particles from direct contact with flowing liquid is detrimental to reactor performance, this nonideal situation being incomplete wetting of the catalyst. As indicated before, the fraction of catalytic bed effectively wetted is a function of the initial distribution of liquid at the top of the reactor and its length-to-diameter ratio. This is more evident in the case of commercial reactors with large diameter, where the chances of channeling, low effective distribution, and so on, are greatly increased. About this last point, Baker et al. (1935) reported the need of some device for uniform initial distribution, since the flow from a single stream does not become uniform until the condition $L_B > 5d_R$ is achieved.

The situation of liquid maldistribution showed in Figure 2.3 can be sustained if the flow of liquid is dictated largely by the force of gravity and influenced very little by frictional forces. In the opposite situation, the high pressure drop will force the liquid to spread over every interstitial channel available, so that the liquid will flow more uniformly through the entire bed cross section, thus improving contact with the catalytic surface. Hence, the condition for even irrigation in trickle flow may be stated as the predominance of impedance to flow over gravity, as expressed by Eq. (2.38).

Adequate wetting or even irrigation of catalyst may be obtained by assuming that $\varepsilon_L = 0.15$; therefore, the condition for even irrigation can be rewritten as in Eq. (2.39). W, known as the *wetting number*, compares frictional and gravitational forces, and its minimum value agrees very well with that reported by Gierman (1988) for a large quantity of experimental data [Eq. (2.40)]. It can then be established that the main variables dictating the uniformity of catalyst irrigation are the superficial liquid velocity, the particle diameter, and the kinematic viscosity of the liquid ($\nu_L = \mu_L/\rho_L$), being a particle size of particular importance. For less viscous feeds, small reactors may be operated under the condition of an optimally diluted catalyst bed.

For commercial-scale reactors, Henry and Gilbert (1973) and Sie and Krishna (1998) suggested that operating at Re << 10 will produce poor wetting. About 100% of catalyst utilization is found in industrial hydroprocessing reactors when they are operated at high mass velocities (>0.7 g/s·cm^2). For pilot-plant reactors, mass velocities are much lower (0.007 to 0.07 g/s·cm^2), which implies an operation with less than 100% catalyst utilization. According to Gianetto and Specchia (1992), a contacting effectiveness factor of less than unity is almost certain when $u_L < 4$ to 5 mm/s.

A criterion to show whether the reaction under consideration is affected by partial wetting ($f_w < 1$) was derived by Lee and Smith (1982) for catalyst particles in TBRs. Since the criterion is expressed in terms of observable quantities, it is applicable to limiting reactant in the liquid phase. The external surfaces of catalytic pellets are partially covered by liquid ($f_w < 1$) when the criterion given by Eq. (2.41) is fulfilled. This criterion is valid for isothermal first-order reactions in PFRs when the reaction is diffusion limited ($\Phi_j^L >> \mathrm{Bi}_{mi}^L$) and is based on the fact that $\eta_j > \eta_j^L$ when $f_w < 1$. If this is not the case, the criterion given by Eq. (2.42) based on the Thiele modulus must be applied.

In cases when the reaction is limited by either the liquid- or gas-phase reactant, the influence of partial external wetting is considerably different. For first-order reactions limited by nonvolatile reactant in the liquid phase, as in hydrotreating operations, a criterion at the catalytic pellet level was formulated by Cassanello et al. (1992). When the external wetting fraction is less than unity, the effectiveness factor may be lower than the factor corresponding to a completely liquid-covered pellet ($\eta_{j,1}$). Based on the fact that the effect of partial wetting is larger than 5% when $\eta_j/\eta_{j,1} < 0.95$, the criterion establishes that the impact of partial wetting is less than 5% and Eq. (2.43) is fulfilled. For liquid-limited reactions when the rate constant is unknown, the criterion given by Eq. (2.44) was proposed by Cassanello et al. (1992).

Naturally, for a limiting reactant in the liquid phase, if $f_w \geq 0.95$, there is no influence of external wetting. Thus, Eq. (2.44) is applied when the external wetting efficiency is less than 0.95. Although Eq. (2.44) was developed by taking the observed reaction rate from a single catalyst pellet into account, its use can be extended to integral reactors by calculating R^*_{exp} with Eq. (2.45). As Eq. (2.45) can be written in terms of LHSV, the left-hand side of Eq. (2.44) is then developed to give Eq. (2.46).

It is important to point out that when the design of the distribution tray in commercial HDT reactors leaves a significant percentage of the top of the bed unwetted, it is expected that radial dispersion mixing of the liquid compensates for maldistribution by wetting the entire cross-sectional bed area (Jacobs and Milliken, 2000). Some reports have showed that diluting lab-scale multiphase FBR with fines can be effective for decoupling the kinetic behavior of the catalyst pellets from the reactor hydrodynamics. However, Tsamatsoulis et al. (2001) and Ramírez et al. (2004) found that dilution of catalyst particles with inert fines does not guarantee full external catalyst wetting (not poor wetting) at all superficial liquid flow rates, because maldistribution of diluent particles may influence the conversion negatively, due to local bypassing of the catalyst pellets by the reactant stream.

The fraction effectively wetted is a function of the initial distribution of liquid in the reactor, superficial liquid mass velocity, and its length-to-diameter ratio, particularly in the case of commercial reactors. Simulations and experimental observations showed that nonuniform coverage of particles by liquid occurs particularly when liquid loads are low, as in the case of laboratory reactors. As we discuss in the next section, achieving complete uniform liquid wetting of the catalyst is desirable because it increases catalyst utilization and reduces the potential impact of any channeling, resulting in operations at a lower reactor temperature and thus a longer run length.

2.2.5 Wall Effects

Both hydrodynamics and intrinsic chemical kinetics affect the reaction rates. This means that when sizing a reactor, if its length is reduced, the representativeness of kinetic measurements obtained in the laboratory reactor will be lost. The only dimension that can be reduced a priori is the reactor width, but care has to be taken when the wall effect begins to become important. In downward-flow TBR, the liquid tends to move toward the reactor walls, which are not catalytically active. This excess of liquid flow near the wall gives rise to an appreciable spread in the residence time of the reactant.

The most common criteria to ensure uniform liquid distribution and hence to eliminate wall flow along a catalytic bed are based on a minimum value of reactor-to-particle diameter ratio. Table 2.5 summarizes some of these criteria. The wide variation in the values of d_R/d_{pe} ratio reported by different authors could be an effect of particle orientation in the bed (i.e., the method of catalyst loading affects the liquid distribution). It has also been reported that the lower the surface tension and density of liquid, the lower the wall flow (Saroha et al., 1988).

How small the d_R/d_{pe} ratio should be is not easy to know, since the smaller the reactor diameter, the more important the wall effect. The overall characteristics of the packed bed are different as the reactor diameter is reduced because the packing of particles is more open near the wall (high bed porosity) only by the presence of the wall surface, provoking a larger fluid velocity (channeling) and a lower catalyst bulk density. Near the wall the fluid velocities are greater and the conversions are lower as a result of the lower resistance to flow, causing an apparent lower activity for all the bed, which can be improved by better radial transfer. Therefore, the better isothermality in reduced-diameter reactors appears to more than compensate for the slight channeling loss to d_R/d_{pe} ratios as low as 4. If diffusion in the radial direction is sufficiently fast, the effect of the velocity profile might be diminished. Since diffusivities of liquids are too low to wipe out the effect of radial flow profiles, and the liquid phase rather than the gas phase determines HDT reactor performance, it is not possible to consider a beneficial effect of radial diffusion with low d_R/d_{pe} ratios. Furthermore, since the reduction in effective bed activity by the wall effect is more serious in trickle flow, larger ratios must be used for this operation.

To study the kinetics of highly exothermic heterogeneous catalytic reactions, dilution of catalyst with an inert material has frequently been used to work in the isothermal mode. A good practice is to dilute the catalyst with inert particles as small as $d_{pe}/10$, so that fluid dynamics are dictated largely by the packing of the small inert particles and by the chemical kinetics of the active catalyst particles. Van den Bleek et al. (1969) proposed a general criterion in order to examine the possible influence of the amount of inert material (i.e., overdilution) on conversion in FBRs, on the basis that the influence of dilution on conversion can be neglected if the relative experimental error in the conversion is an order of magnitude larger than the dilution effect. This criterion makes it possible to determine the allowable degree of dilution in isothermal reactors with uniform dilution, which avoids the reaction mixture bypassing the catalyst, hence for allowable neglect of the dilution effect on the conversion in irreversible isothermal reactions [Eq. (2.47)]. For this criterion it is important to point out that care must be taken in estimating the real value of the dilution parameter (ζ), especially in cases where the diameter of an inert particle is smaller than that of a catalyst particle.

For reactors larger than 1 in. in diameter, Ross (1965) has reported that liquid maldistribution is an important problem. However, Henry and Gilbert (1973) have demonstrated no significant liquid distribution effect in their experiments with reactors up to 4 in. in diameter. If an isothermal reactor operation mode is desired, the heat transfer effect is often more significant than the mass transfer effect. The resulting axial temperature gradient can only be avoided if the heat is transferred sufficiently faster from the catalyst particles to the reactor. This necessarily creates a radial gradient that will impose an upper limit on the diameter of the reactor. However, there will also be a lower limit on the reactor diameter because at some point the liquid flow close to the reactor wall will be more significant with respect to the flow at the center of the bed, so according to Doraiswamy and Tajbl (1974), if the radial aspect ratio $d_R/d_{pe} > 4$, good liquid distribution can be assumed and there are no adverse channeling and heat transfer effects at the reactor wall. Satterfield (1975) has reported that hydrodynamic problems associated with liquid distribution and wall flow decrease for a reactor diameter-to-particle diameter ratio above 10. Gierman (1988) has suggested a minimum value of 16 for this ratio, which would have been thought to be sufficient to ensure minimal liquid maldistribution and wall effects for intermediate petroleum feedstocks (°API > 20). However, heavy oils (20 > °API > 10) and bitumens (°API < 10) may exhibit coning and incomplete wetting of the catalyst even when the reactor diameter-to-particle diameter is 16 (Kwak, 1994).

The change in voidage near the reactor wall has been the subject of many experimental as well as theoretical investigations. Chu and Ng (1989) calculated local bed porosities and local specific surface areas using computer-generated model packings, which are columns randomly filled with uniform spheres generated by following the computer-simulated packing approach of Zimmerman and Ng (1986). Their main finding was that the average bed voidage (or bed porosity) near the reactor wall reaches a value of unity, and the outer surface area of the particles per bed volume unit approaches a higher value than the average value in the interior of the bed if the wall surface is also considered. Because the greater average permeability in the wall zone tends to enhance the flow and the reactor wall itself produces flow retardation, the radial velocity profile presents a maximum at a location of one to a few particle diameters away from the reactor wall. These physical characteristics (bed voidage and specific area) influence the local fluid velocity and contribute thereby to deviations from the ideal plug flow when the ratio $d_R/d_{pe} < 25$.

By taking the data published by Herskowitz and Smith (1978a,b) and other authors into account, Gianetto and Specchia (1992) used the dimensionless group $D_r^L L_B / d_R^2$ considered previously by Hoftyzer (1964), and established that an acceptable liquid distribution could be obtained when Eq. (2.48) is fulfilled. If fluid dynamics and reaction kinetics are so closely interlinked that their effects on the conversion are inseparable, the only way to scale down a commercial-size reactor is to reduce the diameter while keeping the length unchanged. As long as the diameter is not reduced to the extent that wall effects begin to cause the packing to deviate from the unperturbed packing in a large-diameter reactor, the packing structure, the particle Reynolds number, and therefore the fluid dynamics should be identical to those in the commercial-size reactor. This approach is based on the notion that in a well-designed commercial reactor the situation in a horizontal plane within the bed is the same everywhere. Thus, the smaller pilot reactor is a hypothetical narrow vertical column cut out from the large catalyst bed. The minimum diameter of the pilot reactor is determined by the catalyst particle diameter, and this size dimension, together with the same commercial length, gives the smallest truly representative reactor that it is possible to obtain. Further reduction in length is possible as far as criteria to avoid axial dispersion effect allow. These considerations are implied to a lesser extent in the sizing of a typical pilot plant for fixed-bed processes: an L_B/d_R ratio of 50 to 250. From measurements of residence-time distribution in microreactors operated with gas, it has been confirmed that because of fast radial diffusion, the contribution of the wall effect is relatively insignificant, so its effect is negligible compared with axial molecular diffusion.

In summary, the wall effect (bed channeling) is caused by nonuniformities in the catalyst bed properties, which can be distinguished from other forms of liquid maldistribution (e.g., poor distribution tray design, sediment deposition on a catalyst bed). The liquid maldistribution occurring at the top of bed can be corrected with the bed depth, provided that the catalyst bed properties are uniform. Bed channeling works against this corrective action, however. As a result, reactor depths are often limited to 457 to 1067 cm to mitigate the impact of bed channeling on catalyst utilization. Therefore, the reactor wall will withdraw part of the liquid from the packing, introducing a complication both in the interpretation of experimental data on the distribution of the liquid and in performing calculations about the course of the liquid distribution. However, fortunately the liquid bypassing effect due to wall flow is not a significant concern in commercial TBRs, as the reactor-to-particle diameter ratios are very large, making this effect small enough to treat it as a correction to the distribution of the liquid over the cross-sectional area. Those effects of the wall in narrow reactors with full-sized catalyst particles and the axial dispersion in all the catalyst beds at low linear velocities are likely to remain important. Hence, particularly at high conversions and for reactions of high order, it is doubtful whether substantial downscaling of reactors will be possible without loss of accuracy and meaningfulness of results. The axial dispersion formed by wall flow will be significant only for cases where the conversion per pass through the reactor is high, and many errors introduced by this dispersion will result in an observed reaction rate expression that slightly underestimates the true reaction rate, leading to a slightly oversized commercial reactor design. The methods to reduce the wall flow are to control the d_R/d_{pe} ratio (see Table 2.5), to dilute liquid feedstocks in order to obtain low surface tension and density, and to increase the liquid and gas flow, which was also found to improve liquid distribution within the reactor.

TABLE 2.5. Rule of Thumb for Wall Effects

d_R/dp_e	Application	Reference
>4	For good liquid distribution without channeling and heat transfer effects at the reactor wall	Kumar et al. (2001)
	To gain isothermality	Mears (1971)
	For cutting down radial temperature gradients	Doraiswamy and Tajbl (1974)
>10	Practical rule to avoid the wall influence on the position of the dumped packing	Hoftyzer (1964)
	Decrease of hydrodynamic problems associated with liquid distribution and wall flow	Satterfield (1975)
>11.5	For particles 2 mm in diameter	Larachi et al. (1991a,b)
>12	From large and hollow particles with water–air flow at conditions of absorption column operation	Baker et al. (1935)
>12–14	Co-current trickle gas–liquid flow regime (0.1–10 MPa)	Attou et al. (1999)
>10–20	To control radial temperature gradients	Butt and Weekman (1974)
>16	To avoid wall flow	Gierman (1988)
>18	Gas continuous and trickling flow regimes, air–water at 25°C, 1 atm pressure; $d_R/d_{pe} < 30$, catalyst particle sizes from 0.26 to 1.1 cm with granular, spherical, and cylindrical shapes; two columns of 4.08 and 11.4 cm I.D., L_B=0–26 in a small column, 0–70 cm in a larger column; u_L=0.1–0.5 cm/s, u_G=0.1–5.0; results with 10% of reproducibility	Herskowitz and Smith (1978a,b)
>20	To minimize the nonuniformity of liquid flow distribution at elevated pressures	Al-Dahhan and Duduković (1994)
	Value proposed from the dependence of overall bed porosity and permeability on the ratio between bed and particle diameters	Sie and Krishna (1998)
20–25	From large and hollow particles at flow conditions of absorption column operation	Porter et al. (1968; Porter and Templemen, 1968)
>20–25	To avoid liquid bypassing along the wall	Froment and Bischoff (1990)
>23	Uniform liquid flow at equilibrium; glass beds 0.64-0.98 cm in size; trickling flow regime	Specchia et al. (1974)
>25	Wall flow less than 10%; pellets 0.64–0.98 cm in size (shape unknown); trickling flow regime without gas flow	Prchlik et al. (1975a,b)
	Nonreactive system with an incompressible single-phase (water) flow in porous media; pseudosteady (laminar flow) operation mode; spheres with almost uniform size from 0.024 to 0.160 cm I.D. and a density of 2.5 g/cm^3; experimental tubes with 0.273, 0.491, 0.802, 0.957, and 1.89 cm I.D; value obtained from measured radial velocity profiles assuming that there is a reasonably large inert part of the bed where flow is even	Chu and Ng (1989)
	To avoid bypassing (e.g., wall effects) to become significant	Sie (1991)
>100	Negligible heat transfer resistance at the wall	Mears (1971)

The gas fraction in the flow feed bypass of the gas increases due to a segregated gas flow pattern. This is critical considering that under industrial conditions, sulfur-containing hydrocarbons may have fractions of up to 75% in the gas phase. Therefore, the gas rate should be kept low in order to limit the degree to which it bypasses the liquid. On the other hand, from micro-packed-bed experiments at low gas flow rates it was shown that gas and liquid move through the column together, and evaporation and condensation did not affect the residence time of the component of interest, it being prudent to perform the kinetic experiments at lower gas-to-liquid flow ratios than those typically employed in TBRs (0.06 to 0.1).

2.3 KINETIC MODELING APPROACHES

Various approaches to kinetic modeling of reactions that take place in the petroleum refining industry have been reported in the literature. On the one hand, kinetic studies considering each compound and all the possible reactions are complex due to the huge number of hydrocarbons involved. However, they permit a mechanistic description based on detailed knowledge of the mechanism of the various reactions. Most of the time, applying this method to reactions with real feeds is difficult because of analytical complexity and computational limitations. The situation is clear: The more compounds a model includes intrinsically, the more kinetic parameters that need to be estimated; and consequently, the more experimental information that is required. On the other hand, the problem can be simplified to consider the partition of the species into a few equivalent classes, called *lumps* or the *lumping technique*, and then assume that each class is an independent entity (Wei and Kuo, 1969). These two approaches are very well known as being the two extreme cases for kinetic modeling of complex mixtures. The second approach is most used nowadays due to its simplicity. There are other models that can be considered as a combination of these two methods; of course, their complexity is based on the experimental information available.

In the following sections, a detailed description of the approaches to kinetic modeling of petroleum refining reactions is reported. Hydrocracking has been chosen as a base for discussing the various kinetic models. The particular kinetics of other reactions, such as hydrotreating, reforming, and catalytic cracking, is described with more detail in subsequent chapters. For better organization of this section, the kinetic model approaches have been classified as (1) models based on the lumping technique, (2) models based on continuous mixtures, and (3) structure-oriented lumping and single-event models.

2.3.1 Traditional Lumping

Models Based on Fractions with a Wide Distillation Range The kinetics of hydrocracking of gas oil was studied by Qader and Hill (1969) in a continuous fixed-bed tubular flow reactor. These authors found that the rate of hydrocracking is of first order with respect to feed concentration, with an activation energy of 21.1 kcal/mol. The kinetic data were obtained at 10.34 MPa pressure, 400 to 500°C temperature, 0.5 to 3.0 h^{-1} space velocity, and a constant H_2/oil ratio of 500 standard m^3/m^3. The liquid product was distilled into gasoline (IBP to 200°C), middle distillate (200 to 300°C), and diesel (300°C+). This seems to be the first experimental study in which kinetics of hydrocracking of real feed was reported.

Callejas and Martínez (1999) studied the kinetics of Maya residue hydrocracking. They used a first-order kinetic scheme involving a three-lump species: atmospheric residuum (AR; 343°C+), light oils (343°C−), and gases. The experiments were conducted continuously in a stirred-tank reactor (1 L) in the presence of a NiMo catalyst supported on γ-Al_2O_3. All tests were carried out at 12.5 MPa hydrogen pressure at temperatures of 375, 400, and 415°C, and WHSV in the range 1.4 to 7.1 L/g_{cat}·h. The total liquid products from each experiment were analyzed by simulated distillation using the ASTM D-2887 method, which was employed to estimate the boiling distribution of the oil samples. Rate constants at various temperatures are listed in Table 2.6. The authors reported that experimental data at 375 and 400°C are in agreement with the model proposed ($r > 0.82$), but at 415°C the fits were bad ($r < 0.70$). Ancheyta et al. (2005) recalculated the kinetic parameters by nonlinear regression and found some inconsistencies with the values reported by Callejas and Martínez (1999), which were attributed to the individual estimation of k_1, k_2, and k_0 by lineal regression. Better agreement between experimental and calculated yields was reported with optimized k_i values, particularly for gas lumps. The original recalculated values of activation energies are reported in Table 2.6.

Aboul-Gheit (1989) determined the kinetic parameters of vacuum gas oil (VGO) hydrocracking, expressing composition in molar concentration. The experiments were carried out at 400, 425, and 450°C, 0.5 to 2 h^{-1} LHSV, and 12 MPa pressure. Two different NiMo catalysts with HY zeolite supported on silica–alumina matrix were used. He proposed that VGO reacts to form gases, gasoline, and middle distillates. Kinetic parameters and activation energies are summarized in Table 2.6. The same problem as that seen in the previous model was observed, which was also due to individual determination of each parameter by lineal regression. The exact values of k_0 are reported in parentheses in Table 2.6, which are very close to the original values. Consequently, the activation energies determined with the two series of k_0 values are similar.

Another kinetic model for gas oil hydrocracking was proposed by Yui and Sanford (1989), who performed experiments in a pilot plant with a trickle-bed reactor at different operating conditions (350 to 400°C, 7 to 11 MPa, 0.7 to 1.5 h^{-1} LHSV, and an H_2/oil ratio of 600 std m^3/m^3). They used Athabasca bitumen–derived coker and hydrocracker heavy gas oils (HGOs) as feed and two different commercial NiMo/Al_2O_3 hydrotreating catalysts. A three-lump model was considered [HGO, LGO (light gas oil), and naphtha], which can

TABLE 2.6. Kinetic Data Reported for Various Lump Models

Kinetic Data Reported by Callejas and Martínez (1999) and Ancheyta et al. (2005)

	375°C	400°C	415°C	E_A
k_0 (L/g_{cat}·h) Callejas and Martínez (1999)	1.13	3.26	9.20	45.32
Ancheyta et al. (2005)	1.13	3.18	7.22	41.32
k_1 (L/g_{cat}·h) Callejas and Martínez (1999)	0.07	0.25	1.52	64.40
Ancheyta et al. (2005)	0.30	0.46	1.45	32.57
k_2 (L/g_{cat}·h) Callejas and Martínez (1999)	0.21	1.5	5.12	70.43
Ancheyta et al. (2005)	0.79	2.72	5.77	43.90

Kinetic Data Reported by Aboul-Gheit (1989)

	Catalyst 1				Catalyst 2			
	400°C	425°C	450°C	E_A	400°C	425°C	450°C	E_A
k_1 (h^{-1})	0.286	0.500	0.688	17.51	0.469	0.612	0.916	13.09
k_2 (h^{-1})	0.040	0.083	0.140	24.02	0.111	0.216	0.350	22.23
k_3 (h^{-1})	0.026	0.048	0.069	18.67	0.040	0.074	0.106	18.96
k_0^* (h^{-1})	0.352	0.631	0.897	18.14	0.620	0.902	1.372	15.35
	(0.333)	(0.667)	(1.059)	22.51	(0.714)	(1.125)	(1.75)	17.15

Kinetic Data Reported by Yui and Sanford (1989)

	Coker Feed		Hydrocracker Feed	
Parallel Scheme (k_3=0)	A	E_A	A	E_A
k_1+k_2	8.754×10^4	17.75	4.274×10^4	17.24
k_1	8.544×10^3	15.02	3.775×10^3	14.32
k_2	1.780×10^8	29.78	6.847×10^8	32.17
Consecutive scheme (k_2=0)				
k_1	8.754×10^4	17.75	4.274×10^4	17.24
k_3	6.206×10^7	26.96	2.711×10^5	20.46

*Values in parentheses correspond to $k_1+k_2+k_3$. A in h^{-1}, E_A in kcal/mol.

follow parallel, consecutive, and combined reaction schemes. The model includes first-order reactions and considers the effects of partial pressure (in MPa), temperature (in °C), and space velocity on the total liquid products yield [Eqs. (2.49), (2.50), and (2.51), Table 2.7]. Fitted parameters are $Y_0 = 1.0505$, $a = 0.2517$, $b = 0.0414$, and $c = -0.0163$ for the coker gas oil; and $Y_0 = 1.0371$, $a = 0.1133$, $b = 0.0206$, and $c = -0.0134$ for the hydrocracker gas oil. The kinetic parameters are presented in Table 2.6. According to the authors, it was not possible to fit a set of parameters for the combined reaction scheme.

The kinetics of hydrocracking of vacuum distillates from Romashkin and Arlan crude oils was studied by Orochko (1970) in a fixed-bed reactor over an alumina–cobalt molybdenum catalyst using a first-order kinetic scheme

TABLE 2.7. Equations for Kinetic Models Based on Traditional Lumping

Criterion	Eq.	Criterion	Eq.
$Y = Y_0\left(\frac{T}{400}\right)^a\left(\frac{P_{H_2}}{10}\right)^b \text{LHSV}^c$	(2.49)	$T^* = \frac{\text{FBP}_f - T}{\text{FBP}_f - T_{50}}$	(2.59)
$\frac{dC_{\text{HGO}}}{dt} = -(k_1 + k_2)C_{\text{HGO}}$	(2.50)	$f = \frac{1}{2} + \frac{1}{2}\text{erf}\left[\frac{1-T^*}{(2T^*)^{0.5}}\text{Pe}^{0.5}\right]$	(2.60)
$\frac{dC_{\text{LGO}}}{dt} = k_1 C_{\text{HGO}} - k_3 C_{\text{LGO}}$	(2.51)	$T_{50,\tau} = T_{50,f}\exp(-k_{50}\tau)$	(2.61)
$\alpha\tau = \ln\frac{1}{1-y} - \beta y$	(2.52)	$\text{Pe} = 20.125 - 0.175P$	(2.62)
$z = \frac{(1-y)^{k'} - (1-y)}{1-k'}$	(2.53)	$\frac{d(T_{50,\tau}/T_{50,f})}{d\tau} = -k_{50}(T_{50,\tau}/T_{50,f})^n$	(2.63)
$x = k'\frac{(1-y)^{k''} - (1-y)^{k'}}{(1-k')(k'-k'')} + k'\frac{(1-y) - (1-y)^{k''}}{(1-k')(1-k'')}$	(2.54)	$n = 1.9 - 0.0015P$	(2.64)
$g = y - (z + x)$	(2.55)	$k_{50} = 0.4 - 0.003P$	(2.65)
$C_{\text{HGO}_{\text{out}}} = C_{\text{HGO}_{\text{in}}}\exp\left(-\frac{k_H}{SV}\right)$	(2.56)	$k_i(T) = k_{365}K_i$	(2.66)
$k_H = a[\text{S}] + b[\text{PA}] + c(1 + d[\text{I}])^{-n}$	(2.57)	$K_i = 0.494 + 0.52\times10^{-2}\text{TBP}_i - 2.185\times10^{-5}\text{TBP}_i^2 + 0.312\times10^{-7}\text{TBP}_i^3$	(2.67)
$k_H = k_{\text{HT}} + k_{\text{HC}}$	(2.58)	$k_{365} = 4.273\times10^3\exp\left(\frac{-2.11\times10^4}{RT}\right)$	(2.68)

involving four lumps. This model is similar to that proposed by Aboul-Gheit (1989). The rate of a first-order heterogeneous catalytic reaction was expressed by the Eq. (2.52) (Table 2.6), where α is the rate constant, τ the nominal reaction time, γ the total conversion, and β the inhibition factor of the process by the reaction products formed and absorbed on the active surface of the catalyst and also by their effect on mass transfer in the heterogeneous process. These authors indicate that in this case the consecutive reactions predominate, the parallel reactions in the calculations being comparatively minor and negligible to a first approximation. All experiments were carried out at 5 and 10.13 MPa of hydrogen pressure and temperatures of 400, 425, and 450°C. For the case of Arlan petroleum vacuum distillate at 425°C and 10.13 MPa, a value of $\beta = 1$ was reported. Rate constants and activation energies based on the experimental data reported by others are given in Table 2.8. The kinetic model is represented by Eqs. (2.53), (2.54), and (2.55) for diesel, gasoline, and gases, respectively (Table 2.6). In these equations, k' and k'' are kinetic factors with a meaning similar to that of the rate constants, which are determined from the experimental data and are dependent on the equivalent kinetic temperature of the process and the catalyst activity. For the Romashkin petroleum vacuum distillate at 10.13 MPa, the values of k' and k'' are 1.3 and 2.0, respectively.

Botchwey et al. (2004) studied overall conversion kinetic models within specified, short-range temperature regimes for the hydrotreating of bitumen-derived heavy gas oil from Athabasca over a commercial $NiMo/Al_2O_3$ catalyst in a trickle-bed reactor. All experiments were carried out at various reaction temperatures between 340 and 420°C, 8.8 MPa of pressure, LHSV of $1\,h^{-1}$, and a H_2/oil ratio of 600 standard m^3/m^3. The oil samples (feed and products) were grouped into four different boiling cuts with temperature ranges of D (IBP to 300°C), C (300 to 400°C), B (400 to 500°C), and A (500 to 600°C). The boiling-point distribution was derived from gas chromatography simulated distillation. It should be noted that the product analyses were limited to liquid samples, because negligible amounts of gaseous hydrocarbon products were formed from mass balances. The proposed kinetic model included the four lumps (A, B, C, and D) and five kinetic parameters ($k_1,\ldots,k_5$). The low-severity temperature regime was considered to be that at the lowest operating temperature range (340 to 370°C), and the reactions A to C and C to D were negligible. In the intermediate-severity temperature regime (370 to 400°C), only the reaction A to C was negligible. The value of k_5 is equal to zero in both kinetic schemes derived from both temperature regimes. The high-severity temperature regime covered the most severe operating temperature range (400 to 420°C). All kinetic parameter values for the three regimes are tabulated in Table 2.8.

Aoyagi et al. (2003) studied the kinetics of hydrotreating and hydrocracking of conventional gas oils, coker gas oils, and gas oils derived from Athabasca bitumen. They were interested in studying the influence of feed properties on product yield and composition. The experiments were fixed as follows: a temperature of 380°C, an operating pressure at 13.8 MPa, a liquid hourly space

TABLE 2.8. Activation Energies Reported for Various Kinetic Models

Activation Energies Reported by Orochko (1970)

Feedstock	Total Pressure (MPa)	E_A (kcal/mol)
Romashkin petroleum vacuum distillate	5.06	56.7
	10.13	63.8
Arlan petroleum vacuum distillate	5.06	63.0
	10.13	64.8

Activation Energies Reported by Botchwey et al. (2004)

	k_1	k_2	k_3	k_4	k_5
Low-severity temperature regime (340–370°C)					
E_A	33.94	40.15	8.84		
$\ln A$	25.2	28.9	5.4		
Intermediate-severity temperature regime (370–400°C)					
E_A	24.14	21.03	16.01	31.79	
$\ln A$	17.3	14.2	10.7	21.9	
High-severity temperature regime (400–420°C)					
E_A	25.57	26.53	22.46	29.16	28.44
$\ln A$	17.9	18.2	15.4	20.1	19.9

Activation Energies Reported by Sánchez et al. (2005, 2007)

h^{-1}	400°C	E_A (kcal/mol)	h^{-1}	400°C	E_A (kcal/mol)
k_1	0.147	48.5	k_6	0.007	37.1
k_2	0.022	44.2	k_7	0	
k_3	0.020	38.0	k_8	0.003	53.7
k_4	0.098	27.3	k_9	0	
k_5	0.057	39.5	k_{10}	0	

velocity of $0.75\,h^{-1}$, and a H_2/oil ratio of 400 std m^3/m^3. The feeds with different properties were obtained mixing hydrotreated gas oils with gas oil without hydrotreating. A kinetic model was developed and the parameters were adjusted with experimental data from a system with two reactors in series, each with a different catalyst. In the first reactor, a commercial NiMo/γ-Al_2O_3 catalyst was used, and in the second reactor, a commercial hydrocracking catalyst with NiMo/boria USY was employed. The model considers that in the first hydrotreating reactor, the modifications in molecular weight are due to reactions of hydrodesulfurization and hydrogenation of polycyclic aromatic compounds. Hydrocracking is the most important reaction in the second reactor. The model uses a first-order expression to describe the rate of disappearance of heavy gas oil (HGO), given by Eq. (2.56), where k_H is the overall hydrocracking rate constant. Its value depends on both hydrotreating and

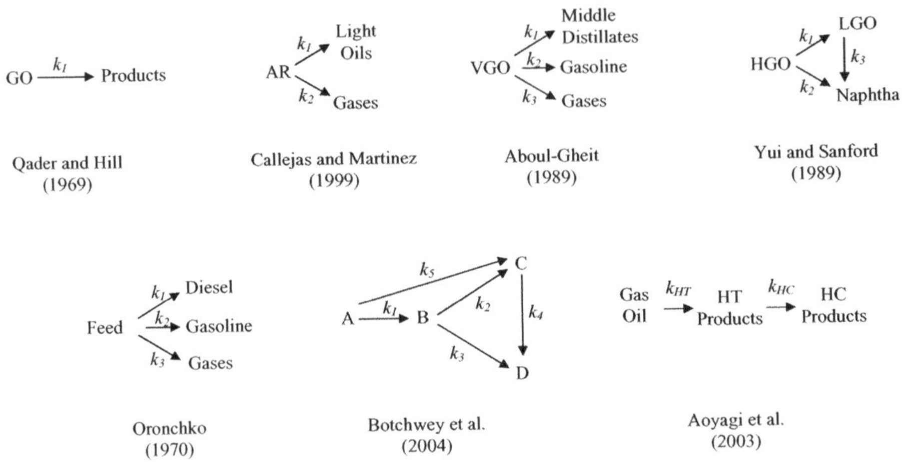

Figure 2.8. Reaction schemes for hydrocracking models with two to four lumps.

hydrocracking reactions and is calculated with Eqs. (2.57) and (2.58), in which the last term includes the nitrogen content's inhibitor effect. HGO_{in} and HGO_{out} are the inlet and outlet concentrations of heavy gas oil; and [S], [PA], and [I] are the contents of sulfur, polycyclic aromatic compounds, and inhibitors. The best set of model parameters reported by the authors is $a = 9.5 \times 10^{-4}$, $b = 1.8 \times 10^{-3}$, $c = 0.32$, $d = 9.1 \times 10^{-4}$, and $n = 2$. Figure 2.8 presents the reaction schemes proposed for developing the previous kinetic models, which contain no more than four lumps.

Another reaction pathway was proposed by Botchwey et al. (2003). The pathways describe the conversion of gas oil to products via heteroatom removal, aromatics saturation, and hydrocracking. Typical hydrotreating reactions are represented by solid lines and cracking reactions are shown by dashed lines. These authors consider conversion to take place according to different regimes: the hydrotreating regime (reactions 1 to 7) at temperatures of 340 to 390°C and the mild hydrocracking regime (reactions 1 to 9) at 390 to 420°C. They arrived at this conclusion after performing experiments in a micro-trickle-bed reactor. The study covered a pressure range between 6.5 and 11 MPa at temperatures of 360, 380, and 400°C. The liquid hourly space velocity and the H_2/oil ratio were maintained constant at 1 h^{-1} and 600 std m^3/m^3, respectively. However, kinetic expressions and rate constants are not given.

Mosby et al. (1986) reported a model to describe the performance of a residue hydrotreater using lumped first-order kinetics which divides residue into lumps that are easy and difficult to crack. This lumping scheme was used by Ayasse et al. (1997) to fit experimental product yields from catalytic hydrocracking of Athabasca bitumen obtained in a continuous-flow mixed reactor over a NiMo catalyst at 430°C and 13.7 MPa. To develop the model, stoichiometry concepts of a complex reacting mixture were applied. The resulting compact model was fitted to data from single-pass hydrocracking and used to predict the performance of multipass experiments. The liquid product was distilled into four cuts: naphtha (IBP to 195°C), middle distillates (195 to 343°C), gas oil (343 to 524°C), and residue (>524°C). Residue fraction was then distilled under vacuum to obtain the gas oil and residue fractions using the ASTM D-1160 procedure. After all the data had been utilized to estimate the parameters of the general lumped model, it was found that the model was overdetermined. The number of parameters was too large, and it was concluded that seven lumps are not required to give the experimental data a satisfactory fit. Afterward, three new models were proposed, two with six lumped components and one with five lumps, which were considered to be adequate to describe the data with an equivalent sum of squared residuals. In model 1, hard and soft residues were lumped as a single component under "hard residue." The initial concentration of lump 2 was zero, and the kinetic parameters of this lump (k_2, s_{24}, s_{25}, s_{26}, and s_{27}) were not determined. In model 2, all the gas oil, whether it originated with the feed or was formed by cracking of the residue, was lumped as a single component under "product gas oil." The initial concentration of lump 3 was zero, and the kinetic parameters of this lump (k_3, s_{35}, s_{36}, and s_{37}) were not determined. Consequently, the simplest model that could capture this chemistry was a five-lump model (model 3), consisting of one residue lump (hard residue), one gas oil lump (product gas oil), middle distillates, naphtha, and light ends. The resulting five-lump model had seven independent parameters (two rate constants and five independent stoichiometric coefficients). After determining the optimal parameter values, it was found that model 1 overpredicted the yield of middle distillates and underestimated the yield of naphtha at high residue conversion in experiments with bitumen as feed. Model 1 was therefore satisfactory for fitting yields over a wide range of residue conversion. Model 2 was inferior to model 1 in predicting the products, with large errors in the proportions of naphtha and gas oil. However, model 3 underestimated the yield of middle distillates and tended to overpredict the yield gas oil. The models with six and seven lumps are unnecessarily complex for these data, whereas the simpler five-lump model is satisfactory.

Recently, Sánchez et al. (2005, 2007) proposed a five-lump kinetic model for moderate hydrocracking of heavy oils: (1) unconverted residue (538°C+), (2) vacuum gas oil (VGO: 343 to 538°C), (3) distillates (204 to 343°C), (4) naphtha (IBP to 204°C), and (5) gases. The model includes 10 kinetic parameters which were estimated from experimental data obtained in a fixed-bed downflow reactor, with Maya heavy crude and a NiMo/γ-Al_2O_3 catalyst at a 380 to 420°C reaction temperature, 0.33 to 1.5 h^{-1} LHSV, a H_2/oil ratio of 890 m^3/m^3, and a 6.9 MPa pressure. Activation energies reported by these authors are given in Table 2.8. The kinetic model was developed for basic reactor modeling studies of a process for hydrotreating of heavy petroleum oils, which, among several characteristics, operates at moderate reaction

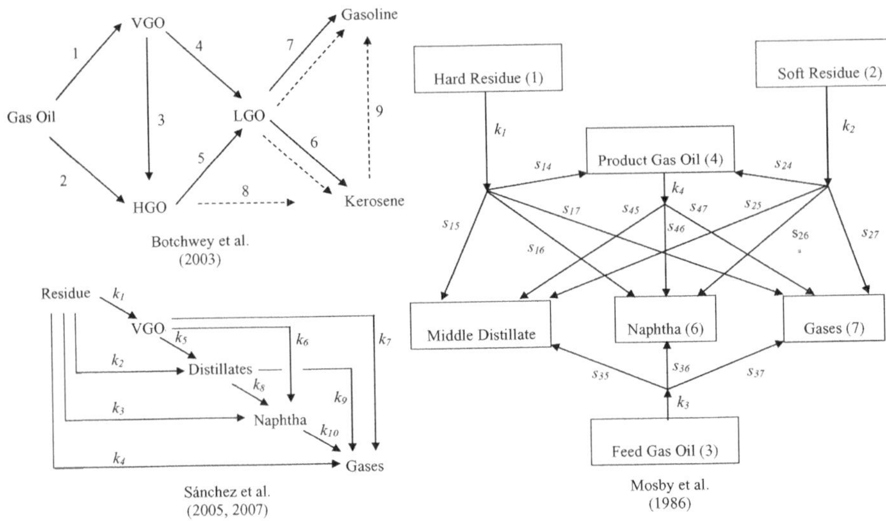

Figure 2.9. Reaction schemes for hydrocracking models with more than four lumps.

conditions and improves the quality of the feed while keeping the conversion level low. Figure 2.9 presents the kinetic models reported by Mosby et al. (1986), Botchwey et al. (2003), and Sánchez et al. (2005).

Models Based on Pseudocomponents: Discrete Lumping Krishna and Saxena (1989) reported a detailed kinetic model with seven lumps in which different cut temperatures are considered. The lumps are sulfur compounds, heavy and light aromatics, naphthenes, and paraffins. The pseudocomponents are considered light if they are formed from fractions with boiling points lower than the cut temperature (T_{cut}). Sulfur compounds are assumed to be a heavy lump. Experimental data reported by Bennett and Bourne (1972) were used to test the model; the values of the 60 kinetic parameters are presented in Table 2.9. The authors proposed a second model based on the analogy between reactions of hydrocracking and the phenomena of axial dispersion of a tracer in a flow; this model used only two parameters. Figure 2.10 shows the reaction scheme proposed and a comparison of the experimental data with the results predicted. The dispersion model is based on a study of the TBP (true boiling point) curves of hydrocracking products. An increment in residence time causes a reduction in the average molecular weight of the product and a drop in the distillation curve's middle boiling point (T_{50}). TBP curves at different residence times are normalized to obtain values of T^* according to Eq. (2.59), where FBP_f is the final boiling point of the feed. Krishna and Saxena (1989) used Bennett and Bourne's (1972) pilot-plant experimental data to develop the model. The normalized temperature data of feed and products, shown in Figure 2.11, can be described roughly by Eq. (2.60); the solid line is the repre-

TABLE 2.9. First-order Rate Constants for the Kinetic Model Proposed by Krishna and Saxena (1989) and Comparison of Calculated and Plant Data Obtained by Mohanty et al. (1991)

Kinetic Constant (h^{-1})	T_{cut} (°C) (Krishna and Saxena, 1989) 371	225	191	149	82	0
k_0	8.3000					
k_1	1.2633	0.4943	0.4799	0.4624	0.4345	0.4000
k_2	0.6042	0.1809	0.1105	0.0397	0.0034	0.0000
k_3	0.0421	0.3131	0.2719	0.2593	0.2501	0.2302
k_4	0.5309	0.0211	0.0096	0.0095	0.0095	0.0095
k_5	0.0397	0.0383	0.0249	0.0131	0.0086	0.0000
k_6	1.1855	0.2772	0.2134	0.1117	0.0073	0.0000
k_7	0.1619	0.0474	0.0275	0.0275	0.0275	0.0275
k_8	0.4070	0.2391	0.1993	0.1518	0.0978	0.0299
k_9	0.2909	0.5434	0.5219	0.4509	0.4391	
k_{10}	0.0818	0.0740	0.0709	0.0618	0.0608	

	Mohanty et al. (1991) Calculated	Plant Data	Error (%)
Total feed to second stage (kg/h)	183, 236	183, 385	−0.08
Hydrogen consumption (kg/h)			
first stage	2816	3267	−13.8
second stage	1196	1363	−12.2
Reactor outlet temperature (°C)			
first stage	693.3	714 (max.)	
second stage	677.7	700 (max.)	
Diesel (wt%)	48.79	50.5	−3.46
Jet fuel (wt%)	30.53	29.4	+3.83
Naphtha (wt%)	16.17	15.8	+2.51
Butanes and lights (wt%)	4.51	4.5	+0.22

sentation of the axial dispersion model with Pe = 14. The middle boiling-point temperature is obtained using Eq. (2.61), which assumes a first-order decay function.

Furthermore, Krishna and Saxena (1989) developed empirical correlations to predict the values of the decay rate of T_{50} (k_{50}) with respect to residence time (τ) and Peclet number (Pe). Both parameters are functions of the paraffin content in the feedstock (P). Equations (2.62) to (2.65) permit the estimation of these parameters considering an n-order decay function.

Stangeland (1974) developed a kinetic model for predicting hydrocracker yields using correlations based on the boiling point of each of the pseudocomponents that characterize the cut. The model includes four parameters: k_0 and A quantify each pseudocomponent's reaction rate, C gives the butane yield

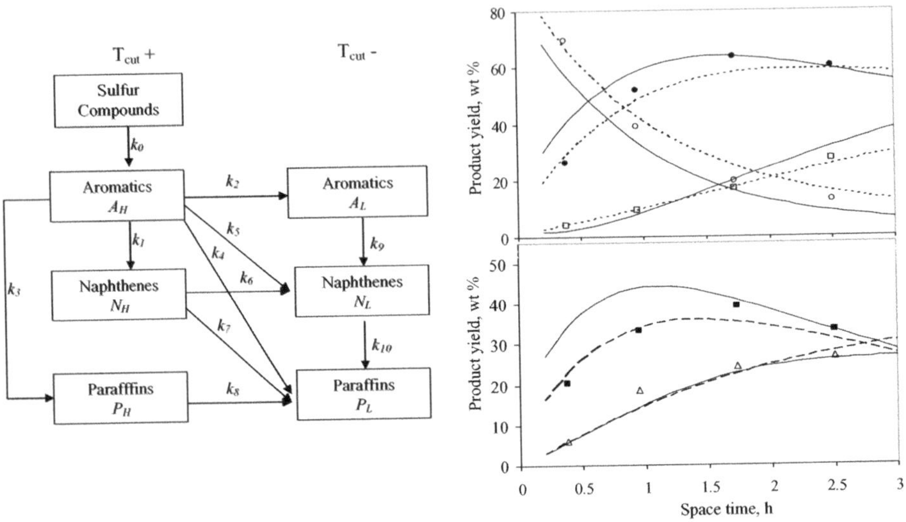

Figure 2.10. Reaction scheme for hydrocracking proposed by Krishna and Saxena (1989), and comparison of calculated and experimental yields (○, 371°C+; ●, 149–371°C; □, 149°C–; ■, 149–225°C; △, 225–371°C; —, dispersion model; ---, kinetic model).

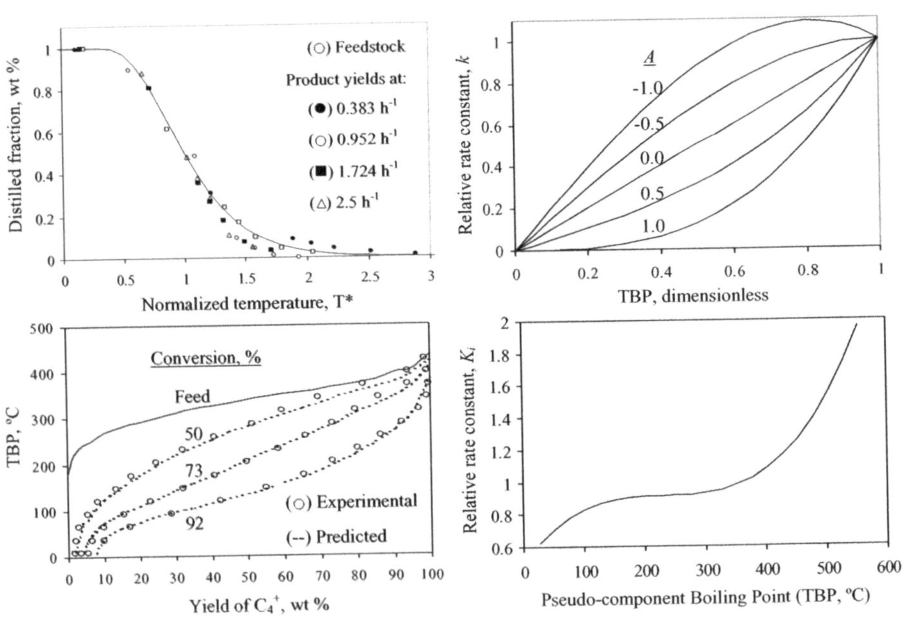

Figure 2.11. Normalized TBP curves, cracking rate function, and comparison of yields for once-through hydrocracking (Stangeland, 1974) and the relative rate constant function (Mohanty et al., 1991).

magnitude, and B varies with both the type of feed (naphtenic or paraffinic) and the type of catalytic process (random or selective). Parameters B and C determine the shape of the yield curve. Although A usually lies in the range 0 to 1.0, it can take negative values. Parameter A determines the shape of the reactivity curve, which varies from a linear to a cubic function, as shown in Figure 2.11. The complete set of equations, presented in Table 2.10, permits calculation of the formation of component i due to the decomposition of heavy components j.

Sets of data at three conversion levels are illustrated in Figure 2.11 for hydrocracking of raw California gas oil in once-through liquid operation. The yields predicted based on these parameters are shown as dashed lines for conversions of 50, 73, and 92% (288°C–). In general, the agreement with experimental data is quite good and the differences are probably within experimental error. The major disadvantage of this approach is that a change in hydrocracker product specifications, or in the number of products, requires reformulating the model and refitting the data.

Mohanty et al. (1991) implemented Stangeland's kinetic model in a computer model for a two-stage commercial-scale VGO hydrocracker. The feed and products were lumped into 23 pseudocomponents for the hydrocracking reactions, and pseudohomogeneous first-order reactions were assumed. Estimation of the hydrocracking kinetic constants for the other pseudocomponents that comprise the VGO was done with Eq. (2.66), where K_i was adjusted with plant data using Eq. (2.67) (Figure 2.11). A hydrocracking kinetic constant of vacuum gas oil with an average boiling point of 365°C reported by Qader and Hill (1969) was employed [Eq. (2.68)]. Calculated yields, hydrogen consumption, and outlet temperatures with this model are tabulated in Table 2.9. The model was validated against plant data and the agreement was generally good. It is important to indicate that with the parameters reported by Mohanty et al. (1991), the mass balance closure in each individual hydrocracking reaction is not satisfied.

Dassori and Pacheco (2002) established a link between the stoichiometric coefficient of the hydrocracking reactions and the parameters P_{ij} of Stangeland's kinetic model; this analogy imposes a constraint on the values that the P_{ij} matrix can take. Such a constraint is given by the mass balance closure in each hydrocracking reaction and would require determination of the values of parameters B and C (Table 2.10). It was noticed that only with these two parameters is it not possible to rearrange the product distribution to satisfy the mass balance for each reaction. These authors modified the model proposed by Stangeland by adding two additional parameters, B_2 and ω, so that the mass balance in each hydrocracking reaction is satisfied. They used a second-order hydrocracking rate constant to quantify the effect of hydrogen partial pressure on the rate of cracking. The kinetic constants are determined from the pseudo-first-order constants reported by Qader and Hill (1969). This model was applied to the hydrocracking of VGO in a commercial reactor described by Mohanty et al. (1991).

TABLE 2.10. Equations of the Kinetic Models for Hydrocracking Based on TBP of Pseudocomponents and on Continuous Mixtures

Eq.	Kinetic Models	
	Model Proposed by Stangeland (1974)	
(2.69)	$\frac{d}{dt}F_i(t) = -k_iF_i(t) + \sum_{j=1}^{i-1} P_{ij}k_jF_j(t)$	Mass balance
(2.70)	$k(T) = k_0[T + A(T^3 - T)]$	Cracking rate constant function
(2.71)	$PC_{ij} = [y_{ij}^2 + B(y_{ij}^3 - y_{ij}^2)](1 - [C_4]_j)$	Liquid product distribution function
(2.72)	$[C_4]_j = C\exp[-0.00693(1.8TBP_j - 229.5)]$	Weight fraction of butane
(2.73)	$y_{ij} = \frac{TBP_i - 2.5}{(TBP_j - 50) - 2.5}$	Normalized boiling-point temperature (TBP)
(2.74)	$P_{ij} = PC_{ij} - PC_{i-1,j}$	Actual fraction of lighter component
	Stangeland(1974)Model Modified by Dassori and Pacheco (2002)	
(2.75)	$\sum_{i=1}^{j-2} P_{i,j}MW_i = (MW_j - n_jMW_{H_2})$	Mass balance for each individual reaction
(2.76)	$PC_{ij} = [y_{ij}^2 + B_1y_{ij}^3 - B_2y_{ij}^2](1 - [C_4]_j)$	Modified product distribution function (with B_2)
(2.77)	$[C_4]_j = C\exp[-\omega(1.8TBP_j - 229.5)]$	Modified weight fraction of butane (with ω)
	Model of Laxminarasimhan and Verma (1996)	
(2.78)	$\theta = \frac{TBP - TBP(l)}{TBP(h) - TBP(l)}$	
(2.79)	$\frac{k}{k_{max}} = \theta^{1/\alpha}$	
(2.80)	$D(k) = \frac{N\alpha}{k_{max}^{\alpha}}k^{\alpha-1}$	
(2.81)	$\frac{dC(k,t)}{dt} = -kC(k,t) + \int_k^{k_{max}} [p(k,K)KC(K,t)D(K)]dK$	
(2.82)	$p(k,K) = \frac{1}{So\sqrt{2\pi}}\left\{\exp\left[-\left(\frac{(k/K)^{a_0} - 0.5}{a_1}\right)^2\right] - A + B\right\}$	
(2.83)	$A = e^{-(0.5/a_1)^2}$	
(2.84)	$B = \delta\left(1 - \frac{k}{K}\right)$	
(2.85)	$So = \int_0^K \frac{1}{\sqrt{2\pi}}\left\{\exp\left[-\left(\frac{(k/K)^{a_0} - 0.5}{a_1}\right)^2\right] - A + B\right\}D(K)dk$	

2.3.2 Models Based on Continuous Mixtures

Laxminarasimhan and Verma (1996) developed a kinetic model for hydrocracking of a petroleum mixture based on the continuous theory of lumping. The model considers properties of the reaction mixture, the underlying pathways, and the associated selectivity of the reactions. The parameter of characterization is the TBP temperature. During the reaction of a particular feed, the mixture's distillation curve changes continuously inside the reactor, and as the residence time increases, most of the heavier components are converted into lighter components. A normalized TBP as a function of an index (θ) is used instead of the TBP. Normalized TBP is defined by Eq. (2.78) (Table 2.10). The reactivity is considered to be monotonic and can be represented by a simple power-law type of function [Eq. (2.79)], where k is the reaction rate of a particular compound, k_{max} is the reaction rate of the compound of higher TBP, and α is a model parameter. The model equations are formulated as a function of reactivity following a procedure proposed by Chou and Ho (1989). To express the equation with k as the independent reactivity, a transformation operator is required, which is approximated by Eq. (2.80). $D(k)$ can be considered as a species-type distribution function, where N is the number of compounds in the mixture and tends toward infinitum in a heavy fraction of oil. A material balance of species of reactivity k, the core of the kinetic model, can be expressed with integrodifferential equation (2.81).

$p(k,K)$ is ideally the yield distribution function that describes the formation of compounds of reactivity k from hydrocracking of compounds of reactivity K. This function is approximated in this model by a skewed Gaussian-type distribution function obtained from experimental data on the reactivity of several model compounds [Eqs. (2.82) to (2.85)]. The parameters a_0, a_1, and δ are specific for each system and are used for model tuning.

This model was employed successfully to experimental data published previously. Bennett and Bourne (1972) reported product yields from the hydrocracking of Kuwait vacuum gas oil at four different residence times: 0.383, 0.952, 1.724, and 2.5 h. The set of parameters was tuned using data obtained at 2.5 h of residence time. These parameters are $\alpha = 1.35$, $k_{max} = 1.35\,h^{-1}$, $a_0 = 6.41$, $a_1 = 28.15$, and $\delta = 2.6667 \times 10^{-5}$. Figure 2.12 shows a typical $p(k,K)$ function, in this case, $k = k_{max}$. Comparisons of experimental and estimated values are also shown in Figure 2.12.

El-Kady (1979) reported another set of experimental data at several reaction temperatures and residence times for the hydrocracking of a vacuum gas oil. In Figure 2.12, comparisons of these experimental data and estimated values at 390°C are presented. The set of fitted parameters of the model reported by Laxminarasimhan and Verma (1996) is $\alpha = 0.77$, $k_{max} = 0.88\,h^{-1}$, $a_0 = 3.67$, $a_1 = 22.86$, and $\delta = 0.77 \times 10^{-9}$ for experimental data at 390°C.

Extensions of the Laxminarasimhan and Verma (1996) model, in which the reacting mixture is divided into continuous mixtures of paraffinic, naphthenic, and aromatics components, have been published by the same research group

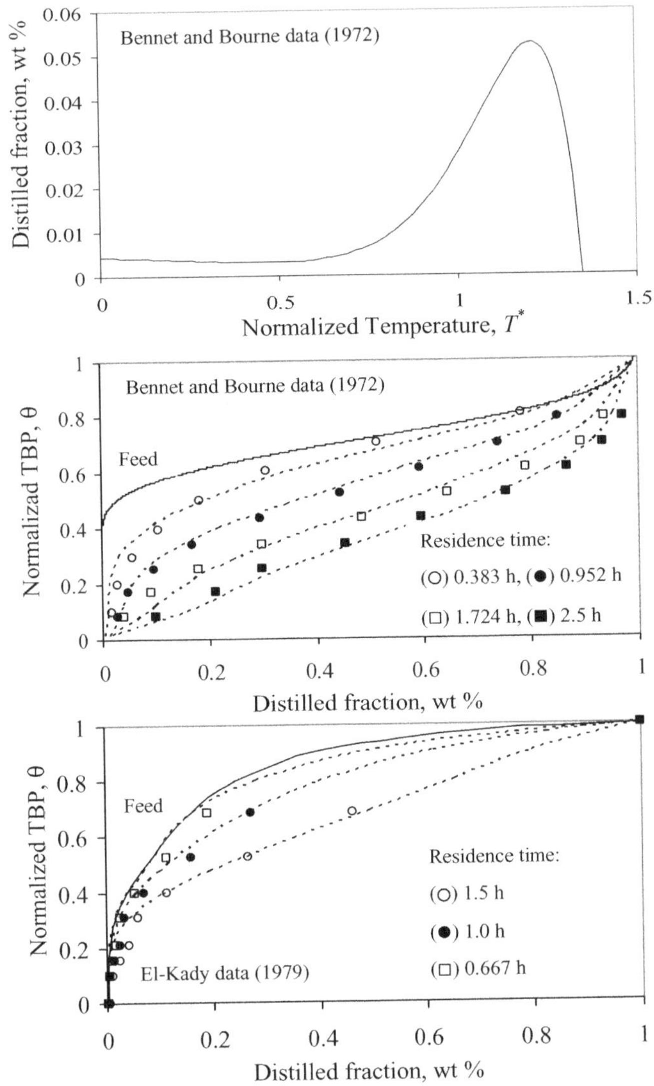

Figure 2.12. Yield distribution function $p(k,K)$ and comparison of estimations (lines) with Laxminarasimhan and Verma model (1996) and experimental data (symbols).

(Narasimhan et al., 1997; Basak et al., 2004). In addition to the reactions of hydrocracking that form compounds in the same family, the formation reactions of paraffins from naphthenes, paraffins from aromatics, and naphthenes from aromatics were considered. Therefore, the models require the definition of a concentration function, a reactivity function, and a species distribution function for each family of compounds, as well as six different product distribution functions. The models were validated with the pilot-plant experimental data reported by Bennett and Bourne (1972). However, parameters of the model functions were not reported. Elizalde et al. (2009) recently reported application of the continuous lumping approach for modeling of hydrocracking of heavy crude oil at moderate reaction conditions. The model parameters were estimated from experiments obtained in an isothermal fixed-bed reactor at 380 to 420°C, 0.33 to $1.5\,h^{-1}$ LHSV at constant pressure (9.8 MPa), and an H_2/oil ratio of $5000\,ft^3$/bbl. More details were provided about values and estimation of model parameters. Comparisons between experimental data and predictions using the continuous lumping kinetic model were reported to show good agreement with an average absolute error of less than 5%.

2.3.3 Structure-Oriented Lumping and Single-Event Models

Structure-oriented lumping kinetic models, which employ most of the information obtained through modern analytical techniques for model reaction modeling at a molecular level, have been proposed for some catalytic processes. The lumps are defined according to the structure of the compounds in the reacting mixture.

Liguras and Allen (1989) utilized contribution group concepts, which provide a mechanism for making use of pure compound data in modeling complex reactions. They describe the conversion of vacuum gas oil in terms of a relatively large number of pseudocomponents, most of which are lumps in their own right. Quann and Jaffe (1992) developed a procedure to describe molecules and reactions with a notation of vectors which allows a computer program to represent the reaction networks. These authors expressed the chemical transformations in terms of a typical structure of the molecules without completely eliminating lumps and rate parameters, which depend on the feedstock composition.

Martens and Marin (2001) reported a model for the hydrocracking of hydrogenated vacuum gas oil based on theoretical and mechanistic considerations. The reaction mechanism is described by a set of single events, each of which can be ascribed a rate equation or a term in a single rate equation. The model considers the reaction rules for carbenium ion of the secondary and tertiary types. A computer algorithm was used for generating the reaction networks.

Froment (2005) has recently reviewed the single-event approach, which retains the full details of the reaction pathways of the individual feed components and reaction intermediates. This approach is illustrated by means of the methanol-to-olefins and catalytic cracking of oil fraction reactions. It is also highlighted the fact that other important processes with complex feedstocks, such as catalytic reforming, hydrocracking, alkylation, and isomerization, can be modeled by means of the single-event concept. As can be seen, all these approaches have been used successfully for some complex reaction systems. However, hydrocracking kinetics of heavy oil fractions with structure-oriented lumping modeling or the single-event approaches has not been reported in the open literature.

Lump models have been used for several years for kinetic modeling of complex reactions. In fact, some commercial catalytic process design is still being performed with this type of approach. Catalyst screening, process control, basic process studies, and dynamic modeling, among others, are areas in which lump kinetic models are employed extensively. The main disadvantages of lump models are their simplicity in predicting product yields, the dependency of kinetic parameters on feed properties, and the use of an invariant distillation range of products, which, if changed, necessitates further experiments and parameter estimation.

Models based on continuous mixtures (continuous theory of lumping) overcome some of these deficiencies by considering the properties of the reaction mixture, the underlying pathways, and the associated selectivity of the reactions. The common parameter of characterization is the true boiling-point temperature, since during reaction it changes continuously inside the reactor as the residence time increases. However, the dependency of model parameters on feed properties is still present. Distillation curves, either chromatographic or physical, also present some difficulties when analyzing heavy oils, since initial and final boiling points are not accurate during experimentation. In fact, for many purposes, 10% and 90% boiling points are commonly utilized instead of IBP and FBP, respectively.

Structure-oriented lumping models are more detailed approaches that express the chemical transformations in terms of typical molecular structures. These models describe reaction kinetics in terms of a relatively large number of pseudocomponents, and hence they do not completely eliminate lumps. In addition, dependency of rate parameters on feed properties is present.

The single-event concept,which uses elementary steps of cation chemistry, consists of a limited number of types of steps involving a series of homologous species. The number of rate coefficients to be determined from experimental information can be reduced and are modeled based on transition-state theory and statistical thermodynamics. With this approach, parameter values are not dependent on feed properties. However, even though the number of parameters can be diminished, detailed and sufficient experimental data are necessary.

The complexity of real feedstocks suggests that models based on lumping theory will continue to be used for the study of hydrocracking reaction kinetics. However, more sophisticated and accurate approaches need to be studied with more detail for a better understanding and representation of heavy oil hydrocracking kinetics.

2.4 REACTOR MODELING

2.4.1 Classification and Selection of Reactor Models

For decades, different models have been developed to understand, design, simulate, or optimize the performance of the reactors used in the petroleum refining industry. This development has been parallel to the need for more detailed prediction capability, motivated by continuous changes in process conditions (i.e., nature of feedstocks, new reactor and catalyst design, reaction conditions, etc.). In this section, different classifications of reactor models as reported in the literature are described. The classification is based on the various levels of sophistication of each model reported. For example, Shinnar (1978) proposed a simple classification that distinguishes between learning and predictive models. *Learning models* are of neuronal network type, and *predictive models* can be subclassified as deterministic and stochastic models, as illustrated in Figure 2.13.

On the one hand, in *deterministic models*, continuous models (i.e., Fickian, dispersion, diffusion, or effective transport models using Fick's and Fourier's laws of mass and heat dispersion, respectively) are represented by differential equations that have one or more independent variables; discrete models are finite-stage models described mainly by capillary, sphere packing, or cell models, the latter introduced by Deans and Lapidus (1960). According to these authors, interstices between packing elements are idealized as perfectly stirred tanks in order to represent the dispersion behavior. *Stochastic models* are characterized by taking into account the random arrangement of particles and void spaces within randomly packed beds. Purely stochastic models could be employed only if sufficient statistical information concerning the structure of packed beds, void spaces, and fluid discrete paths is available. Its application in cell models, for example, may be developed by varying the cell size and choosing cell sizes to conform to the size distribution of void spaces in the packed bed. However, its use seems doubtful, due to the computational cost. An alternative option within the stochastic model type could be the use of deterministic models based on differential equations while considering void spaces in the packing as sources of perturbations in the patterns of concentration, temperature, and fluid velocity through the bed. In these models, also known as *deterministic models with random perturbation*, the particles and void spaces can be positioned randomly throughout the bed, and statistical properties of the patterns can be calculated. Whereas purely stochastic models change the original conception of dispersion models, the introduction of randomly distributed parameters in discrete models produces a large number of model variations without changing the original character of the model. Since deterministic models are described by differential equations, their application using model parameters distributed randomly is only appropriate for macroscale studies, where the effect of individual void spaces on the patterns is negligible.

Crine et al. (1980) defined another classification, based on successive levels of modeling at various observation planes:

- The microscopic level, which corresponds to large volume elements at the molecular scale but small elements at the bed granulometry. The transport processes are described theoretically by differential equations of the continuum type.

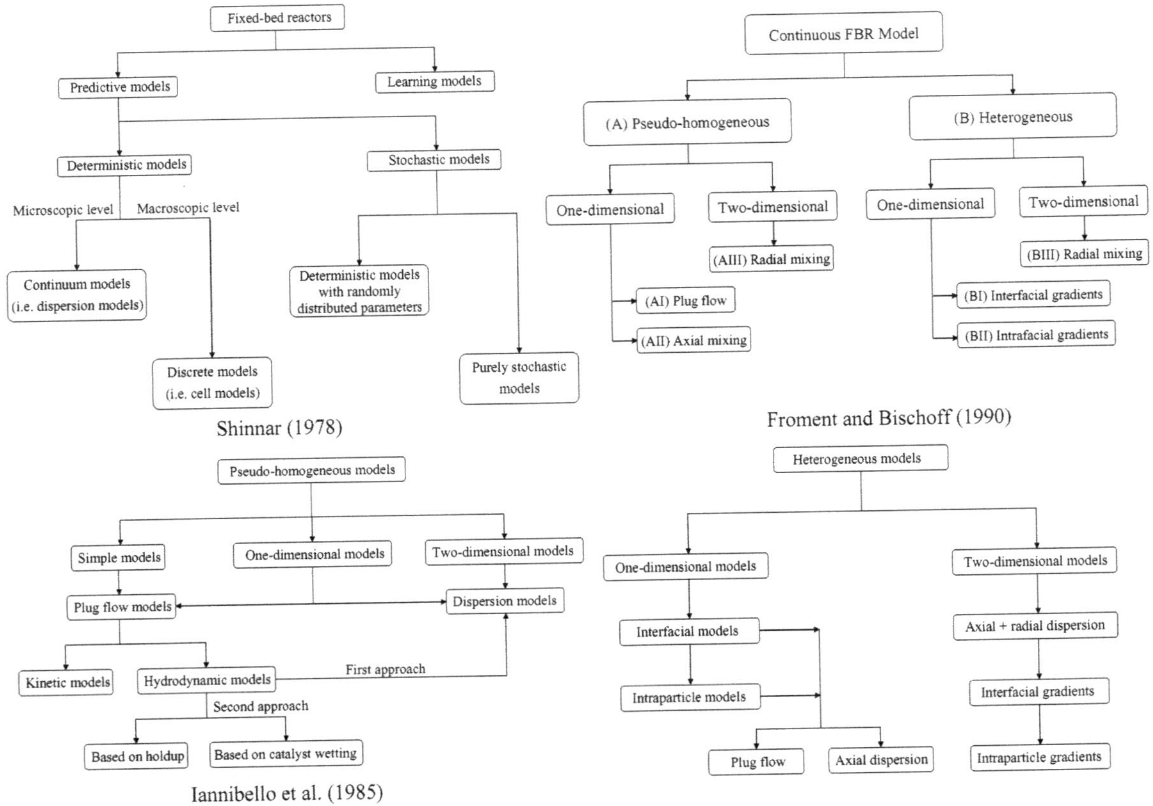

Figure 2.13. Various classifications of catalytic reactor models.

- The first macroscopic level, corresponding to local phenomenological observations, which assumes that the volume elements are large enough to consider the bed as being locally homogeneous. At this level, the regionalized variables can be defined (e.g., fluid hold-ups, irrigation rate), and all the transport processes can be described using the concept of an elementary transport cell based on a simplified representation of the packing.
- The second macroscopic level, characterizing the bed as a whole, where liquid maldistribution (due to random clustering of the transport cells) has to be considered.

Froment and Bischoff (1990) have introduced perhaps the most popular classification of continuous models for adiabatic and nonadiabatic fixed-bed reactors (FBRs), which is shown in Figure 2.13. They considered that if gradients of concentration and temperature across the phase boundaries cannot be neglected, the continuum concept can be narrowed to the phases present in the reactor (heterogeneous continuum models), while if the heterogeneous fluid–particle system is regarded as a single pseudohomogeneous phase, the modeling of the FBR is drastically simplified to the state variables of a single isotropic continuum (pseudohomogeneous continuum models).

According to Iannibello et al. (1985), the simple plug-flow pseudohomogeneous model proposed in the literature to predict the behavior of hydrotreating trickle-bed reactors can be grouped into two types: kinetic models and hydrodynamic models (Figure 2.13). *Kinetic-based models* are generally functions of intrinsic rates and do not account for the influence of hydrodynamics and related phenomena on the conversion rate. *Hydrodynamic-based models* attempt to incorporate the influence of hydrodynamics on catalyst utilization, generally assuming plug flow and introducing an apparent kinetic rate constant (k_{app}). Two approaches have been followed in the development of hydrodynamic models: one relating to the reactor overall efficiency to the external liquid mixing, and the other relating to the liquid–solid contacting efficiency, which can be determined with the liquid hold-up or the irrigation rate of the bed (Crine et al., 1980).

The pseudohomogeneous models consider the bed as a pseudocontinuum, while heterogeneous models distinguish between temperatures and concentrations in the bulk gas phase and at the surface of the catalyst. Each category can be considered with one- or two-dimensional models to account in less or more detail for temperature and concentration gradients inside the reactor (Figure 2.13).

The level of sophistication and complexity of a reactor model depends mainly on the purpose of the investigation and the need for prediction capabilities. On the one hand, the simplest models assume either perfect mixing or plug flow (i.e., the well-known extreme cases of ideality). Deviations from such ideal flow patterns are frequently accounted for using axial dispersion coefficients. On the other hand, the most sophisticated models resolve the fluid dynamics clearly with direct numerical solution of the Navier–Stokes equations and superimpose the kinetics on it (e.g., computational fluid dynamics models).

Some general rules regarding the level of reactor model sophistication have been reported. For example, Feyo De Azevedo et al. (1990) established that a model should not be more detailed than absolutely necessary for the particular purpose involved. Glasscock and Hale (1994) reported 80% of benefit with only 20% of model complexity and concluded that the obligation to develop complex models may be avoided as the only way to properly simulate the behavior of a reactor. Duduković et al. (2002) stated that in the definition of the sophistication level for reactor modeling, the flow pattern and mixing should be commensurate with the level of modeling used to understand the kinetics. Whenever that is not the case, the modeling effort yields less than maximum benefits, since kinetics ultimately drive reactor performance.

There is not a magic rule for establishing the complexity of a model; however, the best practice is to consider the simplest model with all the main relevant phenomena and then add complexity to reduce the error between experimental and calculated data (Andrigo et al., 1999). The model equation complexity is determined strongly by flow conditions, but for FBRs the plug-flow hypothesis (i.e., relatively simple equations) is often satisfactory (Froment, 1986). In general, to describe all physical and chemical phenomena in a PBR, a model should have the following characteristics:

1. It must be a conservative system.
2. It should not predict backmixing of material over large distances.
3. It should produce the correct asymptotic (steady-state) solution.

2.4.2 Description of Reactor Models

The analysis of multiphase catalytic fixed-bed reactors is a challenging task, as the reactor performance in most cases depends not only on the chemical reaction rate but also strongly on fluid dynamics and several transport processes. To combine these factors quantitatively, reactors models with various levels of sophistication have been developed. The simplest globally model for taking fluid dynamics of the flowing liquid in trickle-bed reactors into account is the one-dimensional dispersion model (PD). This model has been used by numerous authors for determining the particle Bodenstein number (Bo_p) for the spreading of a tracer as a function of operating conditions. In general, Bo_p defined in this way is significantly lower for trickling conditions than in single-phase flow through packed beds (Gierman, 1988) (i.e., the dispersion in the liquid phase is higher). The PD model does not explicitly take *stagnant zones* into account, which makes them noticeable by marked tailing in tracer experiments, particularly with porous catalysts. Therefore, a model in which the liquid phase is divided into a stagnant (inactive) and a free-flowing (active)

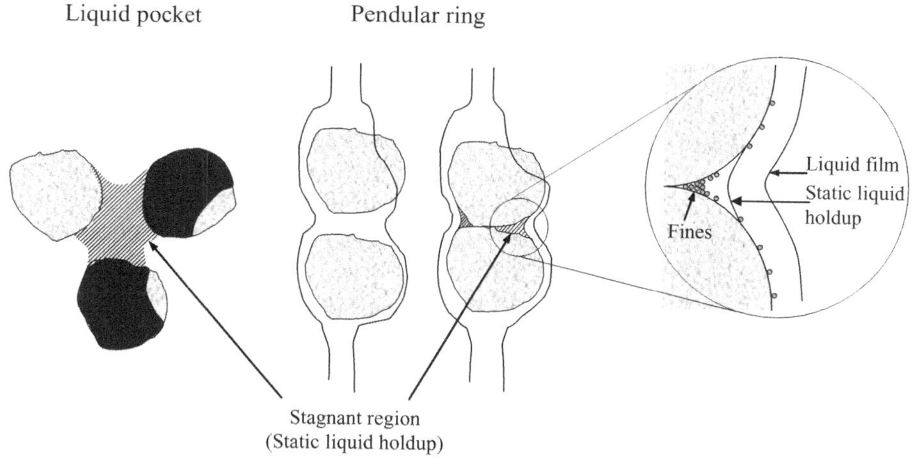

Figure 2.14. Stagnant liquid zones between catalyst particles in TBRs.

fraction is considerably more realistic. Such a model is known as a cross-flow (PE) or cross-flow dispersion (PDE) model. Figure 2.14 shows the presence of stagnant zones in adjacent catalyst particles in TBRs. It is also shown in the zoom of this figure that the stagnant regions contribute to the capture of fines.

Despite all the information available on fluid dynamics and mass transfer, the a priori design of TBR is still far removed from a chemical reaction engineering routine. For robust modeling, correlations for the dependence of model parameters on operating variables are still lacking, as well as the performance of model complexity to predict the degree of conversion with sufficient accuracy. In this situation, to make optimal use of the data available, extensive use of analogies and similarities among the various models must be made (Hofmann, 1978).

Non-steady-state methods permit more detailed kinetic analysis of elementary steps. Reaction rate constants and mechanisms have frequently been determined from an analysis of the transient behavior of multiphase reactors (Marroquín de la Rosa, 2002). In the past, catalyst development drove the selection of an appropriate multiphase reactor type. This sequential approach is increasingly being replaced by a parallel method for catalyst and reactor selection. This approach requires quantitative models for the flow patterns, phase contacting, and transport in various multiphase reactor types. Proper selection of the reactor type and its efficiency of operation greatly affects the total capital and manufacturing costs of the entire process (Duduković et al., 2002). Since the reactors used in the petroleum refining industry for catalytic hydrotreating are complex in nature and difficult to model due to the presence of three phases, much attention has been directed in this chapter to describing such models for hydrotreating application.

Simple Pseudohomogeneous Models The earliest models reported in the literature for HDT reactors were of pseudohomogeneous type, which may be considered to be very simple since they only take into account the reactor inlet and outlet data. Of course, they were not used alone; experimental and commercial experiences with similar reactors were and continue to be employed for such purposes. These models are based on various assumptions that lead to drastically simplified equations. In some cases the models were so oversimplified that today, books of chemical reactor design at the undergraduate level use them only for academic purposes.

The pseudohomogeneous models can be categorized into two types: kinetic and hydrodynamic (Iannibello et al., 1985). *Kinetic models* do not consider the influence of hydrodynamics and related phenomena on conversion and are generally based on intrinsic rates of reactions (k_{in}). They describe the performance of the process generally in terms of first- or nth-order kinetics. On the other hand, *hydrodynamic models* attempt to incorporate the influence of hydrodynamics on catalyst utilization. These models emphasize some other aspects of the reactor, such as external liquid holdup, catalyst wetting, and axial dispersion. They generally assume a plug-flow pattern with first-order kinetics and introduce an apparent kinetic rate constant (k_{app}) in place of the intrinsic rate constant to account for the effects of hydrodynamics. It has been shown that incorporation of a hydrodynamic parameter in reaction rate equations improves the performance of the model in terms of data fitting, thus providing a more appropriate basis for the scale-up of pilot-plant data by adding chemical and physical complexities in the kinetic analysis of HDT reactions. Crine et al. (1980) assumed that the development of hydrodynamic models followed two approaches: in one, the overall reactor efficiency is related to external liquid mixing (i.e., D_a^f, D_r^f, λ_a^f, λ_r^f, etc.), whereas in the other, it is related to the liquid–solid contacting efficiency (i.e., ε_{TL}, ε_L, f_w, η_{CE}, etc.).

Models Based on Kinetics Many authors have reported that pore diffusion effects can be taken into account within the framework of an effective or apparent reaction rate constant (i.e., multiplying the intrinsic reaction rate constant by the effectiveness factor), in order to formulate a pseudohomogeneous basic plug-flow model, which is sufficient to describe the progress of chemical reactions in the liquid phase of a TBR (Henry and Gilbert, 1973; Paraskos et al., 1975; Satterfield, 1975; Hofmann, 1978). The suppositions of this model are:

1. Plug-flow pattern of the liquid phase
2. No evaporation or condensation from or into the liquid phase
3. No mass or heat transfer limitations between gas–liquid and liquid–solid interfaces
4. Liquid saturated with gas at all times

5. First-order isothermal, irreversible reaction with respect to the liquid reactant
6. No homogeneous reaction
7. Gaseous reactant present in large excess
8. Reactions occurring only at the catalyst surface

The analysis of TBR performance under such ideal circumstances and the assumption of simple first-order power-law kinetics can easily be approximated by an expression analogous to that of the well-known and widely used piston-flow reactor design equation with a single reactant phase. This approach provides a quick initial analysis of real cases. The final integrated equation is represented by a logarithmic dependence of the degree of conversion on the apparent rate constant and mean liquid residence time or space–time ($\tau_L = 1/\text{LHSV}$). The derivation of such an equation is typically reported in chemical reaction engineering textbooks (Smith, 1960; Hill, 1977; Froment and Bischoff, 1990) and can be obtained from the following mass balance of the limiting reactant in the reaction (say sulfur, S, in an HDS reaction) in a differential element of the reactor volume:

$$\text{input} = \text{output} + \text{disappearance by reaction} + \text{accumulation} \tag{2.86}$$

At steady state, the term of accumulation is equal to zero; thus,

$$[F_S] = [F_S + dF_S] + [(-r_S)dV] \tag{2.87}$$

Since $F_S = C_S^L \upsilon_L$,

$$(-r_S)dV = -dC_S^L \upsilon_L \tag{2.88}$$

after separating variables, integration, and inclusion of the space-velocity concept ($\text{LHSV} = \upsilon_L/V$); for example, for first-order kinetics, $-r_S = \eta k_{\text{in}} C_S^L$, the final equation is

$$\ln \frac{(C_S^L)_0}{(C_S^L)_f} = \frac{\eta k_{\text{in}}}{\text{LHSV}} \tag{2.89}$$

The chemical complexity of the reaction may reasonably be taken into account by assuming nth-order kinetics $[-r_S = \eta k_{\text{in}} (C_S^L)^n]$, with $n \neq 1$. The final equation for this case is

$$\frac{1}{n-1}\left[\frac{1}{(C_S^L)_f^{n-1}} - \frac{1}{(C_S^L)_0^{n-1}}\right] = \frac{\eta k_{\text{in}}}{\text{LHSV}} \tag{2.90}$$

Frye and Mosby (1967) derived an isothermal reactor equation (or kinetic equation) for desulfurization of light catalytic cycle oil applicable over a wide range of conditions. The equation has one adjustable parameter, which depends on the physical properties, assuming liquid feed vaporization and phase equilibrium first-order with respect to each sulfur compound and hydrogen, and on the effect of H_2S and aromatic hydrocarbon adsorption. The model was able to predict the total HDS percent of the feedstock with acceptable accuracy by using the equation developed for model compounds. This approach could be enhanced by considering more than a lump for prediction of HDS. Further analysis about the effect of the gas/oil ratio on the vapor–liquid equilibrium (VLE) in a deep diesel HDS reactor was performed by Hoekstra (2007) using the model of Frye and Mosby (1967).

Papayannakos and Georgiou (1988) presented a simple kinetic model for hydrogen consumption during residue catalytic HDS, including the effects of reactor conditions and catalyst type, size, and age. The intrinsic reaction rates for hydrogen consumption were described using a second-order kinetic equation, and the intraparticle diffusional effects were discussed by means of the effective diffusivity. Strong pore diffusion limitations were observed for commercial-size catalyst particles. The differential mass balance equation for the remaining hydrogen demand and sulfur compounds was given by Eq. (2.90), where the total hydrogen reaction rate was found to be of second order ($n = 2$), while the reaction order for the HDS reaction was 2.5.

Many other authors have used either Eq. (2.89) or (2.90) to model experimental isothermal reactors at different scales for different purposes, such as catalyst screening, the effect of feedstock properties, or the evaluation of commercial catalysts. Some of these studies have been summarized by Ancheyta et al. (1999, 2001, 2002).

Kinetic models are usually employed for catalyst testing on a laboratory scale and also to obtain intrinsic parameters of any rate of reaction. Several experimental techniques, such as reduction in catalyst size, variation in the amount of catalyst and flow rate needed to maintain a specific LHSV, or filling up the catalyst bed with inert fines, are necessary to properly compare different catalysts. The main drawback of PBRs for kinetic studies is the fact that they are integral reactors; that is, the concentrations gradients may be significant. The only way to obtain kinetic information is to assume a kinetic model and adjust its parameters by comparing model results with experimental results (Pitault et al., 2004). It must be performed iteratively in order to find the best set of parameters. Having the effective constant rate, the intrinsic parameters can be obtained to provide an accurate effectiveness factor. In kinetic studies, the main advantage of this type of reactor over a batch reactor is that the former makes it possible to determine the deactivation of catalyst, although it is difficult to decouple the kinetic model from deactivation phenomena (Perego and Paratello, 1999).

Models Based on Hydrodynamics To account for hydrodynamics and other physical effects, an apparent kinetic rate constant can be introduced: $k_{app} = k_{in} f(\text{hydrodynamics})$. Therefore, to obtain the hydrodynamic model (e.g., for $n = 1$), Eq. (2.89) can be rewritten (note that k_{app} is used instead of ηk_{in})

$$\ln \frac{(C_S^L)_0}{(C_S^L)_f} = \frac{k_{app}}{\text{LHSV}} \tag{2.91}$$

where k_{app} depends on catalyst utilization. The interparticle and intraparticle physical phenomena may be accounted for separately by means of the equation

$$k_{app} = k_{in}(1-\varepsilon_B)\eta\psi(u_L) \tag{2.92}$$

where k_{in} is based on catalyst pellet volume, and hence a factor $(1 - \epsilon_B)$ appears, which is the fraction of reactor volume occupied by the catalyst undiluted, η the catalyst effectiveness factor, and $\psi(u_L)$ a function of the superficial liquid velocity (u_L) that considers the variation of the degree of utilization of the catalyst due to hydrodynamic phenomena (Iannibello et al., 1985). Substituting Eq. (2.92) into Eq. (2.91), the following general hydrodynamic model is obtained:

$$\ln \frac{(C_S^L)_0}{(C_S^L)_f} = \frac{k_{in}(1-\varepsilon_B)\eta\psi(u_L)}{\text{LHSV}} \tag{2.93}$$

Satterfield (1975) reported a simple first-order kinetic model for analysis of TBR performance employed in the HDT of gas oils under ideal conditions. The model only takes the liquid phase into consideration as a single homogeneous phase, assumes a plug-flow pattern for the liquid flow rate, no mass transfer limitation, an irreversible reaction with respect to the liquid reactant, no reaction heat effects, no homogeneous reaction, total wetting of catalyst pellets, and a pseudo rate constant for reactor design. For a more realistic approach, this author suggested including the effect of liquid–solid contact effectiveness (f_w *or* η_{CE}), defined as k_{app}/k_{in}, using a correlation presented by Bondi (1971). It was found that k_{app} increased, approaching k_{in}, as the liquid flow rate tends to infinity. Furthermore, in accordance with experimental observations, the value of the exponent α in the relationship given by Eq. (2.94) proposed by several authors was found to vary substantially depending on the flow-rate region being considered.

$$\ln \frac{(C_S^L)_0}{(C_S^L)_f} = \frac{\kappa k_{in}(1-\varepsilon_B)\eta(L_B)^{\alpha}}{(\text{LHSV})^{1-\alpha}}(d_{pe})^{\beta}\left(\frac{\mu_L}{\rho_L}\right)^{\gamma}\left(\frac{\sigma_c}{\sigma}\right)^{\omega} \tag{2.94}$$

At substantially high superficial liquid velocities, where liquid contact becomes essentially complete and liquid maldistribution is no longer a problem, the exponent α should approach zero (Montagna and Shah, 1975). The relationship between inlet and outlet concentrations of the liquid reactant is given by Eq. (2.91).

1. MODELS BASED ON LIQUID HOLDUP Since it is normally assumed that reactions take place in all the catalyst particles inside the reactor, the total volume of catalyst is employed to calculate the LHSV. Otherwise, there must be some means for correcting this inefficiency if the true conversion severity needs to be determined. To solve this problem it has been considered that catalyst utilization depends on the liquid volume within the reactor (Ross, 1965; Henry and Gilbert, 1973), which is called *liquid holdup*.

Ross (1965) was the first author to publish holdup data for commercial reactors, more than 40 years ago. He measured the liquid residence-time distributions by using the pulse technique in different-size trickle hydrotreaters containing the same reactants and catalyst and operating under the same reaction conditions. It was found that liquid holdup in commercial reactor was only about two-thirds that of the pilot reactor, which indicated that those commercial units were less efficient than the pilot-plant reactors, despite using higher linear liquid and gas velocities. This seemed to have been caused by poorer liquid distribution over the catalyst bed (poor catalyst utilization) and/or by mass transport of reactants through the liquid film on the catalyst. It was reported that for reactors larger than about 1 in. in diameter, liquid distribution becomes an important problem (Henry and Gilbert, 1973).

The uniformity of liquid distribution improves considerably with increasing liquid velocity. The greater the liquid and gas velocities in commercial units, the greater the turbulence in the liquid film. This, in turn, would increase the transfer of reactants through the film. On the basis of these results it was considered that the total (external plus intraparticle) liquid holdup (ε_{TL}) could be used as a measure of liquid–catalyst contact. Therefore, when analyzing data from commercial and pilot-plant HDS reactors, k_{app} was assumed to be proportional to ($k_{in}\,\varepsilon_{TL}$), which means that $(1 - \varepsilon_B) = 1, \eta = 1$, and $\psi(u_L) = \kappa\varepsilon_{TL}$, as indicated in Eq. (2.93), in which liquid space–time (or the total residence time of the liquid) is the basic parameter in reactor performance. Therefore, this model takes into account both external and internal wetting of catalyst (Duduković, 1977; Iannibello et al., 1985).

Henry and Gilbert (1973) extended the model reported by Ross (1965) to use (ε_L/LHSV) as the space–time for correlating pilot-plant and full-scale hydrotreating performances. Their modified plug-flow model is based on the external holdup of the liquid (ε_L) with negligible backmixing effects (assuming that holdup effects were controlling), and it was used to analyze the kinetic data with the undiluted catalyst bed obtained by Mears (1971). The external liquid holdup is considered a parameter that, although empirical, accounts for the external effective wetting of the catalyst pellets.

This model, also known as the *holdup model*, can be employed to correlate catalyst activity with parameters such as liquid mass velocity, liquid hourly space velocity, catalyst size, and catalyst bed length. It has been shown that the liquid velocity and the catalyst bed length have important effects on the performance of the reactor. These effects (i.e., maldistribution of liquid) have been explained in terms of volume of liquid in the reactor or holdup. For example, a higher liquid holdup (higher space time) in the catalyst bed appears to be the key to increased catalyst utilization.

When the reaction rate is indeed proportional to the free-drainage holdup (or dynamic liquid holdup) and this holdup is proportional to $u_L^{1/3}$ (Crine et al., 1980) for the laminar film model, ε_L is proportional to $d_{pe}^{-2/3}$ and to $\nu^{1/3}$, where $\nu = \mu/\rho$; thus, the Henry and Gilbert (1973) correlation would become that given by Eq. (2.94) with $\alpha = 1/3$, $\beta = -2/3$, $\gamma = 1/3$, and $\omega = 0$. The latter equation predicts that decreasing catalyst size will increase conversion, but the same general effect would be produced by varying catalyst size if diffusion limitations within catalyst particles were significant.

It has been pointed out by Henry and Gilbert (1973) and Satterfield (1975) that for certain combinations of liquid and gas flow rates, the gas flow reduces the liquid holdup. However, very high gas flow rates could be detrimental to catalyst utilization. In the development of Henry and Gilbert's model, the influence of gas flow rates over catalyst utilization was not analyzed; therefore, the expression derived can be used only for a constant gas flow. Furthermore, because this model does not consider the effect of the pore diffusion limitations at high temperatures, it cannot be reliable for predictions under the typical conditions employed in HDT reactions.

Paraskos et al. (1975) evaluated the effects of backmixing and flow behavior (e.g., liquid holdup, incomplete catalyst wetting) on the mass transfer resistance in hydrotreating of gas oils in a pilot-plant TBR. The effect of varying LHSV on conversions of sulfur, metals, and nitrogen showed that the percentage of HDS should be dependent on catalyst bed length, backmixing, and liquid holdup; incomplete catalyst wetting reduces the efficiency of the TBR, so it was established that greater catalyst bed length minimizes the effects of liquid holdup or incomplete catalyst wetting.

The function $\psi(u_L)$, on an empirical basis, was assumed to be proportional to the liquid–solid contact effectiveness (η_{CE}) in order to correlate the experimental results of HDT. The following correlation was used for the liquid–solid contact effectiveness:

$$\eta_{CE} = \kappa (u_L)^{\alpha} \tag{2.95}$$

Equation (2.95) correlates the liquid–solid contact efficiency with the superficial liquid velocity. Using Eq. (2.94), it was also found that the power-law coefficient in the holdup or the effective catalyst wetting–LHSV relationship (α) might be dependent on the reaction conditions (e.g., temperature, LHSV, H_2 partial pressure) as well as on the nature of the feed and reaction. First-order kinetics represents well HDS, hydrodenitrogenation (HDN), and hydrodemetallization (HDM) reactions when either liquid holdup or catalyst wetting effects are taken into account. However, when these effects are neglected, the apparent order of reaction undergoes a change as the conversion level increases. For small deviations from plug flow with first-order reactions, the relations between outlet and inlet concentrations were expressed by

$$\ln\frac{(C_S^L)_0}{(C_S^L)_f}=\frac{k_{in}(1-\varepsilon_B)\eta}{\mathrm{LHSV}}-\frac{[k_{in}(1-\varepsilon_B)\eta]^2}{(\mathrm{LHSV})^2}\frac{d_{pe}}{L_B\cdot \mathrm{Pe}_d} \tag{2.96}$$

when the Peclet number is related to LHSV and L_B by an empirical correlation of the following type:

$$\mathrm{Pe}_d=\kappa(\mathrm{LHSV})^{\alpha}L_B^{\alpha} \qquad \text{with } 1>\alpha\geq 0.5 \tag{2.97}$$

Equation (2.96) becomes

$$\ln\frac{(C_S^L)_0}{(C_S^L)_f}=\frac{k_{in}(1-\varepsilon_B)\eta}{\mathrm{LHSV}}-\frac{[k_{in}(1-\varepsilon_B)\eta]^2 d_{pe}}{\kappa(\mathrm{LHSV})^{2+\alpha}L_B^{1+\alpha}} \tag{2.98}$$

2. MODELS BASED ON CATALYST WETTING Experiments in bench-scale TBRs have shown that distribution of liquid over a catalyst particle bed can be extremely nonuniform at the low-liquid space velocities prevailing in bench-scale reactors compared with commercial-scale reactors. This liquid maldistribution within the catalyst bed causes an ineffective use of catalyst active sites, also known as *incomplete catalyst wetting*. This effect can be reduced considerably by improving the uniformity of liquid distribution with increasing superficial liquid velocity and reducing catalyst particle size. In catalyst wetting–based models, catalyst utilization is assumed to be proportional to the fraction of the outside catalyst surface effectively wetted by the flowing liquid, also known as *effective catalyst wetting*, which is defined as the ratio of external wetted area to the total area of catalyst particle.

Murphree et al. (1964) presented the application of liquid residence-time distribution studies in order to examine the deviations of downflow two-phase FBRs from plug flow by calculating the *contacting efficiency*, defined as the ratio of real reactor performance to ideal plug-flow reactor performance. The ability of the model to estimate the reaction rate constant was also reported. These authors concluded that any difference in conversion of these two units under the same operating conditions must be attributed to differences in contact between catalyst and fluid existing in the two units. The model presented by the authors seems to be the first attempt to separate chemical and physical effects on HDS conversion, in an easy manner, in any reactor by measuring the effects of fluid–catalyst external contact efficiency.

Bondi (1971) presented a simple procedure by which it is possible to separate chemical kinetics from physical conversion resistances that are present in TBRs. The procedure to calculate chemical reaction rate constants from conversion data on bench- and pilot-scale TBRs was reported to be useful for any other reactor system. An empirical parameter is introduced which characterizes the conversion resistance depending on liquid and gas velocities, and is valid only for the particular experiments. The low flow rates which are characteristic of experimental reactors can magnify poor oil–catalyst contact and cause low conversion rates that vary with liquid velocity.

The empirical relationship developed by Bondi (1971) can be used to reduce the gap between conversion data obtained in batch reactors and those from small-scale TBRs, which often perform poorly. His empirical correlation for HDS of heavy gas oil, relates the space–time required to achieve 50% conversion (or conversion half time, $\tau_{1/2}$), to the analogous space–time at complete wetting ($\tau_{1/2,c}$), and to the linear superficial liquid velocity:

$$\tau_{1/2} = \tau_{1/2,c} + \frac{A'}{u_L^{b'}} \tag{2.99}$$

where A' and b' are empirical constants. According to Satterfield (1975), this relationship can also be expressed in terms of reaction rate constants:

$$\frac{1}{k_{\text{app}}} = \frac{1}{k_{\text{in}}} + \frac{A'}{G_{mL}^{b'}} \tag{2.100}$$

It is important to point out that Bondi (1971) also found an insignificant positive gas flow effect which was neglected in his model.

The effect of backmixing in the presence of liquid holdup and incomplete catalyst wetting was discussed by Mears (1974), who proposed a relationship between $\ln\left[(C_S^L)_0/(C_S^L)_f\right]$ and L_B based on the effective catalyst wetting effects in order to show that for a bench-scale TBR, the liquid velocity and the catalyst bed length have important effects on the performance of the reactor; in other words, he postulated the hypothesis that the fraction of catalyst (and hence k_{app}) utilized is proportional to the true constant of completely wetted catalyst (k_{in}), to the catalyst effectiveness factor (η) and the contact efficiency η_{CE} (or f_w), that is, to the fraction of the external catalyst area wetted by liquid:

$$k_{\text{app}} = k_{\text{in}}\eta f_w \tag{2.101}$$

where f_w (or a_w/a_S) is the fraction of the external pellets area that is effectively wetted. By incorporating the correlation of the effectively wetted area (a_w) proposed by Puranik and Vogelpohl (1974), which was developed for incomplete contact in absorption towers using different packing size and shape, Mears (1974) arrived at Eq. (2.94) with $\alpha = 0.32$, $\beta = 0.18$, $\gamma = -0.15$, and $\omega = 0.21$.

If Onda's correlation is used, the following model equation can be derived (Mears, 1974):

$$-\log\frac{(C_S^L)_f}{(C_S^L)_0}=\frac{k_{in}\eta}{\text{LHSV}}\left[1-\exp\left(-\kappa L_B^{0.4}\text{LHSV}^{0.4}\right)\right] \tag{2.102}$$

Mears' approach is more acceptable from a physical point of view than that of Henry and Gilbert (1973), since it considers the reaction rate to be proportional to the effectively (freshly) wetted area of the catalyst instead of to the liquid volume (Gianetto et al., 1978). Mears also found that even when a fraction of his data (Mears, 1971) was explained satisfactorily by the holdup model of Henry and Gilbert (1973), data with a diluted bed could not be evaluated with this model.

Duduković (1977) suggested that the catalyst effectiveness factor and partial surface-wetting effects, being coupled local phenomena in TBRs, are a function of the Thiele modulus for nonvolatile liquid reactants in liquid-phase reactant-limited reactions, considering both incomplete external wetting and fractional pore fill-up (or internal partial wetting). Fractional pore fill-up will depend on the catalyst pore structure and physical properties (particularly on the surface tensions) of the gas–liquid–solid system involved. This trickle-bed effectiveness factor model is based on the following formulation for partially wetted catalyst pellet in TBR (η_{TB}), with a reaction occurring only in the liquid-filled pore region of the pellet:

$$\eta_{TB}=\eta_i\eta^* \tag{2.103}$$

where η_i represents the fraction of the particle internal volume wetted (pore filling) and η^* is the effectiveness factor of a pellet partially wetted (inside and outside), defined as

$$\eta^*=\frac{\tanh(\Phi_{TB})}{\Phi_{TB}}=\frac{\tanh[(\eta_i/\eta_{CE})\Phi_T]}{(\eta_i/\eta_{CE})\Phi_T} \tag{2.104}$$

The effectiveness factor in TBRs is then obtained by substituting Eq. (2.104) into Eq. (2.103):

$$\eta_{TB}=\eta_{CE}\frac{\tanh[(\eta_i/\eta_{CE})\Phi_T]}{\Phi_T} \tag{2.105}$$

Equation (2.105) reduces to the following relationship as that used by Mears (1974) only if $\Phi_T >> 1$ (very fast reaction) or $\eta_i/\eta_{CE}\approx 1$:

$$\eta_{TB}=\eta_{CE}\eta \tag{2.106}$$

For very low values of the Thiele modulus, that is, very slow reactions, by expanding tanh in a Maclaurin series and dropping out the terms with order higher than cubic, Eq. (2.105) is reduced to

$$\eta_{TB} \approx \eta_i \left[1 - \frac{1}{3}\left(\frac{\eta_i}{\eta_{CE}} \Phi_T\right)^2\right] \tag{2.107}$$

The reactor design equation for first-order reactions based on assumed plug flow of the liquid is equal to Eq. (2.93), considering $\eta\psi(u_L) = \eta_{TB}$.

Crine et al. (1980) presented a phenomenological description of the hydrodynamic and mass transfer processes occurring in a TBR. The model proposed, which accounts for the random and discontinuous nature of the packed bed, has been verified with experimental hydrotreating data. These data may be correlated using a single wetting parameter (L_m) introduced on physical grounds, which is independent of the temperature and nature of the reaction system investigated. It varies only with the fluid properties and catalyst size, and exhibits a logical dependence with the operating conditions. Crine's model was developed to take into account the discontinuous and random nature of the bed, relating the TBR global efficiency (η_G) to the particle effectiveness factor (η) as (Duduković, 1977)

$$\eta_G = \eta_E \eta_{CE} \frac{\tanh[(\eta_i/\eta_{CE})\Phi_T]}{\Phi_T} \tag{2.108}$$

Due to capillary forces and the high molecular weight of the liquid reactants, η_i is commonly assumed to be unity (Gianetto et al., 1978; Callejas and Martínez, 2002). The relative contributions of the films and of the liquid pockets contacted are taken into account in η_E and η_{CE}. The proposed reactor design model is equal to Eq. (2.93), considering that $\eta\psi(u_L) = \eta_G$.

Iannibello et al. (1983) observed that pore filling of the catalyst can be considered as total, even at very low liquid flow rates and hence very low intraparticle holdup, and at relatively high temperatures. The partial catalyst utilization is probably due to intraparticle diffusivity phenomena rather than to partial pore filling. The rate of reaction of large molecules containing sulfur and metals was reported to be strongly affected by intraparticle mass transport phenomena. The decrease in the apparent kinetic constant may be ascribed to the reduction in the apparent intraparticle diffusivity of reactants. These authors validated the correlation for predicting the contact effectiveness suggested by Mills and Duduković (1981) and proposed that it may be useful to evaluate the hydrodynamic conditions where the kinetic rate constant may be independent of hydrodynamics.

The same group (Iannibello et al., 1985) employed four models that take into consideration the physical and chemical complexity of three-phase systems in order to interpret the results obtained from a pilot trickle-bed reactor in which the removal of sulfur and metals from a heavy residual oil was carried out. The external holdup (EH), total holdup (TH), apparent diffusivity (AD), and second-order kinetic models were tested with different catalysts, and it was observed that the AD and EH models gave almost the same result in terms of data fitting and a slightly better than second-order model. The TH model seemed less attractive from an engineering point of view because the reaction order calculated differed considerably from one catalyst to another or from one reaction to another. Iannibello's group concluded that a model based on the contact efficiency (AD model) seems to be theoretically stronger than the others. The authors also analyzed a first-order reaction model that incorporates the parallel reactions of reactive and refractory fractions provided that there is a good interpretation of experimental data concerning alumina catalyst. The model is

$$\frac{\left(C_S^L\right)_f}{\left(C_S^L\right)_0} = \alpha e^{-k'_\alpha/\mathrm{LHSV}} + \beta e^{-k'_\beta/\mathrm{LHSV}} \tag{2.109}$$

where $\beta = 1 - \alpha$, $k'_\alpha = \eta_{\mathrm{CE}} k_\alpha$, and $k'_\beta = \eta_{\mathrm{CE}} k_\beta$.

This two-lump model—namely, reactive and refractory fractions—has been used more recently in the hydrocracking of asphaltenes, providing the best data fitting (Trejo et al., 2007). On the other hand, due to sieving effects caused by fine pores present in alumina-based catalyst, appropriate modeling of a hydrotreating reactor using this type of catalyst probably must take into account the particular catalyst pore structure connected with the feed that was treated.

Kumar et al. (1997) conducted pilot-scale HDT experiments on straight-run diesel using commercially available $CoMo/Al_2O_3$ catalyst. The kinetics of HDS and HDN were studied using a plug-flow model together with the external holdup and apparent diffusivity models reported by Iannibello et al. (1985), which considered the physical and chemical complexities in the kinetic analysis of hydrotreating reactions in the three-phase system. The chemical complexity of the HDS and HDN reactions were taken into account by assuming that nth-order kinetics ($n > 1$), hydrodynamics, and other physical effects were incorporated through the apparent kinetic rate constant, and because the authors considered the total reaction pressure effects in the reaction system, the rate of reaction was expressed as

$$-r_S = k_{\mathrm{app}} P^m \left(C_S^L\right)^n \tag{2.110}$$

Therefore, the hydrodynamic model resulted in

$$\frac{1}{n-1}\left[\frac{1}{\left(C_S^L\right)_f^{n-1}} - \frac{1}{\left(C_S^L\right)_0^{n-1}}\right] = \frac{k_{\mathrm{app}} P^m}{\mathrm{LHSV}} \tag{2.111}$$

The pilot-plant reactor has deviations from plug flow and they can be accounted for effectively by incorporating external holdup and catalyst wetting in evaluation of the apparent kinetic parameters. Therefore, for the EH model,

$$k_{\mathrm{app}} = (k)_{\mathrm{EH}}\, \varepsilon_L \tag{2.112}$$

The AD model, on the other hand, uses the following representation:

$$k_{\text{app}} = (k)_{\text{AD}}\, \eta_{\text{CE}} \tag{2.113}$$

where $(k)_{\text{EH}}$ and $(k)_{\text{AD}}$ are pseudokinetic rate constants that should be independent of hydrodynamics when an appropriate n value is used. It was observed that the EH model gave the best data fit among the three models employed.

3. MODELS BASED ON AXIAL DISPERSION Perfect piston flow (i.e., an ideal plug-flow pattern) will never occur with Newtonian fluids, as there will be always some axial mixing, due to viscous effects and molecular or eddy diffusion. The deviations from piston flow behavior caused by restricted axial mixing have traditionally been characterized by means of residence-time distribution curves. Some researchers have proposed that deviation from the plug flow of trickle-bed reactors is caused by axial dispersion and have recommended that the hydrodynamic effect must be accorded proper consideration in trickle-bed reactors through axial mixing (Danckwerts, 1953; Wehner and Wilhelm, 1956).

Mears (1971) established that axial eddy dispersion or backmixing (deviations from plug flow) appears to be responsible for the adverse mass velocity effects observed in isothermal laboratory-, bench-, and pilot-scale TBRs for petroleum processing. To examine the possibility that axial dispersion might be responsible for decreased reactor efficiency at low mass velocities, backmixing in the liquid phase was described by a one-dimensional plug-flow model with longitudinal dispersion superimposed. Possible channeling or holdup effects in the reactor were neglected by assuming the superficial liquid velocity along the catalyst bed to be constant and the catalyst effectiveness factor to be independent of temperature. Thus, the differential equation describing the steady-state concentration profile in an isothermal reactor is

$$D_a^L \frac{d^2 C_i^L}{dz^2} - u_L \frac{dC_i^L}{dz} - r_j = 0 \tag{2.114}$$

Using perturbation solutions of Eq. (2.114) obtained by Burghardt and Zaleski (1968) for appropriate boundary conditions, considering small deviations from plug flow (large Peclet number) and a first-order reaction, and substituting the Pe_d number by an empirical correlation proposed by Hochman and Effron (1969) and Sater and Levenspiel (1966), Eq. (2.98) is obtained.

Schwartz and Roberts (1973) presented the application of liquid residence-time distribution (RTD) studies for determining the performance of a downflow two-phase fixed-bed reactor (contacting efficiency and reaction rate constant). The RTD of the liquid external to the catalyst pores is the desired information; however, sometimes the use of a tracer may include some contribution from the internal holdup as well (Satterfield, 1975). From these studies it was concluded that any difference in conversion at the same operating conditions between pilot and commercial units must be attributed to differences in liquid–solid contact between the two units, and that liquid-phase deviation from plug flow has an insignificant effect on conversion in commercial-scale TBRs (Duduković, 1977).

With an appropriate redefinition of parameters, it was found that the cross-flow model is mathematically equivalent not only to the modified mixing-cell model of Deans (Schwartz and Roberts, 1973) but also to a probabilistic time-delay model developed by Buffham et al. (1976). Results over a range of typical trickle-bed conditions showed that predictions based on the simpler dispersion model differ slightly from those obtained using the more complex cross-flow model. Important differences between these models occur only at a high degree of backmixing (short reactors) and at a high reactant conversion; therefore, the assumption of plug flow of liquid represents the TBR behavior quite well, and when it is required to account for liquid backmixing, the dispersion model [Eq. (2.96)] can be selected as an adequate representation, which may also be more conservative in general. The dispersion model is, in addition, appropriate for making initial estimates (i.e., whether or not deviation from plug flow will be significant in any specific case) (Satterfield, 1975).

Montagna and Shah (1975) investigated, both experimentally and theoretically, the backmixing effect on the performance of a pilot-plant HDS reactor with atmospheric residue as feedstock when both gas and liquid are passed co-currently and upward through the reactor. For upflow operation, an increase in gas and liquid flow rates (at constant temperature and pressure) decreases the HDS, HDM, and hydrodeasphaltenization (HDAsph) reaction rates, due to backmixing in the reactor. For a very shallow catalyst bed it was also found that upflow operation yields better conversions for all reactions than does typical downflow operation under the same reaction conditions. The performance of a co-current upflow (both gas and liquid) was compared with that of a co-current downflow HDT reactor using experimental data and the back-mixing (or axial dispersion) model reported by Paraskos et al. (1975). By assuming that $\mathrm{Pe}_d = \kappa (G_{mL})^{\beta}$, the dispersion model [Eq. (2.96)] becomes

$$\ln \frac{\left(C_{\mathrm{S}}^{L}\right)_{f}}{\left(C_{\mathrm{S}}^{L}\right)_{f,P}} = \frac{\kappa k_{\mathrm{in}}^{2} \eta^{2} (G_{mL})^{\beta}}{(\mathrm{LHSV})^{2}} \tag{2.115}$$

The effect of catalyst bed length at constant LHSV or of the liquid flow rate on the performance of a TBR in the HDS of atmospheric residue was also studied by Montagna and Shah (1975). The experimental data were evaluated on the basis of the axial dispersion model of Mears (1971), the holdup model of Henry and Gilbert (1973), and the effective catalyst wetting model of Mears (1974) in order to validate their applicability when explaining the catalyst bed length (or superficial liquid velocity) effects on the removal of nitrogen, sulfur, metals, and asphaltenes from atmospheric residue.

For the axial dispersion model [Eq. (2.96)] it was first necessary to determine the kinetic constants and effectiveness factors for the various reactions.

The Peclet number was expressed according to Hochman and Effron (1969) as $\mathrm{Pe}_d = \kappa L_B^{\alpha}$ with $1 > \alpha \geq 0.5$, and the following equation was derived:

$$\ln \frac{(C_S^L)_f}{(C_S^L)_{f,P}} = \frac{\kappa k_{\mathrm{in}}^2 \eta^2 d_{pe} (L_B)^{-(1+\alpha)}}{(\mathrm{LHSV})^2} \tag{2.116}$$

It was found that Mears' (1971) criterion predicts significantly larger values of $L_{B,\min}$ than those obtained experimentally. From experimental data it was concluded that backmixing, liquid holdup, and effective catalyst wetting all appear to be strongly dependent on the catalyst particle size or on the viscosity of the feedstock.

Shah and Paraskos (1975) outlined an approximate solution to the governing differential equations for an adiabatic hydroprocessing TBR operating in the presence of axial dispersion effects. With this approximation the following criteria for significant axial dispersion effects were obtained: (1) at a high conversion rate, adiabatic operation produces a larger axial dispersion effect than that of isothermal operation, and (2) at a low conversion rate the opposite results are obtained.

Since axial dispersion effects tend to reduce conversion, it is important to design and operate pilot-plant reactors under conditions where this effect is minimal. Under that premise, Mears' criterion was extended by Shah and Paraskos (1975) to the case of pilot-scale adiabatic trickle-bed hydroprocessing reactors using the mass balance given by Eq. (2.116) for a reactant undergoing an nth-order irreversible reaction in the slower-moving liquid phase. The criteria derived by Mears (1971) and Gierman (1988), among others, are used to evaluate the order of magnitude of the Peclet number required to avoid the axial dispersion effects in pilot-scale adiabatic reactors for (a) residual HDS, (b) the HCRs of gas oils, and (c) the hydrodenitrogenation of shale oils. The results indicate that the axial dispersion effect is less important in case (c) than in cases (a) and (b).

Empirical Correlations Since the existence of different phases is neglected in pseudohomogeneous models, and catalytic reaction rates can be described only with concentrations in the liquid bulk, the following empirical approaches can fit in the simple pseudohomogeneous models classification, as they do not need to recognize variations of mass and heat between phases.

Nowadays, many desulfurized middle distillates, although relatively low in total sulfur, contain high concentrations of β-DBTs [dibenzothiophenes (DBTs) with sulfur atoms in the 4 and 6 positions]. This means that the typical HDS technology may be viewed as a pretreater of raw middle distillates. The challenge, then, is to effectively reduce the sulfur from these prehydrotreated distillates to less than 10 to 15 ppmw. The solution may be the concept of two-stage process, where the liquid effluent from the first stage is fed to a second hydrotreater aimed to desulfurize residual β-DBTs.

For refiners the most important consideration in ultradeep HDS is feedstock quality, since they constantly face the question of which feedstocks from widely different origins and preprocessing histories are most attractive. Therefore, the development of phenomenological property–reactivity correlations in terms of readily measurable properties can guide the feedstock selection and blending. These correlations can also be useful for designing model compound experiments, kinetic data interpretation, process modeling, and economics and planning studies.

Although various correlations in the literature try to "predict" product properties (mainly sulfur content) as a function of feed properties and composition, reaction conditions, and so on, they are highly empirical in nature, and not many attempts have been made to develop suitable and well-supported correlations. Here, the most significant efforts to represent the effect of feed properties on an HDS reaction rate are described briefly.

Some of these correlations, developed by Tsamatsoulis et al. (1991), were obtained from results in bench-scale TBRs using atmospheric heavy residue as a feed and commercial $CoMo/Al_2O_3$ catalyst. The empirical correlations shown in this work, called design equations, interrelate the chemical reactions considered (HDS, HCR, hydrogen consumption rate, asphaltenic fraction desulfurization, and nonasphaltenic fraction desulfurization) and relate the characteristic properties of the products (density, viscosity, Ramsbottom carbon residue, API gravity, and Conradson carbon residue) to the severity level of the HDS and HCR reactions. One important finding of this work is that in the range of temperatures used in the experiments (350 to 465°C), the relation between HCR and HDS was not affected by reaction temperature and residence time.

The relationships between percentages of sulfur removal and metal removal during the HDT process were investigated by Callejas and Martínez (1999) employing a residue coming from Maya crude. Typical conditions—pressures of 10 to 15 MPa and temperatures ranging from 375 to 415°C—were employed. At the lowest temperature studied, 375°C, a heavy dependency of hydrodemetallization of nickel (HDNi) and vanadium (HDV) on HDS conversion was observed. Finally, empirical linear equations showing the dependence with HDS conversion were obtained at 375 and 400°C for demetallization reactions. It was observed that pressure only affects the HDS rate constant along the range of pressures studied, but it is necessary to keep in mind the possible effect of pressure over HDN and HDM reactions for values of this variable out of the range reported.

Ho (2003) developed a correlation for the single-stage HDS of raw distillates at low H_2 pressures over a sulfided $CoMo/Al_2O_3$ catalyst, which has the general form

$$k_{\mathrm{HDS}} \propto (\mathrm{API})^{\alpha} \left(C^{L}_{\mathrm{DBTs}}\right)^{\beta} \left(C^{L}_{\mathrm{N}}\right)^{\gamma} \tag{2.117}$$

This correlation may not be applicable to prehydrotreated distillates, which are more likely to be desulfurized at relatively high H_2 pressures over sulfided

NiMo or NiW catalyst because of their higher hydrogenation functionality compared with sulfided CoMo catalysts. The property–reactivity correlation for prehydrotreated distillates was obtained from a simple competitive adsorption model based on the Langmuir isotherm:

$$k_{\text{HDS}} \propto k_{\text{app}}\left[1-\lambda K_N \left(C_{\text{N}}^L\right)_0\right] \tag{2.118}$$

where k_{app} is a phenomenological rate constant for HDS, λ the extent of the HDN reaction, K_N an inhibition constant, and $\left(C_{\text{N}}^L\right)_0$ the feed nitrogen content. The latter expression indicates that the feed nitrogen content should be used as an approximate overall indicator of the feed reactivity, measured by means of the overall rate constant k_{HDS}.

Ho and Markley (2004) also proposed a property–reactivity correlation for hydrodesulfurization of prehydrotreated distillates as diesel fuels. It was found that HDS reactivity of such prehydrotreated distillates decreases primarily according to the feed nitrogen content in a linear fashion.

To optimize process conditions employing a minimum of experiments, Ferdous et al. (2006) performed a statistical design that involves the effect of intensive parameters—LHSV, pressure, and temperature—which have significant effects over HDS and HDN of a heavy gas oil derived from Athabasca bitumen employing an NiMo/alumina catalyst modified with boron. Typical operating conditions for HDTs were studied: 340 to 420°C, 6.1 to 10.2 MPa, and 0.5 to 2.0 h^{-1} in a micro-TBR. The expressions obtained for HDN and HDS were second-order polynomial models (i.e., they showed no straight-line dependence of conversion on the optimization parameters). The kinetic studies were also performed in order to have a tool available for prediction of the effect of catalyst activity over process variables. Two types of expressions were employed: power-law and Langmuir–Hinshelwood models, obtaining good agreement between predictions and experimental data. These authors also have reported that pressure does not have any effect on sulfur conversion during the HDT of heavy gas oil using $NiMo/Al_2O_3$ catalyst containing boron. This result contradicts that reported by Jiménez et al. (2007a,b), who found an appreciable positive effect of high pressure for HDS and HDN reactions during HDT of the heaviest fractions of vacuum gas oils (VGOs). However, according to other researchers (Berger et al., 1996; Shokri et al., 2007), that observation, where no appreciable effect on HDS reactions was found, could be correct only in narrow ranges of pressure. Shokri et al. (2007) pointed out that the viscosity of liquid tends to increase as the pressure does, resulting in lower diffusivity and hence mass transfer. Therefore, all these reports encourage planning experiments to highlight the effects of higher-pressure conditions on sulfur conversion.

Continuous Pseudohomogeneous Models

Steady-State Continuous Pseudohomogeneous Models Among different approaches, steady-state continuous pseudohomogeneous models have been widely reported in the literature. This is due to their reliability and simplicity. A pseudohomogeneous model generally assumes a power-law kinetic type, although sometimes Langmuir–Hinshelwood expression has been employed. Use of the power law for kinetics has been questioned due to its meaningless; that is, it does not permit following the phenomena occurring intrinsically, such as the reaction mechanism and inhibitory effects, however, for a certain range of temperature and composition, the power-law kinetic model has been employed successfully for preliminary designs and to explore related phenomena, such as hydrogen consumption, catalytic deactivation, quench studies, and dynamic behavior, among others.

The reactor is modeled assuming no gradients of mass or temperature between two adjacent phases. Generally, one-dimensional analysis and few reports of two-dimensional modeling reactors for HDT have appeared in the literature. The main contributions toward modeling TBR systems applied to an HDT process by pseudohomogeneous models are described briefly below.

Shah et al. (1976) discussed the proper location for a quench in an exothermic, time-dependent catalyst activity system. The system chosen for study was a trickle-bed reactor, and the feedstock was residue oil. Empirical catalyst activity functions for HDS and HDM were developed from pilot-plant data. Differential mass balances for irreversible first-order reactions of HDS and HDM were written assuming ideal plug-flow conditions, whereas the energy balance was formulated under adiabatic conditions. They concluded that the value of the maximum cycle life and the quench position depend significantly on reaction variables such as feed temperature, feed concentrations of sulfur and metals, activation energies of sulfur and metals removal reactions, residence time, and the sulfur conversion level. This model seems to be the first attempt to predict, although empirically, deactivation in a TBR system sustaining HDT reactions. Although this report is interesting in itself, almost all the studies performed with quench systems take the maximum allowable temperature as a criterion for the localization of quench, that is, the temperature at which the quality of products may become undesirable, and they propose that the quench location is dependent on maximum catalyst life.

Kodama et al. (1980) developed a simulation model of residue HDS reaction based on a catalyst deactivation model. To represent the fouling process, an improved model that included both interaction of the coking reaction and vanadium removal of the pore plugging was proposed. Both rate equations of desulfurization and vanadium removal were expressed by second-order reactions, and they were assumed to be proportional to the hydrogen concentration in the liquid phase. The material balances for sulfur and metals in a plug-flow reactor were carried out, the energy balance was developed considering an adiabatic reactor, and the heat of reaction was attributed only to the HDS reaction. Mass transfer in the porous catalyst was taken into account through the effectiveness factor. The model allows for prediction of the actual operations of bench-scale fixed- and moving-bed reactors. This model was validated with enough data, and thus its predictions can be expected to be reliable.

By using plug-flow reactor model and power-law kinetics, Akgerman et al. (1985) showed the effect of liquid volatility on the conversion, arriving at the conclusion that the difference between predictions of models, assuming either volatile or nonvolatile liquid phases, is significant. For the first-order case, differences range from 24 to 38%. At high conversions, the difference between the models, volatile and nonvolatile, diminishes, due to depletion of the limiting reactant. This can be attributed to a change in concentration in the liquid phase. In another work of Akgerman and Netherland (1986), several equations of state for prediction of partial vaporization of feed in reactor performance were compared. Although VLE has to be performed at each step of integration through the length of the reactor, these authors bypassed this feature, supposing linear variation of equilibrium constant between the inlet and outlet conditions. They assumed almost complete wetting, and no appreciable influence of vaporization effects was observed in the conversion. Further studies in this direction have confirmed the importance of taking into consideration the volatility of the light feedstock in HDT reactions, because it can cause incomplete wetting and thus poor performance of the catalytic bed, increase of conversion of refractory species, and depletions on conversion of reactive species and related phenomena.

Döhler and Rupp (1987) performed laboratory-scale experiments with the same feed and catalyst as those in an industrial VGO hydrotreating unit, in order to simulate the adiabatic behavior of the industrial reactor using a plug-flow pseudohomogeneous one-dimensional reactor model. The model was validated only with HDS, HDN, and hydrodearomatization (HDA) reaction data. They pointed out that calculation of the weight-average bed temperature (WABT) in an adiabatic reactor having a ΔT value of 55°C or higher does not agree well with the isothermal temperature of experimental reactors because there is a nonlinear relationship between the temperature and the rate of reaction.

Used oil hydrotreating in a pilot TBR was simulated by Skala et al. (1991) employing a pseudohomogeneous model with a power term for LHSV, where HDS, hydrodeoxygenation (HDO), and HDM reactions were used for validation. Those reactions were described by first-order power-law kinetic models, which were then used for the simulation of an industrial TBR. Catalyst deactivation by coke and metals was also simulated according to the model of Shah et al. (1976), and a similar model was used to predict the pressure drop dependence on decrease of the bed porosity. Good agreement between the model and industrial data of the pressure drop was reported. The model for pressure drop dependent on catalyst activity could be useful for industrial analysis of reactor performance affected by continuous plugging of a catalyst bed.

Tsamatsoulis and Papayannakos (1998) employed real feeds and operating conditions such as those encountered in hydroprocessing of heavy VGO and a set of four nonporous catalysts to derive a correlation for predicting the Bodenstein number (Bo) as a function of bed characteristic and Reynolds number. Two-thirds of their data fell within the Bo range given by Gierman (1988). In another work of the same authors, the effects of liquid dispersion in a bench-scale HDT on the determination of the intrinsic desulfurization kinetics and on reactor performance for three porous catalysts with different activity were studied. Porous catalysts were used. Only two reactions, hydrodesulfurization and hydrogen consumption, were considered. Activation energies for HDS and HCON were almost the same for each catalyst, but higher values were estimated when an axial dispersion model was used instead of plug flow, although the difference was negligible. Some observations given by these authors were that the plug-flow model can be employed successfully for HDS and hydrogen consumption predictions when the conversion is kept low, but that for higher values such as deep desulfurization (>95%), deviation of up to 40% can be estimated when axial dispersion effects are incorporated in the plug-flow model. Therefore, the influence of dispersion effects on reaction kinetics must be taken into account when using data at high conversions.

A commercial kero-HDS reactor was simulated successfully by Sau et al. (1997) by means of a pseudohomogeneous plug-flow model. The novel continuum theory of lumping was employed for kinetics, and very good predictions were observed. This work is a good example of how, with a simple reactor model but following the chemistry closely, it is possible to achieve reliable predictions with an important reduction in the total number of model parameters.

The HDS, HDN, and olefin hydrogenation (HGO) reactions were simulated in a commercial diesel HDT reactor by Cotta and Maciel Filho (1996) employing a one-dimensional pseudohomogeneous model. Each reaction was described by a power-law kinetic model because they found the Langmuir–Hinshelwood model to be inconsistent with their results. They observed higher experimental values than were calculated for HDS, while the opposite effect was found for HDN. This behavior could be attributed to the fact that the model does not take into account the inhibiting effect of H_2S.

A deterministic quasi-steady-state model of the reaction section of the atmospheric residue desulfurization unit was developed by Lababidi et al. (1998) to simulate the long-term behavior of the catalyst bed. A single fixed-bed experimental reactor was first considered, followed by an industrial-scale reactor. An appropriate correlation was used to determine the dissolved hydrogen concentration in the oil. Simulation results of the single-bed reactor showed a perfect match with Kodama et al. (1980) work, which validate the main assumptions of the model proposed. After validation, a series of four industrial-scale reactors were simulated. Their conclusions were that actual industrial profiles of concentration and temperature with respect to time were very similar to the profiles predicted. Deviations were observed at start-of-run (SOR) and end-of-run (EOR), whereas the model was capable of predicting perfectly the middle-of-run (MOR). According to the authors, the simulation program developed might be useful for predicting the life of the catalyst if the product temperature is considered as an acceptable measure.

To select the best rate expression to predict the industrial reactor behavior, an adiabatic diesel hydrotreating trickle-bed reactor packed with commercial NiMo catalyst was simulated by Cotta et al. (2000). The HDS, HDN, and HDO reactions were considered. Power-law kinetics was employed in this system, and parameters for HDN and HDS were obtained from an experimental isothermal downflow pilot-plant fixed-bed reactor, while for HDO, kinetic parameters were obtained from the literature. A one-dimensional pseudohomogeneous model was employed in this work. On the basis of their results, the authors determined that it is necessary to use the most severe processing conditions (pressure of about 95 atm and temperature of 390°C) to increase HDN conversion, and the best model to represent HDN and HDS processes is a power-law kinetic instead of a Langmuir–Hinshelwood kinetic under typical conditions. Due to the complex composition of different feedstocks, the intrinsic kinetics assumed for estimating HGO conversion might not be adequate, and results obtained from the model might not be reliable.

Mejdell et al. (2001) modeled an experimental plug-flow TBR reactor for the HDS of oil products based on a discretization of the entire spectrum of sulfur components into small pseudocomponents of only 1°C boiling-point range (132 pseudocomponents), and identifiable components with low reactivity such as 4-Me-DBT and 4,6-DMe-DBT (six real components) were modeled separately. A Langmuir–Hinshelwood kinetic type of expression was used. Experimental data to estimate the 277 kinetic parameters were obtained on a reactor operated in upflow mode, employing light gas oil (LGO) as feed. Predictions of conversion were carried out and results were compared with experimental results, showing good agreement. An observation derived from this work was its utility for simulating the HDS process at high conversions because it permits prediction of the conversion of pseudocomponents with high reactivity and also that of refractory components which suffer from large deviations from a TBP–reactivity tendency. This approach may also have a certain generality for other feedstocks if one assumes that the reactivity for the lumps is the same for other oils. Although the authors have reported the implementation of this model in an industrial TBR, and they claimed very accurate predictions in conversion, they did not give any evidence of such a study.

Bellos and Papayannakos (2003) studied the HDS and hydrogen consumption kinetics of a straight-run heavy gas oil in a microreactor loaded with a diluted bed of commercial catalyst which was simulated by means of two models, one of them a plug-flow pseudohomogeneous model assumming no liquid evaporation, and the other an improved model that took into account feed evaporation and gas- and liquid-phase equilibrium along the reactor axis. The former was developed only to derive the initial values for the kinetic parameters of the improved model. Predictions of gas- and liquid-phase equilibrium were carried out at each step of integration over the entire length of the reactor. Miscalculation in the mass balance of the improved model was observed when the catalyst mass was taken as a constant value, since the mass of catalyst was also a function of the bed length.

Melis et al. (2004) employed a pseudohomogeneous axial dispersion reactor model for the interpretation of the HDA reaction during gas oil HDT. This model only considers the HDA reaction by means of a lumped scheme for aromatics composition in gas oil and assumes that hydrogenation and dehydrogenation reactions occur according to the Langmuir–Hinshelwood mechanism. The model developed was capable of predicting experimental results with different types of feed containing different concentrations of aromatics.

A process for LGO HDS via catalytic distillation was proposed by Vargas-Villamil et al. (2004). It was compared with an optimized conventional HDS process using similar flow conditions, which represented an industrial plant. A compromise was established among the production of diesel and naphtha and the operating costs in order to optimize the conventional HDS process. The kinetics of HDS employed was represented by a Langmuir–Hinshelwood equation, using DBT as a representative of all sulfur compounds to HDS via two parallel pathways, hydrogenolysis and hydrogenation. A pseudohomogeneous plug-flow model of an industrial TBR was developed and incorporated in an HDS unit, which was modeled using commercial software. The energy balances and the distribution of the components between the phases were defined by isoenthalpic equilibrium. An effectiveness factor was included to describe an industrial-size catalyst, which accounts for the intraparticle transport phenomena. Some remarks were made with respect to the use of catalytic distillation, such as the possibility of improving the quality of products to a level even higher than that of a conventional process, keeping lower fixed and operational costs. This technology is very prominent in meeting future requirements in specifications of low sulfur contents in diesel fuel.

A simple one-dimensional pseudohomogeneous plug-flow reactor model for a multicatalyst system was developed by Kam et al. (2005) to study the deactivation mechanisms of hydroprocessing catalysts in atmospheric residue desulfurization (ARDS) units due to coking and metal deposition. Three different stages of deactivation are considered: SOR, MOR, and EOR. The reactions considered were HDS, HDM (the removal of vanadium and nickel were considered separately), and HDAsph, the latter accounting for catalyst deactivation. The equations for the mass and heat balances are pseudo-steady-state because of catalyst deactivation. The model was applied further to a parametric study that examines the effects of LHSV, temperature, and maximum capacity on the performance catalyst systems.

A steady-state pseudohomogeneous plug-flow model to predict HDS conversions in an experimental TBR was developed by Sertić-Bionda et al. (2005). The simple reactor and kinetic models proposed in this work were used to investigate the influence of some reaction parameters (i.e., H_2/oil ratio, pressure, and LHSV) on HDS, using atmospheric gas oil and LCO from FCC as feeds.

Toulhoat et al. (2005) have presented a plug-flow pseudohomogeneous model to predict the performance and cycle length of fixed-bed residue hydroprocessing units. The model simulates catalyst activity in a pseudo-steady-state regime and resistance to deactivation by metals and coke deposition. Both HDS and HDAsph reactions described by pseudo Langmuir–Hinshelwood kinetics were considered; coke deposition was assumed to be first order with respect to a driving force equal to the difference between actual and equilibrium coke concentrations in the solid phase.

By taking the model developed by Kam et al. (2005) and adding hydrothermal treatment as a term modeled as a power law, Juraidan et al. (2006) simulated the long-term behavior of a catalyst and reactor considering the same reactions studied previously by Kam et al. (2005) and also carried out the same parametric study. The additional term (coefficient and exponent) was obtained from the results of blank experiments in a laboratory-scale reactor (i.e., experiments carried out with inert materials without a catalyst). Other conditions were the same as those employed by Kam et al. (2005). Kinetic parameters for HDM (HDV and HDNi) and HDasph reactions using Boscan crude were estimated. The model was utilized to predict the complete accelerated test run of experimental results obtained from a pilot plant after verification. Simulated results from this model matched quite well with those of the pilot plant. A marked improvement over the original model of Kam et al. (2005) was achieved.

A two-stage micro-TBR for HDT of heavy gas oil derived from Athabasca bitumen was simulated by Botchwey et al. (2006). A one-dimensional pseudohomogeneous mass transfer model and a two-dimensional heat transfer model were developed. Kinetic models for HDS and HDN reactions used in simulations were based on the Langmuir–Hinshelwood approach. This paper represents an earlier work on the modeling of a two-stage micro-TBR for HDT with interstage H_2S removal. It was observed that removing H_2S improved the levels of HDN and HDS.

Galiasso (2006) developed a simplified pseudohomogeneous plug-flow model for isothermal TBR and gas- and liquid-phase reactors to optimize a scheme of reactors and to minimize investment. The effect of adding reactor volume to existing units to produce a low-emission diesel fuel was compared by using the new scheme of reactors and the conventional TBR. The model reproduced HDS, HDA, and HDN reactions. It was shown that by using the new gas- and liquid-phase reactors, the aromatic hydrogenation and hydrogenolysis reactions can be enhanced. Simplified kinetic rate models (Langmuir–Hinshelwood type) in the gas and liquid phases for simple lumps of HDA reactions were used in the simulations, and kinetics and fluid-dynamic-related parameters were calculated previously through an optimization algorithm.

Dynamic Continuous Pseudohomogeneous Models Since perturbations can occur in the various HDT processes due to changes in composition of reactants, flows, inlet temperatures, and so on, it is highly desirable to account for a robust model capable of predicting the performance of the reactor system under such sudden changes. In this direction, some work has been reported in the literature; the main contributions are summarized below.

Chao and Chang (1987) showed a one-dimensional pseudohomogeneous model incorporating the effects of mass and heat dispersion, mass and heat transfer resistance inside catalyst particles, and catalyst deactivation in order to investigate the dynamic behavior of an adiabatic residue HDS trickle-bed pilot reactor system. The reactions taken into account were HDS, HDV, and the coking deposition rate on catalyst. This dynamic model was validated using the experimental data of Kodama (1980) and producing step changes on feed composition, feed rate, and inlet temperature. This rigorous model could be used only for off-line studies, because it involves a large number of equations, and as a consequence its solution requires a huge amount of time.

Oh and Jang (1997) presented a rigorous modeling and simulation of commercial naphtha HDS reactor in the dynamic regime. The mathematical model is two-dimensional pseudohomogeneous and uses a kinetic model of Langmuir–Hinshelwood type to describe an HDS reaction. They have also studied the influence of changing the hydrogen flow rate by 10%, showing how it influences conversion and temperature. The agreement between predictions and design data can be attributed to well-established correlations for gas–solid systems.

Chen et al. (2001) proposed a pseudohomogeneous two-dimensional reactor model to describe the dynamic and steady states of a fixed-bed pilot-plant hydrotreater used for the hydrotreating of partially stabilized light-coker naphtha; therefore, the reaction system was gas–solid. The rate reaction parameters were obtained in an experimental pilot-plant reactor, and kinetics was assumed as nth-order power. Dynamic behavior was induced by changes in hydrogen volumetric flow rate. As a main conclusion, it was reported that the thermowell can provoke heat conduction within the reactor; thus, temperature measurements in the thermowell could differ from those of the bed, and due to that, special care must be taken when interpreting pilot-plant data.

Heterogeneous Models

Steady-State Heterogeneous Models

1. CONTINUOUS MODELS The main reason for developing heterogeneous models (i.e., models that distinguish the phases in a trickle-bed reactor) is to account for inhibitory effects. In the literature, the majority of reports have supposed no significant resistance to mass transfer from the gas phase to the gas–liquid interface. On the other hand, several researchers have modeled heterogeneous adiabatic systems based on an isothermal bed catalyst, due to the lack of proper correlations accounting for this feature. Generally, the energy balance is carried out by supposing pseudohomogeneous behavior even though material balances are considered heterogeneous. Different features remain for discussion, such as the influence of axial dispersion in countercurrent operation, the level of vaporization during HDT reactions, and the degree of saturation of liquid-phase and phase equilibria, among other rele-

vant aspects. Some important contributions considering different aspects of this type of model applied to HDT processes are reviewed below.

Van Parijs and Froment (1984) simulated an adiabatic reactor for hydrodesulfurization of naphtha, using a one-dimensional heterogeneous reactor model, Hougen–Watson kinetic expressions, and internal concentration gradients. The thiophene was chosen as a model sulfur compound for HDS reaction. A review of equations accounting for interfacial and intraparticle gradients was presented by Froment (1986), who also recommended a Hougen–Watson approach for expressing rates of catalytic reactions, since power-law equations account insufficiently for interaction of the reacting species with the catalyst. It was also pointed out that kinetics and transport phenomena have to be treated separately to simulate and design the reactor successfully. These models seem to be the first rigorous heterogeneous models presented in the literature for an HDT process.

Trambouze (1990) carried out comparative simulations of co-current and countercurrent fixed-bed heterogeneous reactors. The criterion selected to make the comparison was the conversion of one of the reactants, with the quantity of catalyst employed used as a reference. It was remarked that a countercurrent reactor requires less catalyst than the co-current reactor to obtain the same conversion in irreversible reactions, equilibrium reactions, or those inhibited by one of the reaction products that are typical cases of hydrogenation of aromatics and hydrotreating of petroleum fractions. It is known that axial dispersion is more important when operating in the countercurrent mode compared with the co-current downflow mode; however, Trambouze neglected this feature, probably because the objective of his work was only to show a potential for countercurrent operation, although some miscalculation can affect the results quantitatively by taking into account, or overlooking, axial dispersion in a real system.

A one-dimensional heterogeneous model was also employed by Froment et al. (1994) to simulate diesel HDS using a kinetic model for HDS of DBT and alkyl-substituted dibenzothiophenes based on structural contributions. This kinetic approach, which retains the details of the complex reaction network of every feed component, allowed us to reduce significantly the number of parameters with respect to the molecular approach and satisfactorily represented experimental data of HDS. It was proposed that the kinetic approach can also be applied to nitrogen-containing compounds. This approach gives good results, but the model is complex and involves extended analytical work to identify the components (Mejdell et al., 2001).

A set of differential first-order equations was solved by Korsten and Hoffmann (1996) to simulate the performance of a pilot trickle-bed reactor. The main reaction was the desulfurization of VGO, which was assumed to be saturated with hydrogen at the inlet of the reactor bed. Mass transfer coefficients, pressure drop, and physical properties were estimated with correlations reported in the literature, and kinetic parameters of Langmuir–Hinshelwood type were obtained from pilot-plant experiments. Although the correlations employed were developed for nonreacting systems and ambient pressure and temperature, the mathematical formulation showed reasonably good agreement with experimental results. On the basis on their observations, these authors pointed out that scale-up of pilot-plant data to an industrial trickle-bed reactor can yield some miscalculation, due to differences in mass-superficial velocity, which strongly affect the contact effectiveness between the fluid phase and the catalyst. It seems that the correlation used for the solubility of H_2S in oil is not applicable at other conditions because it neglects the influence of pressure.

An attempt to address the main requirements by relaxing many of the assumptions used in previous models was proposed by Khadilkar et al. (1999), who also reported three models. The first model, at the pellet-scale level, assumed power-law kinetics; the second, at the reactor-scale level, considered dry and wet zones but without a distinction between external and internal wetting of catalyst pellets; and the third was a combination of both levels: rigorous multicomponent mass and energy balances at the reactor scale and its extension to the pellet scale. The model was formulated by a set of steady-state one-dimensional differential equations and tested with data available for cyclohexene hydrogenation, giving accurate predictions of conversion and temperature profiles at the reactor scale. Additional features, such as capillary effects, incomplete catalyst filling, and evaporation, were incorporated in the third-level model. They recommended their rigorous approach, level three, for future models with complex reaction systems and volatiles. This model could be implemented for the HDS of diesel when considerable volatilization occurs.

Van Hasselt et al. (1999) developed a novel model for the countercurrent three-levels-of-porosity reactor and for the internally finned monolith reactor and compared them with traditional co-current reactor model in the hydrodesulfurization of VGO. To develop the simulation, a combination of continuous approach and discrete cells were employed; the former approximation was used to simulate the reactions occurring in a cell package and the latter to simulate gas–liquid contact through channels existing through a packed bed conceived as quench; hence modeling a reactor can be visualized as a combination of continuous and discrete models. For comparison, the TBR model was simulated with a one-dimensional heterogeneous model, and equations were written for mass and energy balances. Deep conversion was chosen as 98%. It was observed that the catalyst volume required for countercurrent flow is lower than that for co-current flow, but the main disadvantage of countercurrent flow was observed to be cooling because it is less effective since hydrogen flows from high to low temperature areas. Due to the high degree of freedom for developing this new model, packing could be adapted to satisfy demands imposed by mass transfer mechanisms.

A one-dimensional heterogeneous model for simulation of commercial trickle-bed reactor was presented by Lopez and Dassori (2001). The fluid pattern in the gas and liquid phases was approximated by plug flow. Kinetics was of the Langmuir–Hinshelwood type for the main reactions considered:

hydrodesulfurization and hydrodenitrogenation. The model incorporated catalyst deactivation caused by metal deposition, coking, and the decrease in effective diffusivity. Parameters of the model were obtained from the literature as well as information compiled from runs in VGO hydrotreater units for HDN and frequency factor, activation energies, absorption equilibrium constant, and catalyst deactivation coefficient. Also, they have reported on an ammonia profile within the reactor but did not report the correlation employed for this prediction.

Bhaskar et al. (2004) used a three-phase heterogeneous model to analyze the performance of a pilot-plant trickle-bed reactor employed for the hydrodesulfurization of an atmospheric gas oil fraction and to show the influence of intrinsic kinetics and hydrodynamics. Effects of pressure, temperature, space velocity, and H_2/oil ratio were discussed on a model results basis. The simulation showed good agreement with the experiments carried out in a wide range of operating conditions.

A one-dimensional heterogeneous model was employed by Vanrysselberghe and Froment (2002) to illustrate the performance of an industrial hydrotreating reactor. The continuity, energy, and momentum equations were formulated, and appropriate correlations were employed to determine physical properties. A synthetic diesel mixture was chosen, and detailed Hougen–Watson kinetics based on structural contributions was used. Predictions on the evolution of the content of a number of sulfur components and on the molar flux of hydrogen in the liquid phase were shown.

A heterogeneous adiabatic plug-flow model reactor for trickle-bed reactor based on previous works (Korsten and Hoffmann, 1996; Vanrysselberghe and Froment, 2002) was employed by Marroquín et al. (2002) to represent diesel hydrodesulfurization and hydrogen consumption. Model compounds were chosen and kinetic parameters were taken from the literature, although some changes were finally necessary to fit monoaromatics in the bench-scale data and sulfur content in the industrial diesel product.

Avraam and Vasalos (2003) employed a steady-state model for a trickle-bed reactor to simulate the hydroprocessing of light oil feedstocks. Plug-flow conditions and uniform pellet conditions were assumed. Four general chemical processes were modeled: HDS, HDN, HGO, and hydrogenation of mono-, di-, and tri-aromatics, taking into account equilibrium aromatic and inhibition by hydrogen sulfide, ammonia, and aromatics. This is an important contribution and seems to be the first one to consider changes in liquid and gas holdup along an HDT reactor due to the volatility of light oil compounds. Excellent agreement was found between the results predicted and pilot-plant results.

Chowdhury et al. (2002) investigated the desulfurization and dearomatization of diesel oil in an experimental isothermal trickle-bed reactor. A one-dimensional reactor model based on Korsten's model was developed for a two-phase flow reactor considering both mass transfer and chemical reaction, and the kinetics for HDS and hydrogenation of three types of aromatics were established. Nonactive zones packed with inert particles, which are located before and after the catalytic bed (the active zone), were also modeled in order to simulate the hydrogen mass transfer from gas to liquid. The correlation between experimental and predicted data was higher than 0.9.

Pedernera et al. (2003) studied the influence of oil fraction composition on the conversion of sulfur compounds in laboratory-scale TBR. The reactor model was used to evaluate various configurations of the desulfurization process with straight-run gas oil as feed, as no advantage was found when separated treatments of individual oil fractions were used. Hydrogen consumption was ascribed to the conversion of sulfur and nitrogen, hydrogenation of aromatics, and hydrocracking. Additionally, liquid distribution and wetting efficiency were determined using a magnetic resonance imaging technique. The model used by these authors was an extension of that presented by Chowdhury et al. (2002), which includes modeling of the heat balance in an adiabatic industrial reactor. This paper illustrated the use of new techniques for flow pattern characterization, which highlight the trends for hydrodynamic studies in the future.

Bhaskar et al. (2004) developed a one-dimensional heterogeneous reactor model to simulate the performance of pilot-plant and industrial TBRs applied to the HDS of diesel fractions. It employed a three-phase heterogeneous model based on two-film theory. The major HDT reactions were modeled: HDS, HDN, HDA, HGO, and HCR. The kinetic parameters were obtained from pilot-plant experiments. The authors reported that the model was capable of successfully reproducing industrial profiles of temperature and the concentration of impurities. This work is one of the first that simulates most HDT reactions.

Cheng et al. (2004) investigated the performance of a fixed-bed reactor in co-current and countercurrent flows to remove sulfur and aromatics in diesel fuel. The model presented by this group is one-dimensional heterogeneous and accounts for HDS and HDA reactions to simulate the concentration profiles of the reactants and products in the gas, liquid, and solid phases. Superior performance for removing sulfur was observed when an experimental reactor was operated in countercurrent mode with respect to co-current mode. These authors have expressed adequate HDA reaction rates, compared with Chowdhury et al. (2002) and Bhaskar et al. (2004).

Froment (2004) illustrated a fundamental approach for kinetic modeling of HDS, accounting to the maximum extent for the information provided by the physical–chemical characterization. This structural contribution approach considers detailed feedstock compositions but also transfer limitations inside the catalyst. To validate this approach, an adiabatic commercial reactor for the HDS of a synthetic diesel mixture was simulated using a heterogeneous plug-flow model. Rate equations were considered for the conversion of thiophene, (substituted) benzothiophene, and (substituted) DBT. The results of simulations showed an improvement in the removal of the majority of refractory sulfur components by intermediate flashing of H_2S.

The effect of different catalyst particle shapes on HDS reaction was studied by Macías and Ancheyta (2004). They employed an isothermal heterogeneous reactor model, which was validated with experimental information obtained from a small HDS reactor using straight-run gas oil as feed. This study provides a series of formulas for calculating characteristic factors of the catalytic bed involved in the development of the reactor model.

Rodríguez and Ancheyta (2004) have extended the model of Korsten and Hoffmann (1996) to include mathematical expressions for the rate of HDS, HDN, and HDA reactions. The HDS reaction was described by kinetic equations of the Langmuir–Hinshelwood type; HDN was modeled as a consecutive reaction scheme in which nonbasic compounds are hydrogenated first to basic nitrogen compounds (HDN_{NB}), which undergo further reactions to eliminate the nitrogen atom from the molecule (HDN_B); and HDA was represented by a first-order reversible reaction. The model was validated with experimental information obtained during the HDT of VGO in a pilot-plant reactor operated under isothermal conditions. The commercial reactor was simulated and temperature and concentration profiles were obtained.

Yamada and Goto (2004) also used the model proposed by Korsten and Hoffmann (1996) to simulate and compare the HDS of VGO in a TBR for both modes of operation, cocurrent and counter-current. Pilot and industrial scales were simulated with both modes of operation. The hydrogen velocity was also varied in both reactor scales to observe its effect on the outlet sulfur concentration. They assumed almost no resistance between the gas and liquid phases. It was recognized that more research is necessary for correct simulation of the countercurrent mode of operation because it could involve significant axial dispersion.

To optimize a cost function representing the essential economical parameter of the HDT process, Al-Adwani et al. (2005) employed a reactor model described by Lababidi et al. (1998), including a deactivation model. The model was time dependent, which means that all operating variables were time variant. However, since catalyst deactivation is a slow process, the mathematical model was considered a quasi-steady-state model. Heavy residuum was used as a feedstock. This study was focused on conversion, throughput, and catalyst life. An industrial-scale atmospheric residue HDS process was selected as a typical HDT unit to demonstrate the capabilities of the optimization model. This study showed that the optimum cost is affected strongly by the catalyst cost and the monetary benefit of lower-sulfur products.

Jiménez et al. (2005, 2006, 2007a,b) illustrated the use of a steady-state one-dimensional heterogeneous TBR model with both gas and liquid phases in plug flow and upflow, based on data obtained at a pilot-plant scale to predict the quality of products during the HDT of VGO and demetallized oil (DMO) over commercial $CoMo/\gamma\text{-}Al_2O_3$ and $NiMo/Al_2O_3$ catalysts. The model involved HDS, HDN, and HDA (mono-, di-, and triaromatic) reactions, and combined the Froment et al. (1994) and Korsten and Hoffmann (1996) models. The HDS reaction was described by the kinetic model of Broderick and Gates (1981) for DBT, while HDN and HDA reactions used the kinetic models proposed by Avraam and Vasalos (2003). Two types of sequential design of experiments were used: for optimal model discrimination and for optimal parameters estimation during kinetic investigation. In most recent papers (Jiménez et al., 2007a,b), the HDS process was simulated using the mathematical model developed in previous work (Jiménez et al., 2005, 2006) and several kinetic models were reported in the literature. The best kinetic model and its optimal parameters estimation were selected by means of sequential design of experiments (SDE). They also reported that water markedly enhanced the capacity to remove sulfur and nitrogen compounds during the HDT of the heaviest fractions of VGOs (Jiménez et al., 2007a).

Mostoufi et al. (2005) developed a one-dimensional plug-flow heterogeneous model in order to simulate the two-stage pyrolysis gasoline hydrogenation process to obtain a C_6–C_8 cut suitable for extraction of aromatics. The first hydrogenation stage was performed in the liquid phase in an adiabatic TBR over a Pd/Al_2O_3 catalyst in which hydrogenation of diolefins was the main reaction. The second hydrogenation stage took place in a two-compartment adiabatic fixed-bed reactor in series loaded with $NiMo/Al_2O_3$ and $CoMo/Al_2O_3$ catalysts, and operating in the vapor phase. Hydrogenation of monoolefins took place in the first compartment, and sulfur was removed in the second compartment. Simulations for HGO and HDS reactions in the second-stage reactor were carried out considering model compounds such as cyclohexene and thiophene, respectively. The model proposed considered hydrodynamic parameters: pressure drop, liquid holdup, and catalyst wetting efficiency.

Stefanidis et al. (2005) presented a study on the improvement of representative operating temperature from temperature profiles of an industrial adiabatic reactor, which is used to simulate reactor performance by laboratory-scale isothermal reactors. To validate the temperature estimated, a steady-state pseudohomogeneous plug-flow model with no resistance to mass and heat transfer was developed to describe mass balances of sulfur, hydrogen sulfide, hydrogen consumption, and hydrogen, as well as the heat balance in the adiabatic HDT reactor, with feeds ranging from heavy gas oil to diesel. The main disadvantage of this technique is the need of three experimental points: inlet, middle, and outlet, while the main advantage is its reliable prediction when deep desulfurization is performed.

Nguyen et al. (2006) developed a one-dimensional heterogeneous model at a steady-state regime with axial dispersion to analyze the influence of fluid dynamic nonidealities on the HDS performance of gas oils in isothermal bench-scale reactors. A Langmuir–Hinshelwood type of rate model was used to represent the HDS rate of reaction. Recently, Shokri and Zarrinpashne (2006) developed a two-phase (liquid–solid) heterogeneous model for the effectiveness factor of an HDS reaction with DBT as representative of sulfur compounds in gas oil. The mathematical model is at particle-scale conditions because it was based only on the mass balance equations inside a catalyst particle. However, more recently, Shokri et al. (2007) reported a hybrid model, the previous model with a plug-flow one-dimensional heterogeneous model that was validated with gas oil HDS pilot data. The model was implemented in Hysys commercial software through Fortran codes. The rate of chemical reactions was described by kinetics of the Langmuir–Hinshelwood–Hougen–Watson type, with DBT representing all the sulfur compounds in the feedstock.

Murali et al. (2007) developed a one-dimensional heterogeneous model in order to simulate the performance of bench- and commercial-scale HDT reactors. The HDS, HDA, and HGO reactions were taken into account in the model. The HDS reaction kinetics were described by a single lumped model for total sulfur similar to the Langmuir–Hinshelwood-type rate equation used by Korsten and Hoffmann (1996), whereas the kinetic model for HDA reactions was taken from Chowdhury et al. (2002). In the simulations, a significant amount of feed vaporization (20 to 50%) was found under normal operating conditions of HDT, which suggested that partial-feed vaporization during simulations needs to be considered. The model was validated with pilot-plant data obtained from an upflow operating mode, near ULS levels, to account properly for feed vaporization in heat balance equations. It was mentioned that diesel vaporization is very important in heat balance equations for adiabatic plant simulation because it consumes a significant amount of energy, but it is normally neglected in models reported in the literature. Therefore, the most important contribution of this work in the simulation of HDT reactors was a consideration of whether diesel vaporization and the temperature–H_2/oil ratio were dependent on the liquid specific heat capacity.

A one-dimensional heterogeneous plug-flow model, which accounts for intraparticle transport of the compounds by Fickian diffusion inside the catalyst pellets, was developed by Verstraete et al. (2007) to predict the performance of fixed-bed hydrotreating units. The feedstock of this study was vacuum residue, and experimental data were obtained from an isothermal fixed-bed reactor unit. The model predicts the evolution of concentration profiles of gas, saturates, aromatics, resins, and asphaltenes, their atomic composition in terms of C, H, S, N, O, Ni, and V, and the hydrotreating performances throughout the reactor. It was remarked that it is necessary to take intraparticle diffusion into account when modeling residue hydrotreating processes.

Alternative methods of quenching a trickle-bed reactor were analyzed by Alvarez and Ancheyta (2008) by employing a one-dimensional heterogeneous model and correlations reported in the literature. The HDS, HDN, and HDA were modeled and an energy balance procedure was performed in order to predict profiles of temperature along the reactor bed when quenching was employed.

Liu et al. (2008) proposed a novel methodology to understand the dynamic behavior of an HDT process. The new methodology, known as the *system dynamics* (SD) *model*, can predict the influence of operating conditions on the conversion efficiencies of HDS, HDN, HDA, and consumption of H_2. In this work, the SD methodology was applied for the first time in HDT process modeling with the intention of simulating individual sulfur, nitrogen, and aromatics compounds separately and achieving successful simulation. The methodology was validated with experimental LCO HDT data. The methodology consists of two steps; in the first step it is necessary to develop a dynamic loop diagram showing how a change in one variable modifies other variables, which in turn affects the original variable, and so on. The second step consists of developing a mathematical model, usually shown as a stock-flow diagram, which captures the model structure and the interrelationships between variables. This last diagram is translated to a system of ordinary differential equations, which in this case represented a steady-state one-dimensional heterogeneous model. Liu et al. (2008) also reported a similar study using the SD methodology to simulate the HDS process of LCO, including the nitrogen and aromatic compound inhibition effects on HDS activity.

2. COMPUTATIONAL FLUID DYNAMICS MODELS Although great effort has been made to incorporate hydrodynamics in modeling trickle-bed reactors through correlations derived from empiricism, more fundamental approximations must be made to account for suitable predictive models. The fundamental approximation in modeling, from a rigorous point of view, must be performed by solving conservation equations called *Navier–Stokes equations*, a set of nonlinear partial differential equations, whose solution is possible only for a few simple flows in simple geometries; however, the analysis of fluid dynamics is mathematically complex for actual packed beds. Additionally, constitutive relations that govern a material's internal response to external effects must be introduced into the conservation laws. Constitutive relations are derived from correlations; therefore, appropriate selection and validation of those relations are extremely important. Since constitutive equations are established by experimental data, experiments are fundamental in the study of fluid mechanics.

Realistic problems in fluid mechanics can be solved quite effectively by using both computational methods, called CFD models, which solve conservation equations, and experimental information. Some assumptions must be made to reduce the complexity of conservation equations, such as a consideration of the lack of effect of viscosity.

The CFD models can be employed as both a competitor and a natural complement to experimentation. For many problems, computational fluid dynamics provide a cost-effective alternative to experimental fluid mechanics. Various physical effects can be turned off, thus providing the opportunity to partially study the phenomena. The simulation of fluid dynamics can help us to understand the hydrodynamics of trickle-bed reactors and hence to perform scale-up and scale-down properly.

Duduković et al. (1999) have reported the use of CFD models for hydrodynamics, highlighting the two approaches commonly employed: Euler–Euler formulation and the Euler–Lagrange approach. Although the second approach seems to be more fundamental, it contains the tuning of parameters, which in turn must be validated with experimental information. Moreover, no clear advantages of one over the other formulation has been documented. An application of CFD was reported by Gunjal and Ranade (2007), who simulated the fluid dynamics of trickle-bed reactors in order to understand its interaction with chemical reactions in laboratory- and commercial-scale reactors. The model was employed to understand the influence of porosity distribution, particle characteristics, and scale on the overall reactor performance. The CFD model for TBR consisted of two main parts: (1) implementation of porosity distribution in the bed, and (2) flow equations for each phase (mass and momentum equations), which are based on Eulerian–Eulerian multifluids models. The model was applied to the HDS and HDA of diesel oil, and configuration and operating conditions were similar to those reported by Chowdhury et al. (2002). It was pointed out that CFD-based models with appropriate validation can be helpful in reducing the gap that exists on prediction between laboratory scales and commercial reactors. It was also reported that CFD models overpredicted conversions because they use apparent kinetic parameters reported in the literature, which have previously lumped hydrodynamics and intrinsic kinetic parameters together. Therefore, when those apparent kinetics parameters are used again in the CFD model, which takes into account the prediction of hydrodynamic parameters, the hydrodynamic effects are being estimated twice (i.e., liquid holdup effects). However, the authors mentioned that the CFD model was used only to understand the influence of reactor scales on its performance. The CFD simulations indicated that porosity distribution is an important parameter when estimating hydrodynamic variables (i.e., pressure drop, liquid holdup, wetting efficiency, etc.), which needs to be taken into account for proper prediction of reactor performance (Duduković et al., 1999; Gunjal and Ranade, 2007). The authors also recognized that H_2S solubility in oil fractions is not predicted correctly by empirical correlation at different operating conditions; hence it is advisable to use an equation of state (EoS) in order to improve the estimation. The use of an EoS makes it possible to include the effects of both temperature and pressure; however, suitable interaction parameters could be the limiting factor.

3. DISCRETE MODELS Instead of employing the continuum theory (i.e., modeling a TBR with a set of differential equations), some relaxations have been proposed, such as the supposition that the system can be treated as a number of connected cells. This assumption allows for simplifying the problem of complex reactor system modeling and also favors the use of commercial simulators which accurately predict the results of light petroleum fractions.

a. Cell Models A trickle-bed reactor was modeled by Sánchez et al. (1995) as a group of consecutive cells, consisting of a CSTR reactor coupled with a separator, in order to take into account vapor–liquid equilibrium existing in the reactor. The flash calculations were performed using a commercial simulator, while proper correlations were taken from the literature for simulating pressure drop and catalyst wetting fraction. The pseudocomponent evaluations were calculated by lumping a set of 500 different molecules into three compound types: paraffins, naphthenes, and aromatics. The reactions of hydrogenation and hydrocracking were selected and a low conversion level was maintained. First-order irreversible reactions were assumed. The plug-flow model was reached by using 25 cells. It was observed that increasing the number of cells had minor effects on product-distribution simulations. The main conclusion was that using proper thermodynamic properties and compound class lumping can be an effective method of trickle-bed reactor modeling and kinetic parameter estimation of a complex reaction network. This work shows how available tools such as commercial simulators (Aspen, PRO/II, Hysys, etc.) can be used to save time and effort when simulating multiphase catalytic reactors.

Guo et al. (2008) developed one- and two-dimensional mixing-cell reaction network models to simulate the steady-state behavior of TBRs using the highly exothermic benzene HDT reaction to validate the model. The model was based on a network of CSTRs. Each cell was designed to consider the contribution of interphase mass transfer, reaction kinetics, heat transfer, and vaporization effects. This model was developed with the intention of handling multiphase flow and reaction rates, as well as external wetting efficiency, liquid holdup, and temperature change due to both phase transition and flow maldistribution for a TBR. The model was shown to be suitable and efficient to predict temperature runaway in a catalyst bed, and it could also be applied in the scale-up of FBRs.

b. Stage Models Jakobsson et al. (2004) modeled the co-current and countercurrent operations of an HDS reactor using a mixture consisting of DBT, 4,6-dimethyldibenzothiophene (4,6-DMDBT), H_2, H_2S, and *n*-eicosane as a solvent. Previous models were used for simulation of the co-current (Toppinen et al., 1996) and countercurrent (Taylor et al., 1994) modes of operation. Countercurrent operation was studied to demonstrate the separation of H_2S during the HDS process. Since H_2S inhibits the HDS reaction, the countercurrent operation was proposed to be used to protect high-performance catalysts. The modeling of countercurrent operation used a *rate-based stage model*, in which the reactor is modeled as a series of rate-based segments (or stages) and each rate-based segment can be indentified as a segment of a packed bed, with direct consideration of diffusion, heat transfer, and multicomponent interaction effects on the calculated segment. These segments are connected by means of mass and heat balance equations to form the reactor model.

To demonstrate the benefits of countercurrent contacting of gas oil with H_2 over conventional co-current contact in a TBR for HDS, Ojeda and Krishna (2004) used the equilibrium-stage model of Taylor and Krishna (1993) in HDS reactions in the liquid phase. DBT was selected to represent the most refractory sulfur compounds in a liquid feed of *n*-hexadecane, which represented a diesel fraction. The reaction rate for DBT was described as a Langmuir–Hinshelwood type, and a plug-flow pattern for both gas and liquid phase flow was assumed. It was observed that increasing BT concentration in the feed leads to a lower sulfur concentration in the reactor effluent. That finding was attributed to a higher heat of reaction liberated, which provoked higher temperatures and hence larger conversions. Therefore, and in addition to the fact that the countercurrent gas phase cools down the liquid phase, accurate modeling of thermal effects in the reactor during HDS process must be carried out. However, it was observed that profiles of sulfur content in the liquid phase along the reactor do not match those from continuous models.

Dynamic Heterogeneous Models Reliable three-phase reactor modeling and simulation should be based on true dynamic heterogeneous models, which can be used not only for scale-up, startup, shutdown, and operability studies, but also to obtain a meaningful continuity path to the steady state of the reactor and to investigate the existence of exotic phenomena such as oscillations and steady-state multiplicity, since dynamic models provide a realistic description of the transient states of three-phase reactors. Study of the dynamic behavior of the three-phase reactor also makes it possible to design the best system control in order to obtain a safe, efficient, and profitable operation. Dynamic models, although more complicated to formulate and solve, should be preferred over steady-state models because the numerical solution strategy of dynamic models is more robust than the solution of steady-state models (Wärnå and Salmi, 1996; Salmi et al., 2000). Some important reports using such models are described in the following sections.

1. CONTINUOUS MODELS The hydrogenation reaction of toluene to methylcyclohexane, which occurs in a three-phase trickle-bed reactor with countercurrent and co-current gas and liquid flow, was simulated by Wärnå and Salmi (1996) by means of a dynamic three-phase reactor model. The model equations for the gas, liquid, and catalytic phases consisted of ODEs and parabolic PDEs, which were solved using numerical methods. The reactor was assumed to operate adiabatically and nonisothermally. The reaction rate for toluene hydrogenation was of first order and kinetic parameters were obtained in an isothermal laboratory-scale co-current trickle-bed reactor at total pressure of 4 MPa and temperature ranging from 65 to 125°C. It was observed that countercurrent operation gave slightly higher toluene conversion than did co-current operation. This work showed that the dynamic approach provides a meaningful path to the steady state of the reactor and gives valuable information on reaction dynamics. Because no mass transfer resistances inside the catalyst were considered, the model is applicable only for nonporous particles.

The dynamic modeling principles for fixed (trickle) beds were described by Salmi et al. (2000). An axial dynamic heterogeneous model was applied for the hydrogenation of aromatics simulation. The kinetics was conveniently measured in a laboratory-scale autoclave. It was proposed that dynamic models should be preferred to steady-state models, since the former provide a realistic description of the transient states of three-phase reactors and the numerical solutions of dynamic models are more robust than those of steady-state models. The case studies revealed the importance of internal mass transfer resistance in catalyst particles as well as the dynamics of various phases in three-phase reactors. This study confirmed the disadvantages of the Wärnå and Salmi (1996) model, where intraparticle mass transfer resistances were not considered. A two-dimensional model for temperature and concentration was applied by Hastaoglu and Jibril (2003) to simulate gas–solid reactions in a desulfurization fixed-bed reactor. Three levels of process space were used: bed, pellet, and grain. Steady-state experimental naphtha HDS data of a fixed-bed reactor were used to validate the bed model for concentration, whereas thermal behavior was validated transiently. The model was tested by generating the transient concentration of each component, and profiles of system parameters were obtained, giving good insight into the behavior of the system variables. However, since the model was developed for a gas–solid system, it does not include all the mass and energy transfer terms that should be present in a three-phase reactor model to simulate a TBR.

Vogelaar et al. (2006) derived a plug-flow model to describe coke formation and metal deposition profiles in catalyst pellets found in hydroprocessing as a function of position in the isothermal reactor and to predict catalyst deactivation behavior due to pore blocking at the reactor level. A lab-scale HDM experiment was simulated as a case study. The model is based on three levels of scale: the reactor level, the catalyst particle level, and its active phase. The modeling of this process provides a better insight into the deactivation mechanism of hydroprocessing catalysts and can be used to predict their deactivation behavior in industrial reactors. At the particle level, the effective Fickian diffusivity (D_{ei}^{f}) of a molecule inside a porous structure was estimated assuming friction between the solute and pore walls by a restrictive factor due to that friction with the pore wall.

The deposition process of fine particles under chemical reaction conditions in a high-pressure, high-temperature TBR was analyzed theoretically by Iliuta et al. (2006) using a dynamic multiphase flow deep-bed filtration model coupled with heat and mass species balance equations in the liquid, gas, and solid (catalyst + solid deposit) phases. This deep-bed filtration model incorporated the physical effects of porosity and effective specific surface area changes due to fines deposition and detachment, gas and suspension inertial effects, and coupling effects between the filtration parameters and interfacial momentum exchange force terms. The three-phase heterogeneous model developed in this work to simulate TBR performance incorporated the intraparticle mass transfer resistance and solid deposits by fine particles that lead to porosity reduction and bed plugging. It was found that fine particle deposition does not influence TBR performance appreciably. The only undesirable consequence of the fine particle deposition process was reflected in an almost exclusive hydraulic effect of bed plugging and the increase in resistance to gas–liquid flow.

Ho and Nguyen (2006) developed a four-parameter plug-flow one-dimensional heterogeneous model that gave more quantitative insight into how sulfur, nitrogen, and the catalyst surface interact on many widely dissimilar time scales. The theory on which the model is based is applicable to reaction systems where catalyst poisoning dynamics is driven by nonequilibrium adsorption. Modeling of nitrogen-competitive adsorption phenomenon effects in the HDS of oil fractions at the catalyst surface level was addressed with special attention to the design of robust catalyst-deactivation-compensation operating strategies in the deep HDS of middle oil fractions. The experiments were carried out in a co-current fixed-bed reactor operated isothermally in the upflow mode. The model was capable of reproducing the observed inhibiting effect of nitrogen species on the HDS of hindered heterocyclic sulfur compounds.

Mederos et al. (2006) developed a dynamic heterogeneous one-dimensional model to simulate the behavior of TBRs used for catalytic HDT of oil fractions on the pilot and commercial scales. It considered the main reactions present in the HDT process of oil fractions: HDS, HDN, and HDA (total aromatics). The model was validated with experimental data obtained in an isothermal pilot reactor during the HDT of VGO over a commercial NiMo catalyst. After validation of the dynamic model with pilot-plant data, it was employed to predict the dynamic behavior of a commercial HDT reactor. The start-run simulation of the commercial HDT reactor showed the "wrong-way" behavior in the temperature axial profiles before steady state was reached, a phenomenon reported in earlier papers. The combining of heterogeneous mass balance and pseudohomogeneous heat mass balance, as reported by Rodríguez and Ancheyta (2004), seems to be inconvenient; however, Mederos et al. (2006) demonstrated that this assumption is correct only if predictions of concentration and temperature profiles at steady state are necessary. In other contributions by the same authors (Mederos and Ancheyta, 2007), the effects of co-current downflow and countercurrent flow operation modes on HDS, HDN, and HDA were analyzed by employing the same model (Mederos et al., 2006). An important finding was that higher HDT conversion in the countercurrent mode of operation is obtained with respect to the co-current flow mode, which justifies the development of new reactor internals to improve the performance of TBR operating in a countercurrent mode.

2. CROSS-FLOW MODELS The cross-flow model seems to be more realistic than others because it assumes a stagnant zone and a dynamic zone, which is a reasonable supposition for trickle-bed reactors. Only one work dedicated to the HDT process with this assumption is available in the literature.

Tsamatsoulis and Papayannakos (1995) employed a cross-flow model to investigate the nonideal behavior of the liquid flow in a dynamic regime in a bench-scale TBR under HDT operating conditions. The development consists of two first-order partial differential equations to model the static and dynamic regions, which were solved analytically. This study provides information on how a catalyst bed should be diluted with inert particles so that the plug-flow pattern describes the liquid flow in an experimental trickle-flow hydrotreater in order to derive kinetics. The main disadvantage of this model was the use of nonporous grains.

Learning Models An artificial neural network (ANN) builds an internal model of the governing relationships embedded in the database used for training. The basic method of neural networking refers to implementing in the computer, by software or by the special hardware of processing nodes, neurons, which are linked to each other by variable-strength connection weights. Causal relations between each model input and output may be calculated from an analysis of the trained ANN structure.

The ANN must learn about the problem under study, and this learning stage is commonly called the *training process*. Once an ANN is trained, it can be used for proper simulation of an HDT unit, the effect of the type of catalysts evaluated, and feedstocks on unit performance, for control of an operation, for unit optimization, and so on. Since the ANN approach presents user friendliness and simplicity, suppressing the difficulties and complexities associated with first-principle models, it is not necessary to have sufficient mathematical and programming expertise to formulate complex objective functions and constraints.

ANNs was used by Berger et al. (1996) to model hydrodesulfurization of atmospheric gas oil in a mini-pilot-plant trickle-bed reactor as a function of temperature, pressure, LHSV, inlet sulfur concentration, and staging. The hidden layer contained three neurons. Inputs were normalized to give equal importance to each input and to reduce the effect of outliers in the database. The database, which contained 25 examples, was randomly divided into learn and test sets containing 17 and 8 examples, respectively. The results calculated by the ANN model were compared with the experimental data and an average relative error of 10% was observed. The causal index (CI), which determines the relative effect of each input variable on the model outputs, was applied to the five variables tested in the HDS system and the relative significance of LHSV and temperature over HDS was observed. Almost linear dependence was observed for the sulfur outlet as a function of LHSV; however, this behavior does not correspond to experimental data trends. It is probably necessary to input more data to the model in order to do better learning at low space velocities.

Lopez et al. (2001) proposed different structured and trained models based on process data and laboratory analysis obtained from a commercial VGO hydrotreater unit. The authors showed the power of a three-layered perceptron ANN used as an analysis tool for the optimization of several existing functions between important process variables controlling continuous operation of a VGO unit. Those different ANN models were used to predict the following operating conditions: feedstock composition (paraffins, naphthenes, total aromatics, and mono-, di-, tri-, and tetraaromatic compounds), feedstock and liquid product quality properties (sulfur and metals content, API gravity, TBP at 50 vol%, and refractive index), and process operating variables (product flow rate and average reactor temperature). After comparison of results predicted from correlation modeling and ANN, better data fitting was observed for the last approach. This feature could be attributed to the fact that the ANN model globally takes all effects occurring in the reactor, while correlations only allow for predicting specific parameters, ignoring some effects, such as transport problems within the reactor.

Commercial data were used by Bollas et al. (2004) to develop predictive models for the integration of two units, HDS and FCC, and to examine the economical benefits of their optimization. The 350 data series were randomly split into training and validation sets, consisting of 225 and 125 data series, respectively. The HDS kinetics derived from pilot-plant studies was first simulated by a predictive model and then operation of the commercial unit. Vacuum gas oil was considered as a feed to the HDS and the liquid product obtained from this process was fed to an FCC unit. The main product, a gasoline fraction, was subject to maximization and restrictions. The neural network was a multilayer perception (MLP) consisting of three layers: an input layer with as many nodes as the input variables, a hidden layer with the number of nodes varying from 1 to 5, and an output layer with as many nodes as output variables. Simulated trends agreed well with the existing experience, although the model performance deteriorated for predictions of sulfur in gasoline fractions.

A hybrid neural network model, a deterministic pseudohomogeneous mathematical code coupled with a neural network, was presented by Bellos et al. (2005). This model was used to predict the catalyst deactivation rate and the dependence of catalyst activity on the liquid feed quality. The reactions taken into account to validate the industrial HDT reactor model were HDS and HCON. Part of the kinetic parameters was obtained from industrial reactor operation data and also from experiments carried out in a small-scale reactor using industrial catalyst size and representative feeds.

Salvatore et al. (2005) used a hybrid approach based on ANNs together with a postprocessing classification algorithm to detect faults in a simulated HDT unit. An HDT model to represent the real unit was also developed. The modeling equations were chosen so that the process showed a dynamic similar to that of existing units, by means of concentration and temperature profiles through the catalytic beds. The model of the reactor was built assuming that the reactor is composed of *n* CSTR cells (12 stages) in series with equations describing mass and energy balances in each stage.

Zahedi et al. (2006) presented an ANN model for the simulation of an industrial HDT unit based on measured plant data. The model proposed predicts hydrogen demand for HDS, outlet API, and sulfur content as a function of inlet API and sulfur content in weight percent for seven different feedstocks: kerosene, furnace oil, diesel, coker gas oil, cat cycle oil, thermal cycle oil, and virgin gas oil. Eighty-three data sets were used for training, and then 40 data sets were predicted and compared with those collected from operating plants. Optimum ANN architecture was determined to achieve good generalization. The ANN model results were compared with those predicted by a conventional simulator, and it was observed that the ANN model accuracy outperforms the traditional simulator.

Recently, Lukec et al. (2008) developed ANN models to determine the sulfur content in the hydrotreating product of LGO and VGO. The models were trained using the process and laboratory data of routine refinery production. As the models were shown to be simple and easy to use, with good predictability they were used in practice for accurate continuous process monitoring, continuous online predictions, process fault detection, estimation of unmeasured states and parameters, to point out a measurement error to the hardware analyzer, and for process regulation, adaptive control, real-time optimization, and efficient product quality control. This work emphasizes the main advantage of the neural network models because they can estimate the kinetic parameters for different feedstocks, which depends primarily on the number of data sets used during the training process.

Advantages and Disadvantages of Reactor Models A kinetic model based on a detailed description can only be described by a large system of deferential and algebraic equations, implying a huge number of physical and physicochemical parameters. Due to this feature, some simplifications have been proposed. Among the different approaches in the kinetics of petroleum fractions employed during the last two decades, the most common formulation is that based on a single lump by employing a power law or Langmuir–Hinshelwood expression, although it seems that it is most appropriate to divide the reactant mixture into two lumps: easy to convert and refractory. This approach has been employed because of its easy numerical implementation and its dependence on global parameters, which in turn can easily be measured. However, a single lump is not valid for high conversions because it does not take into account variations in the composition of different feedstocks; moreover, due to future requirements of lower impurities contents in fuels, the power-law model is no longer reliable. Other approaches have been proposed following different criteria. As main contributions, the widely cited structural approach of Froments' work is a novel way of lumping more rationally the huge number of sulfur species contained in a real feedstock, although considerable analytical work must be performed to obtain the parameters involved. Recently, Froment et al. (2008) have reported the application of this theory to different feedstocks for which the parameters are available by taking almost the same catalyst system and performing a few experiments. Thus, data obtained together with the numerical parameters from previous work were employed for reproducing the overall conversion of some compounds. It has highlighted the necessity of establishing a catalog of invariant feed elements for different commercial catalyst to apply this approach to routine analysis. Another approach cited by Te et al. (2003) is based on computational quantum chemistry, supposing a linear relationship between the reaction rate and the equilibrium constant. Few reports using this approach in real systems are available; however, with the accelerated development of more efficient computers and friendly-use quantum chemical software, this approach could be, in the next years, an important trend in exploring the kinetics of real feedstocks.

It seems that although accurate, these approaches are not used for exploratory studies, because they demand considerable analytical work; they must be worked extensively in the future to account for more fundamentals and feedstock invariants. The most recent promissory approach employed is *continuum kinetic lumping*. This theory assumes the mixture as a continuum, where reactants are distributed over the entire mixture and reactivity of agglomerates of molecules decreases monotonically with molecular weight or another index. In this same direction, a novel gamma function distribution has been proposed to predict accurately the deep desulfurization of diesel (Inoue et al., 2000). The continuous kinetic approach appears accurate and involves only a few parameters. Moreover, it has been applied to real feedstocks with close agreement between predictions and experimental data (Sau et al., 1997). It seems that a continuum kinetic model is enough to describe an industrial process because it is possible to predict the apparent order of reaction accurately and provides a tool for the prediction of physical properties through the length of the reactor, which can be used for the prediction of transport and thermodynamic properties. Easy adaptation of a power-law or Langmuir–Hinshelwood expression can be incorporated in a continuous theory of mixtures. One disadvantage of the continuum approach observed by Mejdell et al. (2001) was the fact that some components have large deviations from the general TBP–reactivity tendency, such as substituted benzothiophenes. Due to that, these authors recommend modeling these compounds separately from the rest of spectrum, especially for high-conversion kinetics. The same observation, but employing discrete lumping, has been pointed out by Hu et al. (2002): using multiple lumps based on types of sulfur compounds for explaining desulfurization kinetics at ultralow sulfur levels. This observation is in agreement with Murali et al. (2007), who have pointed out that at low LHSV, a better match could be obtained if sulfur speciation is considered. However, such approaches require the support of advanced analytical tools to identify the various sulfur compounds present in the feedstock.

Recent trends in the production of heavy crude oils and the need to refine them and their distillates create the need to develop suitable reactor models in order to make preliminary calculations in the process design of new refining units. Currently, due to the continuous changes in feedstock composition, only average properties can be obtained, and kinetic studies are only carried out by employing the simplest expression (i.e., the power-law or Langmuir–Hinshelwood approach with adjustable parameters and a single lump). A better kinetic approach is not usual because of the unavailability of characterization techniques for heavy crude oils and residua. The same happens for the reactor system, since no correlations are available to evaluate all the parameters of a detailed model. For such cases it has been better and more convenient to employ simple kinetic and reactor models. However, these models cannot be used for modeling the complex hydrodynamics existing in an HDT reactor. Recently, Guo et al. (2008) have proposed a sequential approach for reducing the gap between CFD simulation and a complex reaction system based on the cell network model. Even if one employs a simplistic reactor model, detailed kinetics coupled with such a reactor model is not well suited for online analysis or optimization and control. Again, depending on the purpose of the study, one could sacrifice the chemistry in order to study more realistic systems such as those of a simple (lumped) reaction sustained in a reactor in order to analyze the fluid dynamics by means of CFD models instead of performing cold-flow experiments. Even with the powerful computers now available, modeling hydrodynamics of complex systems can only be performed for simple kinetic models (Ho, 2008).

A deep discussion of kinetic model selection is beyond the scope of this review. More detailed treatment of these kinetic approaches is summarized elsewhere (Te et al., 2003). More comprehensive details and limitations of continuum lumping have been revised by Ho (2008).

Regarding phase equilibria calculations, two different approaches have been proposed: One involves measurement of bulk properties employing EoS, considering the phases as a single compound, and the other one is based on continuous thermodynamics, which is rarely used in real systems such as petroleum distillation.

Akgerman et al. (1985) reported the influence of feed volatility on conversion in TBRs, arriving at the conclusion that a very different level of conversion is predicted if volatility is included with respect to the case when it is not included. Frye and Mosby (1967) correlated the level of HDS for light catalytic cycle oil with the liquid vaporization at the entrance of reactor, supposing that appropriate reaction rate constants are provided. The effect of species volatility on deep desulfurization of diesel has been explained by Hoekstra (2007), arguing that light compounds are striped from the liquid and the remaining sulfur compounds increase its concentration, favoring reaction rates. Avraam and Vasalos (2003) have showed the effect of volatilization through the length of reactor by plotting the variations of liquid and gas holdup as a function of dimensionless position through the reactor and emphasized the importance of volatilization on energy balances. The same conclusion was brought out by Murali et al. (2007), who only calculated the vaporization at the entrance of reactor, however. These authors have also recognized the need for an accurate kinetic model coupled with vaporization effects to predict the performance of a reactor for deep desulfurization. Chen et al. (2009) conducted a VLE study with LCO to investigate the influence of vaporization of feedstock on the operating regime of a pilot-plant hydrotreater, although they remark that results observed at this small scale cannot be extrapolated directly to a commercial plant.

More research in this area is necessary, particularly for HDT of light fractions of petroleum. It is also necessary to incorporate the continuous thermodynamic approach since it permits the description of realistic systems, such as petroleum fractions, which could be considered as a mixture of an infinite number of components (Cotterman and Prausnitz, 1985; Cotterman et al., 1985). As can be inferred, the continuous descriptions, together with separate modeling of some identifiable compounds, can provide an accurate explanation of either kinetics or thermodynamics, although a system described by these approaches could be too complex.

It is not strictly necessary to take into account vaporization effects for the modeling of hydroteatment of heavy crude oils and residua, because almost all reactive compounds remain in the liquid phase even at the high temperatures employed for these processes, as observed in a recent study (i.e., less than 1% of the mole fraction of VGO in the gas phase at typical reaction conditions) (Alvarez and Ancheyta, 2008). This simplification can contribute to reducing the complexity of such a model and favors the exploration of other features, such as the chemistry or related phenomena.

Some researchers have established that probabilistic models can be fitted to the experimental data for TBRs more flexibly than deterministic models, which suggests that a probabilistic description of TBRs corresponds more closely to reality than does a deterministic description (Hofmann, 1977). However, still further research is required to reach a final conclusion: for example, the usefulness of such complex models. The advantages and disadvantages of the various models reported in the literature to simulate HDT reactors are described below.

Pseudohomogeneous models

Based on kinetics

Advantages

- Currently used for testing and evaluating a catalyst in bench-scale reactors.
- When the reaction being studied is first-order or pseudo-first-order, a residence-time distribution curve can be used to calculate the intrinsic reaction rate constants, which allows determining contacting efficiency.
- Easy and fast application to systems where the rate of reaction is limited only by intrinsic reaction kinetics.

Disadvantages

- Based on a priori assumption of appropriate kinetics and weak underlying theory.
- Do not account for the influence of hydrodynamics and related phenomena (i.e., mass transfer) on conversion.

- The use of these models for comparison of different catalysts may suffer from uncertainty, because of superimposition of kinetic and hydrodynamic effects.
- Kinetic models have very restricted applications for deep desulfurization calculations.
- Sometimes k_{in}'s obtained with kinetic models do not really come from intrinsic kinetics because these are frequently masked by transport limitations, and for this reason they are also known as "effective" rate constants ($k_{e,j}$).

Based on hydrodynamics

Advantages

- A flow regime in which the reaction occurs is taken into account.
- A contact effectiveness factor and liquid holdup are incorporated.
- Results from experimental reactors and industrial plants can be correlated.
- Predictions of these models are superior at low levels of conversion.

Disadvantages

- Group various phenomena in a few parameters.
- No theoretical justification for assuming that reaction rate is proportional to total liquid holdup has been found.
- Representing deep conversion as in the case of HDS reaction is not possible using hydrodynamic models.
- It may not be able to explain the performance of TBRs satisfactorily because of the existence of stagnant zones, particularly in the case of a porous catalyst.
- Since the catalyst effectiveness factor and incomplete wetting are strongly coupled phenomena, they cannot necessarily be expressed in the regime of interest by a single product of catalyst effectiveness for a completely wetted pellet and a fraction of the external area wetted as suggested for this type of model.
- Physical reality is very far from these empirical descriptions, which cannot account for characteristic phenomena such as channeling or hot-spot formation; the latter is formed mainly by poor solid–liquid contact.
- Parameters could vary substantially depending on the flow-rate region being considered.
- Assumption of reaction order a priori.

Continuous model

Advantages

- A unique phase is considered, reducing the number of variables, making it easier to reach a solution for the equation balance.
- Useful for simulating industrial-scale equipment when the volume of the reactor is too large compared to that of an individual pellet.
- Rapid means for obtaining an estimate of the reactor size necessary to achieve any given conversion and examining the influence of several design variables on the reactor's behavior.
- Especially suitable for steady-state analysis.

Disadvantages

- The model neglects interfacial resistance; thus it would attribute concentrations different from that actually contacting the catalyst.
- Inhibitory effects provoked by gaseous components such as ammonia and hydrogen sulfide are ignored.
- Hydrogen pressure variation is not considered.
- Not able to predict effects of volatized fractions.
- One-dimensional homogeneous models do not provide information about the possibility of achieving an excessive temperature at the center of the reactor that can be markedly different from the mean temperature at the same longitudinal position.

Empirical correlations

Advantages

- Easy to predict the quality properties of HDT products from data of feed characteristics and process conditions.
- Polynomial expressions resulting from statistic regression analysis can be employed to perform optimization studies.

Disadvantages

- Correlations are valid only within the range of experimental results that were used for its development; extrapolation can be performed only within a very narrow range beyond the extreme values of the experimental framework.
- There is not generality of the equations, even for similar reacting systems (reactor dimensions, size, shape and type of catalyst, feedstock, operating conditions, etc.).

Heterogeneous models
Continuous model

Advantages

- Incorporating mass and heat transfer in the three phases, phenomenological effects can be modeled separately and added to the reactor model to improve predictions.
- Adequate for purposes of scale-up and scale-down reactors.
- Suitable for kinetic studies and for heat and mass parameter estimation.
- Catalyst incomplete wetting and liquid wall flow are taken into account.
- Intraparticle diffusion can be considered.

Disadvantages

- Validation data using different reactors and reaction systems are scarce.
- There are too many unknown and uncertain parameters involved whose correlations reported in the literature, especially to evaluate mass and energy transfer coefficients, are developed under low pressures and temperatures, which differ from typical conditions employed in the HDT process.
- Data for gas–liquid mass transfer for small trickle-bed reactors (low Reynolds number) do not exist in the literature.
- The main obstacle when using models that account for evaporation is the difficulty of incorporating it into reactor simulation codes of VLE for the petroleum cut, which include a great number of components.
- Unsuitable representation of radial temperature profiles within the solid phase when a chemical reaction takes place.
- The slow response of the model limits its use as a control tool for online use in industrial practice.

CFD

Advantages

- Can be used to reduce empiricism in scale-up/scale-down and optimization of the TBRs.
- Reduced time of programming and accuracy solutions are provided.
- Detailed information such as local velocities and local hot-spot formations can be obtained since the structure heterogeneity of the packed bed is taken into account.
- It is possible to use CFD in the scale-up of packed-bed reactors in which the flow distribution is significantly affected by complex reactions, since

the nonreactive CFD model can be combined with other reactive models, such as the cell reaction network model.

Disadvantages

- User correlations are employed to describe phenomena related to a particular reaction system, hydrodynamics, and energy and mass transfer, so that empiricism is still present.
- CFD models are normally computationally expensive, especially when a huge number of geometrical details or small-scale spatial variations need to be taken into account.
- Because the general reactive simulation with a CFD model is conducted for the macroscopic reactor level and tends to track all catalytic reactions and hydrodynamic phenomena in a simultaneous approach, it is not easy to identify the origin for any numerical difficulty present when kinetics of multiple reactions are highly coupled and nonlinear and/or when reactions are highly exothermic.

Cross-flow model

Advantages

- Axial dispersion defined in this model is practically independent of liquid loading.
- Both stagnant and free-flowing zones in a TBR are considered, which is considerably realistic.
- Experimental response curves, even those with a strong tail, are well reproduced.

Disadvantage

- The uncertainties associated with determination of third higher moments, which are employed to calculate parameters.

Cell model

Advantages

- Since each cell can be simulated employing an ideal reactor and flash calculations are performed by tools included in a commercial simulator, modeling a TBR by this method can lead to a considerable saving of time.
- An easy way to simulate vaporization in a TBR.

- Chemical composition information of light fractions is retained in the model.
- Suggested for mass transfer analysis since only one steady-state solution is predicted.
- This model type is an option when other models fail to simulate two-dimensional systems with complex kinetics and enormous reaction heat released.
- Due to the sequential approach of the cell models, convergence is faster because it is obtained at one geometric position at a time.

Disadvantages

- Application of this model to heavy petroleum fractions depends on the accuracy of correlations to calculate parameters employed in the simulator.
- Excessive computational effort.
- Fails to reproduce backmixing behavior.

Stage model

Advantage

- Considerable work exists in the literature regarding equilibrium between phases, which is a supposition in stage models.

Disadvantages

- Fails to account properly for the influence that chemical equilibrium has on VLE, and vice versa.
- No shortcut procedures are available for modeling reactors with the equilibrium stage model.

Learning model

Advantages

- Used when a deterministic model cannot describe a system adequately.
- Easily creates scenarios for optimizing purposes.
- Ability to analyze nonlinear processes.
- Successful in process fault detection and diagnosis.
- Noise tolerance.
- Online adaptability for industrial use.

- Ability to adapt and continue learning (continuous update) to improve performance and extend its applicability.
- High degree of robustness or fault tolerance.
- Comparing the traditional computational time to convergence, ANNs are much faster than traditional modeling.
- Capable of generalizing the answers.
- Acceptable results for unknown samples.
- This type of model may make use of heterogeneous models as a complement to accomplish the optimization and online control of commercial HDT units.
- Coupled with deterministic models, it is possible to predict the catalyst deactivation rate in HDT units treating different feeds.

Disadvantages

- Requires an enormous number of data obtained from experiments or coming from a deterministic simulator for training.
- The success of an ANN model depends on the quality of process and laboratory data used.
- Cannot be used to extrapolate operating conditions out of the data framework employed for training.
- No reports of using ANNs for scaling HDT purposes.
- Since ANNs are empirical models, all influences of the HDT complex system cannot be included.

2.4.3 Generalized Reactor Model

When developing a generalized reactor model, nothing should be neglected a priori, but all the resistances and others terms must be included in mass and heat balance equations (Wärnå and Salmi, 1996). However, such a model can be very complex and difficult to solve, even supposing that all the parameters involved are available; thus some assumptions are still needed. The assumptions, of course, have to be well supported and preferably validated with experimental data. The mass and heat balance equations in the case of the generalized reactor model for hydroprocessing are detailed in Tables 2.11 and 2.12, respectively, which have been developed with the following assumptions: liquid and gas properties (superficial velocities, mass and heat dispersion coefficients, specific heats, holdups, and densities), catalyst properties (porosity, size, activity, effectiveness, etc.), wetting efficiency, and bed void fraction are constant along the entire catalytic bed. Inside the catalyst particle, mass and heat effective diffusivity coefficients may also be assumed constant. Under these considerations, those parameters can be put out of the partial derivatives with respect to axial and radial spatial coordinates. For the case of

TABLE 2.11. Generalized Mass Balance Equations (M)

Mass Balance \ Term	Accumulation (1)		Convective (2)		Axial Dispersion (3)		Radial Dispersion (4)
(A) Gas phase ($i=H_2$, H_2S, NH_3, LHC)	$\frac{\varepsilon_G}{RT_G Z}\frac{\partial p_i^G}{\partial t} =$	$\pm$	$\frac{u_G}{RT_G Z}\frac{\partial p_i^G}{\partial z}$	+	$\frac{\varepsilon_G D_a^G}{RT_G Z}\frac{\partial^2 p_i^G}{\partial z^2}$	+	$\frac{\varepsilon_G D_r^G}{RT_G Z}\left(\frac{\partial^2 p_i^G}{\partial r^2}+\frac{1}{r}\frac{\partial p_i^G}{\partial r}\right)$
(B) Liquid phase ($i=H_2$, H_2S, NH_3, LHC)	$(1-f_{St})\varepsilon_L\frac{\partial C_i^L}{\partial t} =$	−	$u_L\frac{\partial C_i^L}{\partial z}$	+	$\varepsilon_L D_a^L\frac{\partial^2 C_i^L}{\partial z^2}$	+	$\varepsilon_L D_r^L\left(\frac{\partial^2 C_i^L}{\partial r^2}+\frac{1}{r}\frac{\partial C_i^L}{\partial r}\right)$
(C) Liquid phase ($i=H_2$, S, N, A, O, GO, WN, Ni, V)	$(1-f_{St})\varepsilon_L\frac{\partial C_i^L}{\partial t} =$	−	$u_L\frac{\partial C_i^L}{\partial z}$	+	$\varepsilon_L D_a^L\frac{\partial^2 C_i^L}{\partial z^2}$	+	$\varepsilon_L D_r^L\left(\frac{\partial^2 C_i^L}{\partial r^2}+\frac{1}{r}\frac{\partial C_i^L}{\partial r}\right)$
(D) Stagnant liquid ($i=H_2$, H_2S, NH_3, LHC, S, N, A, O, GO, WN, Ni, V)	$f_{St}\varepsilon_L\frac{\partial C_{Sti}^L}{\partial t} =$						

Mass Balance \ Term	Accumulation (1)		Intraparticle Diffusion (9)
(E) Solid phase, wet surface ($i=H_2$, H_2S, NH_3, LHC, S, N, A, O, GO, WN, Ni, V)	$\varepsilon_{pL}(1-\varepsilon_B)\frac{\partial C_{SLi}^S}{\partial t} =$		
(F) Solid phase, dry surface ($i=H_2$, H_2S, NH_3, LHC, S, N, A, O, GO, WN, Ni, V)	$\varepsilon_{pG}(1-\varepsilon_B)\frac{\partial C_{SGi}^S}{\partial t} =$		
(G) Solid phase, wet inner ($i=H_2$, H_2S, NH_3, LHC, S, N, A, O, GO, WN, Ni, V)	$\varepsilon_{pL}\frac{\partial C_{Li}^S}{\partial t} =$	+	$\frac{D_{ei}^L}{\xi^2}\frac{\partial}{\partial \xi}\left(\xi^2\frac{\partial C_{Li}^S}{\partial \xi}\right)$
(H) Solid phase, dry inner ($i=H_2$, H_2S, NH_3, LHC, S, N, A, O, GO, WN, Ni, V)	$\varepsilon_{pG}\frac{\partial C_{Gi}^S}{\partial t} =$	+	$\frac{D_{ei}^G}{\xi^2}\frac{\partial}{\partial \xi}\left(\xi^2\frac{\partial C_{Gi}^S}{\partial \xi}\right)$

nonisothermal reactor models, physicochemical and thermodynamic properties must be evaluated at the temperature of each discretized point in the mathematical model.

Although some correlations have been reported in the literature to predict variation (radial and axial) of bed porosity (Cotterman and Prausnitz, 1985; Stefanidis et al., 2005), it is difficult to incorporate them in conventional continuum models. Hence, the effects of porosity distribution and subsequent local velocity variations are also neglected. This assumption could lead to inaccuracies in the predictions of any reactor model, as shown by Gunjal and Ranade (2007) using a CFD model. These authors reported that the difference in HDS conversion, considering uniform porosity distribution in the bed, was

G-L Transfer (5)	G-S Transfer (6)	L-S Transfer (7)	Flowing-Stagnant Liquid Transfer (8)
$-\; K_{Li}a_L\left(\frac{p_i^G}{H_i}-C_i^L\right)$	$-\; (1-f_w)k_i^{GS}a_S\left(\frac{p_i^G}{RT_GZ}-C_{SGi}^S\right)$		
$+\; K_{Li}a_L\left(\frac{p_i^G}{H_i}-C_i^L\right)$		$-\; f_wk_i^Sa_S(C_i^L-C_{SLi}^S)$	$-\; k_i^m a_S(C_i^L-C_{Sti}^L)$
		$-\; f_wk_i^Sa_S(C_i^L-C_{SLi}^S)$	$-\; k_i^m a_S(C_i^L-C_{Sti}^L)$
			$+\; k_i^m a_S(C_i^L-C_{Sti}^L)-k_i'^S(C_{Sti}^L-C_{SLi}^S)$

	G-S Transfer (6)	L-S Transfer (7)	Generation (10)
		$+\; f_wk_i^Sa_S(C_i^L-C_{SLi}^S)+k_i'^S(C_{Sti}^L-C_{SLi}^S)$	$+\; \rho_B\zeta\sum_{j=1}^{N_{RL}}\upsilon_{ij}^L\eta_j^L r_j'^L(C_{SLi}^S,T_S^S)$
	$+\; (1-f_w)k_i^{GS}a_S\left(\frac{p_i^G}{RT_GZ}-C_{SGi}^S\right)$		$+\; \rho_B\zeta\sum_{j=1}^{N_{RG}}\upsilon_{ij}^G\eta_j^G r_j'^G(C_{SGi}^S,T_S^S)$
			$+\; \rho_S\sum_{j=1}^{N_{RL}}\upsilon_{ij}^L r_j'^L(C_{Li}^S,T_S)$
			$+\; \rho_S\sum_{j=1}^{N_{RG}}\upsilon_{ij}^G r_j'^G(C_{Gi}^S,T_S)$

about 15% higher than in the case where nonuniform bed porosity was considered. The results demonstrated the necessity of coupling the continuous models with rigorous hydrodynamic models to take into account the bed porosity, which strongly influences the hydrodynamic performance of PBRs.

Generalized Mass Balance Equations (M) In the mass balance equations shown in Table 2.11, each phase is assumed to be a continuum and is represented by an Eulerian–Eulerian framework model (Duduković et al., 1999). Figure 2.15 shows graphically the interphase (gas–interface, interface–liquid, and liquid–solid) and intraphase concentration profiles in a TBR, which are represented mathematically by the terms of the generalized mass balance

TABLE 2.12. Generalized Heat Balance Equations (H)

Head Balance \ Term		Accumulation (1)	Convective (2)	Axial Dispersion (3)	Radial Dispersion (4)	
(A)	Gas phase	$\varepsilon_G \rho_G Cp_G \dfrac{\partial T_G}{\partial t} =$	$\pm\ u_G \rho_G Cp_G \dfrac{\partial T_G}{\partial z}$	$+\ \varepsilon_G \lambda_a^G \dfrac{\partial^2 T_G}{\partial z^2}$	$+\ \varepsilon_G \lambda_r^G \left(\dfrac{\partial^2 T_G}{\partial r^2} + \dfrac{1}{r}\dfrac{\partial T_G}{\partial r} \right)$	$-$
(B)	Liquid phase	$\varepsilon_L \rho_L Cp_L \dfrac{\partial T_L}{\partial t} =$	$-\ u_L \rho_L Cp_L \dfrac{\partial T_L}{\partial z}$	$+\ \varepsilon_L \lambda_a^L \dfrac{\partial^2 T_L}{\partial z^2}$	$+\ \varepsilon_L \lambda_r^L \left(\dfrac{\partial^2 T_L}{\partial r^2} + \dfrac{1}{r}\dfrac{\partial T_L}{\partial r} \right)$	$+$

Heat Balance \ Term		Accumulation (1)		Axial Dispersion (3)	Radial Dispersion (4)	
(C)	Solid phase, isothermal	$\varepsilon_S \rho_S Cp_S \dfrac{\partial T_S^S}{\partial t} =$		$+\ \varepsilon_S \lambda_a^S \dfrac{\partial^2 T_S^S}{\partial z^2}$	$+\ \varepsilon_S \lambda_r^S \left(\dfrac{\partial^2 T_S^S}{\partial r^2} + \dfrac{1}{r}\dfrac{\partial T_S^S}{\partial r} \right)$	$+$
(D)	Thermowell	$\rho_W Cp_W \dfrac{\partial T_{TW}}{\partial t} =$		$+\ \lambda_{TW} \dfrac{\partial^2 T_{TW}}{\partial z^2}$	$+\ \lambda_{TW} \left(\dfrac{\partial^2 T_{TW}}{\partial r^2} + \dfrac{1}{r}\dfrac{\partial T_{TW}}{\partial r} \right)$	

Heat Balance \ Term		Accumulation (1)		Intraparticle Transfer (12)
(E)	Solid phase, nonisothermal	$\rho_S Cp_S \dfrac{\partial T_S}{\partial t} =$	$+$	$\lambda_e^S \left(\dfrac{\partial^2 T_S}{\partial \xi^2} + \dfrac{2}{\xi}\dfrac{\partial T_S}{\partial \xi} \right)$

equations. In the following sections we describe all the terms of these equations and the assumptions under which they were derived.

Gas Phase (MA) The mass balance equation in the gas phase for nonvolatile components is ignored, so that those compounds with negligible vapor pressure (i.e., for i = S, N, A, O, GO, WN, Ni, and V) are excluded from the equation MA, where "M" represents the mass balance equation given in Table 2.11 and "A" is row A in the same table (Sater and Levenspiel, 1966; Wärnå and Salmi, 1996; Duduković et al., 1999). The accumulation term M1 (Table 2.11, column 1) gives the dynamic (non-steady-state or transient) behavior of a TBR reactor. This term is of great interest for the modeling and simulation of TBRs for HDT of oil fractions in dynamic conditions, as reported elsewhere (Ho and Nguyen, 2006).

f-L Transfer (5)		Conductive (6)		f-S Transfer (7)		f-W Transfer (8)
$h_{GI}a_L(T_G - T_I)$	+	$\sum_{i=1}^{N_{CG}}\left\{K_{Li}a_L\left(\frac{p_i^G}{H_i}\right)-C_i^L \cdot [Cp_{Gi}(T_I - T_G) - \Delta H_{vi}]\right\}$	−	$(1-f_w)h_{GS}a_S(T_G - T_S^S)$	−	$(1-f_w)h_{GW}\frac{A_W}{V}(T_G - T_W)$
$h_{IL}a_L(T_I - T_L)$	+	$\sum_{i=1}^{N_{CG}}\left\{K_{Li}a_L\left(\frac{p_i^G}{H_i}\right)-C_i^L \cdot [Cp_{Li}(T_I - T_G) + \Delta H_{vi}]\right\}$	−	$f_w h_{LS}a_S(T_L - T_S^S)$	−	$f_w h_{LW}\frac{A_W}{V}(T_L - T_W)$

G-S Transfer (9)		L-S Transfer (10)		Generation (11)
$(1-f_w)h_{GS}a_S(T_G - T_S^S)$	+	$f_w h_{LS}a_S(T_L - T_S^S)$	+	$\rho_B\zeta\left[\sum_{j=1}^{N_{RL}}(-\Delta H_{Rj}^L)\eta_j^L r_j'^L(C_{SLi}^S, T_S^S) + \sum_{j=1}^{N_{RG}}(-\Delta H_{Rj}^G)\eta_j^G r_j'^G(C_{SGi}^S, T_S^S)\right]$

	Generation (11)
	$\rho_S\left[\sum_{j=1}^{N_{RL}}(-\Delta H_{Rj}^L)r_j'^L(C_{Li}^S, T_S) + \sum_{j=1}^{N_{RG}}(-\Delta H_{Rj}^G)r_j'^G(C_{Gi}^S, T_S)\right]$

According to Lopez and Dassori (2001), for the HDS process a reliable representation of reactor dynamics must be accompanied by a deep study of the reactivity of the various sulfur compounds present in the feed, the kinetic mechanism, and the overall effect of upgraded products over the catalyst performance. Once these tasks have been achieved and the kinetic models of the reacting system have been established, it is possible to propose a reactor model.

The term M2 represents the convective flow, which is considered to be of plug-flow type. Use of this term implies that concentration and temperature gradients occur only in the axial direction. The signs "–" and "+" in the term MA2 refer to co-current and countercurrent flow, respectively. Figure 2.16 shows the axial concentration profiles of reactants and products in the co-current and countercurrent operation mode of TBRs.

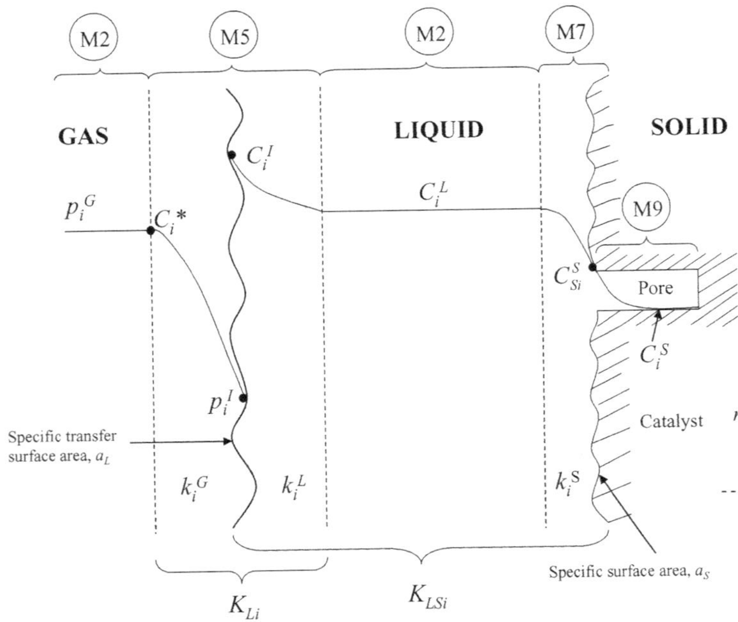

Partial pressure and concentration profiles (for $i = H_2$) in a TBR

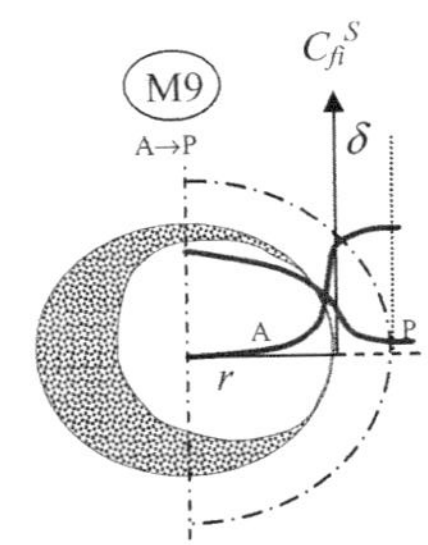

Concentration profile in a spherical catalyst pellet with mass transfer resistance

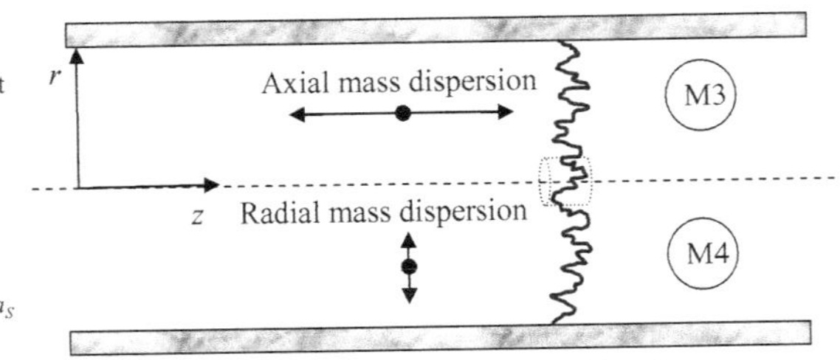

Radial concentration profile resulting from axial and radial mass dispersion phenomena

Figure 2.15. Graphical representation of some terms of the generalized mass balance equation.

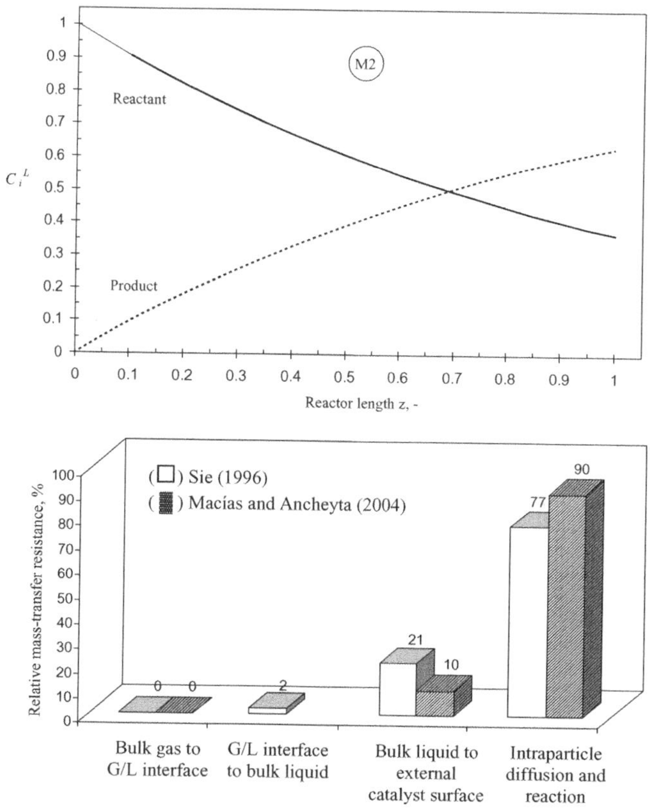

Figure 2.16. Axial average liquid molar concentration profiles and relative mass transfer resistances found in a typical oil fraction HDT process.

Column M3 deals with effective transport in the axial direction. This term may be neglected during modeling of isothermal TBRs according to the criteria reported by Mears (1971) and Gierman (1988), and for adiabatic pilot or commercial TBRs as reported by Shah and Paraskos (1975). Minimization of fluid flow dispersion in commercial reactors may also be ensured because of the high gas and liquid velocities employed. Mears (1971) showed that the axial dispersion effect is much more important in TBRs than in single vapor-phase reactors; therefore, for the HDS of naphtha, for example, this term may be neglected. According to Salmi et al. (2000), the term MA3 can be neglected since the gas phase is closer to the plug-flow pattern, which implies that its axial dispersion effects can be discarded ($D_a^G = 0$).

The term accounting for effective mass radial (or transversal) dispersion is represented in column M4. When this term is used together with the M2 term, and in some cases with M3, the model is called 2D; if the M4 term is neglected, the model is termed 1D. It has been reported that the term M4 can be neglected when the d_R/d_{pe} ratio is higher than 25 since radial porosity variation within the reactor is negligible. Figure 2.15 shows a basic scheme of axial and radial mass dispersion phenomena and the resulting concentration profile inside TBR. The radial mass dispersion term is especially interesting for liquid distributor studies, since it defines the necessary drip point density for adequate liquid distribution and describes how a badly designed distributor would affect the bed. For commercial TBRs if maldistribution of the liquid is not present, the trickle flow regime is described satisfactorily by plug flow (perfect radial mixing) for both phases (Froment, 2004). In summary, when axial and radial mass dispersions are neglected along the TBR, one-dimensional plug flow has been recommended for use in both the gas and liquid phases (Jiménez et al., 2007b).

Taking into account the assumptions described above, when a commercial reactor is modeled, terms MA3 and MA4 of equation MA are normally neglected because of high L_B/d_{pe} and d_R/d_{pe} ratios. Since the gas temperature (T_G) is also a variable dependent of z coordinate, the resulting MA equation should be written

$$\frac{\varepsilon_G}{RZ}\frac{\partial}{\partial t}\left(\frac{p_i^G}{T_G}\right) = \pm\frac{u_G}{RZ}\frac{\partial}{\partial z}\left(\frac{p_i^G}{T_G}\right) - K_{Li}a_L\left(\frac{p_i^G}{H_i} - C_i^L\right) - (1-f_w)k_i^{GS}a_S\left(\frac{p_i^G}{RT_GZ} - C_{SGi}^S\right) \tag{2.119}$$

Developing the partial derivatives for both sides of Eq. (2.119), the following expression is obtained:

$$\frac{\varepsilon_G}{RT_GZ}\frac{\partial p_i^G}{\partial t} - \frac{\varepsilon_G p_i^G}{RT_G^2Z}\frac{\partial T_G}{\partial t} = \pm\frac{u_G}{RZ}\left(\frac{1}{T_G}\frac{\partial p_i^G}{\partial z} - \frac{p_i^G}{T_G^2}\frac{\partial T_G}{\partial z}\right) - K_{Li}a_L\left(\frac{p_i^G}{H_i} - C_i^L\right) - (1-f_w)k_i^{GS}a_S\left(\frac{p_i^G}{RT_GZ} - C_{SGi}^S\right) \tag{2.120}$$

However, this expression is very difficult to solve when it is linked with the complete set of partial differential equations (PDEs). Hence, this expression is only recommendable for steady-state simulations as it was used by Murali et al. (2007).

Column M5 represents the mass transfer from gas phase to liquid phase. The mass transfer resistance is described by the two-film theory, in which the interface is assumed to be in thermodynamic equilibrium and no additional resistance to mass transfer is present. When the overall gas–liquid mass transfer coefficient based on the liquid phase is used (K_{Li}), a fictitious liquid-phase concentration is employed, which for gaseous compounds (H_2, H_2S, NH_3, and light hydrocarbons) is the concentration that would be in equilibrium with the corresponding bulk partial pressure and is represented by the following relationship:

$$C_i^* = \frac{p_i^G}{H_i} \tag{2.121}$$

where H_i (Henry's constant) is an equilibrium relation that in some way represents the solubility of the gaseous compound i in the oil fraction. A higher mass transfer rate for term M5 is only attributed to the enlarged transfer area between the gas and the liquid (a_L) and not to a higher degree of turbulence (Hofmann, 1977).

The gas–solid mass transfer term (column M6) may be neglected when the wetting efficiency is complete ($f_w = 1$) and the molar flow rate of component i in the gas phase can only change by transfer to or from the liquid phase, since no contact exists between gas and solid (Froment et al., 1994).

Gaseous Compounds in the Liquid Phase (MB) The dynamic liquid-phase holdup in the liquid mass balance equation (row MB in Table 2.11) varies strictly along the reactor because of the partial volatilization of lighter oil fractions. Avraam and Vasalos (2003) have reported that there is a decrease in the liquid holdup along the bed reactor due to volatilization of the lighter oil feedstock, whereas the dynamic gas holdup increases.

In the modeling of small TBRs when the catalyst bed is diluted with inert material of smaller size, in order to prevent axial dispersion, the term in column M3 may be neglected if criteria of Mears (1971) are fulfilled and/or that of Gierman (1988) using the smaller inert particles' average diameter as the design parameter. Due to the dilution of the catalyst bed, it is also reasonable to assume no concentration gradients along the reactor radius, leading to the fact that the radial dispersion term in column M4 may also be neglected (Botchwey et al., 2006).

The term in column M7 corresponds to the mass transfer at the stationary liquid film of the liquid–solid interface. It has been observed that the lower superficial mass velocities normally found in smaller reactors result in incomplete catalyst wetting ($f_w < 1$), which can result in less contaminant removal than that expected from commercial reactors (Bhaskar et al., 2004). Because of well-designed distributors and high superficial mass velocities, in commercial reactors complete wetting of catalyst particles is normally assumed. Therefore, mass transfer between the liquid and solid phases is a function of the liquid flow in contact with the external area of the catalytic particles, which originates in a gradient between the two phases ($C_i^L - C_{SLi}^S$). An increase in this gradient means that the reactant is not being transferred totally to the external area of the particle, which apart from the liquid flow, depends on the shape and size of the particle. Small particles minimize these external concentration gradients as a result of the larger solid surface area and the higher mass transfer coefficients (Hofmann, 1977; Macías and Ancheyta, 2004).

Nonvolatile Compounds in the Liquid Phase (MC) For the mass balance equation of nonvolatile compounds (row MC), partial vaporization of feedstock is assumed to be negligible under HDT conditions. Under these conditions, organic sulfur, nitrogen, aromatics, olefins, gas oil, wild naphtha, nickel, and vanadium components can then be considered as nonvolatile (Bhaskar et al., 2004).

Stagnant Liquid Phase (MD) The flowing–stagnant liquid mass balance equation (row MD) assumes that mass can be transferred between the stagnant and flowing portions of liquid in the catalytic bed. As the fraction of stagnant fluid approaches zero ($f_{st} \rightarrow 0$), the TBR flow pattern becomes plug flow (Schwartz and Roberts, 1973). The term representing the mass transfer between flowing and stagnant liquid zones is given in column M8. It has been observed that if liquid feed to a TBR is interrupted and the liquid present in the packed bed is allowed to flow off freely, not all of the external liquid holdup (liquid outside the catalyst's pores) drains off freely but a definite fraction of the external liquid remains stagnant in the dead zones. Most of these stagnant liquid zones are present in the contact points between particles, mainly at the top of the reactor (Hofmann, 1977). When the effects of stagnant zones are neglected ($f_{st} \rightarrow 0$) the following terms must be eliminated from the generalized model: MB8, MC8, row MD, and the second term in ME7.

Surface of the Solid Phase (ME-MF) The mass balance on the external surface of solid phase considering partial wetting is given in row ME, whereas for the external surface of a dry catalyst, the mass balance is given in row MF. In the equations given in rows ME and MF, internal mass gradients in the solid phase are evaluated by a catalyst effectiveness factor in the respective generation terms (ME10 and MF10). These gradients inside particles are the product of an effective diffusion, which depends mainly on catalyst porosity and the size of the molecules being diffused through the pores. The effectiveness factor (η_j^f) in a commercial HDS catalyst ($\sim\frac{1}{20}$ to $\frac{1}{8}$ in. size) has been reported to be in the range 0.4 to 0.6. These low values of η_j^f give $\Phi_j^f > 1$, which implies that the HDS reaction may be considered to be within strong internal diffusional limitations in commercial applications. Because internal diffusion, also depends on external diffusion, which in turn depends on flow rates, to obtain maximum catalyst effectiveness the reactor should operate with no interphase liquid–solid mass transfer limitations. As in the case of external gradients (gas–liquid and liquid–solid), a reduction in particle size yields an increase in the particle effectiveness factor because the path lengths at the interior of the particle are reduced. If the catalyst is crushed, the particles are assumed isoconcentrational and the effectiveness factor is considered to be unity ($\eta_j^f = 1$). However, loading commercial reactors with smaller catalysts will increase ΔP; thus, more attention must be given when designing catalyst size (Marroquín de la Rosa et al., 2002; Macías and Ancheyta, 2004).

The results of dynamic simulations when using ME and MF equations may be affected by inaccuracies in an estimation of some model parameters, such as the bed void fraction, since the reader could chose a correlation to calculate the void fraction bed for an undiluted bed when the real system under study was diluted. This parameter is needed to estimate the specific solid-phase fraction ($\varepsilon_S = 1 - \varepsilon_B$) (Chen et al., 2001).

Interior of the Solid Phase (MG-MH) The major fraction of mass transfer resistances exist inside the catalyst particle: mainly, when large commercially shaped catalysts are used, as shown in Figure 2.16, where the relative percentage of the various mass transfer resistances typically found in the HDT of oil fractions is reported (Sie, 1991; Macías and Ancheyta, 2004). Thus, to account for internal mass transfer limitations, mass balance equations MG and MH are used since they can affect the kinetics of the process (Hofmann, 1977; Botchwey et al., 2006). Figure 2.15 shows the concentration profile inside a catalyst particle with mass transfer resistance.

The term for intraparticle transport by effective diffusion (column M9) accounts for the concentration gradients inside the catalyst particles because chemical reactions are assumed to take place at the inner surface of the solid catalyst (i.e., inside the pores of the catalyst). Those pores are considered to have uniform properties and to be filled completely with liquid ($\eta_i = 1$) or gas ($\eta_i = 0$) (Froment et al., 1994; Jiménez et al., 2005; Iliuta et al., 2006).

For the equations inside solid catalyst particles, a mass balance over an infinitesimal volume element in a porous catalyst particle was assumed, which can be written as

$$\varepsilon_{pf}\frac{\partial C_{fi}^{S}}{\partial t} = -\xi^{-S}\frac{\partial N_i^f \xi^S}{\partial \xi} + \rho_S \sum_{j=1}^{N_{Rf}} \upsilon_{ij}^f r_j'^f \left(C_{fi}^S, T_S\right) \tag{2.122}$$

where N_i^f is the molar flux of component i and s in ξ^s is the shape factor, being ξ the radial spatial coordinate (for slab $s = 0$, for infinitely long cylinder $s = 1$, and for sphere $s = 2$). Nonideal geometries can be treated with noninteger values. The further development of Eq. (2.122) depends on which diffusion model is used. Strictly speaking, multicomponent diffusion should be described with Stefan–Maxwell equations, according to which all of the fluxes are related to all concentration gradients (Salmi et al., 2000):

$$N^f = -F\frac{kC_f^s}{d\xi} \tag{2.123}$$

where F is a coefficient matrix consisting of binary diffusion coefficients (Fott and Schneider, 1984). However, a comparison between different diffusion models has shown that a simpler approach based on Fick's law is sufficient, provided that the diffusion coefficients are described in an approximate manner (e.g., by using the Tyn–Calus correlation for a liquid-phase molecular diffusion coefficient. Thus, by applying Fick's law and assuming the catalyst to be a spherical particle, Eq. (2.122) can be rewritten as those reported in rows MG and MH of Table 2.11 (Salmi et al., 2000).

The generation terms MG10 and MH10 represent the appearance and disappearance of products and reactants by catalytic reactions that take place only at the active sites inside pores of catalyst particles. These terms give the nonlinear behavior in HDT reactors, because of the nonlinear interactions between mass, thermal, and kinetic processes present in the reacting system (Jiménez et al., 2005).

Generalized Heat Balance Equations (H) Commercial HDT reactors operate under nonisothermal adiabatic conditions, and since reactions are mostly exothermic, average reactor temperature would always increase along the catalyst bed (Figure 2.17) (Lopez and Dassori, 2001; Bhaskar et al., 2004). On the other hand, experiments for catalyst screening and process studies are mostly conducted in microreactors on the bench- and pilot-plant scales. These systems commonly operate under the same conditions as those reported for commercial units but keeping the reaction temperature more or less constant (operation in isothermal mode), and hence the heat balance can be omitted for small reactor modeling. However, since commercial HDT reactors do not operate isothermally, experimental information generated from small reactors does not represent the commercial operation exactly. Therefore, to predict the real behavior of commercial reactors using experimental data from small reactors, it is necessary to add the energy balance in industrial HDT reactor modeling (Rodríguez and Ancheyta, 2004).

In the case of HDT of heavy oils, reaction temperature is the most important operating parameter. It is regularly used to adjust the desired degree of HDS or to compensate for catalyst deactivation, and consequently, for catalyst lifetime. For such processes it is of vital importance to properly determine the optimum temperature at the entrance of the reactor so that the desired conversion is achieved for either SOR or EOR conditions. This optimum temperature, known as the *setpoint temperature*, is the primary variable to be used in terms on process control (Al-Adwani et al., 2005).

For modeling purposes it is usually sufficient to consider only the liquid and solid phases (rows HB and HC or HE in Table 2.12), since the heat capacity of the gas phase is much lower than those of the solid and liquid phases (Salmi et al., 2000).

Gas Phase (HA) Table 2.12 reports all the terms that need to be included in an energy balance equation. The signs "–" and "+" in term HA2 refer to cocurrent and countercurrent flow, respectively. Commercial HDT reactors are normally considered to operate adiabatically because energy losses from the reactor to its surroundings are usually negligible compared with the energy generated by the reaction (Shah and Paraskos, 1975; Froment et al., 1994; Vanrysselberghe and Froment, 2002; Froment, 2004). Therefore, terms in column H4 can be neglected because a commercial reactor is isothermal only in the radial direction ($\lambda_r^G = 0$), and terms in column H8 must be neglected because there is no heat transfer from the fluid phase to the reactor wall (Jiménez et al., 2005; Kam et al., 2005).

In nonadiabatic operation, the terms in column H3 (except HD3 term) can be neglected due to the insignificance of this type of compared with radial thermal conductivity (Botchwey et al., 2006). The basic scheme of radial temperature profile for a nonadiabatic TBR is shown in Figure 2.17.

The terms in column H5 are the fluid phase–interface convective energy transfer, where the driving force is the temperature difference between the

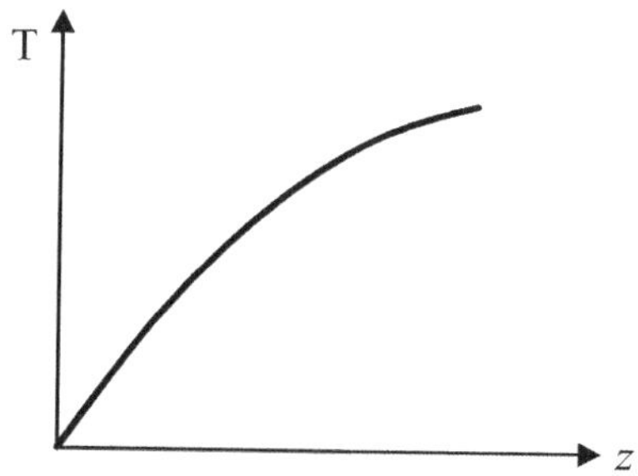

Axial reactor temperature profile
given by the convective term (H2)

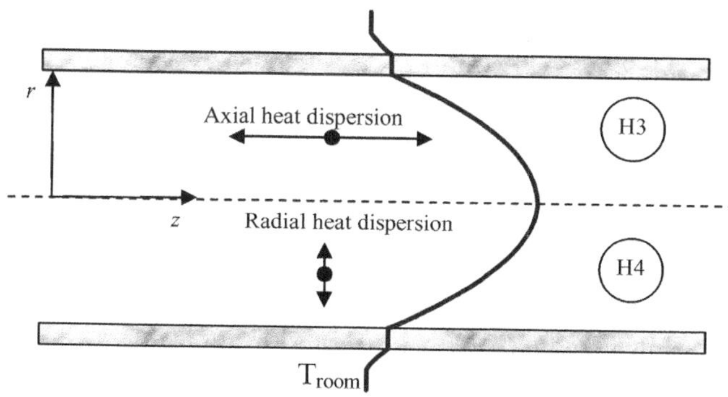

Radial temperature profile in a
nonadiabatic TBR

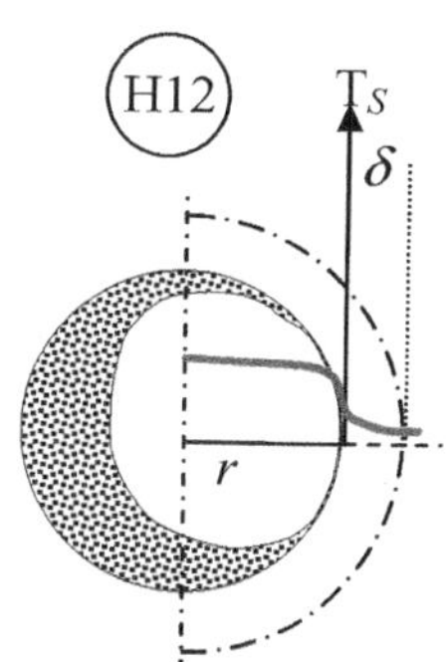

Temperature profile in a spherical catalyst
pellet with heat transfer resistance.

Figure 2.17. Graphical representation of some terms of the generalized heat balance equation.

bulk gas phase and the interface temperature (T_I) in the case of the HA5 term and the temperature difference between the bulk liquid phase and the interface temperature in the HB5 term (Froment et al., 1994; Marroquín de la Rosa, 2002; Vanrysselberghe and Froment, 2002).

The term HA6 corresponds to the conductive heat flux in the gas-film side at the gas–liquid interface, due to the transport of enthalpy by the interfacial mass transfer. The driving force for the conductive heat flux is also the temperature difference between the gas bulk phase and the interface temperature (Froment et al., 1994; Marroquín de la Rosa, 2002; Vanrysselberghe and Froment, 2002). This term also takes into account the flux of heat by vaporization or condensation between the gas and liquid phases. As the partial vaporization of oil fractions is sometimes assumed to be negligible under typical HDT process conditions, the mass transport of nonvolatile components on the gas–liquid interface is neglected as well (Bhaskar et al., 2004).

The terms in column H7 are the convective heat flux from gas to solid external surface (HA7 term) and the convective heat transfer from liquid to solid external surface (HB7 term) (Froment et al., 1994; Vanrysselberghe and Froment, 2002).

Liquid Phase (HB) As with the gas phase, if the reactor is considered as adiabatic, terms HB4 and HB8 must be neglected. The heat of vaporization or condensation is also accounted for in the liquid-phase heat balance equation by means of the HB6 term. The latent heat (ΔH_{vi}) represents the heat consumed by vaporization only for reaction products, where the negative sign indicating heat removal is given by the concentration gradient, since $C_i^L > C_i^*$ for $i = H_2S, NH_3$, and light hydrocarbons (LHCs). On the other hand, the latent heat represents the heat gained by condensation only for H_2 (Vanrysselberghe and Froment, 2002).

The energy balance given by equations HA, HB, and HC in Table 2.12 considers the heat generation on the solid surface and its transfer to the liquid phase, and finally from the liquid phase to the gas phase by convection and mass transfer. This implies that $T_S^S > T_L > T_I > T_G$.

Isothermal Solid Phase (HC) The catalyst particles may be assumed to be isothermal due to the usually low concentration of the oil fraction in the inlet gas–liquid mixture (Sertić-Bionda et al., 2005). If the catalyst particles are assumed to be isothermal (internal energy transfer is carried out without resistance), equation HC is used for modeling of TBR, but utilizing concentrations and temperature at a solid surface (Froment et al., 1994; Vanrysselberghe and Froment, 2002).

In the generation term (HC11) the sign of the reaction heat is negative ($-\Delta H_{Rj}^f$), since the major HDT reactions are exothermic. However, at temperatures higher than ~420°C, thermal cracking reactions are more important, these reactions being endothermic by their chemical nature; then the sign of this reaction heat becomes positive ($+\Delta H_{Rj}^f$) (Chen et al., 2001).

Thermowell (HD) According to Chen et al. (2001), at the outlet of the catalyst bed the temperature difference between the thermowell and the bed could be as high as 60 K. In this case, the temperature measured in the thermowell cannot represent the temperature in the bed. Therefore, if ignored, it might lead to an incorrect interpretation of pilot-plant data. The reason for this difference is that the thermowell, a stainless steel tube, has much higher heat conductivity than that of the catalyst particles of the bed. Hence, heat can be transferred much more easily from the higher-temperature zone to the lower-temperature zone along the thermowell than along the catalyst bed. That back transfer of heat to the inlet of the reactor along the thermowell and reactor wall enhances heat dispersion in the catalyst bed, leading to a flattened axial temperature profile. This effect is quite pronounced in pilot-plant reactors that can sustain highly exothermic reactions because of a the relatively small diameter of the catalyst bed and the use of a thermowell to measure temperatures inside the bed.

Nonisothermal Solid Phase (HE) The film resistances are very important for heat transfer, while the heat transfer inside the solid catalyst is usually fast. Heat transfer in a porous particle may be described by Fourier's law, leading to a PDE with respect to temperature for a nonisothermal catalyst, as shown in row HE of Table 2.12 (Salmi et al., 2000). Figure 2.17 presents the temperature profile developed inside a catalyst particle and across an external boundary layer of thickness δ.

Boundary Conditions The mass and energy balance equations have boundary conditions that relate the surface properties to the bulk properties of the reacting system. For these balance equations, Danckwerts' boundary conditions are generally used (Danckwerts, 1953), especially for models with axial dispersion (Wärnå and Salmi, 1996), but in the case of numerical difficulties in the proximity of the reactor exit, Salmi and Romanainen (1995) have proposed a modified semiempirical boundary condition. This suggests that there are a lot of alternative boundary conditions proposed in the literature for the reactor inlet, for the reactor outlet, and for the transfer of heat between the catalyst and the reactor wall. Therefore, since the generalized model in Tables 2.11 and 2.12 is a system of PDEs, it is necessary to define the initial ($t = 0$) and boundary ($t > 0$) conditions per equation, which are shown in Tables 2.13 and 2.14. Hence, to fix boundary conditions at the reactor inlet and outlet, points $z = 0$ and $z = L_B$ on the axial coordinate are assumed to be the entrance and exit of the reactor, respectively (Lopez and Dassori, 2001).

For some authors at the boundary condition $z = 0$, gas and liquid molar flow is normally assumed to be in physical equilibrium at the reactor inlet temperature and pressure (oil fraction is saturated with the gaseous compound), and the following expression is used (Korsten and Hoffmann, 1996; Pedernera et al., 2003; Kumar and Froment, 2007):

TABLE 2.13. Initial Conditions (t = 0) of Generalized Mass and Heat Balance Equations

		Gas Phase	Liquid Phase		Solid Phase		Stagnant Zone
Condition	Operation Mode	($i=H_2, H_2S, NH_3$, LHC)	($i=H_2, H_2S, NH_3$, LHC)	(i=S, N, A, O, GO, WN, Ni, V)	Surface (i=All Compounds)	Interior (i=All Compounds)	(i=All Compounds)
$z=0$, $0 \le r \le R$	Co-current	$p_i^G=(p_i^G)_0$	—	$C_i^L=(C_i^L)_0$	$C_{Sfi}^S=(C_{Sfi}^S)_0$	$C_{fi}^S=(C_{fi}^S)_0$	
	Countercurrent	$p_i^G=0$	$C_i^L=0$	$C_i^L=(C_i^L)_0$	$C_{Sfi}^S=(C_{Sfi}^S)_0$	$C_{fi}^S=(C_{fi}^S)_0$	
	Co-current/ countercurrent	$T_G=(T_G)_0=T_0$	—	$T_L=(T_L)_0=T_0$	$T_S^S=(T_S^S)_0=T_0$	$T_S=(T_S)_0=T_0$	$C_{Sti}^S=0$
$0 \le z \le L_B$, $0 \le r \le R$	Co-current/ countercurrent	$p_i^G=0$	—	$C_i^L=0$	$C_{Sfi}^S=0$	$C_{fi}^S=0$	$C_{Sti}^S=0$
		$T_G=T_0$	—	$T_L=T_0$	$T_S^S=T_0$	$T_S=T_0$	
$z=L_B$, $0 \le r \le R$	Co-current	$p_i^G=0$	—	$C_i^L=0$	$C_{Sfi}^S=0$	$C_{fi}^S=0$	
	Countercurrent	$p_i^G=(p_i^G)_{LB}$	—	$C_i^L=0$	$C_{Sfi}^S=0$	$C_{fi}^S=0$	
	Co-current/ countercurrent	$T_G=(T_G)_{L_B}=T_0$	—	$T_L=(T_L)_{L_B}=T_0$	$T_S^S=(T_S^S)_{L_B}=T_0$	$T_S=(T_S)_{L_B}=T_0$	$C_{Sti}^S=0$
$0 \le \xi \le d_{pe}/2$, $0 \le z \le L_B$, $0 \le r \le R$	Co-current/ countercurrent	—	—	—	—	$C_{fi}^S=0$ $T_S=T_0$	

$$C_i^f = \left(C_i^f\right)_0 = \frac{\left(p_i^G\right)_0}{H_i} \tag{2.124}$$

Others researchers, in order to simulate co-current and countercurrent operation of pilot TBRs for HDT of oil fractions, have considered that the oil is not saturated with H_2; that is, the initial H_2 concentration in oil is zero $[(C_{H_2}^L)_0 = 0]$ for $t \geq 0$ at $z = 0$ (Yamada and Goto, 2004; Mederos and Ancheyta, 2007).

When a high-purity hydrogen stream without gas recycle is used, such as in the case of some laboratory and bench-scale HDT reactors, or when the gas recycle has been subject to the purification process in commercial units, values of partial pressure (p_i^G) and liquid molar concentrations (C_i^L) of H_2S, NH_3, and LHC at the entrance of the catalytic bed ($z = 0$ and $z = L_B$ for co-current and countercurrent operation, respectively) are equal or very close to zero. For commercial HDT reactors without the high purification of a gas recycle stream, values of partial pressures (p_i^G) and liquid molar concentrations (C_i^L) of H_2S, NH_3, and LHC at the entrance of the catalytic bed ($z = 0$) differ from 0 (Mederos et al., 2006).

The axial and radial dispersion terms of mass and heat result in a second-order differential equation for all phases; consequently, two boundary conditions are necessary. According to Danckwerts (Wehner and Wilhelm, 1956), the Danckwerts' boundary condition at $z = 0$ is

$$-\varepsilon_f D_a^f \left.\frac{\partial\left(C_i^f\right)}{\partial z}\right|_{z=0^+} = u_f\left[\left(C_i^f\right)_0 - \left(C_i^f\right)_{z=0^+}\right] \tag{2.125}$$

$$-\varepsilon_f \lambda_a^f \left.\frac{\partial T_f}{\partial z}\right|_{z=0^+} = u_f \rho_f Cp_f\left[\left(T_f\right)_0 - \left(T_f\right)_{z=0^+}\right] \tag{2.126}$$

which can be simplified to

$$C_i^f = \left(C_i^f\right)_0, \qquad T_f = \left(T_f\right)_0 \tag{2.127}$$

These boundary conditions are true because the axial dispersion of mass and heat is relatively small and the concentration and temperature gradients at the reactor inlet are quite flat (Chen et al., 2001).

The Danckwerts' boundary condition at $z = L_B$ is

$$\frac{\partial C_i^f}{\partial z} = \frac{\partial T_f}{\partial z} = 0 \tag{2.128}$$

The boundary condition

$$z \to \infty, \qquad T_f \to T_W \tag{2.129}$$

TABLE 2.14. Boundary Conditions ($t > 0$) of Generalized Mass and Heat Balance Equations

		Gas Phase (i=H_2, H_2S, NH_3, LHC)		Liquid Phase (i=H_2, H_2S, NH_3, LHC)	
Condition	Operation Mode	Dispersion	No Dispersion	Dispersion	No Dispersion
$z=0,\ 0\le r\le R$	Co-current	—	$p_i^G=(p_i^G)_0$	—	$C_i^L=(C_i^L)_0$
		—	$T_G=(T_G)_0$	—	$T_L=(T_L)_0$
	Countercurrent	$\frac{\partial p_i^G}{\partial z}=\frac{\partial T_G}{\partial z}=0$	—	—	$C_i^L=0$
					$T_L=(T_L)_0$
$z=L_B,\ 0\le r\le R$	Co-current	$\frac{\partial p_i^G}{\partial z}=\frac{\partial T_G}{\partial z}=0$	—	$\frac{\partial C_i^L}{\partial z}=\frac{\partial T_L}{\partial z}=0$	—
	Countercurrent		$p_i^G=(p_i^G)_{L_B}$	$\frac{\partial C_i^L}{\partial z}=\frac{\partial T_L}{\partial z}=0$	—
			$T_G=(T_G)_{L_B}$		
$r=0,\ 0<z<L_B$	Co-current	$\frac{\partial p_i^G}{\partial r}=\frac{\partial T_G}{\partial r}=0$	—	$\frac{\partial C_i^L}{\partial r}=\frac{\partial T_L}{\partial r}=0$	—
	Countercurrent	$\frac{\partial p_i^G}{\partial r}=0$	—	$\frac{\partial C_i^L}{\partial r}=0$	—
		$-\lambda_r^G\frac{\partial T_G}{\partial r}=h_{GW}(T_G-T_W)$		$-\lambda_r^L\frac{\partial T_L}{\partial r}=h_{LW}(T_L-T_W)$	
$\xi=0,\ 0\le z\le L_B,\ 0\le r\le R$	Co-current/ countercurrent	—	—	—	—
$\xi=\frac{d_{pe}}{2},\ 0\le z\le L_B,\ 0\le r\le R$	Co-current/ countercurrent	—	—	—	—

is the real condition at the outlet of the reactor, but as it is an infinite type of boundary condition, it is difficult to apply under most numerical methods. Because of this, many researchers prefer to use Danckwerts' boundary conditions, where the temperature gradients at the exit of the reactor are zero. It must be pointed out that unless the reactor is infinitely long, this exit condition will not be true because the fluid will not reach equilibrium with its surroundings. This lack of equilibrium at the exit would be even more pronounced at the high temperatures frequently experienced in TBRs. The boundary condi-

(*i*=S, N, A, O, GO, WN, Ni, V)		Solid Phase	
Dispersion	No Dispersion	Surface (*i*=All Compounds)	Interior (*i*=All Compounds)
—	—	$C_{Sfi}^S=(C_{Sfi}^S)_0$	
—	—	$T_S^S=(T_S^S)_0$	
—	$C_i^L=(C_i^L)_0$	$C_{Sfi}^S=(C_{Sfi}^S)_0$	
—	—	$T_S^S=(T_S^S)_0$	
$\frac{\partial C_i^L}{\partial z}=\frac{\partial T_L}{\partial z}=0$	—	$\frac{\partial T_S^S}{\partial z}=0$	
$\frac{\partial C_i^L}{\partial z}=\frac{\partial T_L}{\partial z}=0$	—	$\frac{\partial T_S^S}{\partial z}=0$	
$\frac{\partial C_i^L}{\partial r}=\frac{\partial T_L}{\partial r}=0$	—	$\frac{\partial T_S^S}{\partial z}=0$	
$\frac{\partial C_i^L}{\partial r}=0$	—	$-\lambda_r^S\frac{\partial T_S^S}{\partial r}=h_{SW}(T_S^S-T_W)$	
$-\lambda_r^L\frac{\partial T_L}{\partial r}=h_{LW}(T_L-T_W)$			
—	—	—	$\frac{\partial C_{fi}^S}{\partial \xi}=\frac{\partial T_S}{\partial \xi}=0$
—	—	—	$-D_{ei}^L\frac{\partial C_{fi}^S}{\partial \xi}=f_w k_i^S a_S(C_{SLi}^S-C_i^L)k_i'^S(C_{SLi}^S-C_{SLi}^L)$ $=\rho_B\zeta\sum_{j=1}^{N_{RL}}\upsilon_{ij}^L\eta_j^L r_j'^L(C_{SLi}^S,T_S^S)$ $-D_{ei}^G\frac{\partial C_{Gi}^S}{\partial \xi}=(1-f_w)k_i^{GS}a_S(C_{SGi}^S-C_i^G)$ $=\rho_B\zeta\sum_{j=1}^{N_{RG}}\upsilon_{ij}^G\eta_j^G r_j'^G(C_{SGi}^S,T_S^S)$ $(1-f_w)h_{GS}a_S(T_S^S-T_G)+f_w h_{LS}a_S(T_S^S-T_L)$ $=-\lambda_e^S\frac{\partial T_S}{\partial \xi}$

tions proposed by Young and Finlayson (1973) avoid this abnormality. They derived inlet and outlet conditions similar to those of Danckwerts for the case in which both axial and radial dispersion is present. However, the outlet condition was reported with a nonzero gradient, which is reduced to Danckwerts form when radial dispersion is neglected. The boundary condition $\partial T_S^S/\partial z=0$ for $z \geq L_B$ is a Danckwerts type also, but as the solid phase is commonly nonexistent for $z > L_B$, it is generally regarded as a valid exit condition. For the same reason, as the reaction does not take place without the solid catalyst, the concentration gradients can be assumed to be zero at the exit. Therefore, the Danckwerts boundary conditions are used when there is no mass (or heat) dispersion outside the reactor (Chao and Chang, 1987; Feyo De Azevedo et al., 1990).

The boundary condition at $\xi = d_{pe}/2$ is sometimes considered to be $C_{fi}^S = C_{Sfi}^S$ (Shokri and Zarrinpashne, 2006; Shokri et al., 2007). According to Chen et al. (2001), the following boundary conditions will be employed for heat transfer in the thermowell:

$$\lambda_{TW} \frac{\partial T_{TW}}{\partial z} = h_{air}\left[(T_{TW})_0 - T_{room}\right] \quad \text{at } z = 0 \tag{2.130}$$

$$-\lambda_{TW} \frac{\partial T_{TW}}{\partial z} = h_{air}\left[(T_{TW})_f - T_{room}\right] \quad \text{at } z = L_B \tag{2.131}$$

$$\frac{\partial T_{TW}}{\partial r} = 0 \quad \text{at } z = 0 \tag{2.132}$$

$$\lambda_{TW} \frac{\partial T_{TW}}{\partial r} = h_{TW}(T_f - T_{TW}) \quad \text{at } r = R_{TW} \tag{2.133}$$

The resulting set of PDEs coupled with the respective initial and boundary conditions are then solved simultaneously using an appropriate numerical method (Melis et al., 2004).

Example of Simplification of the Generalized Model Sometimes, to simplify heat transfer modeling in HDT reactors, the three processes involved (heat transfer in the solid, liquid, and gas phases) can be lumped into only one equation with a pseudohomogeneous heat balance. In the following, an example of how to obtain a pseudohomogeneous heat balance from the generalized heat balance presented in Table 2.12 is shown.

The general heat balance for the gas phase is

$$\text{HA1} = \text{HA2} + \text{HA3} + \text{HA4} + \text{HA5} + \text{HA6} + \text{HA7} + \text{HA8} \tag{2.134}$$

The general energy heat balance for the liquid phase is

$$\text{HB1} = \text{HB2} + \text{HB3} + \text{HB4} + \text{HB5} + \text{HB6} + \text{HB7} + \text{HB8} \tag{2.135}$$

Assuming no temperature gradient within the catalyst particles (isothermal catalyst), the general heat balance for the solid phase is

$$\text{HC1} = \text{HC3} + \text{HC4} + \text{HC9} + \text{HC10} + \text{HC11} \tag{2.136}$$

When modeling commercial HDT reactors it is generally accepted that axial dispersion can be omitted; then HA3 = HB3 = HC3 = 0. If adiabatic operation is also assumed, then HA4 = HB4 = HC4 = 0 (isothermal reactor in the radial direction) and HA8 = HB8 = 0 (no heat transfer between the fluid phase and the reactor wall).

The pseudohomogeneous model is based on the fact that the temperature difference among the gas, liquid, and catalyst at any particular axial position of the reactor is negligible. Hence, $T_G = T_I = T_L = T_S^S = T$, and in consequence the following terms can be neglected: HA5 = HA7 = HB5 = HB7 = HC9 = HC10 = 0. The temperature gradients $T_G - T_I$ and $T_I - T_L$ in the HA6 and HB6 terms are also neglected. The final heat balance equation for the gas, liquid, and solid phases is then

$$\mathrm{HA1} + \mathrm{HB1} + \mathrm{HC1} = \mathrm{HA2} + \mathrm{HB2} + \mathrm{HC6}' + \mathrm{HB6}' + \mathrm{HC11} \qquad (2.137)$$

Equation (2.137) in terms of model parameters for co-current operation gives

$$\begin{aligned}
&\varepsilon_G \rho_G C_{pG} \frac{\partial T}{\partial t} + \varepsilon_L \rho_L C_{pL} \frac{\partial T}{\partial t} + \varepsilon_S \rho_S C_{pS} \frac{\partial T}{\partial t} \\
&= -u_G \rho_G C_{pG} \frac{\partial T}{\partial z} - u_L \rho_L C_{pL} \frac{\partial T}{\partial z} + \sum_{i=1}^{N_{CG}} \left\{ K_{Li} a_L \left(\frac{p_i^G}{H_i} - C_i^L \right) (-\Delta H_{vi}) \right\} \\
&+ \sum_{i=1}^{N_{CG}} \left\{ K_{Li} a_L \left(\frac{p_i^G}{H_i} - C_i^L \right) (\Delta H_{vi}) \right\} + \rho_B \zeta \left[\sum_{j=1}^{N_{RL}} (-\Delta H_{Rj}^L) \eta_j^L r_j'^L (C_{SLi}^S, T) \right. \\
&\left. + \sum_{j=1}^{N_{RG}} (-\Delta H_{Rj}^G) \eta_j^G r_j'^G (C_{SGi}^S, T) \right] \qquad (2.138)
\end{aligned}$$

The final simplified pseudohomogeneous heat balance is

$$\begin{aligned}
(\varepsilon_G \rho_G C_{pG} + \varepsilon_L \rho_L C_{pL} + \varepsilon_S \rho_S C_{pS}) \frac{\partial T}{\partial t} &= -(u_G \rho_G C_{pG} + u_L \rho_L C_{pL}) \frac{\partial T}{\partial z} \\
&+ \rho_B \zeta \left[\sum_{j=1}^{N_{RL}} (-\Delta H_{Rj}^L) \eta_j^L r_j'^L (C_{SLi}^S, T) \right. \\
&\left. + \sum_{j=1}^{N_{RG}} (-\Delta H_{Rj}^G) \eta_j^G r_j'^G (C_{SGi}^S, T) \right] \qquad (2.139)
\end{aligned}$$

If the catalyst wetting efficiency is also assumed to be complete, $f_w = 1$ (e.g., when properly designed liquid distributors and high liquid velocities in commercial reactors are used). Therefore, reactions occur only in the liquid phase; then Eq. (2.139) is rewritten as

$$\left(\varepsilon_G\rho_G C_{PG} + \varepsilon_L\rho_L C_{PL} + \varepsilon_S\rho_S C_{PS}\right)\frac{\partial T}{\partial t} = -\left(u_G\rho_G C_{PG} + u_L\rho_L C_{PL}\right)\frac{\partial T}{\partial z} + \rho_B\zeta\sum_{j=1}^{N_{RL}}\left(-\Delta H_{Rj}^{L}\right)\eta_j^L r_j'^L\left(C_{SLi}^S, T\right) \tag{2.140}$$

According to Feyo De Azevedo et al. (1990), some researchers have included the radiation effects in nonisothermal reactors in the radial direction ($HA4 = HB4 = HC4 \neq 0$) by means of the effective radial thermal conductivity

$$\lambda_{er} = \left(\varepsilon_S\lambda_r^S + \varepsilon_L\lambda_r^L + \varepsilon_G\lambda_r^G\right) = \left(\lambda_{er}\right)_0 + \lambda_{\text{rad}} \tag{2.141}$$

$$\lambda_{\text{rad}} = 4\psi_r\sigma d_{pe}T^3 \tag{2.142}$$

where $(\lambda_{er})_0$ is the conductive plus convective contributions to the effective radial conductivity, λ_{rad} the radiant contribution, σ the Stephan–Boltzmann constant, and ψ_r a radiant transfer factor defined as

$$\psi_r = \frac{2}{(2/e) - 0.264} \tag{2.143}$$

where e is the particle emissivity. Therefore, rearranging Eq. (2.151) as $\lambda_{er}(r) = (\lambda_{er})_0 + K_1[T(r)]^3$ (where $K_1 = 4\psi_r\sigma d_{pe}$), the pseudohomogeneous radial heat dispersion term ($HA4 + HB4 + HC4$ with $T_S^S = T_L = T_G = T$) is expressed as

$$\lambda_{er}\left(\frac{\partial^2 T}{\partial r^2} + \frac{1}{r}\frac{\partial T}{\partial r}\right) = \frac{1}{r}\frac{\partial}{\partial r}\left(r\lambda_{er}(r)\frac{\partial T}{\partial r}\right) = \lambda_{er}(r)\left(\frac{1}{r}\frac{\partial T}{\partial r} + \frac{\partial^2 T}{\partial r^2}\right) + \frac{\partial\lambda_{er}(r)}{\partial r}\frac{\partial T}{\partial r}$$
$$= \lambda_{er}(r)\left(\frac{1}{r}\frac{\partial T}{\partial r} + \frac{\partial^2 T}{\partial r^2}\right) + 3K_1T^2\left(\frac{\partial T}{\partial r}\right)^2 \tag{2.144}$$

where the last term on the right side of Eq. (2.144) represents the radioactive heat transfer, and its exclusion may produce errors in estimation of some other heat transfer parameters.

2.4.4 Estimation of Model Parameters

To solve the set of ordinary differential equations (ODEs) (for the steady-state regime) or the set of PDEs (for the dynamic regime), it is necessary to evaluate several parameters and chemical properties of the system. Those parameters can be estimated with existing correlations, whose accuracy is of great importance for the entire state of robustness of the reactor model.

Effective Diffusivity The description of steady-state diffusion and reaction of a multicomponent liquid mixture in a porous catalyst particle requires appropriate definition of the species fluxes to the particle. Fick's law for equimolar counterdiffusion through an ideal cylindrical pore is given by

$$N_i^f = -D_{ei}^f \nabla C_{fi}^s \tag{2.145}$$

where D_{ei}^f is the effective diffusivity, by means of which the structure (porosity and tortuosity) of the pore network inside the particle is taken into consideration in the modeling. Table 2.15 shows Bosanquet's formula to estimate the effective diffusivity inside the catalyst particle (Bosanquet, 1944), which consists of two diffusion contributions: Knudsen diffusivity (D_{Ki}^f) and molecular diffusivity (D_{Mi}^f). Both Bosanquet's formula and Fick's law can also be applied with sufficient accuracy to cases involving a narrow unimodal pore-size distribution and very dilute mixtures. The restrictive factor $F(\lambda_g)$ accounts for additional friction between the solute and the pore walls. The exponent $\hat{Z}$ is 4 for $\lambda_g < 0.2$ (Iliuta et al., 2006). The tortuosity factor of the pore network, τ, is used in the calculation of D_{ei}^f because the pores are not oriented along the normal direction from the surface to the center of the catalyst particle, and its value generally varies between 3 and 7, but for HDT process it is commonly assumed to be equal to 4 (Satterfield, 1970, 1975; Macías and Ancheyta, 2004; Ancheyta et al., 2005; Iliuta et al., 2006). It is also possible to estimate the tortuosity factor assuming valid the upper bound correlation given by Weissberg (1963) for a packing of random spheres as shown in Table 2.15.

Effectiveness Factor The effectiveness factor of independent reactions can be defined as the ratio of the volumetric average of the reaction rate into the particle to the reaction rate at the surface of the particle as proposed by Thiele (1939) and Zeldovich (1939):

$$\eta_i^f = \frac{(1/V_p)\int r_j'^f (C_{fi}^S, T_S)\, dV_p}{r_j'^f (C_{Sfi}^S, T_S^S)} \tag{2.146}$$

$$\eta_i^f = \frac{r_j'^f (C_{fi}^S, T_S)}{r_j'^f (C_{Sfi}^S, T_S^S)} = \frac{(r_j'^f)_{\text{obs}}}{(r_j'^f)_{\text{in}}} \tag{2.147}$$

Analytical solutions for Eqs. (2.146) and (2.147) are possible only for single reactions and for zero- and first-order rate expressions. The various correlations used in the literature to estimate the catalyst effectiveness factor for isothermal and irreversible reactions are shown in Table 2.16.

For kinetic models other than the power-law approach, such as Langmuir–Hinshelwood–Hougen–Watson (LHHW)-type kinetic expressions, there is no analytical solution of Eq. (2.146). Therefore, an alternative method to avoid the numerical integration of Eq. (2.146) is the Bischoff generalized modulus

TABLE 2.15. Estimation of Effective Diffusivity

Parameter	Gas Phase	Liquid Phase
Molecular diffusivity coefficient	$\frac{1}{D_{Mi}^G} = \frac{1}{1-y_i}\sum_{k\neq i}^{N_{CG}} \frac{y_k}{D_{i,k}}$	$D_{Mi}^L = 8.93\times 10^{-8} \frac{v_L^{0.267} T_L}{v_i^{0.433}\mu_L}$
Knudsen diffusivity coefficient	$D_{Ki}^G = 9700 r_g \left(\frac{T_G}{\mathrm{MW}_i}\right)^{0.5}$	$\frac{1}{D_{Ki}^L} \approx 0$
Effective diffusivity	$D_{ei}^f = \frac{\varepsilon_S}{\tau}\frac{1}{1/D_{Mi}^f + 1/D_{Ki}^f} F(\lambda_g)$	
Restrictive factor	$F(\lambda_g) = (1-\lambda_g)^{\hat{Z}}$	
Ratio of radius of gyration over pore radius	$\lambda_g = \frac{r_{\text{solute}}}{r_g}$	
Mean pore radius	$r_g = \frac{2V_g}{S_g}$	
Tortuosity factor	$\frac{1}{\tau} = \frac{\varepsilon_S}{1-\frac{1}{2}\log(\varepsilon_S)}$	
Binary diffusion coefficient (P in atm)	$D_{i,k} = 0.0018583\sqrt{T_G^3\left(\frac{1}{\mathrm{MW}_i}+\frac{1}{\mathrm{MW}_k}\right)}\frac{1}{P\sigma_{ik}^2\Omega_D}$	
Dynamic liquid viscosity	$\mu_L = 3.141\times 10^{10}(T_L - 460)^{-3.444}[\log_{10}(\mathrm{API})]^a$	
	$a = 10.313[\log_{10}(T_L - 460)] - 36.447$	
Molar volume of solute (i) in liquid phase and liquid solvent (L)	$v_i = 0.285(\upsilon_{ci})^{1.048}$	
	$v_L = 0.285(\upsilon_{cL})^{1.048}$	
Solvent critical specific volume	$\upsilon_{cL} = \upsilon_{cL}^m \mathrm{MW}_L$	
	$\upsilon_{cL}^m = 7.5214\times 10^{-3} T_{\mathrm{MeABP}}^{0.2896} d_{15.6}^{-0.7666}$	

approach (Bischoff, 1965), which enables an analytical solution to any type of rate equation and single reaction.

As mentioned previously, the effectiveness factor in commercial HDS catalysts has been reported to be in the range 0.4 to 0.8 (Satterfield, 1970, 1975; Duduković, 1977; Hofmann, 1977; Glasscock and Hale, 1994; Korsten and Hoffmann, 1996; Macías and Ancheyta, 2004; Marroquín et al., 2005). Expressions for isothermal first-order reactions with irregularly shaped

TABLE 2.16. Estimation of Catalyst Effectiveness Factor

Kinetic Model	Reaction Order	Shape	Thiele Modulus	Effectiveness Factor
Power law	$n=1$	Spheres and crushed	$\Phi_j^f = \frac{V_p}{S_p}\sqrt{\frac{k'_{\text{in},j}\rho_S}{D_{ei}^f}}$	$\eta_j^f = \frac{3\Phi_j^f \coth(3\Phi_j^f)-1}{3(\Phi_j^f)^2}$
	$n=1$	Pellet, cylinder, 2-, 3-lobe, etc.	$0.5 > \Phi_j^f > 10$	$\eta_j^f = \frac{\tanh(\Phi_j^f)}{\Phi_j^f}$
	$2 \le n \le 3$	Any geometry	$\Phi_j^f = \frac{1}{\phi_S}\frac{V_p}{S_p}\sqrt{\left(\frac{n+1}{2}\right)\frac{k'_{\text{in},j}(C_{Sfi}^S)^{n-1}\rho_S}{D_{ei}^f}}$	$\eta_j^f = \frac{1}{\sqrt{(\Phi_j^f)^2+1}}$
	$n \ge 0$	Any geometry	$\Phi_j^f > 3$	$\eta_j^f = 1/\Phi_j^f$
	$n \ge 0$	Any geometry	Supposed to be $\Phi_j^f > 3$	$\eta_j^f = \frac{2D_{ei}^f(S_p/V_p)^2}{(n+1)k'_{\text{app},j}(C_{Sfi}^S)^{n-1}\rho_S}$
LHHW	—	Spheres and crushed	$\Phi_j^f = \frac{V_p\rho_S r_j'^f(C_{Sfi}^S, T_S^S)}{S_p\sqrt{2}}\left[\int_{C_{i,eq}}^{C_{Sfi}^S} D_{ei}^f \rho_S r_j'^f(C_{fi}^S, T_S^S)\,dC_i\right]^{-1/2}$	$\eta_j^f = \frac{3\Phi_j^f \coth(3\Phi_j^f)-1}{3(\Phi_j^f)^2}$
	—	Pellet, cylinder, 2-, 3-lobe, etc.	$\Phi_j^f = \frac{V_p\rho_S r_j'^f(C_{Sfi}^S, T_S^S)}{S_p\sqrt{2}}\left[\int_{C_{i,eq}}^{C_{Sfi}^S} D_{ei}^f \rho_S r_j'^f(C_{fi}^S, T_S^S)\,dC_i\right]^{-1/2}$	$\eta_j^f = \frac{\tanh(\Phi_j^f)}{\Phi_j^f}$

catalysts lead under steady-state conditions to acceptable results with errors not exceeding 20% (Aris, 1975; Duduković, 1977). Bischoff (1965) proposed a general modulus to predict the effectiveness factor for any reaction type within a relatively narrow region. If reactions of order less than one-half are excluded, the spread between all the various curves is about 15%. The mean deviation of values of η calculated from the empirical correlation proposed by Papayannakos and Georgiou (1988) is less than 2.4% from those predicted with the normalized modulus for simple order reactions proposed by Froment and Bischoff (1990) in the range $0.05 < \eta < 0.99$.

Global Gas–Liquid Mass Transfer The gas–liquid interphase mass transfer flux is described in terms of the simple two-film theory:

$$\frac{1}{K_{Li}} = \frac{RT_G Z_{(p,T_G)}}{k_i^G H_i} + \frac{1}{k_i^L} \tag{2.148}$$

The overall external resistance to mass transfer (K_{Li}) is composed by the resistance to mass transfer in the gas (k_i^G) and liquid (k_i^L) films. Estimates of the compressibility factor Z sometimes give values close to 1; therefore, ideal gas law could be used (Mejdell et al., 2001).

For slightly soluble gases such as H_2, the value of Henry's constant (H_i) exceeds unity, and then mass transfer resistance in the gas film can be neglected (Zhukova et al., 1990). Therefore, the total mass transfer is approximately equal to the liquid-side mass transfer coefficient:

$$\frac{1}{K_{Li}} = \frac{1}{k_i^L} \tag{2.149}$$

The liquid film mass transfer coefficient (k_i^L) is calculated using the correlations reported in Table 2.17.

Gas–Liquid Equilibrium The gas–liquid equilibrium along the catalyst bed is represented in the mass balance equations by the Henry's law constant. The constants related to this law for different chemical species available in the system may be defined in two ways, described below:

1. *Solubility coefficients.* Employing solubility coefficients, the following expression is used to estimate Henry's constant:

$$H_i = \frac{v_n}{\lambda_i \rho_L} \tag{2.150}$$

where λ_i stands for the component i solubility and v_N is the molar volume under normal conditions. Using this expression implies knowledge of the gaseous component solubility in the liquid phase considering the process temperature effect. Korsten and Hoffmann (1996) have reported the next correlations to evaluate this parameter only for H_2 and H_2S:

TABLE 2.17. Correlations to Estimate Model Parameters

Parameter	Symbol	References
Holdup gas	ε_G	Calculated from $\varepsilon_B = \varepsilon_L + \varepsilon_G$
Holdup liquid	ε_L	Charpentier and Favier (1975), Satterfield (1969), Specchia and Baldi (1977), Ellman et al. (1990)
Bed void fraction	ε_B	Haughey and Beveridge (1969), Carberry and Varma (1987), Froment and Bischoff (1990)
Catalyst wetting efficiency (or contacting effectiveness)	f_w (η_{CE})	Ring and Missen (1991), Al-Dahhan and Duduković (1995)
Two-phase pressure drop	ΔP	Larkins et al. (1961), Ellman et al. (1988)
Gas–liquid interfacial area	a_L	Iliuta et al. (1999)
Gas–solid interfacial area	a_S	Puranik and Vogelpohl (1974), Onda *et al.* (1967)
Mass transfer gas–solid coefficient	k_i^{GS}	Petrovic and Thodos (1968), Dwivedi and Upadhyay (1977)
Mass transfer liquid–solid coefficient	k_i^S	Duduković et al. (2002), Dwivedi and Upadhyay (1977), Bird et al. (2002), Evans and Gerald (1953), Wilson and Geankoplis (1966), Goto and Smith (1975), Satterfield et al. (1978), Specchia et al. (1974)
Mass transfer coefficient of liquid side at G–L interface	k_i^L	Goto and Smith (1975)
Mass transfer coefficient of gas side at G–L interface	k_i^G	Goto and Smith (1975), Reiss (1967), Yaïci et al. (1988)
Axial dispersion of gas	D_a^G	Hochman and Effron (1969), Sater and Levenspiel (1966), Demaria and White (1960)
Axial dispersion of liquid	D_a^L	Gierman (1988), Hochman and Effron (1969), Sater and Levenspiel (1966), Tsamatsoulis and Papayannakos (1998)
Radial dispersion of gas	D_r^G	Fahien and Smith (1955)
Radial dispersion of liquid	D_r^L	Fahien and Smith (1955), De Ligny (1970), Herskowitz and Smith (1978a,b)
Binary interaction parameter for H_2–oil using PR EoS	$k_{H_2,oil}$	Moysan et al. (1983), Ronze et al. (2002), Riazi (2005), Lal et al. (1999), Magoulas and Tassios (1990)
Binary interaction parameter for H_2S-oil using PR EoS	$k_{H_2S,oil}$	Feng and Mather (1993a,b), Carroll and Mather (1995)
Binary interaction parameter for NH_3-oil using SRK EoS	$k_{NH_3,oil}$	API (1997)
Density of the gas phase	ρ_G	Soave (1972), Peng and Robinson (1976)

Parameter	Symbol	References
Density of the liquid phase	ρ_L	Ahmed (1989)
Catalyst bulk density	ρ_B	ASTM (2003)
Dynamic viscosity of liquid	μ_L	Glaso (1980), Ahmed (1989), Brulé and Starling (1984)
Dynamic viscosity of gas	μ_G	Ahmed (1989), Brulé and Starling (1984)
Diffusion coefficients for gases	D_{Mi}^G	Wilke and Chang (1955)
Diffusion coefficients for liquids	D_{Mi}^L	Tyn and Calus (1975)
Specific heat of gas phase	C_{PG}	Lee and Kesler (1975), Perry et al. (2004)
Specific heat of liquid phase	C_{PL}	Lee and Kesler (1975, 1976), Perry et al. (2004)
Specific heat of solid phase	C_{PS}	Perry et al. (2004)
Gas–liquid heat transfer coefficient	h_{GL}	Marroquín de la Rosa et al. (2002), Chilton and Colburn (1939)
Liquid–solid heat transfer coefficient	h_{LS}	Chilton and Colburn (1939)
Chilton–Colburn *j*-factor for energy transfer	j_H	Froment and Bischoff (1990), Hill (1977), Bird et al. (2002), Gupta et al. (1974)
Effective thermal conductivity radial	λ_{er}^f	Hashimoto et al. (1976)
Effective thermal conductivity axial	λ_{ea}^f	Tarhan (1983), Dixon (1985)
Thermal conductivity of *f* phase	k_f	API (1997), Chung et al. (1988)
Heat of reaction	ΔH_R	Tarhan (1983)
Heat of vaporization/condensation	ΔH_v	Soave (1972), Peng and Robinson (1976)
Thiele modulus	Φ	Bischoff (1965)
Effectiveness factor	η	Froment and Bischoff (1990), Aris (1975)

$$\lambda_{H_2} = -0.559729 - 0.42947 \times 10^{-3} T_L + 3.07539 \times 10^{-3} \frac{T_L}{\rho_L} + 1.94593 \times 10^{-6} T_L^2 + 0.835783 \frac{1}{\rho_L^2} \quad (2.151)$$

for hydrogen and

$$\lambda_{H_2S} = \exp(3.3670 - 0.008470 T_L) \quad (2.152)$$

for hydrogen sulfide. However, this last correlation may not be no adequate to evaluate the solubility of H_2S in oil fractions at the complete temperature range used in commercial units. In that case, it is possible to use an EoS to estimate this Henry's constant.

2. *Equation of state.* It is also possible to obtain Henry's constant assuming local equilibrium at the liquid–gas interface:

$$H_i = \frac{P y_i}{C_{tot}^L x_i} = \frac{P}{C_{tot}^L} K_{eq,i}(x_i, y_i, T_I, P) \quad (2.153)$$

where the equilibrium constant ($K_{eq,i}$) is calculated using an adequate EoS (e.g., Peng–Robinson (PR), Soave–Redlich–Kwong (SRK), *t*-van der Waals, Grayson–Streed). For this expression, it was necessary to calculate a two-phase thermodynamic equilibrium previously at each local point along the catalytic bed to estimate the interface temperature and the molar compositions in liquid and gas phases. The main advantage of this expression is that it takes into account the volatility of the feedstock; however, it also increases the computing time too greatly.

Another way to calculate the Henry's constant of gaseous solute in a solvent is to use the next thermodynamic assumption:

$$H_i = \lim_{x_i \to 0} \frac{f_i^L}{x_i} = \lim_{x_i \to 0} P \varphi_i^L \quad (2.154)$$

where φ_i^L is the fugacity coefficient of a gaseous compound *i* (solute) in the liquid phase (solvent), and its calculation using EoS is addressed below. The SRK and PR equations of state are the most widely used in HDT process modeling, being defined by the generalized expression

$$P = \frac{R T_f}{v^f - b} - \frac{a}{(v^f + \delta_1 b)(v^f + \delta_2 b)} \quad (2.155)$$

For pure compounds, values of parameters *a* and *b* are given by

$$a = a_{ii} = a_i = a_{c,i} \alpha_i (T_f) \quad (2.156)$$

with

$$a_{c,i} = \Omega_a \frac{R^2 T_{c,i}^2}{P_{c,i}} \tag{2.157}$$

$$b = b_i = \Omega_b \frac{RC_{c,i}}{P_{c,i}} \tag{2.158}$$

The generalized temperature function $\alpha_i(T_f)$ was proposed by Soave (1972) to be an equation of the form

$$\alpha_i(T_f) = \left[1 + m_i\left(1 - \sqrt{\frac{T_f}{T_{c,i}}}\right)\right]^2 \tag{2.159}$$

with

$$m_i = M_0 + M_1\omega_i + M_2\omega_i^2 \tag{2.160}$$

When applied to mixtures, the classical mixing rules may be considered to evaluate parameters a and b:

$$a = a_m = \sum_{i=1}^{N_{CL}} \sum_{k=1}^{N_{CL}} x_i x_k a_{ik} \tag{2.161}$$

$$a_{ik} = a_{ki} = \sqrt{a_{ii} a_{kk}}\,(1 - k_{ik}) \tag{2.162}$$

$$b = b_m = \sum_{i=1}^{N_{CL}} x_i b_i \tag{2.163}$$

where k_{ik} are the binary interaction parameters, which may be obtained from the references reported in Table 2.17. It is important to point out that the quality of Henry's constant calculation depends enormously on the accuracy of these interaction parameters. The liquid-phase fugacity coefficient can be derived from Eq. (2.155) to give

$$\ln \varphi_i^L = \frac{b_i}{b}(Z^L - 1) - \ln(Z^L - B) - \frac{A}{B(\delta_2 - \delta_1)}\left(\frac{2\sum_{k=1}^{N_{CL}} x_k a_{ik}}{a} - \frac{b_i}{b}\right)\ln\frac{Z^L + \delta_2 B}{Z^L + \delta_1 B} \tag{2.164}$$

with

$$A = \frac{aP}{(RT_L)^2} \tag{2.165}$$

TABLE 2.18. Parameters of Soave–Redlich–Kwong and Peng–Robinson Equations of State

	EoS	
Parameter	SRK	PR
δ_1	1	$1+\sqrt{2}$
δ_2	0	$1-\sqrt{2}$
Ω_a	0.42748	0.457236
Ω_b	0.08664	0.077796
M_0	0.48	0.37464
M_1	1.574	1.54226
M_2	−0.176	−0.26992
A_Z	1	1
B_Z	−1	$-(1-B)$
C_Z	$A - B - B^2$	$A - 2B - 3B^2$
D_Z	$-AB$	$-(AB - B^2 - B^3)$

$$B = \frac{bP}{RT_L} \tag{2.166}$$

where Z^L is the compressibility factor of the liquid phase at saturation obtained from the solution of Eq. (2.155) expressed in its cubic compressibility factor form:

$$A_Z\left(Z^L\right)^3 + B_Z\left(Z^L\right)^2 + C_Z\left(Z^L\right) + D_Z = 0 \tag{2.167}$$

The values of the universal parameters (δ_1, δ_2, Ω_a, Ω_b, M_0, M_1, M_2, A_Z, B_Z, C_Z, and D_Z) are given in Table 2.18.

Heat Transfer Coefficients Correlations employed for mass transfer can be used to calculate the parameters for energy transfer between phases by employing the Chilton and Colburn (1939) analogy. The Chilton–Colburn j-factor for mass transfer (j_D) is given by

$$j_D = \frac{\text{Sh}}{\text{Re}_f\,\text{Sc}^{1/3}} = f(\text{Re}_f, \text{geometry}, \text{boundary conditions}) \tag{2.168}$$

To evaluate the liquid–solid mass transfer coefficients, for example, Eq. (2.168) must be expressed as

$$j_D = \frac{k_i^S}{C_i^L u_L}\left(\frac{\mu_L}{\rho_L D_{M,i}^L}\right)^{2/3} \tag{2.169}$$

On the other hand, the Chilton–Colburn *j*-factor for energy transfer (j_H) is given by

$$j_H = \frac{\text{Nu}}{\text{Re}_f\ \text{Pr}^{1/3}} = g(\text{Re}_f, \text{geometry}, \text{boundary conditions}) \qquad (2.170)$$

where $g(\text{Re}_f)$ is a correlation that is a function of the Reynolds number of the respective phase f to be evaluated; some of these correlations are shown in Table 2.17. If one needs to estimate the liquid–solid heat transfer coefficient, for example, Eq. (2.169) is rewritten in the expression

$$j_H = \frac{h_{LS}}{C_{p_L} u_L \rho_L} \left(\frac{C_{p_L} \mu_L}{k_L} \right)^{2/3} \qquad (2.171)$$

There are various correlations to estimate the mass transfer coefficients, but the Chilton–Colburn analogy is not usually employed to evaluate them, whereas due to the lack of correlations to estimate the heat transfer coefficients in the gas or liquid film side at the gas–liquid interface and in the liquid film at the liquid–solid interface, it is common to use the Chilton–Colburn analogy, by equating Eqs. (2.168) and (2.170) ($j_D = j_H$), to estimate these coefficients. The physical and geometrical properties involved in the dimensionless numbers must be evaluated at conditions of reaction and for each phase in a heterogeneous reactor.

To use the Chilton–Colburn analogy, it is necessary to consider the following conditions (Bird et al., 2002):

- Constant physical properties
- Small net mass transfer rates
- No chemical reaction
- No viscous dissipation heating
- No absorption or emission of radian energy
- No pressure diffusion, thermal diffusion, or forced diffusion

Theoretical Calculations of Some Parameters Relative to the Catalyst Bed Dilution of the catalyst bed with inert material is a common practice in experimental HDT reactors (Sie, 1996). The following simple formula is employed to calculate the dilution factor:

$$\zeta = \frac{V_c}{V_c + V_i} \qquad (2.172)$$

where V_c is the catalyst volume and V_i is the volume of inert particles, both obtained experimentally.

An equivalent particle diameter (d_{pe}), defined as the diameter of a sphere that has the same external surface (or volume) as the actual catalyst particle, is an important particle characteristic that depends on particle size and shape. For fixed beds with catalyst extrudates of commercial size, the equivalent particle diameter can be calculated according to Cooper et al. (1986):

$$d_{pe} = \frac{6(V_p/S_p)}{\phi_S} \tag{2.173}$$

Bed void fraction (or bed porosity) for undiluted catalyst bed can be calculated with the following correlation reported by Froment and Bischoff (1990), Haughey and Beveridge (1969), and Carberry and Varma (1987):

$$\varepsilon_B = 0.38 + 0.073\left[1 + \frac{(d_t/d_{pe} - 2)^2}{(d_t/d_{pe})^2}\right] \tag{2.174}$$

This correlation was developed for undiluted packed beds of spheres; however, if the equivalent particle diameter concept is used, it can also be employed for nonsphere particles. Once the bed void fraction is determined, the particle density can be calculated as follows (Tarhan, 1983):

$$\rho_S = \frac{\rho_B}{1 - \varepsilon_B} \tag{2.175}$$

Since the continuous models are based on the volume-average form of the transport equations for multiphase systems, equations expressing conservation of volume are (Whitaker, 1973)

$$\varepsilon_B = \varepsilon_L + \varepsilon_G \tag{2.176}$$

$$\varepsilon_L + \varepsilon_G + \varepsilon_S = 1 \tag{2.177}$$

Relationships between phase holdups inside the catalyst solid are given in the following expressions:

$$\varepsilon_S = \varepsilon_{pL} + \varepsilon_{pG} \tag{2.178}$$

$$\varepsilon_{pL} + \varepsilon_{pG} + \varepsilon_{pS} = 1 \tag{2.179}$$

The external surface area of catalyst particles per unit of reactor volume for PBRs can be calculated as

$$a_S = \frac{6(1 - \varepsilon_B)}{d_{pe}} \tag{2.180}$$

Catalyst porosity (ε_S) may be calculated with the following equation from the experimental data for total pore volume (V_g):

$$\varepsilon_S = \rho_S V_g \tag{2.181}$$

In an extreme case where the experimental S_g parameter is not available in order to estimate the average pore radius, one can use the correlation proposed by Macé and Wei (1991):

$$r_g = \frac{4\varepsilon_S}{S'_g} \tag{2.182}$$

where

$$S'_g = 4\pi\rho_S\left(\frac{d_{pe}}{2}\right)^2 \varepsilon_S \tag{2.183}$$

Some parameters that account for bed characterization are experimentally measurable, others are experimental or can be obtained through simulations, and others are empirical. Of course, although it is better to obtain the local porosity experimentally, measurements required the use of advanced techniques. To do that, computational calculations are preferred.

Most of empirical correlations for predicting liquid saturation, pressure drop, and flow regimes are based on experiments performed at atmospheric conditions. Since industrial trickle-bed reactors are operated at high pressures and temperatures, the applicability of these correlations for such operating conditions needs to be investigated (Nguyen et al., 2006). The majority of correlations that have been developed on a laboratory scale may not work for large-scale reactors (operated at high pressure and temperature) due to significant changes in hydrodynamic characteristics with a reactor scale (Gunjal and Ranade, 2007). Although extensive studies in hydrodynamic correlations are available in the literature (Duduković et al., 2002) it seems that many works in modeling reactors for petroleum fractions still employ the classical correlations (i.e., those derived from reasonable assumptions with simple expressions). Some researchers have employed the same correlations of previous papers without checking the accuracy of these expressions or the range in which they are applicable. On the other hand, although the benefit of using correlations based on neural networks has been reported, many data are necessary to employ this approach. To overcome this feature, an alternative is to create a database with different types of crude oils and fractions in order to develop an online program able to determine the various parameters involved in modeling a HDT reactor. At present it is not available. The only valuable effort seems to be that of Larachi et al. (1999), who have developed a simulator based on neural networks for the prediction of some hydrodynamic parameters for trickle-bed reactors. Although neural networks are updated continuously, which favors its use for parameter predictions, it is necessary to develop a fundamental relationship that takes advantage of novel techniques for characterization of hydrodynamic parameters.

The limitations of the various models reported in the technical literature are closely related to the number of parameters involved and to the reliability of the data available. Therefore, appropriate correlations for mass and energy transfer should be employed in order to calculate each of the terms used in a model reactor (i.e., correlations developed under similar conditions, such as the same flow regime, pressure, and temperature, assuming similar liquid system and porous particles). Generalized correlations for the prediction of properties having a broad range of variation should be avoided when modeling a TBR in detail, because they could produce some miscalculations. Constant values assumed a priori, such as tortuosity factor, binary interaction parameters, heat of reaction, and specific heat, should be used as a reference when experimental data are available. To simplify a TBR model, some researchers have ignored the low heat of some reactions because its contribution is not significant in the energy balance, which seems to be a reasonable assumption.

Using different approaches to fluid dynamics, kinetics or thermodynamics can lead to very different conclusions in predictions of reactor performance. For example, Gunjal and Ranade (2007) have reported 15% more conversions considering all parameters to be constant and assuming only uniform porosity (i.e., no nonuniform porosity). Akgerman et al. (1985) has reported 24 to 38% higher conversions when considering volatiles with respect to nonvolatiles, and Inoue et al. (2000) has predicted the use of larger reactors employing an *n*th kinetic model when more accurate kinetic expression has been utilized. Another important finding is the consideration of a thermowell in the energy balance for a pilot plant, as pointed out by Chen et al. (2001). These are the reasons to account for detailed models which allow both for making accurate descriptions of the chemical phenomena and for reliable preliminary calculations when designing a TBR reactor.

REFERENCES

Aboul-Gheit, K. (1989) Hydrocracking of vacuum gas oil (VGO) for fuels: production-reaction kinetics. *Erdoel Erdgas Kohle* 105:319–320.

Ahmed, T. (1989) *Hydrocarbon Phase Behavior*. Gulf Publishing, Houston, TX.

Al-Adwani, H. A. H., Lababidi, H. M. S., Alatiqi, I. M., Al-Dafferi, F. S. (2005) Optimization study of residuum hydrotreating processes. *Can. J. Chem. Eng.* 83(2):281–290.

Al-Dahhan, M. H., Duduković, M. P. (1994) Pressure drop and liquid holdup in high pressure trickle-bed reactors. *Chem. Eng. Sci.* 49(24):5681–5698.

Al-Dahhan, M. H., Duduković, M. P. (1995) Catalyst wetting efficiency in trickle-bed reactors at high pressure. *Chem. Eng. Sci.* 50(15):2377–2389.

Akgerman, A., Netherland, D. W. (1986) Effect of equation of state on prediction of trickle bed reactor model performance. *Chem. Eng. Commun.* 49(1):133–143.

Akgerman, A., Collins, G. M., Hook, B. D. (1985) Effect of feed volatility on conversion in trickle bed reactors. *Ind. Eng. Chem. Fundam.* 24(3):398–401.

Alvarez, A., Ancheyta, J. (2008) Simulation and analysis of different quenching alternatives for an industrial vacuum gasoil hydrotreater. *Chem. Eng. Sci.* 63:662–673.

Alvarez, A., Ramírez, S., Ancheyta, J., Rodríguez, L. M. (2007) Key role of reactor internals in hydroprocessing of oil fractions. *Energy Fuels* 21(3):1731–1740.

Ancheyta, J., Aguilar, E., Salazar, D., Betancourt, G., Leiva, M. (1999) Hydrotreating of straight run gas oil–light cycle oil blends. *Appl. Catal. A* 180(1–2):195–205.

Ancheyta, J., Maity, S. K., Betancourt, G., Centeno, G., Rayo, P., Gómez, M. T. (2001) Comparison of different Ni-Mo/alumina catalysts on hydrodemetallization of Maya crude oil. *Appl. Catal. A* 216(1–2):195–208.

Ancheyta, J., Angeles, M. J., Macías, M. J., Marroquín, G., Morales, R. (2002) Changes in apparent reaction order and activation energy in the hydrodesulfurization of real feedstocks. *Energy Fuels* 16(1):189–193.

Ancheyta, J., Rodríguez, M. A., Sánchez, S. (2005) Kinetic modeling of hydrocracking of heavy oil fractions: a review. *Catal. Today* 109:76–92.

Andrigo, P., Bagatin, R., Pagani, G. (1999) Fixed bed reactors. *Catal. Today* 52(2–3):197–221.

API (1997) *Technical Data Book: Petroleum Refining*. American Petroleum Institute, Washington, DC, Chap. 8, pp. 8–56.

Aoyagi, McCaffrey, W. C., Gray, M. R. (2003) Kinetics of hydrocracking and hydrotreating of coker and oilsands gas oils. *Pet. Sci. Technol.* 21:997–1015.

Aris, R. (1975) *The Mathematical Theory of Diffusion and Reaction in Permeable Catalysts*, Vol. 1, *The Theory of the Steady State*. Clarendon Press, Oxford, UK.

ASTM (2003) *Test Method Bulk Density ("Unit Weight") and Voids in Aggregate*. Test Method c-29. ASTM, Philadelphia.

Attou, A., Boyer, C., Ferschneider, G. (1999) Modelling of the hydrodynamics of the cocurrent gas–liquid trickle flow through a trickle-bed reactor. *Chem. Eng. Sci.* 54(6):785–802.

Avraam, D. G., Vasalos, I. A. (2003) HdPro: a mathematical model of trickle-bed reactors for the catalytic hydroprocessing of oil feedstocks. *Catal. Today* 79–80(1–4): 275–283.

Ayasse, R., Nagaishi, H., Chan, E. W., Gray, M. R. (1997) Lumped kinetics of hydrocracking of bitumen. *Fuel* 76:1025–1033.

Baker, T., Chilton, T. H., Vernon, H. C. (1935) The course of liquor flow in packed towers. *Trans. Am. Inst. Chem. Eng.* 31:296–315.

Basak, K., Sau, M., Manna, U., Verma, R. P. (2004) Hydrocracker model based on novel continuum lumping approach for optimization in petroleum refinery. *Catal. Today* 98:253–264.

Bellos, G. D., Papayannakos, N. G. (2003) The use of a three phase microreactor to investigate HDS kinetics. *Catal. Today* 79–80(1–4):349–355.

Bellos, G. D., Kallinikos, L. E., Gounaris, C. E., Papayannakos, N. G. (2005) Modelling of the performance of industrial HDS reactors using a hybrid neural network approach. *Chem. Eng. Process.* 44:505–515.

Bennett, R. N., Bourne, K. H. (1972) Hydrocracking for middle distillate: a study of process reactions and corresponding product yields and qualities. In: *Proceedings of the ACS Symposium on Advances in Distillate and Residual Oil Technology*, New York, pp. G45–G62.

Berger, D., Landau, M. V., Herskowitz, M., Boger, Z. (1996) Deep hydrodesulfurization of atmospheric gas oil: effects of operating conditions and modelling by artificial neural network techniques. *Fuel* 75(7):907–911.

Bhaskar, M., Valavarasu, G., Sairam, B., Balaraman, K. S., Balu, K. (2004) Three-phase reactor model to simulate the performance of pilot-plant and industrial trickle-bed reactors sustaining hydrotreating reactions. *Ind. Eng. Chem. Res.* 43(21):6654–6669.

Bird, R. B., Stewart, W. E., Lightfoot, E. N. (2002) *Transport Phenomena*. Wiley, Hoboken, NJ.

Bischoff, K. B. (1965) Effectiveness factors for general reaction rate forms. *AIChE J.* 11(2):351–355.

Bischoff, K. B., Levenspiel, O. (1962) Fluid dispersion—generalization and comparison of mathematical models: II. Comparison of models. *Chem. Eng. Sci.* 17:257–264.

Bollas, G. M., Papadokonstantakis, S., Michalopoulos, J., Arampatzis, G., Lappas, A. A., Vasalos, I. A., Lygeros, A. (2004) A computer-aided tool for the simulation and optimization of the combined HDS-FCC processes. *Trans. IChemE A* 82(A7): 881–894.

Bondi, A. (1971) Handling kinetics from trickle-phase reactors. *Chem. Technol.* 1:185–188.

Bosanquet, C. H. (1944) British TA Report BR-507. Sept. 27.

Botchwey, C., Dalai, A. K., Adjaye, J. (2003) Product selectivity during hydrotreating and mild hydrocracking of bitumen-derived gas oil. *Energy Fuels* 17:1372–1381.

Botchwey, C., Dalai, A. K., Adjaye, J. (2004) Kinetics of bitumen-derived gas oil using a commercial $NiMo/Al_2O_3$ catalyst. *Can. J. Chem. Eng.* 82:478–487.

Botchwey, C., Dalai, A. K., Adjaye, J. (2006) Simulation of a two-stage micro trickle-bed hydrotreating reactor using Athabasca bitumen-derived heavy gas oil over commercial $NiMo/Al_2O_3$ catalyst: effect of H_2S on hydrodesulfurization and hydrodenitrogenation. *Int. J. Chem. Reactor Eng.* 4(A20):1–15.

Broderick, D. H., Gates, B. C. (1981) Hydrogenolysis and hydrogenation of dibenzothiophene catalyzed by sulfided $CoO\text{-}MoO_3/\gamma\text{-}Al_2O_3$: the reaction kinetics. *AIChE J.* 27(4):663–673.

Brulé, M. R., Starling, K. E. (1984) Thermophysical properties of complex systems: applications of multiproperty analysis. *Ind. Eng. Chem. Process Des. Dev.* 23(4): 833–845.

Buffham, B. A., Gibilaro, L. G., Rathor, M. N. (1976) A probabilistic time delay description of flow in packed beds. *AIChE J.* 16(2):218–223.

Burghardt, A., Zaleski, T. (1968) Longitudinal dispersion at small and large Peclet numbers in chemical flow reactors. *Chem. Eng. Sci.* 23(6):575–591.

Butt, J. B., Weekman Jr., V. W. (1974) Characterization of the activity, selectivity and aging properties of heterogeneous catalysts. *AIChE Symp. Ser.* 70(143):27–41.

Callejas, M. A., Martínez, M. T. (1999) Hydroprocessing of a Maya residue: intrinsic kinetics of sulfur-, nitrogen-, nickel-, and vanadium-removal reactions. *Energy Fuels* 13(3):629–636.

Callejas, M. A., Martínez, M. T. (2002) Evaluation of kinetic and hydrodynamic models in the hydroprocessing of a trickle-bed reactor. *Energy Fuels* 16(3):647–652.

Carberry, J. J. (1976) *Chemical and Catalytic Reaction Engineering*. McGraw-Hill, New York.

Carberry, J. J., Wendel, M. M. (1963) A computer model of the fixed bed catalytic reactor: the adiabatic and quasi-adiabatic cases. *AIChE J.* 9(1):129–133.

Carberry, J. J., White, D. (1969) On the role of transport phenomena in catalytic reactor behavior. *Ind. Eng. Chem.* 61(7):27–35.

Carberry, J., Varma, A. (1987) *Chemical Reaction and Reactor Engineering*. Chemical Industries Series, Vol. 26. Marcel Dekker, New York.

Carroll, J. J., Mather, A. E. (1995) A Generalized correlation for the Peng–Robinson interaction coefficients for paraffin–hydrogen sulfide binary systems. *Fluid Phase Equilib.* 105(2):221–228.

Cassanello, M. C., Martinez, O. M., Cukierman, A. L. (1992) Effect of the liquid axial dispersion on the behavior of fixed bed three phase reactors. *Chem. Eng. Sci.* 47(13–14):3331–3338.

Cassanello, M. C., Cukierman, A. L., Martinez, O. M. (1994) Actas del V Congreso Latinoamericano de Transferencia de Calor y Materia, IID-5.1.

Cassanello, M. C., Cukierman, A. L., Martinez, O. M. (1996) General criteria to analyze the role of mass transfer and hydrodynamics in trickle-bed reactors. *Chem. Eng. Technol.* 19(5):410–419.

Chao, Y. C., Chang, J. S. (1987) Dynamics of a residue hydrodesulfurization trickle bed reactor system. *Chem. Eng. Commun.* 56(1–6):285–309.

Charpentier, J. C., Favier, M. (1975) Some liquid holdup experimental data in trickle-bed reactors for foaming and nonfoaming hydrocarbons. *AIChE J.* 21(6): 1213–1218.

Chen, J., Ring, Z., Dabros, T. (2001) Modeling and simulation of a fixed-bed pilot-plant hydrotreater. *Ind. Eng. Chem. Res.* 40(15):3294–3300.

Chen, J., Wang, N., Mederos, F., Ancheyta, J. (2009) Vapor–liquid equilibrium study in trickle-bed reactors. *Ind. Eng. Chem. Res.* 48:1096–1106.

Cheng, Z.-M., Fang, X.-C., Zeng, R.-H., Han, B.-P., Huang, L., Yuan, W.-K. (2004) Deep removal of sulfur and aromatics from diesel through two-stage concurrently and countercurrently operated fixed-bed reactors. *Chem. Eng. Sci.* 59(22–23): 5465–5472.

Chilton, T. H., Colburn, A. P. (1939) Mass transfer (absorption) coefficients. *Ind. Eng. Chem.* 26(11):1183–1187.

Chou, Y., Ho, T. C. (1989) Lumping coupled nonlinear reactions in continuous mixtures. *AIChE J.* 35:533–538.

Chowdhury, R., Pedernera, E., Reimert, R. (2002) Trickle-bed reactor model for desulfurization and dearomatization of diesel. *AIChE J.* 48(1):126–135.

Chu, C. F., Ng, K. M. (1989) Flow in packed tubes with a small tube to particle diameter ratio. *AIChE J.* 35(1):148–158.

Chung, T.-H., Ajlan, M., Lee, L. L., Starling, K. E. (1988) Generalized multiparameter correlation for nonpolar and polar fluid transport properties. *Ind. Eng. Chem. Res.* 27(4):671–679.

Cooper, B. H., Donnis, B. B. L., Moyse, B. (1986) Hydroprocessing conditions affect catalyst shape selection. *Oil Gas J.* 84(49):39–44.

Cotta, R. M., Maciel Filho, R. (1996) Kinetic and reactor models for HDT of middle distillates. Presented at the 5th World Congress of Chemical Engineers, San Diego, CA.

Cotta, R. M., Wolf-Maciel, M. R., Maciel Filho, R. (2000) A cape of HDT industrial reactor for middle distillates. *Comput. Chem. Eng.* 24(2–7):1731–1735.

Cotterman, R. L., Prausnitz, J. M. (1985) Flash calculations for continuous mixtures using an equation of state. *Ind. Eng. Chem. Process Des. Dev.* 24(2):434–443.

Cotterman, R. L., Bender, R., Prausnitz, J. M. (1985) Phase equilibria for mixtures containing very many components: development and application of continuous thermodynamics for chemical process design. *Ind. Eng. Chem. Process Des. Dev.* 24(1):194–203.

Crine, M., Marchot, P., L'Homme, G. A. (1980) A phenomenological description of trickle-bed reactors application to the hydrotreating of petroleum fractions. *Chem. Eng. Sci.* 35(1–2):51–58.

Danckwerts, P. V. (1953) Continuous flow systems: distribution of residence times. *Chem. Eng. Sci.* 2(1):1–13.

Dassori, C. G., Pacheco, M. A. (2002) Hydrocracking: an improved kinetic model and reactor modeling. *Chem. Eng. Commun.* 189:1684–1704.

Deans, H. A., Lapidus, L. A. (1960) Computational model for predicting and correlating the behavior of fixed-bed reactors: I. Derivation of model for nonreactive systems. *AIChE J.* 6(4):656–663.

De Ligny, C. L. (1970) Coupling between diffusion and convection in radial dispersion of matter by fluid flow through packed beds. *Chem. Eng. Sci.* 25(7):1177–1181.

Demaria, F., White, R. R. (1960) Transient response study of gas flowing through irrigated packing. *AIChE J.* 6(3):473–481.

Dixon, A. G. (1985) Thermal resistance models of packed-bed effective heat transfer parameters. *AIChE J.* 31(5):826–834.

Döhler, W., Rupp, M. (1987) Comparison of performance of an industrial VGO-treater with reactor model predictions. *Chem. Eng. Technol.* 10(1):349–352.

Doraiswamy, L. K., Tajbl, D. G. (1974) Laboratory catalytic reactors. *Catal. Rev. Sci. Eng.* 10(1):177–219.

Duduković, M. P. (1977) Catalyst effectiveness factor and contacting efficiency in trickle-bed reactors. *AIChE J.* 23(6):940–944.

Duduković, M. P., Larachi, F., Mills, P. L. (1999) Multiphase reactors: revisited. *Chem. Eng. Sci.* 54(13–14):1975–1995.

Duduković, M. P., Larachi, F., Mills, P. L. (2002) Multiphase catalytic reactors: a perspective on current knowledge and future trends. *Catal. Rev. Sci. Eng.* 44(1):123–246.

Dwivedi, P. N., Upadhyay, S. N. (1977) Particle-fluid mass transfer in fixed and fluidized beds. *Ind. Eng. Chem. Process Des. Dev.* 16(2):157–165.

Elizalde, I., Rodríguez, M. A., Ancheyta, J. (2009) Application of continuous kinetic lumping modeling to moderate hydrocracking of heavy oil. *Appl. Catal. A* 365:237–242.

El-Kady, F. Y. (1979) Hydrocracking of vacuum distillate fraction over bifunctional molybdenum–nickel/silica–alumina catalyst. *Indian J. Technol.* 17:176–183.

Ellman, M. J., Midoux, N., Laurent, A., Charpentier, J. C. (1988) A new, improved pressure drop correlation for trickle-bed reactors. *Chem. Eng. Sci.* 43(8):2201–2206.

Ellman, M. J., Midoux, N., Wild, G., Laurent, A., Charpentier, J. C. (1990) A new, improved liquid hold-up correlation for trickle-bed reactors. *Chem. Eng. Sci.* 45(7):1677–1684.

Evans, G. C., Gerald, C. F. (1953) Mass transfer from benzoic acid granules to water in fixed and fluidized beds at low Reynolds numbers. *Chem. Eng. Prog.* 49(3):135–140.

Fahien, R. W., Smith, J. M. (1955) Mass transfer in packed beds. *AIChE J.* 1(1): 28–37.

Feng, G. X., Mather, A. E. (1993a) Solubility of H_2S in *n*-hexadecane at elevated pressure. *Can. J. Chem. Eng.* 71(2):327–328.

Feng, G. X., Mather, A. E. (1993b) Solubility of H_2S in *n*-dodecane. *Fluid Phase Equilib.* 87(2):341–346.

Ferdous, D., Dalai, A. K., Adjaye, J. (2006) Hydrodenitrogenation and hydrodesulfurization of heavy gas oil using $NiMo/Al_2O_3$ catalyst containing boron: experimental and kinetic studies. *Ind. Eng. Chem. Res.* 45(2):544–552.

Feyo De Azevedo, S., Romero-Ogawa, M. A., Wardle, A. P. (1990) Modelling of tubular fixed-bed catalytic reactors: a brief review. *Trans. Inst. Chem. Eng.* 68(A):483–502.

Fott, P., Schneider, P. (1984) Multicomponent mass transport with complex reaction in porous catalyst. In: *Recent Advances in the Engineering Analysis of Chemically Reacting Systems*, Doraiswamy, L. K. (ed.). Wiley Eastern, New Delhi, India.

Froment, G. F. (1986) The kinetics of catalytic processes: importance in reactor simulation and design. *Appl. Catal.* 22(1):3–20.

Froment, G. F. (2004) Modeling in the development of hydrotreatment processes. *Catal. Today* 98(1–2):43–54.

Froment, G. F. (2005) Single event kinetic modeling of complex catalytic processes. *Catal. Rev. Sci. Eng.* 47:83–124.

Froment, G. F., Bischoff, K. B. (1990) *Chemical Reactor Analysis and Design*, 2nd ed. Wiley, New York.

Froment, G. F., Depauw, G. A., Vanrysselberghe, V. (1994) Kinetic modeling and reactor simulation in hydrodesulfurization of oil fractions. *Ind. Eng. Chem. Res.* 33(12): 2975–2988.

Froment, G. F., Castaneda-Lopez, L. C., Marin-Rosas, C. (2008) Kinetic modeling of the hydrotreatment of light cycle oil and heavy gas oil using the structural contributions approach. *Catal. Today* 130(2–4):446–454.

Frye, C. G., Mosby, J. F. (1967) Kinetics of hydrodesulfurization. *Chem. Eng. Prog.* 63(9): 66–70.

Galiasso, T. R. (2006) Diesel upgrading into a low emissions fuel. *Fuel Process. Tech.* 87(9):759–767.

Gianetto, A., Specchia, V. (1992) Trickle-bed reactors: state of art and perspectives. *Chem. Eng. Sci.* 47(13–14):3197–3213.

Gianetto, A., Baldi, G., Specchia, V., Sicardi, S. (1978) Hydrodynamics and solid–liquid contacting effectiveness in trickle-bed reactors. *AIChE J.* 24(6):1087–1104.

Gierman, H. (1988) Design of laboratory hydrotreating reactors scaling down of trickle-flow reactors. *Appl. Catal.* 43(2):277–286.

Glaso, O. (1980) Generalized pressure–volume–temperature correlations. *J. Pet. Technol.* 32(5):785–795.

Glasscock, D. A., Hale, J. C. (1994) Process simulation: the art and science of modeling. *Chem. Eng.* 101(11):82–89.

Goto, S., Smith, J. M. (1975) Trickle-bed reactor performance: I. Holdup and mass transfer effects. *AIChE J.* 21(4):706–713.

Gunjal, P. R., Ranade, V. V. (2007) Modeling of laboratory and commercial scale hydroprocessing reactors using CFD. *Chem. Eng. Sci.* 62(18–20):5512–5526.

Guo, J., Jiang, Y., Al-Dahhan, M. H. (2008) Modeling of trickle-bed reactors with exothermic reactions using a cell network approach. *Chem. Eng. Sci.* 63(3): 751–764.

Gupta, S. N., Chaube, R. B., Upadhyay, S. N. (1974) Fluid-particle heat transfer in fixed bed and fluidized beds. *Chem. Eng. Sci.* 29(3):839–843.

Hashimoto, K., Muroyama, K., Fujiyoshi, K., Nagata, S. (1976) Effective radial thermal conductivity in cocurrent flow of gas and liquid through packed bed. *Int. Chem. Eng.* 16(4):720–727.

Hastaoglu, M. A., Jibril, B. E. (2003) Transient modeling of hydrodesulfurization in a fixed-bed reactor. *Chem. Eng. Commun.* 190(2):151–170.

Haughey, D. P., Beveridge, G. S. (1969) Structural properties of packed beds: a review. *Can. J. Chem. Eng.* 47(2):130–140.

Henry, H. C., Gilbert, J. B. (1973) Scale up of pilot data for catalytic hydroprocessing. *Ind. Eng. Chem. Process Des. Dev.* 12(3):328–334.

Herskowitz, M., Smith, J. M. (1978a) Liquid distribution in trickle-bed reactors: I. Flow measurements. *AIChE J.* 24:439–450.

Herskowitz, M., Smith, J. M. (1978b) Liquid distribution in trickle-bed reactors: II. Tracer studies. *AIChE J.* 24:450–454.

Hill, C. G. (1977) *An Introduction to Chemical Engineering Kinetics and Reactor Design*. Wiley, New York.

Hlavacek, V., Marek, M. (1966) Axialer stoff- und wärmetransport im adiabatischen rohrreaktor: II. Numerische untersuchung-ablauf einer einfachen reaktion bezw. einer folgereaktion. *Chem. Eng. Sci.* 21:501–513.

Ho, T. C. (2003) Property–reactivity correlation for HDS of middle distillates. *Appl. Catal. A* 244(1):115–128.

Ho, T. C. (2008) Kinetic modeling of large-scale reaction systems. *Catal. Rev.* 50(3):287–378.

Ho, T. C., Markley, G. E. (2004) Property–reactivity correlation for hydrodesulfurization of prehydrotreated distillates. *Appl. Catal. A* 267(1–2):245–250.

Ho, T. C., Nguyen, D. (2006) Modeling of competitive adsorption of nitrogen species in hydrodesulfurization. *Chem. Eng. Commun.* 193(4):460–477.

Hochman, J. M., Effron, E. (1969) Two-phase cocurrent downflow in packed beds. *Ind. Eng. Chem. Fundam.* 8(1):63–71.

Hoekstra, G. (2007) The effects of gas-to-oil rate in ultra low sulfur diesel hydrotreating. *Catal. Today* 127(1–4):99–102.

Hofmann, H. (1977) Hydrodynamics, transport phenomena, and mathematical models in trickle-bed reactors. *Int. Chem. Eng.* 17(1):19–28.

Hofmann, H. P. (1978) Multiphase catalytic packed-bed reactors. *Catal. Rev. Sci. Eng.* 17(1):71–117.

Hoftyzer, P. J. (1964) Liquid distribution in a column with dumped packing. *Trans. Inst. Chem. Eng.* 42:T109–T117.

Hu, M. C., Ring, Z., Briker, J., Te, M. (2002) Rigorous hydrotreater simulation. *Pet. Tech. Q.* Spring, pp. 85–91.

Iannibello, A., Marengo, S., Guerci, A., Baldi, G., Sicardi, S. (1983) Performance of a pilot trickle-bed reactor for hydrotreating of petroleum fractions: dynamic analysis. *Ind. Eng. Chem. Process Des. Dev.* 22(4):594–598.

Iannibello, A., Marengo, S., Burgio, G., Baldi, G., Sicardi, S., Specchia, V. (1985) Modeling the hydrotreating reactions of a heavy residual oil in a pilot trickle-bed reactor. *Ind. Eng. Chem. Process Des. Dev.* 24(3):531–537.

Iliuta, I., Ortiz-Arroyo, A., Larachi, F., Grandjean, B. P. A., Wild, G. (1999) Hydrodynamics and mass transfer in trickle-bed reactors: an overview. *Chem. Eng. Sci.* 54(21):5329–5337.

Iliuta, I., Ring, Z., Larachi, F. (2006) Simulating simultaneous fines deposition under catalytic hydrodesulfurization in hydrotreating trickle beds: Does bed plugging affect HDS performance? *Chem. Eng. Sci.* 61(4):1321–1333.

Inoue, S., Takatsuka, T., Wada, Y., Hirohama, S., Ushida, T. (2000) Distribution function model for deep desulfurization of diesel fuel. *Fuel* 79(7):843–849.

Jacobs, G. E., Milliken, A. S. (2000) Evaluating liquid distributors in hydroprocessing reactors. *Hydrocarbon Process.* Nov., pp. 76–84.

Jakobsson, K., Hasanen, A., Aittamaa, J. (2004) Modelling of a countercurrent hydrogenation process. *Trans. IChemE A* 82(A2):203–207.

Jiménez, F., Núñez, M., Kafarov, V. (2005) Study and modeling of simultaneous hydrodesulfurization, hydrodenitrogenation and hydrodearomatization on vacuum gas oil hydrotreatment. *Comput.-Aided Chem. Eng.* 20(1):619–624.

Jiménez, F., Kafarov, V., Nuñez, M. (2006) Computer-aided modeling for hydrodesulfurization, hydrodenitrogenation and hydrodearomatization simultaneous reactions in a hydrotreating industrial process. *Comput.-Aided Chem. Eng.* 21(1):651–657.

Jiménez, F., Ojeda, K., Sánchez, E., Karafov, V., Maciel Filho, R. (2007a) Modeling of trickle bed reactor for hydrotreating of vacuum gas oils: effect of kinetic type on reactor modeling. *Comput. Aided Process Eng.* ESCAPE17, 1–6.

Jiménez, F., Kafarov, V., Nuñez, M. (2007b) Modeling of industrial reactor for hydrotreating of vacuum gas oils: simultaneous hydrodesulfurization, hydrodenitrogenation and hydrodearomatization reactions. *Chem. Eng. J.* 134(1–3):200–208.

Juraidan, M., Al-Shamali, M., Qabazard, H., Kam, E. K. T. (2006) A refined hydroprocessing multicatalyst deactivation and reactor performance model-pilot-plant accelerated test applications. *Energy Fuels* 20(4):1354–1364.

Kam, E. K. T., Al-Shamali, M., Juraidan, M., Qabazard, H. (2005) A hydroprocessing multicatalyst deactivation and reactor performance model-pilot-plant life test applications. *Energy Fuels* 19(3):753–764.

Khadilkar, M. R., Mills, P. L., Dudukovic, M. P. (1999) Trickle-bed reactor models for systems with a volatile liquid phase. *Chem. Eng. Sci.* 54(13–14):2421–2431.

Kodama, S., Nitta, H., Takatsuka, T., Yokoyama, T. (1980) Simulation of residue hydrodesulfurization reaction based on catalyst deactivation nodel. *J. Jpn. Pet. Inst.* 23(5):310–320.

Korsten, H., Hoffmann, U. (1996) Three-phase reactor model for hydrotreating in pilot trickle-bed reactors. *AIChE J.* 42(5):1350–1360.

Krishna, R., Saxena, A. K. (1989) Use of an axial-dispersion model for kinetic description of hydrocracking. *Chem. Eng. Sci.* 44:703–712.

Kumar, H., Froment, G. F. (2007) Mechanistic kinetic modeling of the hydrocracking of complex feedstocks, such as vacuum gas oils. *Ind. Eng. Chem. Res.* 46(18): 5881–5897.

Kumar, V. R., Balaraman, K. S., Rao, V. S. R., Ananth, M. S. (1997) Modelling of hydrotreating process in a trickle-bed reactor. *Pet. Sci. Technol.* 15(1–2): 283–295.

Kumar, V. R., Balaraman, K. S., Rao, V. S. R., Ananth, M. S. (2001) Performance study of certain commercial catalysts in hydrodesulfurization of diesel oils. *Pet. Sci. Technol.* 19(9–10):1029–1038.

Kundu, A., Nigam, K. D. P., Duquenne, A. M., Delmas, H. (2003) Recent developments on hydroprocessing reactors. *Rev. Chem. Eng.* 19(6):531–605.

Kwak, S. (1994) Hydrotreating heavy oils over a commercial hydrodemetallation catalyst. Ph.D. dissertation, University of Utah, Aug.

Lababidi, H. M. S., Shaban, H. I., Al-Radwan, S., Alper, E. (1998) Simulation of an atmospheric residue desulfurization unit by quasi-steady state modeling. *Chem. Eng. Technol.* 21(2):193–200.

Lal, D., Otto, F. D., Mather, A. E. (1999) Solubility of hydrogen in Athabasca bitumen. *Fuel* 78(12):1437–1441.

Larachi, F., Laurent, A., Midoux, N., Wild, G. (1991a) Experimental study of a trickle-bed reactor operating at high pressure: two-phase pressure drop and liquid saturation. *Chem. Eng. Sci.* 46(5–6):1233–1246.

Larachi, F., Laurent, A., Wild, G., Midoux, N. (1991b) Some experimental liquid saturation results in fixed-bed reactors operated under elevated pressure in cocurrent upflow and downflow of the gas and the liquid. *Ind. Eng. Chem. Res.* 30(11): 2404–2410.

Larachi, F., Grandjean, B., Iliuta, I., Bensetiti, Z., André, E., Wild, G., Chen, M. (1999) Excel worksheet simulator for trickle-bed reactors. http://www.gch.ulaval.ca/bgrandjean/pbrsimul/pbrsimul.html.

Larkins, R. P., White, R. R., Jeffrey, D. W. (1961) Two-phase concurrent flow in packed beds. *AIChE J.* 7(2):231–239.

Laxminarasimhan, C. S., Verma, R. P. (1996) Continuous lumping model for simulation of hydrocracking. *AIChE J.* 42:2645–2653.

Lee, B. I., Kesler, M. G. (1975) A generalized thermodynamic correlation based on three-parameter corresponding states. *AIChE J.* 21(3):510–527.

Lee, B. I., Kesler, M. G. (1976) Improve prediction of enthalpy of fractions. *Hydrocarbon Process* 3(Mar.):153–158.

Lee, H. H., Smith, J. M. (1982) Trickle-bed reactors: criteria of negligible transport effects and of partial wetting. *Chem. Eng. Sci.* 37:223.

Levenspiel, O., Bischoff, K. B. (1963) Patterns of flow in chemical process vessels. *Adv. Chem. Eng.* 4:95–198.

Liguras, D. K., Allen, D. T. (1989) Structural models for catalytic cracking: model compound reactions. *Ind. Eng. Chem. Res.* 28:665–673.

Liu, Z., Zheng, Y., Wang, W., Zhang, Q., Jia, L. (2008) Simulation of hydrotreating of light cycle oil with a system dynamics model. *Appl. Catal. A* 339:209–220.

Lopez, R., Dassori, C. G. (2001) Mathematical modeling of a VGO hydrotreating reactor. *SPE Int.* SPE 69499:1–14.

Lopez, R., Perez, J. R., Dassori, C. G., Ranson, A. (2001) Artificial neural networks applied to the operation of VGO hydrotreaters. *SPE Int.* SPE 69500:1–10.

Lukec, I., Sertić-Bionda, K., Lukec, D. (2008) Prediction of sulphur content in the industrial hydrotreatment process. *Fuel Process Tech.* 89(3):292–300.

Macé, O., Wei, J. (1991) Diffusion in random particle models for hydrodemetalation catalysts. *Ind. Eng. Chem. Res.* 30(5):909–918.

Macías, M. J., Ancheyta, J. (2004) Simulation of an isothermal hydrodesulfurization small reactor with different catalyst particle shapes. *Catal. Today* 98(1–2):243–252.

Magoulas, K., Tassios, D. (1990) Thermophysical properties of *n*-alkanes from C1 to C20 and their prediction for higher ones. *Fluid Phase Equilib.* 56:119–140.

Marroquín de la Rosa, J. O., Valencia-López, J. J., Ochoa-Tapia, J. A., Viveros-García, T. (2002) Trickle-bed reactor modeling for diesel hydrotreatment. *Proceedings of ISCRE-17*, Hong Kong, China, Aug. 25–28.

Marroquín, G., Ancheyta, J., Esteban, C. (2005) A batch reactor study to determine effectiveness factors of commercial HDS catalyst. *Catal. Today* 104(1):70–75.

Martens, G. G., Marin, G. B. (2001) Kinetics for hydrocracking based on structural classes: model development and application. *AIChE J.* 47:1607–1622.

Mears, D. E. (1971) The role of axial dispersion in trickle-flow laboratory reactors. *Chem. Eng. Sci.* 26(9):1361–1366.

Mears, D. E. (1974) The role of liquid holdup and effective wetting in the performance of trickle-bed reactors. In: *Proceedings of the 3rd International Symposium or Chemical Reaction Engineering II*, Hulburt, H. M. (ed.) ACS Monograph Series No. 133. American Chemical Society, Washington, DC, p. 218.

Mears, D. E. (1976) On criteria for axial dispersion in nonisothermal packed-bed catalytic reactors. *Ind. Eng. Chem. Fundam.* 15(1):20–23.

Mederos, F. S., Ancheyta, J. (2007) Mathematical modeling and simulation of hydrotreating reactors: cocurrent versus countercurrent operations. *Appl. Catal. A* 332(1): 8–21.

Mederos, F. S., Rodríguez, M. A., Ancheyta, J., Arce, E. (2006) Dynamic modeling and simulation of catalytic hydrotreating reactors. *Energy Fuels* 20(3):936–945.

Mejdell, T., Myrstad, R., Morud, J., Rosvoll, J. S., Steiner, P., Blekkan, E. A. (2001) *A New Kinetic Model for Hydrodesurfurization of Oil Products*. Studies in Surface Science Catalysis, Vol. 133, Froment, G. F., Waugh, K. C. (eds.). Elsevier Science Amsterdam, p. 189.

Melis, S., Erby, L., Sassu, L., Baratti, R. (2004) A model for the hydrogenation of aromatic compounds during gasoil hydroprocessing. *Chem. Eng. Sci.* 59(22–23): 5671–5677.

Mills, P. L., Duduković, M. P. (1981) Evaluation of liquid–solid contacting in trickle-bed reactors by tracer methods. *AIChE J.* 27(6):893–904.

Mohanty, S., Saraf, D. N., Kunzru, D. (1991) Modeling of a hydrocracking reactor. *Fuel Process. Tech.* 29:1–17.

Montagna, A. A., Shah, Y. T. (1975) The role of liquid holdup, effective catalyst weetting, and backmixing on the performance of a trickle bed reactor for residue hydrodesulfurization. *Ind. Eng. Chem. Process Des. Dev.* 14(4):479–483.

Mosby, F., Buttke, R. D., Cox, J. A., Nikolaides, C. (1986) Process characterization of expanded-reactors in series. *Chem. Eng. Sci.* 41:989–995.

Mostoufi, N., Sotudeh-Gharebagh, R., Ahmadpour, M., Eyvani, J. (2005) Simulation of an industrial pyrolysis gasoline hydrogenation unit. *Chem. Eng. Technol.* 28(2): 174–181.

Moysan, J. M., Huron, M. J., Paradowski, H., Vidal, J. (1983) Prediction of the solubility of hydrogen in hydrocarbon solvents through cubic equations of state. *Chem. Eng. Sci.* 38(7):1085–1092.

Muñoz, J. A. D., Alvarez, A., Ancheyta, J., Rodríguez, M. A., Marroquín, G. (2005) Process heat integration of a heavy crude hydrotreatment plant. *Catal. Today* 109(1–4):214–218.

Murali, C., Voolapalli, R. K., Ravichander, N., Gokak, D. T., Choudary, N. V. (2007) Trickle bed reactor model to simulate the performance of a commercial diesel hydrotreating unit. *Fuel* 86(7–8):1176–1184.

Murphree, E. V., Voorhies, A., Jr., Mayer, F. X. (1964) Application of contacting studies to the analysis of reactor performance. *Ind. Eng. Chem. Proc. Des. Dev.* 3(4): 381–386.

Narasimhan, C. S. L., Sau, M., Verma, R. P. (1997) *An Integrated Approach for Hydrocracker Modeling*. Studies in Surface Science Catalysis, Vol. 127. Elsevier Science, Amsterdam.

Nguyen, N. L., Reimert, R., Hardy, E. H. (2006) Application of magnetic resonance imaging (MRI) to determine the influence of fluid dynamics on desulfurization in bench scale reactors. *Chem. Eng. Technol.* 29(7):820–827.

Oh, M., Jang, E. J. (1997) Rigorous modelling and dynamic simulation of an industrial naphtha hydrodesulfurization process. *J. Korean Inst. Chem. Eng.* (in Korean) 35(5):791–798.

Ojeda, J. A., Krishna, R. (2004) In-situ stripping of H_2S in gasoil hydrodesulphurization: reactor design considerations. *Trans. IChemE A* 82(A2):208–214.

Onda, K., Takeuchi, H., Koyama, Y. (1967) Effect of packing materials on the wetted surface area. *Kagaku Kogaku* (in Japanese) 31(2):126–134.

Orochko, D. I. (1970) Applied over-all kinetics of hydrocracking of heavy petroleum distillates. *Khim. Tekhnol. Topliv Masel* 8:2–6.

Papayannakos, N., Georgiou, G. (1988) Kinetics of hydrogen consumption during catalytic hydrodesulphurization of a residue in a trickle-bed reactor. *J. Chem. Eng. Jpn.* 21(3):244–249.

Paraskos, J. A., Frayer, J. A., Shah, Y. T. (1975) Effect of holdup incomplete catalyst wetting and backmixing during hydroprocessing in trickle bed reactors. *Ind. Eng. Chem. Process Des. Dev.* 14(3):315–322.

Pedernera, E., Reimert, R., Nguyen, N. L., van Buren, V. (2003) Deep desulfurization of middle distillates: process adaptation to oil fractions' compositions. *Catal. Today* 79–80:371–381.

Peng, D. Y., Robinson, D. B. (1976) A new two-constant equation of state. *Ind. Eng. Chem. Fundam.* 15(1):59–64.

Perego, C., Paratello, S. (1999) Experimental methods in catalytic kinetics. *Catal. Today* 52(2–3):133–145.

Perry, R. H., Green, D. W., Maloney, J. O. (2004) *Perry's Chemical Engineers' Handbook*. McGraw-Hill, New York, p. 2–193.

Petersen, E. E. (1965) *Chemical Reaction Analysis*. Prentice-Hall, Englewood Cliffs, NJ.

Petrovic, L. J., Thodos, G. (1968) Mass transfer in the flow of gases through packed beds. *Ind. Eng. Chem. Fundam.* 7(2):274–280.

Pitault, I., Fongarland, P., Mitrovic, M., Ronze, D., Forissier, M. (2004) Choice of laboratory scale reactors for HDT kinetic studies or catalyst tests. *Catal. Today* 98(1–2): 31–42.

Porter, K. E., Templeman, J. J. (1968) Wall flow: III. *Trans. Inst. Chem. Eng.* 46:T86.

Porter, K. E., Templeman, J. J., Barnett, V. D. (1968) The spread of liquid over random packings: II. *Trans. Inst. Chem. Eng.* 46:T74.

Prchlik, J., Soukup, J., Zapletal, V., Ruzicka, V. (1975a) Liquid distribution in reactors with randomly packed porous beds. *Collect. Czech. Chem. Commun.* 40:845.

Prchlik, J., Soukup, J., Zapletal, V., Ruzicka, V. (1975b) Wall flow in trickle bed reactors. *Collect. Czech. Chem. Commun.* 40:3145.

Puranik, S. S., Vogelpohl, A. (1974) Effective interfacial area in irrigated packed columns. *Chem. Eng. Sci.* 29(2):501–507.

Qader, S. A., Hill, G. R. (1969) Hydrocracking of gasoil. *Ind. Eng. Chem. Process Des. Dev.* 8:98–105.

Quann, R. J., Jaffe, S. B. (1992) Structure-oriented lumping: describing the chemistry of complex mixtures. *Ind. Eng. Chem. Res.* 31(11):2483–2497.

Ramírez, L. F., Escobar, J., Galván, E., Vaca, H., Murrieta, F. R., Luna, M. R. S. (2004) Evaluation of diluted and undiluted trickle-bed hydrotreating reactor with different catalyst volume. *Pet. Sci. Technol.* 22(1–2):157–175.

Reiss, L. P. (1967) Cocurrent gas–liquid contacting in packed columns. *Ind. Eng. Chem. Proccess Des. Dev.* 6(4):486–499.

Riazi, M. R. (2005) *Characterization and Properties of Petroleum Fractions*. ASTM International Standards Worldwide, Philadelphia.

Ring, Z. E., Missen, R. W. (1991) Trickle-bed reactors: tracer study of liquid holdup and wetting efficiency at high temperature and pressure. *Can. J. Chem. Eng.* 69(4): 1016–1020.

Rodríguez, M. A., Ancheyta, J. (2004) Modeling of hydrodesulfurization (HDS), hydrodenitrogenation (HDN), and the hydrogenation of aromatics (HDA) in a vacuum gas oil hydrotreater. *Energy Fuels* 18(3):789–794.

Ronze, D., Fongarland, P., Pitault, I., Forissier, M. (2002) Hydrogen solubility in straight run gasoil. *Chem. Eng. Sci.* 57:547–553.

Ross, L. D. (1965) Performance of trickle bed reactors. *Chem. Eng. Prog.* 61(10):77–82.

Salmi, T., Romanainen, J. (1995) A novel exit boundary condition for the axial dispersion model. *Chem. Eng. Process.* 34(4):359–366.

Salmi, T., Wärnå, J., Toppinen, S., Rönnholm, M., Mikkola, J. P. (2000) Dynamic modelling of catalytic three-phase reactors for hydrogenation and oxidation processes. *Braz. J. Chem. Eng.* 17(4–7):1023–1035.

Salvatore, L., Pires, B., Campos, M. C. M., De Souza, M. B., Jr. (2005) A hybrid approach to fault detection and diagnosis in a diesel fuel hydrotreatment process. *Trans. Eng. Comput. Technol.* 7:379–384.

Sánchez, M., Shah, Y. T., Dassori, C. G. (1995) Multiphase hydrotreating reactor modeling. Presented at the AIChE Spring Meeting, Houston, TX, Mar. 19–24 (42b).

Sánchez, S., Ancheyta, J. (2007) Effect of pressure on the kinetics of moderate hydrocracking of Maya crude oil. *Energy Fuels* 21:653–661.

Sánchez, S., Rodríguez, M. A., Ancheyta, J. (2005) Kinetic model for moderate hydrocracking of heavy oils. *Ind. Eng. Chem. Res.* 44:9409–9413.

Saroha, A. K., Nigam, K. D. P., Saxena, A. K., Kapoor, V. K. (1988) Liquid distribution in trickle-bed reactors. *AIChE J.* 44(9):2044–2052.

Sater, V. E., Levenspiel, O. (1966) Two-phase flow in packed beds: evaluation of axial dispersion and holdup by moment analysis. *Ind. Eng. Chem. Fundam.* 5(1):86–92.

Satterfield, C. N. (1969) Mass transfer limitations in trickle-bed reactor. *AIChE J.* 15(2):226–234.

Satterfield, C. N. (1970) *Mass Transfer in Heterogeneous Catalysis*. MIT Press, Cambridge, MA.

Satterfield, C. N. (1975) Trickle-bed reactors. *AIChE J.* 21(2):209–228.

Satterfield, C. N., Ozel, F. (1973) Direct solid-catalyzed reaction of a vapor in an apparently completely wetted trickle bed reactor. *AIChE J.* 19(6):1259–1261.

Satterfield, C. N., Van Eek, M. W., Bliss, G. S. (1978) Solid mass transfer in packed beds with downward concurrent gas–liquid flow. *AIChE J.* 24(4):709–717.

Sau, M., Narasimhan, C. S. L., Verma, R. P. (1997) A kinetic model for hydrodesulfurisation. In: *Hydrotreatment and Hydrocracking of Oil Fractions*. Studies in Surface Science Catalysis, Vol. 106, Froment, G. F., Delmon, B., Grange, P. (eds.). Elsevier Science, Amsterdam, p. 421.

Schwartz, J. G., Roberts, G. W. (1973) An evaluation of models for liquid backmixing in trickle-bed reactors. *Ind. Eng. Chem. Process Des. Dev.* 12(3):262–271.

Scott, A. H. (1935) Liquid distribution in packed towers. *Trans. Inst. Chem. Eng.* 13:211–217.

Sertić-Bionda, K., Gomzi, Z., Šarić, T. (2005) Testing of hydrodesulfurization process in small trickle-bed reactor. *Chem. Eng. J.* 106:105–110.

Shah, Y. T., Paraskos, J. A. (1975) Criteria for axial dispersion effects in adiabatic trickle bed hydroprocessing reactors. *Chem. Eng. Sci.* 30(9):1169–1176.

Shah, Y. T., Mhaskar, R. D., Paraskos, J. A. (1976) Optimum quench location for a hydrodesulfurization reactor with time varying catalyst activity. *Ind. Eng. Chem. Process Des. Dev.* 15(3):400–406.

Shinnar, R. (1978) Process control research. In: *An Evaluation of Present Status and Research Needs*. ACS Symposium Series, Vol. 72. American Chemical Society, Washington, DC, p. 1.

Shokri, S., Zarrinpashne, S. (2006) A mathematical model for calculation of effectiveness factor in catalyst pellets of hydrotreating process. *Pet. Coal* 48(1):27–33.

Shokri, S., Marvast, M. A., Tajerian, M. (2007) Production of ultra low sulfur diesel: simulation and software development. *Pet. Coal* 49(2):48–59.

Sie, S. T. (1991) Scale effects in laboratory and pilot-plant reactors for trickle-flow processes. *Rev. Inst. Fr. Pet.* 46(4):501–515.

Sie, S. T. (1996) Miniaturization of hydroprocessing catalyst testing systems: theory and practice. *AIChE J.* 42(12):3498–3507.

Sie, S. T., Krishna, R. (1998) Process development and scale up: III. Scale-up and scale-down of trickle bed processes. *Rev. Chem. Eng.* 14(3):203–252.

Skala, D. U., Šaban, M. D., Orlović, A. M., Meyn, V. W., Severin, D. K., Rahimian, I. G.-H., Marjanović, M. V. (1991) Hydrotreating of used oil: prediction of industrial trickle-bed operation from pilot-plant data. *Ind. Eng. Chem. Res.* 30(9):2059–2065.

Smith, J. M. (1960) *Chemical Engineering Kinetics*. McGraw-Hill, New York.

Soave, G. (1972) Equilibrium constants for a modified Redlich–Kwong equation of state. *Chem. Eng. Sci.* 27(6):1197–1203.

Specchia, V., Baldi, G. (1977) pressure drop and liquid holdup for two phase concurrent flow in packed beds. *Chem. Eng. Sci.* 32(5):515–523.

Specchia, V., Rossini, A., Baldi, G. (1974) Distribution and radial spread of liquid in two-phase concurrent flows in a packed bed. *Quad. Ing. Chim. Ital.* 10(11):171–176.

Stangeland, B. E. (1974) Kinetic model for the prediction of hydrocracking yields. *Ind. Eng. Chem. Process Des. Dev.* 13:71–75.

Stefanidis, G. D., Bellos, G. D., Papayannakos, N. G. (2005) An improved weighted average reactor temperature estimation for simulation of adiabatic industrial hydrotreaters. *Fuel Process. Tech.* 86(16):1761–1775.

Székely, J. (1961) Reaction rate constant may modify the effects of backmixing. *Ind. Eng. Chem.* 53:313–318.

Tarhan, M. O. (1983) *Catalytic Reactor Design*. McGraw-Hill, New York.

Taylor, R., Krishna, R. (1993) *Multicomponent Mass Transfer*. Wiley Series in Chemical Engineering. Wiley, New York.

Taylor, R., Kooijman, H. A., Hung, J. S. (1994) A second generation nonequilibrium model for computer simulation of multicomponent separation processes. *Comput. Chem. Eng.* 18:205–217.

Te, M., Fairbridge, C., Ring, Z. (2003) Various approaches in kinetics modeling of real feedstock hydrodesulfurization. *Pet. Sci. Technol.* 21(1–2):157–181.

Thiele, E. W. (1939) Relation between catalytic activity and size of particle. *Ind. Eng. Chem.* 31(7):916–920.

Toppinen, S., Aittamaa, J., Salmi, T. (1996) Interfacial mass transfer in trickle-bed reactor modelling. *Chem. Eng. Sci.* 51(18):4335–4345.

Toulhoat, H., Hudebine, D., Raybaud, P., Guillaume, D., Kressmann, S. (2005) Thermidor: a new model for combined simulation of operations and optimization of catalysts in residues hydroprocessing units. *Catal. Today* 109(1–4):135–153.

Trambouze, P. (1990) Countercurrent two-phase flow fixed bed catalytic reactors. *Chem. Eng. Sci.* 45(8):2269–2275.

Trejo, F., Ancheyta, J., Sánchez, S., Rodríguez, M. A. (2007) Comparison of different power-law kinetic models for hydrocracking of asphalthenes. *Pet. Sci. Technol.* 25(1):263–275.

Tsamatsoulis, D., Papayannakos, N. (1995) Simulation of non-ideal flow in a trickle bed hydrotreater by the cross-flow model. *Chem. Eng. Sci.* 50(23):3685–3691.

Tsamatsoulis, D., Papayannakos, N. (1998) Investigation of intrinsic hydrodesulphurization kinetics of a VGO in a trickle bed reactor with backmixing effects. *Chem. Eng. Sci.* 53(19):3449–3458.

Tsamatsoulis, D., Koutoulas, E., Papayannakos, N. (1991) Improvement of characteristic reactions and properties of a heavy residue by catalytic hydrotreatment. *Fuel* 70(6):741–746.

Tsamatsoulis, D., Al-Dahhan, M. H., Larachi, F., Papayannakos, N. (2001) The effect of particle dilution on wetting efficiency and liquid film thickness in small trickle beds. *Chem. Eng. Commun.* 185(1):67–77.

Tyn, M. T., Calus, W. F. (1975) Diffusion coefficients in dilute binary liquid mixtures. *J. Chem. Eng. Data* 20(1):106–109.

Van den Bleek, C. M., Van Der Wiele, K., Van Den Berg, P. J. (1969) Effect of dilution on the degree of conversion in fixed bed catalytic reactors. *Chem. Eng. Sci.* 24:681–694.

van Hasselt, B. W., Lebens, P. J. M., Calis, H. P. A., Kapteijn, F., Sie, S. T., Moulijn, J. A., van den Bleek, C. M. (1999) A numerical comparison of alternative three-phase reactors with a conventional trickle-bed reactor: the advantages of countercurrent flow for hydrodesulfurization. *Chem. Eng. Sci.* 54(21):4791–4799.

van Herk, D., Kreutzer, M. T., Makkee, M., Moulijn, J. A. (2005) Scaling down trickle bed reactors. *Catal. Today* 106(1–4):227–232.

Van Parijs, I. A., Froment, G. F. (1984) The influence of intraparticle diffusional limitations in naphtha hydrodesulfurization. *Inst. Chem. Eng. Symp. Ser.* 87: 319–327.

Vanrysselberghe, V., Froment, G. F. (2002) Hydrodesulfurization-heterogeneous, simulation of hydrodesulfurization reactors. In: *Encyclopedia of Catalysis*. Wiley, Hoboken, NJ.

Vargas-Villamil, F. D., Marroquín, J. O., de la Paz, C., Rodríguez, E. (2004) A catalytic distilliation process for light gas oil hydrodesulfurization. *Chem. Eng. Process.* 43:1309–1316.

Verstraete, J. J., Le Lannic, K., Guibard, I. (2007) Modeling fixed-bed residue hydrotreatment processes. *Chem. Eng. Sci.* 62(18–20):5402–5408.

Vogelaar, B. M., Berger, R. J., Bezemer, B., Janssens, J.-P., van Langeveld, A. D., Eijsbouts, S., Moulijn, J. A. (2006) Simulation of coke and metal deposition in catalyst pellets using a non-steady state fixed bed reactor model. *Chem. Eng. Sci.* 61(22): 7463–7478.

Wärnå, J., Salmi, T. (1996) Dynamic modelling of catalytic three phase reactors. *Comput. Chem. Eng.* 20(1):39–47.

Wehner, J. F., Wilhelm, R. H. (1956) Boundary conditions of flow reactor. *Chem. Eng. Sci.* 6(2):89–93.

Wei, J., Kuo, J. C. W. (1969) Lumping analysis in monomolecular reaction systems: analysis of the exactly lumpable system. *Ind. Eng. Chem. Res.* 8:114–123.

Weissberg, H. (1963) Effective difusión coefficient in porous media. *J. Appl. Phys.* 34:2636–2639.

Whitaker, S. (1973) The transport equations for multi-phase systems. *Chem. Eng. Sci.* 28(1):139–147.

Wilke, C. R., Chang, P. (1955) Correlation for diffusion coefficients in dilute solutions. *AIChE J.* 1(2):264–270.

Wilson, E. J., Geankoplis, C. J. (1966) Liquid mass transfer at very low Reynolds numbers in packed beds. *Ind. Eng. Chem. Fundam.* 5(10):9–14.

Yaïci, W., Laurent, A., Midoux, N., Charpentier, J. C. (1988) Determination of gas-side mass transfer coefficient in trickle-bed reactors in presence of an aqueous or an organic liquid phase. *Int. Chem. Eng.* 28(2):299–305.

Yamada, H., Goto, S. (2004) Advantages of counter-current operation for hydrodesulfurization in trickle bed reactors. *Korean J. Chem. Eng.* 21(4):773–776.

Young, L. C., Finlayson, B. A. (1973) Axial dispersion in nonisothermal packed bed chemical reactors. *Ind. Eng. Chem. Fundam.* 12(4):412–422.

Yui, S. M., Sanford, E. C. (1989) Mild hydrocracking of bitumen-derived coker and hydrocracker heavy gas oils: kinetics, product yields and product properties. *Ind. Eng. Chem. Res.* 28:1278–1284.

Zahedi, G., Fgaier, H., Jahanmiri, A., Al-Enezi, G. (2006) Artificial neural network identification and evaluation of hydrotreater plant. *Pet. Sci. Technol.* 24(12):1447–1456.

Zeldovich, Y. B. (1939) On the theory on powders and porous substances. *Acta Phys. Chim.* 10:583.

Zhukova, T. B., Pisarenko, V. N., Kafarov, V. V. (1990) Modeling and design of industrial reactors with a stationary bed of catalyst and two-phase gas–liquid flow: a review. *Int. Chem. Eng.* 30(1):57–102.

Zimmerman, S. P., Ng, K. M. (1986) Liquid distribution in trickling flow trickle-bed reactors. *Chem. Eng. Sci.* 41(4):861–866.

NOMENCLATURE

a	Glaso's coefficient for viscosity correlation; energy parameter in equation of state
a, b, c	Power terms for temperature, hydrogen partial pressure, and LHSV, respectively
a, b, c, d, n	Parameters
$a_0, a_1, \delta, A, B, D$, So	Model parameters
a_L	Gas–liquid interfacial area per unit reactor volume, cm_l^2/cm_r^3
a_S	Liquid (or gas)–solid interfacial area per unit reactor volume, cm_S^2/cm_r^3
a_w	Effectively wetted area
A	Reduced parameter of energy parameter a, preexponential factor
A_W	Bed-wall reactor heat transfer area, cm_r^2
A_Z	Coefficient of the cubic equation of state expressed in Z
A', b'	Empirical constants for Bondi's correlation
API	API gravity
b	Volume parameter in equation of state
B	Reduced parameter of volume parameter b
B_Z	Coefficient of the cubic equation of state expressed in Z

$C(k, t)$	Concentration function
C_{HGO}	Concentration of heavy gas oil, wt%
C_{LGO}	Concentration of light gas oil, wt%
C_Z	Coefficient of the cubic equation of state expressed in Z
C_i^I	Molar concentration of component i in the gas–liquid interface, mol_i/cm_L^3
C_i^L	Molar concentration of component i in the liquid phase, mol_i/cm_L^3
C_i^*	Concentration of compound i in the liquid in equilibrium with the gas phase, mol_i/cm_L^3
C_{fi}^S	Molar concentration of component i inside the solid filled with phase f, mol_i/cm_f^3
C_{pf}	Specific heat capacity of phase f, $J/g_f \cdot K$
C_{pfi}	Specific heat capacity of component i in phase f, $J/mol_i \cdot K$
C_{Sfi}^S	Molar concentration of component i at surface of solid covered by phase f, mol_i/cm_f^3
d_p	Catalyst particle diameter, cm_S
d_{pe}	Equivalent particle diameter, cm_S
d_R	Reactor diameter, cm_r
$d_{15.6}$	Liquid specific gravity at 15.6°C and 1 atm, dimensionless
$D(k)$	Species-type distribution function
D_Z	Coefficient of the cubic equation of state expressed in Z
$D_{i,k}$	Fick's binary diffusion coefficient of components i and j, $cm_G^3/(cm_S \cdot s)$
D_a^f	Mass axial dispersion coefficient of phase f, cm_r^2/s
D_{ei}^f	Effective fickian diffusivity of component i inside a porous catalyst, $cm_f^3/cm_S \cdot s$
D_{Ki}^f	Knudsen diffusion coefficient of component i in phase f, $cm_f^3/cm_S \cdot s$
D_{Mi}^f	Molecular diffusion coefficient of component i in phase f, $cm_f^3/cm_S \cdot s$
D_r^f	Mass radial dispersion coefficient of phase f, cm_r^2/s
e	Particle emissivity, dimensionless
erf	Error function
E_A	Activation energy, kcal/mol
f	Fraction of material in a product that boils below the specified temperature
f_{st}	Fraction of liquid that is stagnant, dimensionless
f_w	Catalyst wetting efficiency, $cm_{S,wet}^2/cm_S^2$
f_i^L	Liquid-phase fugacity of component i, MPa

F	Coefficient matrix for Stefan–Maxwell model
F_i	Molar flow rate of component i, mol_i/s
$F(\lambda_g)$	Restrictive factor due to friction between solute and pore walls, dimensionless
FBP_f	Feed final boiling point, °C
g	Gas yield, wt%
G_{mf}	Superficial mass flow velocity of phase f, $g_f/cm_r^2 \cdot s$
h_{air}	Thermowell–atmosphere heat transfer coefficient, $J/(cm_{TW}^2 \cdot s \cdot K)$
h_{GI}	Heat transfer coefficient at the gas side of the gas–liquid interface, $J/cm_I^2 \cdot s \cdot K$
h_{GS}	Gas–solid heat transfer coefficient, $J/cm_S^2 \cdot s \cdot K$
h_{GW}	Gas–wall reactor heat transfer coefficient, $J/cm_W^2 \cdot s \cdot K$
h_{IL}	Heat transfer coefficient at the liquid side of the gas–liquid interface, $J/cm_I^2 \cdot s \cdot K$
h_{LS}	Liquid–solid heat transfer coefficient, $J/cm_S^2 \cdot s \cdot K$
h_{LW}	Liquid–wall reactor heat transfer coefficient, $J/cm_W^2 \cdot s \cdot K$
h_{TW}	f phase–thermowell heat transfer coefficient, $J/cm_{TW}^2 \cdot s \cdot K$
H_i	Henry's law coefficient of component i, $MPa \cdot cm_L^3/mol_i$
ΔH_{Rj}^f	Heat of reaction j in phase f, J/mol_i
ΔH_{vi}	Heat of vaporization/condensation (or latent heat) of component i, J/mol_i
I	Inhibitor content, wt%
j_D	Chilton–Colburn j-factor for mass transfer, dimensionless
j_H	Chilton–Colburn j-factor for heat transfer, dimensionless
k', k''	Kinetic factors
k_{app}	Apparent reaction rate constant (for first-order reaction), s^{-1}
k_H	Overall rate constant, $kg_{feed}/m_{cat}^3/s$
k_{HT}, k_{HC}	Apparent rate constants for hydrotreating and hydrocracking reactions, $kg_{feed}/m_{cat}^3/s$
k_i	Reaction rate constant
k_{ik}	Binary interaction parameter, dimensionless
k_{in}	Intrinsic reaction rate constant (for first-order reaction), s^{-1}
k_L	Thermal conductivity of liquid phase, $J/cm_L \cdot s \cdot K$
k_α, k_β	First-order kinetic constants for reactive and refractory fraction respectively, s^{-1}

$(k)_{AD}$	Pseudokinetic rate constant referred to apparent diffusivity model, $(mol/cm^3)^{1-n}(s^{-1})$
$(k)_{EH}$	Pseudokinetic rate constant referred to external holdup model, $(mol/cm^3)^{1-n}(s^{-1})$
k_i^G	Mass transfer coefficient from gas phase to gas–liquid interface, $cm_G^3/(cm_I^2 \cdot s)$
k_i^{GS}	Gas–solid mass transfer coefficient, $cm_G^3/cm_S^2 \cdot s$
k_i^L	Mass transfer coefficient from gas–liquid interface to liquid phase, $cm_L^3/cm_I^2 \cdot s$
k_i^m	Flowing-stagnant liquid mass transfer coefficient, $cm_L^3/cm_S^2 \cdot s$
k_i^S	Liquid–solid mass transfer coefficient, $cm_L^3/cm_S^2 \cdot s$
$k_i'^S$	Stagnant liquid–solid mass transfer coefficient, $cm_L^3/cm_r^2 \cdot s$
k_{max}	Rate constant of species with the highest TBP
k_{50}	Decay constant, h^{-1}
k_{365}	Kinetic constant with average boiling point of 365°C, $m_L^6/m_{cat}^6 \cdot kmol \cdot s$
$K_{eq,i}$	Liquid–vapor equilibrium constant of component i, dimensionless
K_i	Adsorption equilibrium constant of component i on catalyst active sites, cm_L^3/mol_i
K_{Li}	Overall gas–liquid mass transfer coefficient of component i in the liquid phase, $cm_L^3/cm_I^2 \cdot s$
K_{LSi}	Overall liquid–solid mass transfer coefficient of component i in the liquid phase, $cm_L^3/cm_S^2 \cdot s$
K_1	Constant
L_B	Length of catalyst bed, cm_r
LHSV	Liquid hourly space velocity, $cm_L^3/cm_{cat}^2 \cdot h$
m	Reaction order
m_i	Slope parameter of component i
$M_{0,1,2}$	Universal constants
MW_i	Molecular weight of component i, g_i/mol_i
MW_L	Molecular weight of liquid phase, g/mol
n	Decay order; reaction order
N	Total number of species, number of CSTR in series
N_{Cf}	Number of components in phase f
N_{Rf}	Number of reactions in phase f
N_i^f	Molar flux of component i in phase f, $mol_i/cm_S^2 \cdot s$
p_i^G	Partial pressure of component i in the bulk gas phase, MPa
p_i^I	Partial pressure of component i in the gas–liquid interface, MPa
$p(k, K)$	Yield distribution function

P	Absolute reactor pressure, MPa; paraffin content in feed, wt%
P_{H_2}	Hydrogen partial pressure, MPa
ΔP	Two-phase pressure drop, MPa/cm_r
PA	Content of polycyclic aromatics compounds, wt%
r	Radial reactor coordinate, cm_r
r_g	Mean pore radius, cm_S
r_i	Rate of disappearance of component i, $mol_i/cm_L^3 \cdot s$
r_p	Radius of particle, cm_S
r_{solute}	Hydrodynamic molecular radius of the solute, cm_i
r_{sph}	Radius of sphere, cm_S
$r_j'^f$	Rate of reaction j per unit of catalyst mass in phase f, $mol_i/g_S \cdot s$
R	Gas law constant, 8.314471 J/mol·K
R_{TW}	Outer radius of thermowell, cm_{TW}
S	Sulfur content, wt%
s_i	Stoichiometric coefficient
S_g	Specific external surface area of catalyst particle, cm_S^2/g_{cat}
S_p	Total geometric external surface area of catalyst particle, cm_S^2
S_g'	Surface area defined by eq. 2.183, cm_S^2/g_{cat}
SV	Space velocity, $kg_{feed}/m_{cat}^3/s$
t	Time, s
T	Absolute temperature, K
T^*	Normalized dimensionless temperature
T_{50}	Mid boiling temperature, °F
$T_{50,f}$	Mid boiling temperature of feed, °C
$T_{50,\tau}$	Mid boiling temperature of feed affected by decay function, °C
TBP	True boiling point, °C
u_f	Superficial velocity of phase f, $cm_f^3/cm_r^2 \cdot s$
v_i	Molar volume of solute i at its normal boiling temperature, cm_i^3/mol_i
v_L	Molar volume of liquid solvent at its normal boiling temperature, cm_L^3/mol_L
v_N	Molar gas volume at standard conditions, Nl_G/mol_G
v^f	Molar volume of phase f, cm_f^3/mol_f
V	Reactor volume, cm_r^3
V_c	Bulk volume of catalyst, cm_{cat}^3
V_g	Total pore volume of catalyst particle, cm_{G+L}^3/g_S
V_i	Bulk volume of diluent, cm_{inert}^3
V_p	Total geometric volume of catalyst particle, cm_S^3
W	Wetting number

x	Yield of gasoline, wt%, conversion
x_i	Mole fraction of the component i in the liquid phase, mol_i/mol_L
y_i	Mole fraction of the component i in the gas phase, mol_i/mol_G, product yield, wt%
Y	Total liquid product yield, m^3/m^3 of feed
Yo	Constant
z	Axial reactor coordinate, cm_r; diesel yield, wt%
Z	Compressibility factor, dimensionless
$\hat{Z}$	Constant employed in Bosanquet's formula, dimensionless

Greek Letters

α	Empirical exponent; rate constant, h^{-1}; model parameter
α_i	Generalized temperature function of component i
β	Empirical exponent; inhibition factor
γ	Empirical exponent
$\delta_{1,2}$	Universal constants in the generalized equation of state
ε_B	Catalyst bed void fraction or catalyst bed porosity, cm^3_{G+L}/cm^3_r
ε_S	Catalyst particle porosity, cm^3_{G+L}/cm^3_S
ε_f	External holdup of phase f, cm^3_f/cm^3_r
ε_{pf}	Holdup of f (= G or L) phase inside catalyst particle, cm^3_f/cm^3_S
ε_{pS}	Holdup of solid phase inside catalyst particle, cm^3_p/cm^3_S
ζ	Catalyst bed dilution factor, $cm^3_{cat}/cm^3_{cat} + cm^3_{inert}$
η^*	Effectiveness factor of a partially external and internal wetted pellet, dimensionless
η_{CE}	External catalyst contacting efficiency, $cm^2_{S,wet}/cm^2_S$
η_E	External efficiency, dimensionless
η_G	Global reactor efficiency, dimensionless
η_i	Wetting efficiency inside catalyst particle, dimensionless
η^f_j	Catalyst effectiveness factor of reaction j in phase f, dimensionless
$\eta_{TB,j}$	Overall catalyst effectiveness factor of reaction j in a TBR, dimensionless
θ	Index for normalized TBP
Φ_T	Thiele modulus of an irregular particle, dimensionless
$\Phi_{TB,j}$	Modified Thiele modulus of reaction j for TBRs, dimensionless

Φ_j^f	Thiele modulus of reaction j in phase f, dimensionless
κ	Proportionality constant
λ	Extent of reaction, dimensionless
λ_{er}	Radial effective thermal conductivity, $J/cm_r \cdot s \cdot K$
λ_g	Hydrodynamic diameter of solute/pore diameter ratio, cm_i/cm_S
λ_i	Solubility coefficient of component i, $Nli/kg_L \cdot MPa$
λ_{rad}	Radiant contribution, $J/cm_r \cdot s \cdot K$
λ_a^f	Axial thermal conductivity of phase f, $J/cm_r \cdot s \cdot K$
λ_r^f	Radial thermal conductivity of phase f, $J/cm_r \cdot s \cdot K$
λ_{TW}	Thermowell thermal conductivity, $J/cm_{TW} \cdot s \cdot K$
μ_f	Dynamic viscosity of phase f, cP
ν	Kinematic viscosity, cSt
ξ	Radial coordinate inside spherical catalyst particle, cm_S
ρ_B	Catalyst bulk (or bed) density, g_S/cm^3_{cat}
ρ_f	Density at process conditions of phase f, g_f/cm^3_f
σ	Surface tension, dyn/cm; Stephan–Boltzmann constant, $5.67 \times 10^{-8}\ J/m^2 \cdot s \cdot K^4$
σ_c	Critical surface tension, dyn/cm
σ_{ik}	Collision diameter, Å
τ	Tortuosity factor for catalyst, cm_f/cm_S; space–time, h
$\tau_{1/2}$	Observed overall conversion half-time, s
$\tau_{1/2,c}$	Conversion half-time extrapolated to infinite liquid velocity, s
υ_L	Volumetric flow of liquid phase, cm^3_L/s
υ_{cL}^m	Critical specific volume of liquid, cm^3_L/mol_L
υ_{ij}^f	Stoichiometric coefficient of component i in reaction j in phase f, dimensionless
ϕ_s	Shape factor (= surface area of a sphere of equal volume/surface area of the particle)
φ_i^L	Liquid-phase fugacity coefficient of component i, dimensionless
$\psi(u_L)$	Function of superficial liquid velocity that considers the degree of utilization of the catalyst due to hydrodynamic phenomena
ψ_r	Radiant transfer factor, dimensionless
ω	Empirical exponent; acentric factor, dimensionless
Ω_D	Collision integral, dimensionless
$\Omega_{a,b}$	Universal constants

Subscripts

app	Apparent
B	Referred to reactor catalytic bed
c	Referred to critical conditions

e	Effective
f	Phase (gas, liquid, or solid); final or outlet condition
G	Gas phase
H_2	Molecular hydrogen
H_2S	Hydrogen sulfide
i, k	Component index
in	Intrinsic
I	Gas–liquid interface
j	Reaction index
L	Liquid phase
MeABP	Mean average boiling point
NH_3	Ammonia
obs	Observed
p	Referred to catalyst particle
P	Referred to plug flow
room	Referred to room conditions
S	Sulfur compound
S	Solid phase; condition at external surface of solid catalyst particle
st	Condition at stagnant zone
tot	Total
TW	Referred to thermowell
W	Referred to reactor wall
0	Initial or inlet condition
*	Equilibrium condition

Superscripts

f	Phase (gas, liquid, or solid)
G	Gas phase; gas side of the gas–liquid interface
I	Gas–liquid interface
L	Liquid phase; liquid side of the gas–liquid interface
S	Solid phase; liquid side of the liquid–solid interface; inside solid catalyst particle

Dimensionless Numbers

Bi	Biot number, hL/k
Bo	Bodenstein number, ud_{pe}/D
Nu	Nusselt number, hd_{pe}/λ
Pe	Peclet number, uL_B/D
Pr	Prandtl number, $C_p\mu/\lambda$
Re	Reynolds number, $uL\rho/\mu$ ($L = L_B$ or d_{pe})
Sc	Schmidt number, ν/D
Sh	Sherwood number, $k^f d_{pe}/D$

3

MODELING OF CATALYTIC HYDROTREATING

3.1 THE HYDROTREATING PROCESS

Catalytic hydrotreating (HDT) is applied extensively in the petroleum refining industry to remove impurities, such as heteroatoms (sulfur, nitrogen, oxygen), PNAs (polynuclear aromatics), and metal-containing compounds (primarily V and Ni). The concentration of these impurities increases as the boiling point of the petroleum fraction increases. S-, N-, O-, and PNA-containing compounds are found in low-molecular-weight feedstocks such as straight-run distillates (naphtha, kerosene, gas oil), while high-molecular-weight feedstocks (vacuum gas oils, atmospheric and vacuum residua) contain the same impurities in higher concentrations, as well as complex V- and Ni-containing compounds and asphaltenes (Mochida and Choi, 2004).

Depending on the nature of the feed and the amount and type of the different heteroatoms (i.e., different reactivities compounds), specific hydrotreating processes have been developed. The reactions occurring during hydrotreating are hydrodesulfurization (HDS), hydrodenitrogenation (HDN), hydrodeoxygenation (HDO), hydrodearomatization (HDA), hydrodemetallization (HDM), and hydrodeasphaltenization (HDAs). In addition, the average molecular weight of the feed is lowered by hydrocracking (HDC), which can happen without a substantial loss in liquid product yield, as in the HDT of light distillates, or with moderate or severe reduction of molecular weight, such as in the case of heavy feeds. To accomplish the current and future

Modeling and Simulation of Catalytic Reactors for Petroleum Refining, First Edition.
Jorge Ancheyta.
© 2011 John Wiley & Sons, Inc. Published 2011 by John Wiley & Sons, Inc.

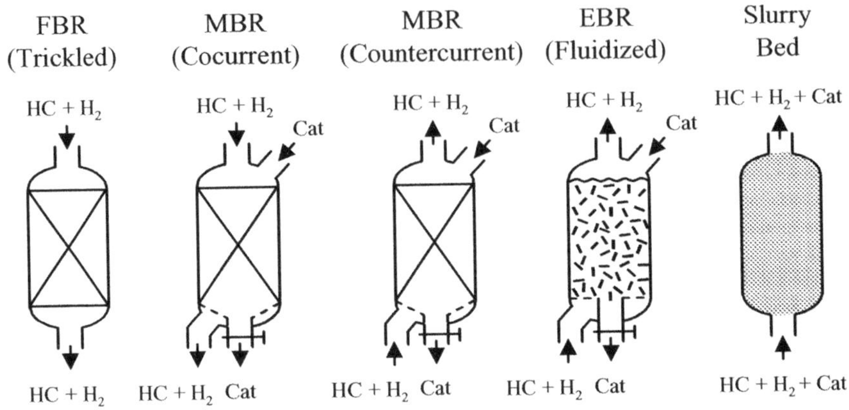

Figure 3.1. Various types of reactors used for catalytic hydrotreating.

stringent environmental regulations to produce clean fuels (e.g., ultralow sulfur fuel), the extent of each reaction needs to be maximized either to obtain the final product or to prepare feeds for subsequent processes. To do that, researchers have focused their attention on the optimization of catalyst properties and composition and also on hydrotreating reactor and process design (Rana et al., 2007). As for reactor and process design, each process is optimized individually according to the nature and boiling range (i.e., physical and chemical properties) of the feed to be hydrotreated, for which reaction conditions and reactor type and configuration are the most important features to be considered. The severity of reaction conditions depends on the type of feed and on the product quality desired. In general, the higher the boiling point of the feed, the higher the reaction severity.

In relation to reactor type and configuration, it should first be mentioned that reactors (as well as catalyst and reaction conditions) used for hydrotreating of heavy feeds are different from those employed for hydrotreating of light feeds (Ancheyta et al., 2002b). In general, HDT reactors operate in fixed beds (FBRs), moving beds (MBRs), expanded or ebullated beds (EBRs), and slurry beds (SBRs). The principle of operation of these three groups of reactors is very similar, but they differ in some technical details (Furimsky, 1998). Figure 3.1 shows schematic representations of the reactors used for catalytic hydrotreating.

In the past, FBR reactors were utilized exclusively only for hydrotreating of light feeds, such as naphtha and middle distillates, but at present they are also used for hydrotreating of heavier feeds, such as petroleum residua. However, when the feed contains large amounts of metals and other impurities (e.g., asphaltenes), the use of FBRs has to be examined carefully according to the catalyst cycle life. Alternatively, MBR and EBR reactors have demonstrated reliable operation with difficult feeds, such as vacuum residua. When hydrotreating petroleum feeds, the life of the catalyst is crucial to retaining its activity and selectivity for some time. Depending on the feed, the catalyst life may vary on the order of months or years. It is then clear that the time scale of deactivation influences the choice of reactor (Moulijn et al., 2001).

3.1.1 Characteristics of HDT Reactors

Figure 3.2 shows examples of the main reactors used for hydrotreating, whose characteristics are described below.

Fixed-Bed Reactors FBRs are the most commonly used reactor systems in commercial hydrotreating operations. They are easy and simple to operate. However, the simplicity of operation limits their use to the HDS of light feeds. For example, in the case of naphtha hydrodesulfurization, the reaction is carried out in two-phases (gas–solid) fixed-bed reactors since at the reaction conditions the naphtha is completely vaporized. On the contrary, for heavier feeds three phases are commonly found: hydrogen, a liquid–gas mixture of the partially vaporized feed, and the solid catalyst. The latter system is called a trickle-bed reactor (TBR), which is a reactor in which a liquid phase and a gas phase flow co-currently downward through a fixed bed of catalyst particles while reactions take place (Rodríguez and Ancheyta, 2004). The gas is the continuous phase, and the liquid is the disperse phase (Quann et al., 1988). A schematic representation of the phenomena occurring in a TBR based on three-film theory is presented in Figure 3.3 (Korsten and Hoffmann, 1996; Bhaskar et al., 2004). It is common to assume that mass transfer resistance in the gas film can be neglected and that no reaction occurs in the gas phase, so that for the reactions to occur, the hydrogen has to be transferred from the gas phase to the liquid phase, whose concentration is in equilibrium with the bulk partial pressure and then adsorbed onto the catalyst surface to react with other reactants. The gas reaction products are then transported to the gas phase, while the main liquid hydrotreated reaction product is transported to the liquid phase.

The operating conditions of TBRs in bench scale and commercial hydrotreating plants show that superficial mass velocities and Reynolds numbers for both liquid and gas phases are always smaller in bench-scale reactors than in commercial reactors. For these reasons, low liquid velocities are used in small-scale reactors in order to match the liquid hourly space velocity (LHSV) of commercial plants, which implies that gas–liquid and liquid–solid mass transfers are much better in commercial HDT reactors. In addition, because of the lower resistance to liquid flow at the wall, the linear velocity next to the wall is greater than that at the center of the reactor. This variation in linear velocity causes an increase in axial dispersion. The extent of this axial dispersion effect depends mainly on the bed length and the conversion (Ancheyta et al., 2002a).

The main undesirable impurity in naphtha is sulfur, and the sulfur compounds present in naphtha are easy to remove. That is the reason that only an HDS catalyst is required for sulfur removal in naphtha. However, when

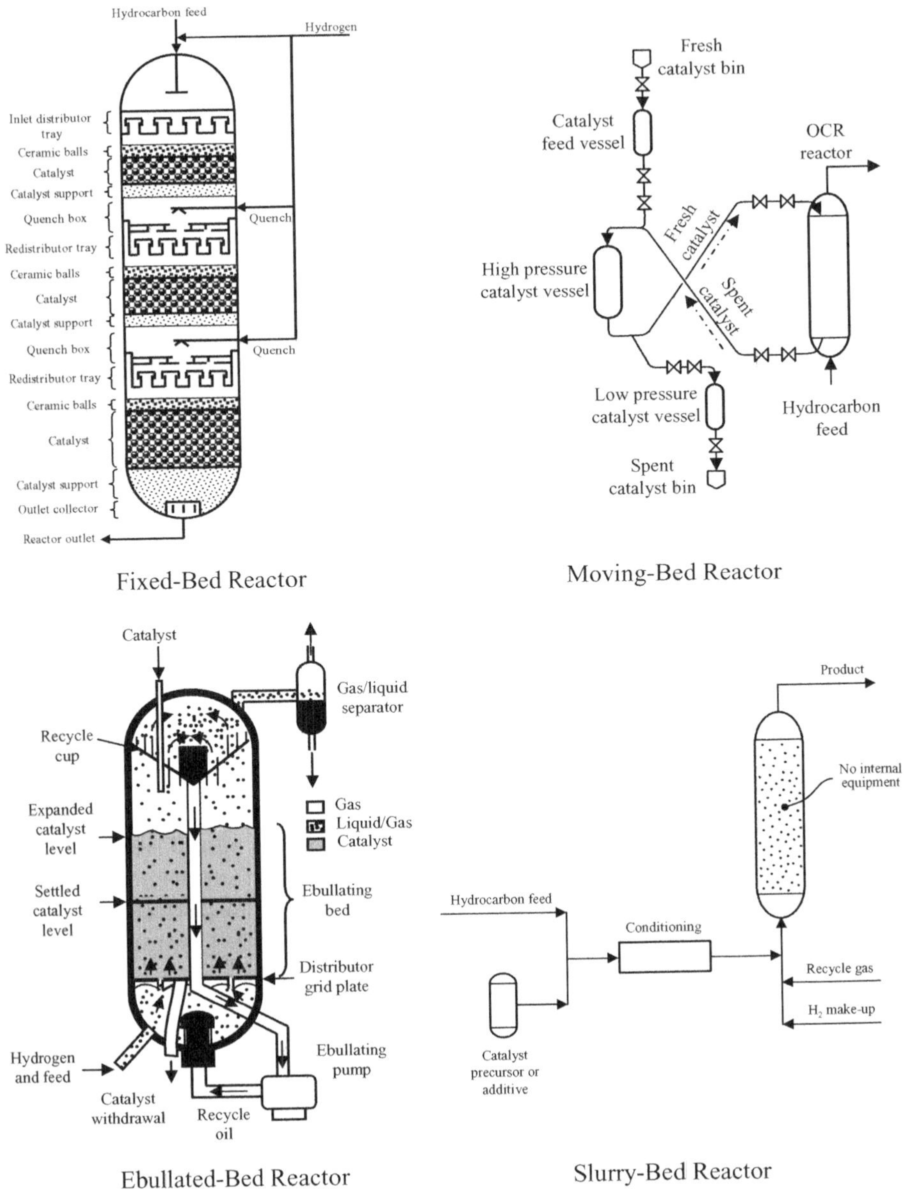

Figure 3.2. Examples of FBR, MBR, EBR, and SPR for catalytic hydrotreating.

processing straight-run gas oil (SRGO), the refractory sulfur compounds are present [4,6-dimethyl-DBT and 4(or 6)-methyl-DBT], making deep HDS for ultralow sulfur diesel (ULSD) production difficult to achieve. In addition, most of the time, SRGO is blended with light cycle oil (LCO) from catalytic

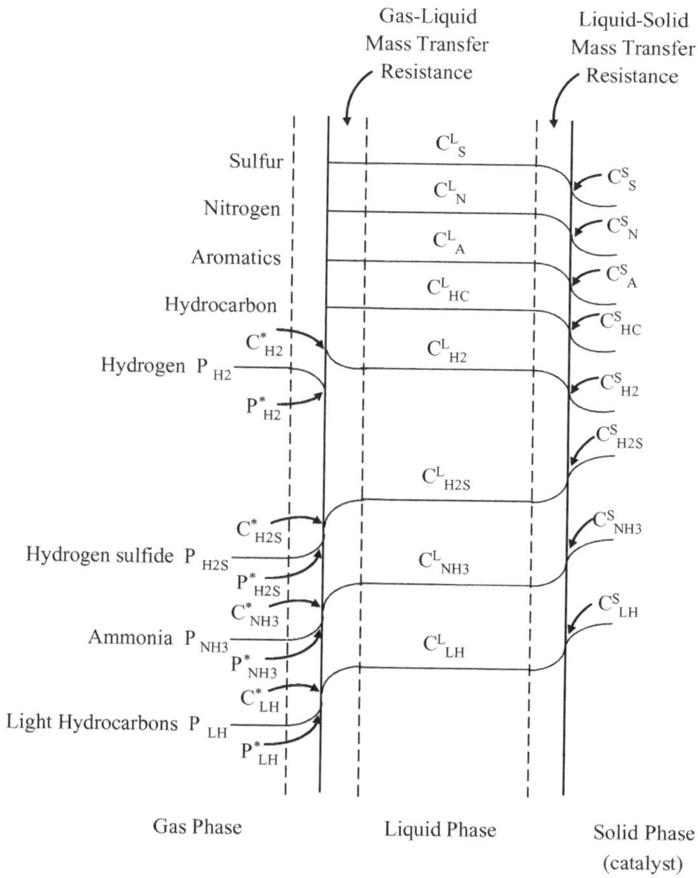

Figure 3.3. Concentration profiles in an HDT reactor. (Adapted from Korsten and Hoffmann, 1996; Bhaskar et al., 2004.)

cracking units (FCC), and both are fed to the hydrotreater. Apart from sulfur, LCO also contains high amounts of nitrogen and aromatics, which make the hydrotreating even more difficult, since they either compete for catalytic active sites or consume large amounts of hydrogen (Ancheyta et al., 1999a). To face this problem, multibed systems with different catalysts have been proposed. Also, hydrogen is introduced between the beds as a quench because of the exothermality of the reaction. The heat released in light feed HDT is relatively low, so that quenching is not necessary and HDT units are designed with just one reactor containing a single catalyst bed. However, for heavier feeds, multiple catalyst beds with cooling in between are used (Robinson and Dolbear, 2006). Multibed configuration with a hydrogen quench system is usually employed for hydrotreating of FCC feeds (a blend of heavy atmospheric gas oil and light and heavy vacuum gas oils) and heavier feedstocks.

In FBR the liquid and gas flow co-currently downward through the catalytic bed, which has unfavorable hydrogen and hydrogen sulfide concentration profiles over the reactor (i.e., a high H_2S concentration at the reactor outlet), which provokes an inhibition effect for removal of the last ppm sulfur compounds (Ancheyta et al., 1999b). A more suitable profile of H_2S concentration can be provided by operating the reaction in countercurrent mode: for example, introducing the feed at the top and hydrogen at the bottom of the reactor. By this means, the possible recombination of hydrogen sulfide with olefins to form small amounts of mercaptans (rebuilding sulfur-containing molecules: $H_2C{=}CH_2 + H_2S \leftrightarrow HS{-}CH_2CH_3$) at the reactor outlet is avoided, since H_2S is removed from the top of the reactor (Babich and Moulijn, 2003). For example, if H_2S is not removed in naphtha hydrodesulfurization units, mercaptans may cause problems for downstream catalytic reforming units. Also, the inhibition effect of H_2S in hydrotreating reactions (e.g., hydrodesulfurization), can be minimized since the final deep HDS is carried out at a low hydrogen sulfide concentration. The reactor employed in SynSat Technology, which combines Criterion's SynSat catalysts and ABB Lummus's reactor technologies, is an example of this approach (Langston et al., 1999). In a SynSat reactor the hydrogen sulfide is removed interstage to prevent the hydrogen sulfide formed in the first part of the reactor from passing through to the final part of the reactor. A very low H_2S partial pressure and low-temperature hydrogenation are enabled by applying a gas–liquid countercurrent operating with an intake of fresh hydrogen gas in the bottom reactor. Another reactor employing this concept comprises the fractionation of the feed into light and heavy cuts that react separately in upper and lower parts of the catalyst bed (Mochida et al., 1996). The feed is introduced at about two-thirds the reactor height in between the top bed and a middle bed of catalyst. Hydrogen is charged from the bottom of the reactor so that hydrogen sulfide inhibition on the heavier fraction hydrotreating can be avoided (Mochida and Choi, 2006). This reactor combines the characteristics of low H_2S partial pressure final-stage hydrotreating, countercurrent operation, and catalytic distillation. One more application of this approach considers the use of a stacked bed of two catalysts, in which the final deep HDS and/or final deep hydrogenation step is performed at low H_2S and high H_2 partial pressures. The hydrocarbon is fed at the middle of the reactor in downflow mode through a conventional HDS catalyst in the bottom bed, while hydrogen is fed at the top of the reactor to the top catalyst bed. The bottom bed catalyst partly desulfurizes the feed at a high H_2S partial pressure, which is then separated from the gas phase and is fed back together with hydrogen to the catalyst in the top bed, where deep HDS and/or deep hydrogenation is performed at low H_2S and high H_2 partial pressures (Sie and de Vries, 1993). Figure 3.4 shows these types of reactors.

When using FBR for hydrotreating of heavy oils and residua, it is common to use guard materials at the top of the reactor or in a separate vessel before the main hydrotreating reactors to catch foulants. Other particulates present in the feed can also settle in the catalysts, causing catalytic bed plugging. As a

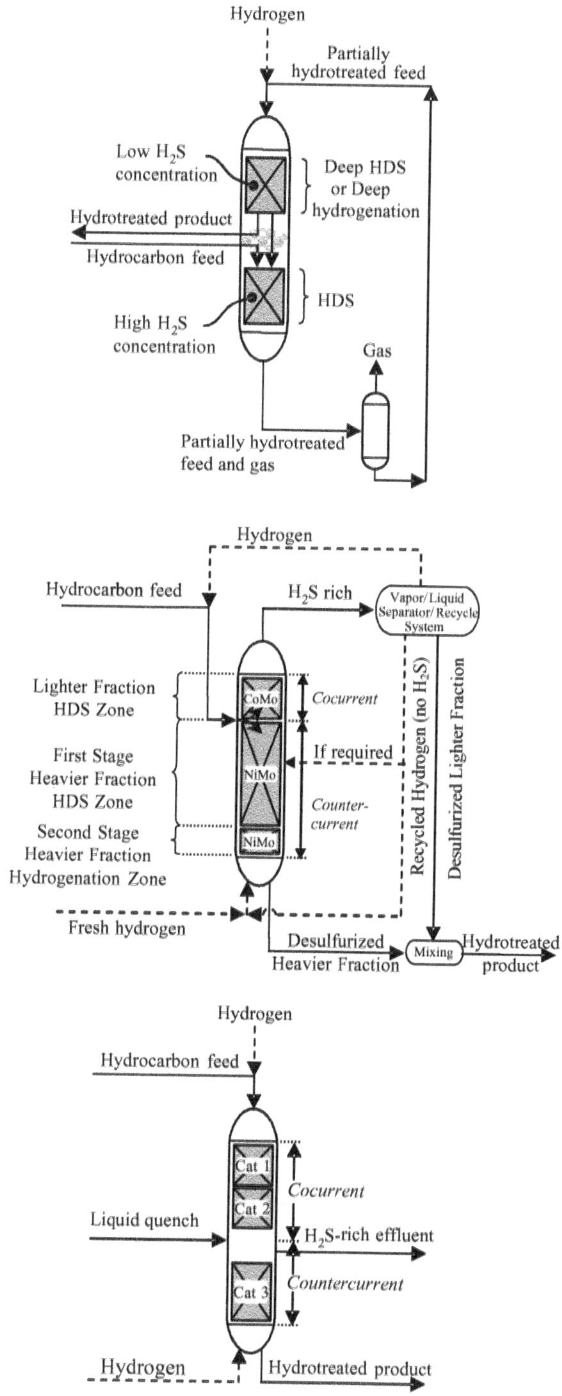

Figure 3.4. Various process approaches for reduction of H_2S partial pressure at the final stage of HDT. (Adapted from Sie and de Vries, 1993; Mochida et al., 1996; Langston et al., 1999.)

result, the pressure drop increases and the reactor performance declines, which ends up in hydrotreating plant shutdown. Pressure drop in hydrotreating units is usually the result of the accumulation of a dense layer of particulates or the formation of gums from reactive species in the feed. In other words, ΔP develops when the particulates layer reduces the void fraction in the catalyst bed over time. The common way to avoid this problem is by using filters to protect the catalyst beds. This solution is partial since filters are ineffective in removing complex polymeric iron sulfide gums (formed from the reaction of soluble iron compounds, e.g., iron porphyrins with sulfur in the feed, or by their dissolution by naphthenic acid) and particulates smaller than 5 to 20 μm, which may pass through the catalyst beds. Another solution is the use of layers of highly macroporous materials graded on the top of the FBR reactors, whose main objective is protection of the catalyst bed from fouling due to the factors mentioned previously. The problem becomes more complicated when hydroprocessing heavy and extraheavy feeds. For such cases, proper bed grading can be used to increase the cycle length and prevent premature shutdown. The control of particulates is done by using catalysts of large diameter and large void fraction to spread out the zone of deposition of particulates. Each layer of catalyst, with a particular shape and size, collects particulates within a certain size range and prevents the formation of a dense layer of particulate accumulation, so that pressure drop due to gum formation is prevented. Macroporous materials and bed grading substantially reduce the pressure drop problems, consequently extending the life of the catalyst bed and eliminating the need for frequent shutdown to skim the head part of the bed to remove crust, plugs, and agglomerates of catalyst particles, and to replace fouled catalyst. The main disadvantage of using bed grading is the loss of reactor volume for loading the catalyst since a part of the grading material can be inert. However, the use of materials with some catalytic activity mitigates these effects. Hence, finding a right balance between fouling prevention and preserving sufficient overall catalyst activity is the key issue in optimizing bed grading (Minderhoud et al., 1999).

The problem of short catalyst cycle life, which is the main reason for declining the use of FBR when feeds have a high amount of metals, may be solved by the association of adequate HDM and HDS catalysts as well as appropriate selection of reaction conditions, which contribute to a strong increase in the performance of new processes for residue refining (Ancheyta et al., 2006).

Moving-Bed Reactors In contrast to FBRs, in an MBR the catalyst goes in downflow through the reactor by gravitational forces. The fresh catalyst enters at the top of the reactor and the deactivated catalyst leaves the reactor at the bottom, while the hydrocarbon goes either in counter- or co-current flow through the reactor. With this moving-bed system, the catalyst can be replaced either continuously or discontinuously (Gosselink, 1998). One example of the application of MBR for hydrotreating of heavy oils and residua is the bunker reactor used in the Hycon process developed by Shell (Van Ginneken et al., 1975; Scheffer et al., 1998). The OCR (on-stream catalyst replacement) process is another option for the hydrotreatment of heavy oils and residua with a significant amount of metals. OCR is a moving-bed reactor operating in a countercurrent mode at high temperature and pressure (Scheuerman et al., 1993). Another alternative of moving systems, with feed and catalyst flowing countercurrent, is the MBR used in the Hyvahl-M process (Euzen, 1991). This technology is part of a series of processes for residua hydrotreating licensed by IFP/Asvahl (Billon et al., 1991): Hyvahl-F, Hyvahl-S, and Hyvahl-M.

In general, MBR catalyst replacement is commonly a batch operation typically done once or twice a week. Catalyst transfer (i.e., adding and removing the catalyst) is the most critical section. The countercurrent mode of operation of MBRs seems to be the best configuration, since the spent catalyst contacts the fresh feed at the MBR bottom while the fresh catalyst reacts with feed almost completely hydrodemetallized at the MBR top, resulting in lower catalyst consumption (Morel et al., 1997).

Ebullated-Bed Reactors Similar to the MBRs, to handle problematic heavy feeds with a large amount of metals and asphaltenes, such as vacuum residua, EBRs are used in process technologies to overcome some of the deficiencies of FBRs. H-Oil, T-Star (an extension of H-Oil), and LC-Fining processes are the commercialized technologies that use EBRs. These hydroprocessing technologies possess very similar features (process parameters and reactor design) but differ in mechanical details (Daniel et al., 1988). In EBRs, hydrocarbon feed and hydrogen are fed upflow through a catalyst bed, expanding and backmixing the bed, minimizing bed plugging, and consequently reducing pressure drop problems. The mixture of gas (makeup and recycle hydrogen) and liquid reactants (feed and recycle oils) enters the reactor plenum chamber and is well mixed through the specially designed gas–liquid mixer, spargers, and catalyst support grid plate. A homogeneous environment is created to hydrotreat and hydrocrack the heavy feedstocks (Kam et al., 1999). Product quality is maintained continually at a high level by intermittent catalyst addition and withdrawal. Reactor features include onstream catalyst addition and withdrawal, eliminating the need to shut down for catalyst replacement as in the case of FBRs.

An EBR is a three-phase system [i.e., gas, liquid, and solid (catalyst)], in which the oil is separated from the catalyst at the top of the reactor and recycled to the bottom of the bed to mix with the new feed. The large liquid recycle causes the reactor to behave as a continuous-stirred-tank reactor. EBR is provided with an ebullating pump, located externally for an H-Oil reactor and internally for an LC-Fining reactor, to maintain liquid circulation within the reactor. This liquid circulation is what maintains the reactor at essentially isothermal conditions, so that there is no need for quenches within the reactor. The liquid recycle rate can be adjusted by varying the ebullition pump speed. The unconverted heavy oils are recycled back to the reactor with a small amount of diluent to improve fluidity and thus overall conversion. The fluidization of the catalyst also results in solid backmixing, which implies that the catalyst removed also contains fresh catalyst particles. That is why this mixture of spent catalyst and fresh catalyst removed from the reactor is also called *equilibrium spent catalyst* or simply *equilibrium catalyst*. Fresh catalyst is added to the top of the reactor, and spent catalyst is withdrawn from the bottom of the reactor. The inventory of catalyst in the reactor is maintained at the desired level by adjusting the catalyst addition rate equal to the withdrawal rate plus any losses. The catalyst replacement rate can be adjusted to suit feed properties or product slate and quality requirements. EBRs are commonly designed to have an expanded catalyst bed of 130 to 150% of the settled catalyst bed, which has been demonstrated to be the bed expansion for achieving uniform fluidization and good contacts among hydrogen, oil, and catalyst.

Slurry-Phase Reactors Slurry-phase reactors (SPRs) can also be used to hydroprocess feeds that have a very high metals content to obtain lower-boiling products using a single reactor. SPR-based technologies combine the advantages of the carbon rejection technologies in terms of flexibility with the high performances peculiar to hydrogen addition processes (Panariti et al., 2000). SPRs achieve a similar intimate contact of oil and catalyst and can operate with a lower degree of backmixing than can EBRs. In contrast to FBRs and EBRs, in SPRs a small amount of finely divided powder is used (typically, from 0.1 to 3.0 wt%), which can be an additive or a catalyst (or catalyst precursors). The catalyst is mixed with the feed (heavy oil) and both are fed upward with hydrogen through an empty reactor vessel. An SPR is free of internal equipment and operates in a three-phase mode. The solid additive particles are suspended in the primary liquid hydrocarbon phase through which the hydrogen and product gases flow rapidly in bubble form. Since the oil and catalyst flow co-currently, the mixture approaches plug-flow behavior (Quann et al., 1988; Speight, 2000).

In an SPR the fresh catalyst is slurried with the heavy oil prior to entering the reactor, and when the reaction finishes, the spent catalyst leaves the SPR together with the heavy fraction and remains in a benign form in the unconverted residue (Furimsky, 1998). Inside the reactor, the liquid–powder mixture behaves as a single phase (homogeneous phase), due to the small size of the catalyst and additive particles. It has been reported that the powder serves primarily as a site where a small amount of coke can deposit, so as to keep walls, valves, and heat exchangers clean, thus maintaining good operability (Schulman and Dickenson, 1991). In other words, the use of a selected catalyst dispersed into the feed inhibits coke formation.

3.1.2 Process Variables

During hydrotreating operation at different scales (laboratory, microreactor, bench scale, pilot scale, and commercially), there are four main process variables that have a great impact on reaction conversion and selectivity as well as on the activity and stability of the catalyst: (1) total pressure and hydrogen

TABLE 3.1. Typical Operating Conditions and Hydrogen Consumption During the HDT of Various Feeds

Type of Feed	Temperature (°C)	Pressure (psig)	LHSV
Naphtha	280–425	200–800	1.5–5.0
Gas oil	340–425	800–1600	0.5–1.5
Resid	340–450	2000–3000	0.2–1.0
	Source	**API Gravity**	**H_2 Consumption (scf/bbl)**
Naphtha	—	—	100–700
Gas oil	—	—	300–800
Resid	—	—	500–2000
AR	Venezuela	15.3–17.2	425–730
VR	Venezuela	4.5–7.5	825–950
AR	West Texas	17.7–17.9	520–670
VR	West Texas	10.0–13.8	675–1200
AR	Khafji	15.1–15.7	725–800
VR	Khafji	5.0	1000–1100
AR	Kuwait	15.7–17.2	470–815
VR	Kuwait	5.5–8.0	290–1200

partial pressure, (2) reaction temperature, (3) H_2/oil ratio and recycle gas rate, and (4) space velocity and fresh feed rate. Typical operating conditions for the hydrotreating of various petroleum fractions are given in Table 3.1.

Total Pressure and Hydrogen Partial Pressure The total pressure of a hydrotreating unit is determined by the reactor design and is controlled by the pressure that is maintained at the high-pressure separator (HPS). Inlet or outlet hydrogen partial pressure is calculated by multiplying the total pressure by the hydrogen purity (H_2 mole fraction) of the recycle gas. Definition of the value of total reactor pressure is decided depending primarily on the nature of the feed and the amount of impurities to be removed (i.e., the quality of the feed and the quality of the product desired). In general, when a hydrotreater is operated at high hydrogen partial pressures, the following main effects are obtained (Mehra and Al-Abdulal, 2005; Gruia, 2006):

- Longer catalyst cycle life
- Capability for processing heavier feeds
- Higher throughput capability
- Higher conversion capability
- Better distillate quality
- Purge gas elimination

Since the catalyst deactivation rate will be increased substantially and catalyst cycle life reduced at low reactor pressure due to coke formation, a hydroprocessing reactor must be operated at H_2 partial pressure very close to the design value. Although it is highly desirable to operate a reactor at the highest allowable pressure, equipment limitations restrict the operation at a pressure close to or a little bit higher than the design value. Given this situation, the only way to increase hydrogen partial pressure is by increasing the purity of the recycle gas, which can be achieved either by increasing the H_2 purity of the makeup hydrogen, or by venting gas off the HPS or reducing the temperature of the HPS (Gruia, 2006).

At higher H_2 partial pressures, the removal of impurities is easier; however, reactors become more expensive and hydrogen consumption increases, which can become a significant cost factor for the refinery. New units are being designed to operate under higher H_2 partial pressure atmosphere by working at higher total pressure.

The performance of any hydrotreating reactor and process is limited by the hydrogen partial pressure at the inlet to the reactor. The higher the hydrogen partial pressure, the better the hydrotreating reactor performance. The overall effect of increasing the partial pressure of the hydrogen is to increase the extent of the conversion (Speight, 2000). This has been confirmed extensively by studies conducted with model compounds for HDS, HDN, HDA, and so on, reactions as well as with real feeds (light distillates, middle distillates, heavy oils, etc.) at microscale, benchscale, and pilot plants. Figure 3.5 exemplifies the effect of hydrogen partial pressure and reaction temperature on sulfur removal and saturation of polyaromatic hydrocarbons (PAHs) (Binghan and Christensen, 2000; Chen et al., 2003). It is clearly seen that PAHs react quite readily, but their conversion is thermodynamically limited, and neither increasing the temperature nor increasing the H_2 partial pressure reduces the PAH content in the product to values lower than 2 wt%.

The presence of heteroatom compounds with different reactivities in a hydrotreating feed makes, for example, HDS of refractory multiring sulfur compounds very difficult, with a high hydrogen demand, a pathway that first goes through prehydrogenation of one of the aromatic rings. If H_2 partial pressure is not at the value required, the following problems will be faced during hydrotreating (Ho, 2003):

- The slow HDN rate of nitrogen compounds blocks off virtually all active sites that are available for HDS.
- The HDS rate of refractory sulfur compounds may be limited by a thermodynamically mandated low hydrogenation rate.
- The catalyst surface may be starved or adsorbed hydrogen.

In commercial operation, hydrogen partial pressure is obtained primarily by feeding the proper amount of makeup gas. The increase in catalyst activity for achieving higher impurities removal and conversion rates would require

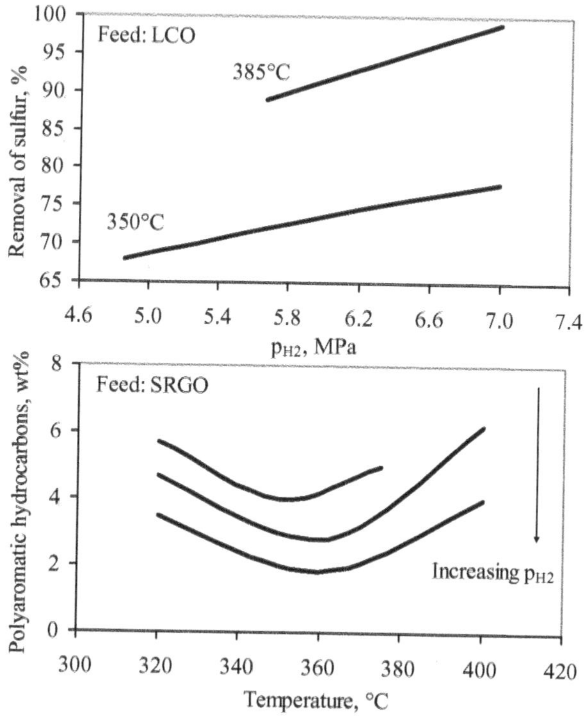

Figure 3.5. Effect of H_2 partial pressure on sulfur removal and aromatic saturation. (Adapted from Bingham and Christensen, 2000; Chen et al., 2003.)

significant modifications in hydrotreating reactor operation, primarily through the use of higher pressure and also by increasing the hydrogen rate and purity, reducing the space velocity, and proper selection of catalyst. Pressure requirements would depend, of course, on the feedstock quality and on the product quality target of each refinery. Also, if all aromatics need to be hydrogenated, a higher pressure is needed in the reactor compared with that of a conventional operating mode. The level of pressure required for such product specifications will be limited by the cost and availability of the technology.

Reaction Temperature Reactor temperature generally determines the types of compounds that can be removed from the petroleum feed and also establishes the working life of the catalyst. Increasing the temperature increases reaction rates and thus the removal of impurities. However, similar to reactor pressure, there are limits to the maximum allowable temperature, since depending on the feed above a certain thermal cracking value of the hydrocarbon constituents becomes more prominent, which can lead to the formation of considerable amounts of low-molecular-weight hydrocarbon liquids and gases, and also to catalyst deactivation much more quickly than at lower temperatures. Thermal cracking also produces olefins, which when hydrogenated, release heat, increasing the temperatures further as well as the thermal cracking rates (hot spots). Finally, this condition inside the reactor provokes temperatures higher than the safe upper limits for the reactor walls (Speight, 2000). The effect of reaction temperature on impurities removal is illustrated in Figure 3.5 together with that of pressure.

Most hydrotreating reactions are exothermic in nature, which makes the commercial reactor temperature increase as feed passes through the catalyst bed. For experimental reactors (e.g., micro- and bench-scale reactors), an isothermal condition is commonly achieved, but for adiabatic (commercial) reactors, the outlet reactor temperature will be higher than the inlet reactor temperature. To determine the average temperature of adiabatic reactors, the weight-average bed temperature (WABT) is typically used. WABT can easily be calculated if the reactor is provided with various temperature indicators (TIs) located in different zones of the catalytic bed, by the following equations (Stefanidis et al., 2005):

$$\text{WABT}_i = \frac{T_i^{\text{in}} + 2T_i^{\text{out}}}{3} \tag{3.1}$$

where WABT_i is the average temperature of each catalytic bed between two TIs and T_i^{in} and T_i^{out} are the inlet and outlet temperatures in each catalytic bed, respectively. The global WABT is calculated with

$$\text{WABT} = \sum_{i=1}^{N} (\text{WABT}_i)(Wc_i) \tag{3.2}$$

where N is the number of catalyst beds and Wc_i is the weight fraction of catalyst in each bed with respect to the total.

Equation (3.1) is used instead of a common arithmetic average since it takes into account the common nonlinear gradient of temperature observed in hydrotreating reactions. In Eq. (3.1) it is assumed that in the last two-thirds of the reactor length the temperature is closer to T_{out}, while in the first one-third of the reactor length the prevailing temperature value is closer to T_{in}.

WABT is frequently used during operation for process control purposes; in such cases Eqs. (3.1) and (3.2) are preferably expressed as only one linear equation as a function of TI values as follows:

$$\text{WABT} = \sum_{i+1}^{N} a_i \cdot \text{TI}_i \tag{3.3}$$

where a_i are constant values determined by solving Eqs. (3.1) and (3.2). Note that the sum of all a_i values must be equal to unity. Equation (3.3) can easily be programmed into the process control system and the WABT can be reported in real time.

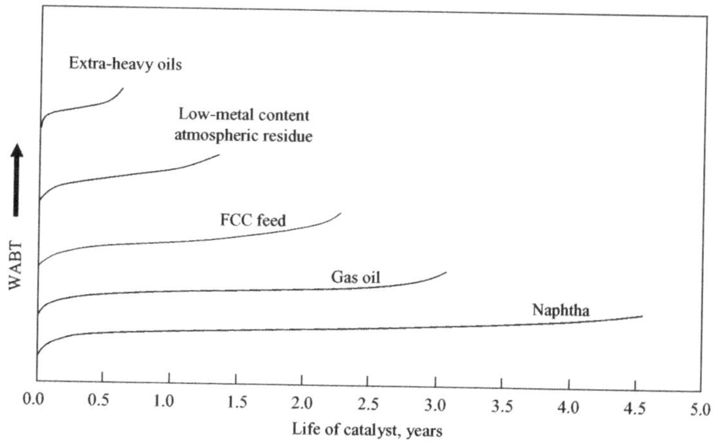

Figure 3.6. Life of catalyst and required increase in WABT for hydrotreatment of various feeds.

A common practice during commercial operation of hydrotreating units is to increase the reactor temperature steadily to compensate for catalyst deactivation and to produce a constant-quality product. This production policy requires the unit to be operated at a different WABT value during time on stream, which are known as start-of-run (SOR) temperature ($WABT_{SOR}$) and end-of-run (EOR) temperature ($WABT_{EOR}$). In fact, when designing a hydrotreating plant, simulations have to be carried out for at least two cases: at SOR and EOR conditions. Properties of the feed, the desired quality of the product, and the reactor design are the main parameters that define the values of $WABT_{SOR}$ and $WABT_{EOR}$ as well as the temperature increase during operation. Typically, $WABT_{EOR} - WABT_{SOR} = 30°C$. High-metal-content feeds require the temperature to be increased more frequently, of course. When WABT reaches a value close to the maximum designed, the catalyst has to be replaced. Figure 3.6 summarizes the typical increases in WABT and catalyst life, depending on the type of feed. For the HDS of naphtha, a long catalyst life is observed and the increase in WABT over time is unimportant. However, during the HDT of heavy oils, WABT has to be increased constantly so that catalyst deactivation is compensated and the product is produced at constant quality.

H_2/Oil Ratio and Recycle Gas Rate The H_2/oil ratio in standard cubic feet (scf) per barrel (bbl) is determined by

$$H_2/\text{oil} = \frac{\text{total hydrogen gas to the reactor, scf/day}}{\text{total feed to the reactor, bbl/day}} [=] \frac{\text{scf}}{\text{bbl}} \tag{3.4}$$

Another unit frequently used to report the H_2/oil ratio is m^3/bbl, obtained by multiplying the H_2/oil ratio (in scf/bbl) by a conversion factor (0.028317). A molar H_2/oil ratio can also be calculated from the volumetric H_2/oil ratio by means of the following equation:

$$\text{molar } H_2/\text{oil} = 1.78093 \times 10^{-7} \left(\frac{H_2}{\text{oil}} \text{ scf/bbl} \right) \frac{MW_{oil}}{MW_{H2}} \frac{\rho_{H2}}{\rho_{oil}} \tag{3.5}$$

where MW_{oil} and MW_{H2} are the molecular weights of the oil to be hydrotreated and of hydrogen, respectively, and ρ_{oil} and ρ_{H2} are the densities of the oil and hydrogen, respectively (ρ_{H2} at 15°C and 1 atm is 0.0898 kg/cm^2).

Apart from economic considerations, gas recycle is used to compensate for hydrogen consumption and hence to maintain the hydrogen partial pressure within the reactor. Use of a high excess of hydrogen (i.e., an elevated H_2/oil ratio) ensures adequate conversion and impurities removal due to efficient physical contact of the hydrogen with the catalyst and hydrocarbon; also, carbon deposition is minimized, which reduces the rate of catalyst deactivation. The latter is actually the main reason to work in a high-hydrogen-concentration atmosphere; otherwise, the catalyst can deactivate faster, due to coking. Another important benefit of operating at high H_2 partial pressure is the reduction in the SOR temperature of the reactor, which increases the cycle life of the catalyst. However, there is a limit in the value of the H_2/oil ratio, since above a certain gas rate, the change in hydrogen partial pressure will be relatively small and no further benefits will be obtained. In fact, higher gas rates than necessary incur extra heating and cooling rates, which may become more important than other advantages.

Increasing the recycle-gas rate increases the H_2/oil ratio and the hydrogen partial pressure in the reactor. Apart from this, the objective of the gas rate is to strip volatile products from the reactor liquids, thus affecting the concentration of various components in the reactive liquid phase. The H_2 partial pressure and H_2/oil ratio must be maintained very close to the design value; otherwise, the catalyst life will be affected adversely.

The hydrogen loop in a hydrotreating unit involves several streams, as can be seen in Figure 3.7. The reactor effluent stream is separated in a high pressure separator (HPS) into liquid hydrotreated products and noncondensable H_2-rich gases (typically, 78 to 83 mol% H_2 plus H_2S and other light gases, CH_4, C_2H_6, C_3H_8, butanes, traces of pentanes). Hydrogen sulfide, either formed via HDS or present in the reactor feed, is commonly separated with an amine contactor to increase the purity of the H_2. Following use of an amine contactor, lighter hydrocarbon gases are still present in the recycle gas stream; part of this stream (10 to 15%) is purged to the fuel gas system or sent to a hydrogen purification process (e.g., PSA: pressure swing adsorption) if the unit is provided with such a plant, from which about 20% is lost to the fuel gas system. The other portion of gases leaving the amine contactor is compressed and recycled to the top of the reactor and/or used for quenching (80 to 85 mol% H_2). The separated H_2-rich gases from the hydrogen purification process are mixed with the makeup hydrogen and recycled to the top of the reactor.

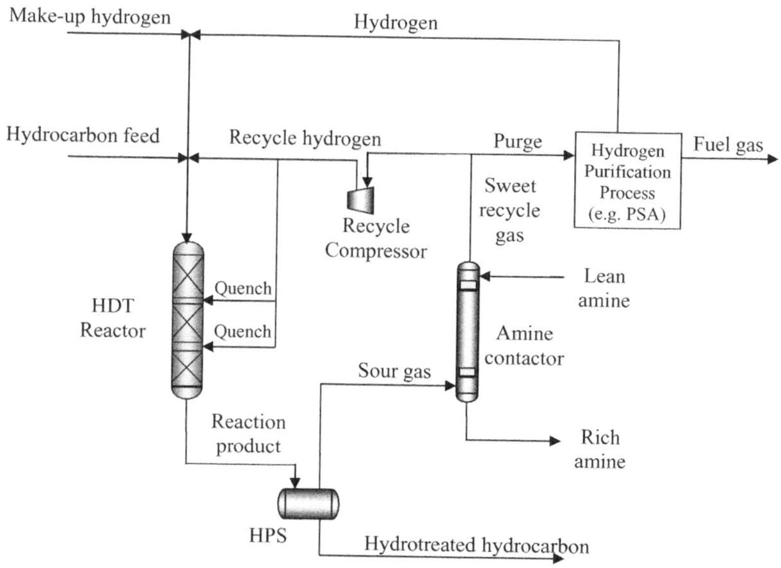

Figure 3.7. Hydrogen loop in a hydrotreating unit.

Depending on the source of makeup hydrogen, it is typically available at 96 to 99.9 mol% H_2 purity. The separated hydrotreated liquids from the HPS are sent to other sections for further processing (Mehra and Al-Abdulal, 2005).

When working under ultralow-sulfur fuel production conditions or with high-sulfur feedstocks, the concentration of H_2S in the recycle gas stream can achieve high values, which consequently reduces the hydrogen purity of the recycle gas stream and hydrogen partial pressure. This highly concentrated H_2S atmosphere inhibits the HDS reaction, as shown in Figure 3.8. According to various authors, HDS activity is reduced around 3 to 5% for each 1 vol% of H_2S in the recycle gas stream, which means roughly that an increase of 3 to 5% in the catalyst amount is needed to balance this situation. Even a low H_2S content results in an increase (e.g., 0.3 mol% can reduce the reaction rate about 5%). The operation of HDS units with 9% H_2S content in the recycle gas stream would require about 15 to 20% more catalyst to achieve the same results than when the H_2S concentration is 0%. In addition, when H_2 purity is increased, the SOR temperature can be lowered about 9°C, and the run length can be extended about 30%.

The recycle gas from the HPS is generally water-washed to remove ammonia, preventing the formation of ammonium sulfide, which might form blockages in the reactor effluent cooler, and is then sent to the sour water plant to remove H_2S. If a scrubbed recycle gas is not available, the reactor temperature must be increased to offset the H_2S inhibition, the effect of which is greater at higher total reactor pressure.

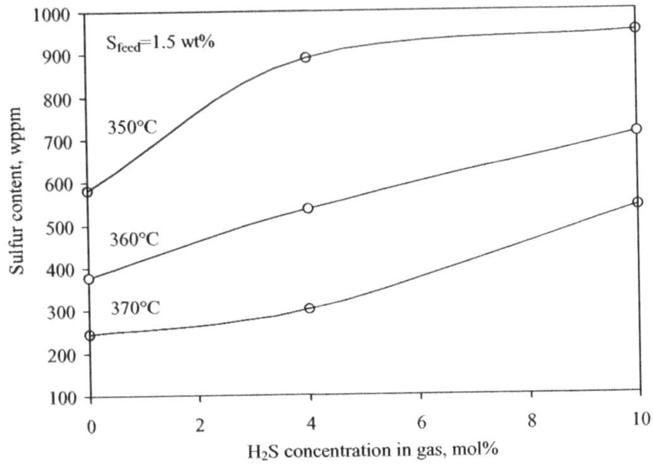

Figure 3.8. Effect of H_2S on product sulfur content during HDT of middle distillates over CoMo/γ-Al_2O_3 ($P = 54\,kg/cm^2$, H_2/oil = 2000 scf/bbl, LHSV = $2\,h^{-1}$).

Hydrogen consumption is another very important process parameter, since it determines the amount of hydrogen makeup. Hydrogen consumption during hydrotreating depends on feedstock properties and impurity removals. As a feed becomes much heavier it will require substantially more addition of hydrogen to reach the product quality desired. Table 3.1 shows typical hydrogen consumption values for hydrotreating of different hydrocarbon feeds.

Total hydrogen consumption is a summation of chemical hydrogen consumption and dissolved hydrogen (calculated from vapor–liquid equilibrium), assuming that any mechanical hydrogen loss is negligible. The most common approach to calculating H_2 consumption not only at the commercial level but also at different experimental scales is by means of a hydrogen balance in gas streams. That is the amount of hydrogen entering the reactor minus the amount of hydrogen exiting the reactor. Another way to do so is by determining the hydrogen content in the liquid feed and products. Liquid products, which have been hydrogenated, must have a higher hydrogen content than that of liquid feed. The difference is the hydrogen added to the feed (i.e., hydrogen consumption). There are also rules of thumb for quick calculations, which employ typical hydrogen consumption values reported in the literature (Edgar, 1993; Speight, 1999), but the values obtained must be taken with care due to their empirical nature, yielding only approximations to the real amount of hydrogen consumed.

Space Velocity and Fresh Feed Rate Space velocity is a process variable normally used to relate the amount of catalyst loaded within a reactor to the amount of feed. Space velocity in normally expressed on a volume basis

(LHSV: liquid hourly space velocity) or a weight basis (WHSV: weight hourly space velocity). LHSV and WHSV are calculated as follows:

$$\text{LHSV} = \frac{\text{total volumetric feed flow rate to the reactor}}{\text{total catalyst volume}} [=]\, h^{-1} \quad (3.6)$$

$$\text{WHSV} = \frac{\text{total mass feed flow rate to the reactor}}{\text{total catalyst weight}} [=]\, h^{-1} \quad (3.7)$$

LHSV and WHSV are related to each other by the equation

$$\text{WHSV} = \frac{\rho_{\text{oil}}}{\rho_{\text{cat}}} \text{LHSV} \quad (3.8)$$

ρ_{oil} and ρ_{cat} are the densities of the hydrocarbon feed and the catalyst, respectively. When using LHSV as a process parameter, ρ_{cat} is not important. However, in the case of WHSV, ρ_{cat} becomes relevant since it can vary depending on how the catalyst was loaded to the reactor. For example, for dense loading, more catalyst is loaded in the same reactor volume and the WHSV value will be different than with nondense loading, although the LHSV value will be the same in both cases.

In certain cases, space velocity is also used as the GHSV, which is calculated as follows:

$$\text{GHSV} = \frac{\text{total volumetric gas flow rate to the reactor}}{\text{total catalyst volume}} [=]\, h^{-1} \quad (3.9)$$

In a hydrotreating process, the space velocity is used as the LHSV. The space velocity is inversely proportional to the residence time. Therefore, an increase in space velocity indicates a decrease in residence time and thus in reaction severity. Figure 3.9 shows the influence of LHSV on the sulfur content of products obtained during the hydrotreating of different middle distillates. It is clearly observed from this figure that a decrease in LHSV results in diminished sulfur content in the product. Operating at a higher space-velocity value (a higher feed rate for a given amount of catalyst) requires a higher reactor temperature to maintain the same impurity removal (i.e., product quality), resulting in an increased deactivation rate, thus reducing the catalyst life.

3.1.3 Other Process Aspects

The operation of hydrotreating reactors is considered to be very close to adiabatic because the heat losses from the reactor are usually negligible compared with the heat generated by the reactions. The exothermality of hydrotreating reactions, predominantly HDS and the hydrogenation of aromatics, can cause

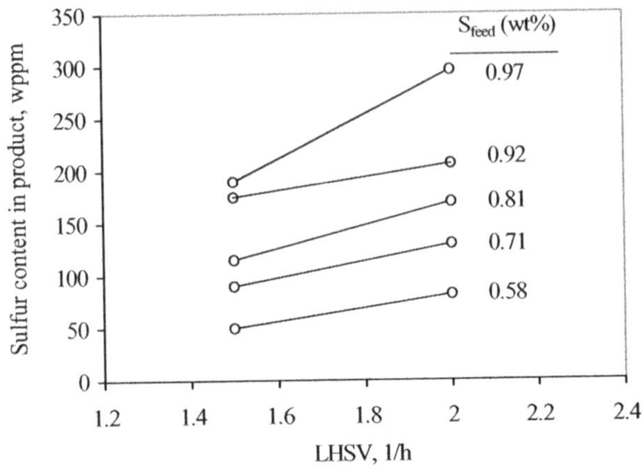

Figure 3.9. Effect of LHSV on product sulfur content during hydrotreatment of various middle distillates ($P = 54\,kg/cm^2$, H_2/oil = 2000 scf/bbl, $T = 360°C$).

an increase in reactor temperature beyond the design limits, depending, of course, on the conversion level, reaction conditions, and feed properties—which is why an appropriate temperature control system is required. Temperature control is essential to achieve an economically acceptable catalyst cycle length and to obtain the product quality required. Hence, in the sense of temperature control of multiple catalytic beds, injection of quench fluids and/or heat integration of the effluent from each bed come together (Alvarez and Ancheyta, 2008).

The design of the quench system is based on establishing an appropriate reactor temperature profile by determining the number of catalyst beds and their respective lengths in order to retain the required product quality and an economically acceptable catalyst cycle life. This is accomplished by solving the reactor mass and energy balances simultaneously until the optimal configuration of the system is obtained. An important aspect when designing a quench system is to consider a maximum allowable temperature which limits the reactor bed length and is commonly about 30°C or less above the inlet reactor temperature. Figure 3.10 illustrates the quench concept for a catalytic fixed-bed reactor with several quenches (Alvarez et al., 2007a).

Commonly, control of the reaction temperature in hydrotreating reactors is achieved by introducing part of the hydrogen recycle stream between the catalytic beds, called *quenching* or *cold shot cooling*. The use of quench liquids has also been reported. Quenching fluids are introduced in the quench zone or quench box, which is typically a mixing chamber where the bed effluent is mixed with the cooling medium. The flow of fluid injected to each quench location is adjusted to achieve the desired temperature profile and is specified to limit the temperature rise below the maximum allowable temperature.

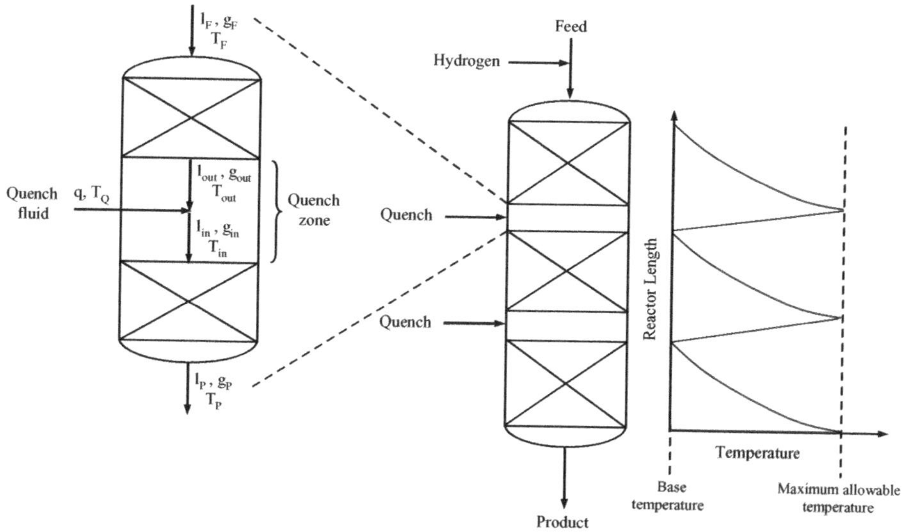

Figure 3.10. General representation of quench in a hydrotreating reactor.

Another vital aspect of fixed-bed hydrotreating reactor performance is the reactor internal hardware design. Reactor internal hardware allows for efficient catalyst utilization by means of effective reactant distribution, quenching performance, and fouling protection. Most of the fixed-bed hydrotreating reactors currently in operation in worldwide petroleum refineries have been built and designed over the past 30 years. These units have been experiencing underperformance with the increasing supply of heavier oils to refineries, the tightening environmental legislation, and poor reactor internal design. Some of these problems were partially solved with increases in reaction severity, which reduced considerably the catalyst cycle life due to enhanced catalyst deactivation. Mechanical constraints in reactor design and product demand were other problems that refiners had to face when trying to increase reactor temperature and reduce feed flow rate (i.e., decreased space velocity). In addition, excessive pressure drops were present due to fouling caused by solids contained in the feed (iron scale, salts, coke fines, etc.) and reaction products (coke and metals).

Over the years, many strategies have been proposed to meet current product specifications and at the same time to keep the catalyst cycle life at acceptable levels. Those strategies are based on the development of new highly active catalysts, tailoring reaction conditions (e.g., temperature, LHSV, hydrogen partial pressure) and designing new reactor configurations (e.g., multibed reactors with interstage quenching, reactors in series, and counterflow reactors); for fouling abatement, improved procedures for catalyst loading, low activity mesoporous materials, and graded-bed designs were developed. However, the experience has shown that improving catalyst performance and maximizing its volume within an existing unit are the most cost-effective options for improving unit performance. Two key parameters have been identified as possible solutions of these problems (Patel et al., 1998): (1) increasing catalyst activity and (2) efficient distribution of the reactants through the catalytic bed by means of proper reactor internal design.

Quench Systems

Conventional Quenching Quenching with hydrogen is used widely in most hydrotreatment units. Hydrogen, being the main reactant in hydrotreating, has the advantage of replenishing some of the chemically consumed hydrogen in the catalytic beds, decreasing the hydrogen sulfide and ammonia partial pressures in the reactor, which reduces the inhibition effect on HDT reactions and keeps the catalyst clean by inhibiting coke formation. The availability of quench hydrogen depends on the H_2/oil ratio along the reactors, which is a design condition that influences product quality. The value of the H_2/oil ratio depends primarily on the compressor capacity within the plant. High H_2/oil ratios improve the product quality and increase the quench availability; for example, a staged hydrocracking unit that operates at a H_2/oil ratio of ~10,000 scf/bbl may have up to five gas injection points. However, high H_2/oil ratios also imply higher hydrogen recycling rates, and therefore larger compressors and equipment in the separation section are required, which increases the investment costs (Muñoz et al., 2005).

Quenching with Liquids Processes that use liquid quench are not as common as gas-based quenching processes. That is why most of the information reported in the literature is related to hydrotreating reactors with hydrogen quench systems. However, quench hydrogen is not always the best option, due to its availability in refineries and compression requirements. In such cases quench liquids may be more attractive, due to their higher heat capacity and lower compression costs; nevertheless, it may require more reactor volume or a lower liquid-hourly space-velocity (i.e., more reactor volume to achieve the same conversion). The way in which a liquid quench is introduced into a reactor is different from that of a gas quench, and special reactor components and liquid quench injection devices are needed to provide an efficient contact between the gas and liquid phases. The processes that use liquid quench streams may be classified in two general categories:

1. *Multiple feed processes.* Processes with multiple feeds are characterized by introducing several liquid hydrocarbon streams of different composition and properties at the top and between the beds of a reactor. Generally, the hydrocarbon feed is first fractionated, then the heaviest fraction is fed at the top of the reactor and lighter fractions are introduced as a side feed (Figure 3.11a). By this approach the side feeds act as quench streams and at the same time are processed together with each bed effluent in the following catalytic bed.

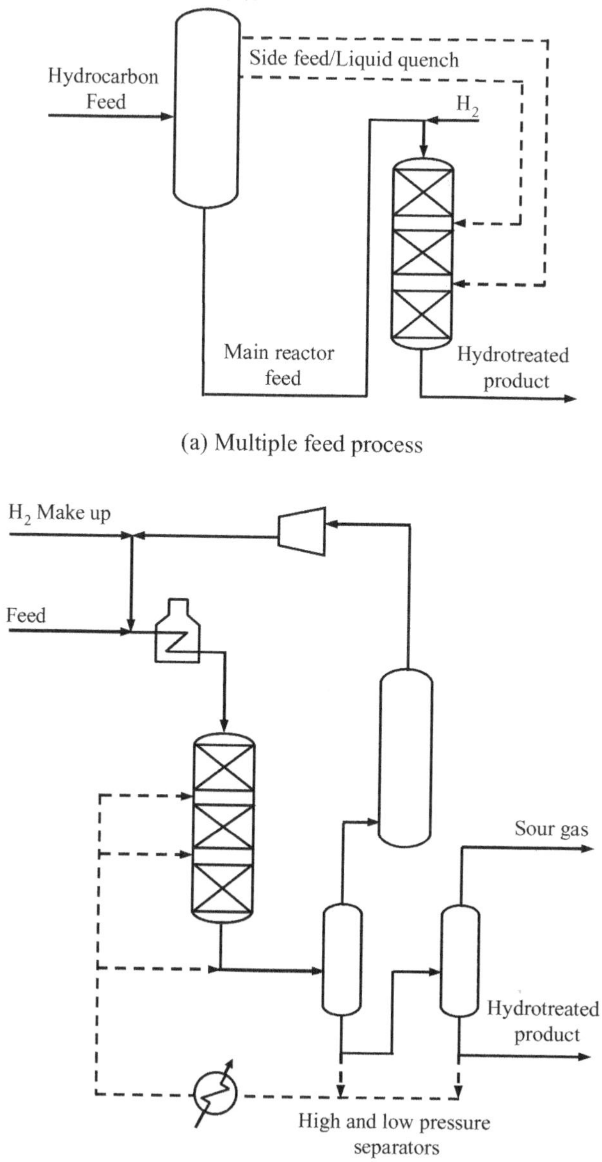

(a) Multiple feed process

(b) Process with product recycle

Figure 3.11. Examples of liquid quench–based processes.

2. *Product recycle processes.* In processes with product recycle, a portion of the reactor effluent is separated and cooled via heat exchange in order to be introduced between beds as quench stream. Generally, the recycle stream used as a quench comes from the bottom of the high- or low-pressure separators, as presented in Figure 3.11b. In this way, the portion of the heaviest treated fraction has a second-pass opportunity through the reaction system.

Comparison of Quench Approaches The use of quench fluids may have either a positive or a negative impact on the process configuration and the product quality, depending on several factors, such as quench fluid properties, flow rate and temperature, and injection points. Table 3.2 summarizes the advantages and disadvantages of using each quench fluid or method. The use of recycle hydrogen as quench always has a positive effect on product quality at the expense of high compression costs. Quenching with liquids may reduce such a disadvantage, but to achieve the same conversion degree will require more reactor volume or a decrease in LHSV. The choice between each alternative requires detailed process studies in order to select the most cost-effective

TABLE 3.2. Advantages and Disadvantages of Quench Fluids

Type of Quench Fluid	Advantages	Disadvantages
Hydrogen	Replenishes consumed hydrogen	Low heat capacity
	Reduces H_2S and NH_3 partial pressures	Increases requirements of equipment for recycling hydrogen
	Reduces coke formation	High pressure drops
	Improves distribution tray performance by increasing gas velocity	Increases height of the reactor
Liquid	High heat capacity	Increases LHSV and decreases reaction severity
	Reduces equipment requirements	Increases height and diameter of the reactor
	Reduces viscosity of the mixture	In case of vaporization, hydrogen partial pressure decreases
	Provides treatment to liquid quench	Increased costs due to fractionation of the feed
	Allows for adjusting the hydrogen distribution in the fractions of the product	May increase heat of reaction
	Provides second-pass opportunity to unreacted species	

option. For example, Bingham and Christensen (2000) evaluated the use of liquid quench versus recycle gas quench as well as other process aspects in order to revamp a two-stage HDS/HDA unit. For this particular unit, they determined that the most cost-effective alternative was the recycling of hydrotreated product from the high-pressure separator as liquid quench combined with state-of-the-art reactor components. Nevertheless, in other types of processes, such as hydrocracking, hydrogen will be the cooling medium of choice due to its effect on the product composition and quality and coke formation.

In some cases it is possible to employ both liquid and gas quenches in the reaction system as described by Bradway and Tsao (2001), who proposed a method for quenching HDT reactions by injecting recycled hydrogen along with liquid hydrocarbons. The proposal combines the multiple feed or product recycle process schemes along with hydrogen recycling in order to reduce the high-pressure drops in the reactor generated by the large gas volumes originally required to cool the reaction zone. Thus, the process scheme proposed makes it possible to enjoy the advantages of each quench method and lessen the disadvantages of each.

Reactor Internals The majority of the hydrotreating units operating at the end of the twentieth century relied on rudimentary reactor internal designs, such as sieve trays, chimney trays, conventional bubble cap trays, and impingement quench boxes; even worse, in some cases reactors may have any components at all. The design of those distributor trays was strongly influenced by hardware employed in fractionation columns, which is not necessarily adequate for trickle-bed reactors. Inappropriate reactor internal design caused poor catalyst utilization, due to flow maldistribution of the reactants at the inlet of the catalyst bed. Flow maldistribution was also enhanced by the increase in reaction severity, which eventually led to its detection by the high radial temperature differences measured at catalyst bed outlets. The main problems generated by flow maldistribution are the overuse of a part of the catalyst inventory and the formation of hot spots; meanwhile, the rest of the catalyst becomes underused, leading to poor product quality and shorter cycle lengths. This fact has increased the awareness of the importance of reactor internal design and its role in efficient catalyst utilization (Alvarez et al., 2007b).

Based on the idea that the efficient catalyst utilization is governed by reactor components, a proper design must perform the following functions (Ouwerkerk et al., 1999):

- Uniform volumetric and thermal distribution of liquid and gas reactants over the cross-sectional area of the catalyst bed
- Inter- and intraphase mixing
- Quench performance
- Fouling resistance

- Space efficiency
- Ease in maintenance and installation

Reactor internal hardware may be located at the reactor inlet, interbed zones, and at the reactor outlet. The hardware at the reactor inlet provides an initial distribution of the reactants and protection against fouling; this is achieved by means of distributor trays together with fouling abatement trays and/or top-bed grading materials. For high-hydrogen-demanding feeds, where a large ΔT value is generated due to reaction exothermality, multibed reactors with interstage quenching are employed to limit the temperature rise; quench zones located between catalyst beds comprise a reactant collection system, a quench fluid-injection device, a chamber for mixing the cooling medium with the hot reactants, and a reactant redistribution tray. Finally, the reactor outlet contains hardware for fluid collection along with catalyst retention.

Figure 3.12 shows the hydrotreating reactor internal fundamentals according to the previous description for a unit with two catalytic beds and one quench (Ancheyta and Speight, 2007). Axial and radial ΔT profiles within the reactor are also illustrated. The axial ΔT represents the typical temperature rise caused by the exothermality of the hydrotreating reactions in the catalytic bed. It allows for establishing the catalytic bed length when the reactor temperature reaches a maximum allowable temperature and for determining the number of beds for a required impurity removal. On the other hand, the radial ΔT value reflects the performance of reactor internals. The figure shows radial ΔT values for good and poor reactor internal performance; the former is characterized by low radial temperature differences after distribution trays and quench zones, and the latter exhibits gradual widening in radial ΔT, which provides evidence of flow maldistribution. It is worth mentioning that maldistribution has a cumulative character if the distributor trays and quench boxes are not working adequately; thus, in multibed reactors the poorest catalyst utilization will be in the last bed, which is reflected in the widest radial temperature differences.

Distributor Trays Certainly, the most relevant reactor internal feature is the distribution system, whose purpose is to establish radial liquid distribution across the catalyst bed, and thus it determines the performance of a trickle-bed reactor. To date, most hydrotreating units have used the original distributor designs, such as sieve trays, chimney trays, and bubble cap trays, the last two being the most successful.

Sieve trays are the most rudimentary systems, being simple and cheap in construction; they comprise a great number of liquid downcomers (perforations) across the tray and sometimes widely spaced chimneys for separating gas flow. This type of tray has been used more as a predistribution system, followed by chimney or bubble cap trays, than as a principal distributor. Chimney trays are basically descendants of sieve trays; their main feature is that of evenly spaced chimneys with lateral apertures for the liquid and a top aperture

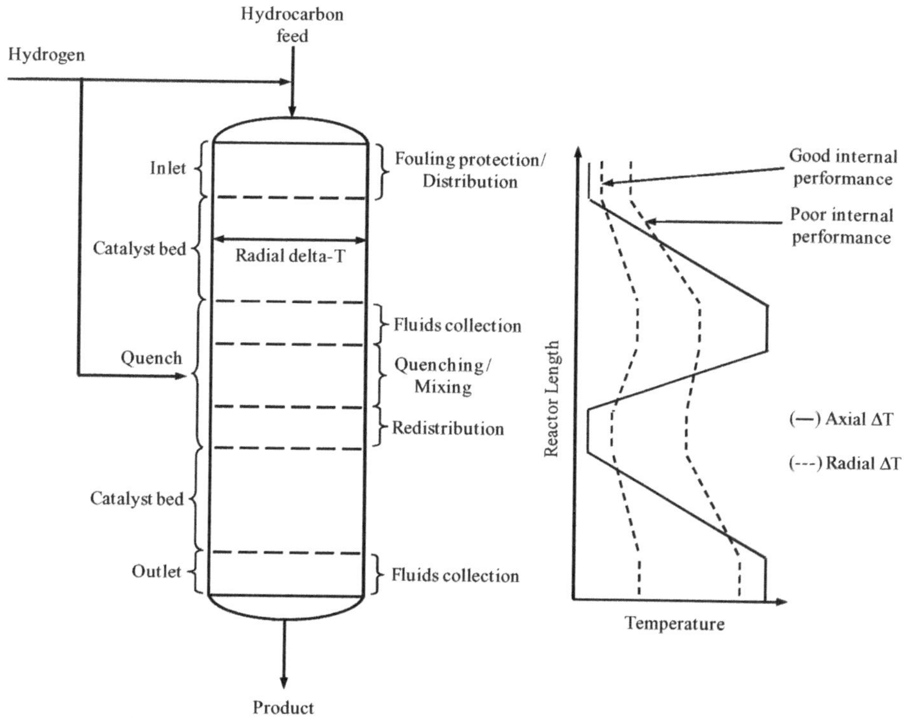

Figure 3.12. Fundamentals of internals of hydrotreating reactors.

for the gas, thus allowing independent flow of both fluids. A great number of chimney designs are available, differing mainly in the number or type of lateral apertures, such as traditional chimney distributors, chimneys with triangular notches, and multiaperture chimneys. An alternative design of chimney trays comprises gas chimneys and triangularly notched liquid chimneys. Bubble cap trays are essentially similar to those employed in distillation columns, but with a different function. They are characterized by the contact and mixture of gas and liquid reactants, creating a mixed phase that flows through the slots of the bubble cap. The wide range of bubble cap trays goes from the early designs to the latest high-performance designs. Many types of the distribution systems described above have been patented over the years; however, they are simply variations of the original systems, delivering little improvement, and in many cases promote liquid maldistribution (Patel et al., 1998). Even though technical information on these designs is available in expired patents, the reasons for their underperformance were not well understood until recently.

The study of distribution systems has been of great interest to mayor oil companies, resulting in the current state-of-the-art distributors, such as Shell GSI's HD (high distribution) tray (Den Hartog and van Vliet, 1997; Altrichter et al., 2004), Topsøe's vapor-lift tray (Yeary et al., 1997; Seidel et al., 2002),

Exxon's spider vortex technologies (Davis, 2002; McDougald et al. 2006), Akzo Nobel's duplex tray (Akzo Nobel, 2003), and Fluor's swirl cap tray (Jacobs et al., 2000). The development of these high-performance distributor trays, using sophisticated high-pressure cold flow units in combination with computational fluid dynamics (CFD), led to a better understanding of the flaws of the original designs. The meticulous evaluation of those designs highlighted the importance of specific parameters, such as a liquid source layout (i.e., tray spacing and wall coverage), discharge pattern, tray levelness, sensitivity to plugging, and flexibility in operation.

1. LIQUID-SOURCE LAYOUT A liquid-source layout is characterized by tray or center-to-center spacing and wall coverage capability. Tray spacing is referred to the distance between the centers of two drip points. This parameter is directly proportional to the catalyst particle diameter and must be optimized so that radial mixing, provided by the grading material, compensates for maldistribution.

Uniform liquid distribution can be achieved closer to the top of the catalyst bed with narrower tray spacing: in other words, a larger number of liquid point sources. On the other hand, wide tray spacing reduces catalyst utilization and requires more bed depth to correct liquid distribution by means of radial dispersion. In this matter, original tray designs do not have optimal tray spacing, as discussed earlier by Patel et al. (1998); for example, bubble cap trays are known for having the worst tray spacing, due to their relatively large size (50 to 100% larger than a chimney).

Wall coverage capability is the other layout parameter that influences reactor performance. Conventional distributors present dead zones without liquid sources near the reactor wall, as in the case of bubble caps. Poor wall coverage together with a disk discharge pattern contributes to flow bypassing, leaving a great percentage of unused catalyst near the reactor wall vulnerable to hot-spot formation.

2. DISCHARGE PATTERN The most important design parameter of distributor trays is perhaps the liquid discharge pattern. Along with tray spacing, it determines the percentage of wetted catalyst across the top of the catalyst bed and, consequently, overall catalyst utilization. For the last decade, distributor tray development has been focused on providing an efficient discharge pattern, which in this context refers to uniform distribution closer to the top of the catalyst bed. Conventional distributors, such as chimney trays and bubble cap trays, produce a disk type of discharge pattern, which wets only the catalytic surface right beneath the discharge point. This type of discharge pattern is very inefficient because it leaves a great percentage of unused catalyst at the beginning of the bed. However, commercial state-of-the-art trays provide a very wide spray discharge pattern, covering almost 100% of the catalyst bed. The liquid discharge pattern is governed by the hydrodynamics present in the discharge points. Liquid flow in sieve and chimney trays is governed by the

overflow principle, where the liquid accumulated over the tray drips down through the sieves or the apertures on the chimneys, generating a disk-type discharge pattern, while gas enters separately through the top of the chimneys. In addition to the inefficient discharge pattern, these designs provide poor vapor–liquid contact, and therefore large temperature gradients may be observed. On the other hand, the gas-assist principle takes advantage of the high gas velocity to drag the liquid held on the tray, forming a dispersed liquid phase which is discharged through a central downcomer, as in the case of bubble caps and state-of-the-art distributors, although for the former this does not produce an efficient discharge pattern. This operation principle provides excellent vapor–liquid contact, reducing interphase temperature differences by up to 90% (Ballard and Hines, 1965).

3. TRAY LEVELNESS Tray tilt or levelness is another important factor that must be considered during installation. When a tray is not leveled properly, the liquid will gravitate toward the lowest area of the tray, resulting in preferential liquid flow, causing poor distribution.

4. LIQUID-LOADING SENSITIVITY A proper distributor design must be able to provide satisfactory performance over a broad range of liquid loads. Variations in liquid loading, such as those presented at start- and end-of-run conditions, may affect the functioning of distributor trays.

Quench Zones Hydrotreating fixed-bed reactors requires a quench system to control the temperature rise caused by the exothermality of the reactions. The main consequences of temperature runaway are hot-spot formation, leading to enhanced catalyst aging by coke formation and sintering, poor product yields due to excessive hydrocracking, and sometimes damage to the reactor vessel. As discussed earlier, controlling reaction temperature in hydrotreating reactors is achieved by introducing quench fluids into the quench zone located between catalytic beds. This interbed zone makes it possible to inject cooling medium, mixed with the hot reactants from the previous bed, and to redistribute the liquid and gas reactants across the following catalyst bed (Ouwerkerk, 1999).

Early interbed hardware designs included an impingement quench box together with a redistribution tray such as those discussed earlier (Ballard and Hines, 1970; Peyrot, 1987). The impingement quench box system comprises a quench tube that allows for injecting cold hydrogen, a liquid collector tray, a mixing box where the fluid impingement occurs, a perforated tray for collecting fluids coming from the mixing box, and a bubble cap redistributor tray. The operating principle of the quench box is based on:

- Division of the downflowing fluids (hot reactants and quench gas) into two streams which enter through the openings of the collector tray to separate chambers of the mixing box

- Directing the flow of the streams by means of baffles located in each chamber, causing the fluids to impinge in a turbulent mixing zone
- Discharge of the mixture toward the redistribution system

However, impingement mixing is known to provide ineffective gas–liquid mixing, due to poor interphase contacting, resulting in large gas–liquid temperature differences. The latter defect, along with an inappropriate redistribution tray, results in poor quench zone performance. This is characterized by wide radial ΔT values after each interbed zone, which persists or in the worst case grows as the fluids move down the reactor.

Nevertheless, the failures of conventional systems were corrected in new interbed designs which are constituted by vortex-type mixers and high-performance distributors [e.g., Shell GSI's UFQ (ultraflat quench), ExxonMobil's spider vortex quench zone, Chevron-Lummus's nautilus reactor technologies, Isomix's internals, Fluor's swirl zone vortex mixer]. The main feature of this type of system is the swirling motion of the fluids generated within the mixing box, which enhances gas–liquid contact. The performance of vortex mixers is explained well by Litchfield et al. (1996) and Pedersen et al. (1995) when describing operation of a proprietary quench zone design. The authors stress the difficulty in achieving effective mixing due to the large density differences between the quench gas and process gases, and the importance of the quench zone configuration in order to maximize intra- and interphase contact. The primary parts of a quench zone are the quench fluid-injection device, which imparts radial and perpendicular mixing of the process fluids and quench gas, and the vortex mixer arrangement, which provides turbulent swirling motion to the fluids. Injection devices include traditional quench pipes for direct injection into the mixing chamber, concentric manifolds with nozzles that surround the mixing chamber (e.g., UFQ quench ring) for radial inward injection, or the spider, which is a small manifold located at the center of the quench zone (spider vortex) for radial outward injection. Vortex mixers vary in the arrangement of vanes and baffles within the chamber, which create passageways and constrictions, which impart a swirling motion and turbulence to the fluids. A different approach to vortex mixers is the Albermarle Q-Plex quench mixer (Albermarle, 2006), where the quench gas and process fluids are passed through a single orificei thus, the constriction provides intimate intra- and interphase mixing. The operation is carried out in three mixers in series.

One important aspect of these systems is their reduced height in comparison to conventional systems. In commercial hydrotreating units it is extremely important to minimize the vertical dimensions of interbed internals to reduce the height of the required reactor vessel, especially in hydrocracking units, which may have more than two quench zones. Large reactors with wall thicknesses of 20 to 40 cm (high-pressure operation) represent considerably heavy reactor vessels, which in return increases the total cost due to the larger supporting structure required and difficult transportation and installation. Of all the commercial technologies, perhaps Shell's technical papers emphasize this aspect most strongly. For example, after installing the UFQ and HD internals in a conventional hydrocracker, where 67% of the reactor vessel volume was occupied by catalyst, catalyst utilization grew up to about 86%, due to the reduced catalyst-to-catalyst distance of the UFQ (1 to 1.4 m) and elimination of the grading material due to the HD tray effectiveness (Swain and Zonnevylle, 2000). On the other hand, Albermarle's Q-Plex has been reported to have a total height of about 0.5 m, which is much smaller than that of most vortex mixers (about 1 m).

However, reduced-height components compromise flexibility for operating at variable liquid–gas loads, especially at high loads, where flooding may occur or the residence time in the mixer is not enough for effective fluid mixing (Litchfield et al., 1996). It has also been reported recently that for optimal fluid mixing over a wide range of liquid–gas loads (33 to 200%) the vortex mixer should be 0.35 to 0.65 times the inner reactor diameter, the total quench zone length being less than 1.5 times the inner diameter (Van Vliet et al., 2006).

3.2 FUNDAMENTALS OF HYDROTREATING

3.2.1 Chemistry

The type and amount of impurities to be removed by catalytic hydrotreating in a petroleum distillate can vary substantially depending on the type and source of the feed. In general, light feeds (e.g., naphtha) contain very little and few types of impurities, while heavy feeds (e.g., residua) possess most of the heavy compounds present in a crude oil. Apart from having a high concentration of heavy compounds, the impurities in heavy feeds are more complex and refractory (i.e., difficult to react) than those present in light feeds. That is why hydrotreating of light distillates is conducted at lower reaction severity, whereas heavy oils require higher reaction pressures and temperatures.

The reactions occurring during catalytic hydrotreating can be classified in two types: hydrogenolysis and hydrogenation. In hydrogenolysis a carbon–heteroatom single bond undergoes "lysis" by hydrogen. The heteroatom is any atom other than hydrogen or carbon present in petroleum, such as sulfur, nitrogen, oxygen, and metals. In hydrogenation, hydrogen is added to the molecule without cleaving bonds. The principal hydrogenolysis and hydrogenation reactions in catalytic hydrotreating are described below.

Hydrogenolysis reactions

- *Hydrodesulfurization* (HDS). Removal of organic sulfur compounds from a petroleum fraction and conversion to hydrogen sulfide (H_2S). Sulfur removal difficulty increases in the following order: paraffins < naphthenes < aromatics. The type of sulfur compounds can be classified as mercaptans, sulfides, disulfides, thiophenes, benzothiophenes, dibenzothiophenes, and substituted dibenzothiophenes. The ease of removal of these sulfur compounds is in the same order, the mercaptans being the easiest to remove and dibenzothiophenes the most difficult.

- *Hydrodenitrogenation* (HDN). Removal of organic nitrogen compounds and conversion to ammonia (NH_3). Removal of nitrogen requires more severe reaction conditions than does HDS. The molecular complexity (five- and six-membered aromatic ring structures), the quantity, and the difficulty of nitrogen-containing molecules to be removed increase with increasing boiling range of the distillate. Nitrogen compounds can be basic or nonbasic. Pyridines and saturated heterocyclic ring compounds (indoline, hexahydrocarabazole) are generally basic, whereas pyrroles are nonbasic.
- *Hydrodeoxygenation* (HDO). Removal of organic oxygen compounds and conversion to water. Similar to HDS and HDN, lower-molecular-weight oxygen compounds are easily converted, while higher-molecular-weight oxygen can be difficult to remove. Phenol is one of the most difficult oxygen compounds to convert.
- *Hydrodemetallization* (HDM). Removal of organometals and conversion to the respective metal sulfides. Nickel and vanadium being the most common metals present in petroleum, hydrodemetallization is frequently subdivided into hydrodeniquelization (HDNi) and hydrodevanadization (HDV). Once metal sulfides are formed, they are deposited on the catalyst and contribute to irreversible deactivation.

Hydrogenation reactions

- *Saturation of olefins.* Conversion to their saturated homologs of organic compounds containing double bonds.
- *Saturation of aromatics, or hydrodearomatization (HDA).* Conversion of aromatic compounds to naphthenes. The aromatic compounds found in petroleum distillates are mono-, di-, tri-, and polynuclear aromatics. Monoaromatics are much more difficult to saturate than the others since their saturation requires more energy.
- *Hydrocracking* (HYC). During hydrotreatment of light and middle distillates, some hydrocracking can occur, but its extent is normally low. However, when processing heavy feeds it can be very high. Hydrocracking is also a hydrogenolysis reaction in which carbon–carbon bonds are broken.

Asphaltenes can undergo both types of reactions (hydrocracking and hydrogenation) depending on reaction conditions. At relatively low or moderate temperatures, the reaction is more hydrogenation dominated during hydrocracking of heavy residue; however, at high temperatures, hydrocracking is more prominent. The overall conversion of asphaltenes is called hydrodeasphaltenization (HDAsp). Examples of some typical reactions occurring during catalytic hydrotreating are presented in Figure 3.13.

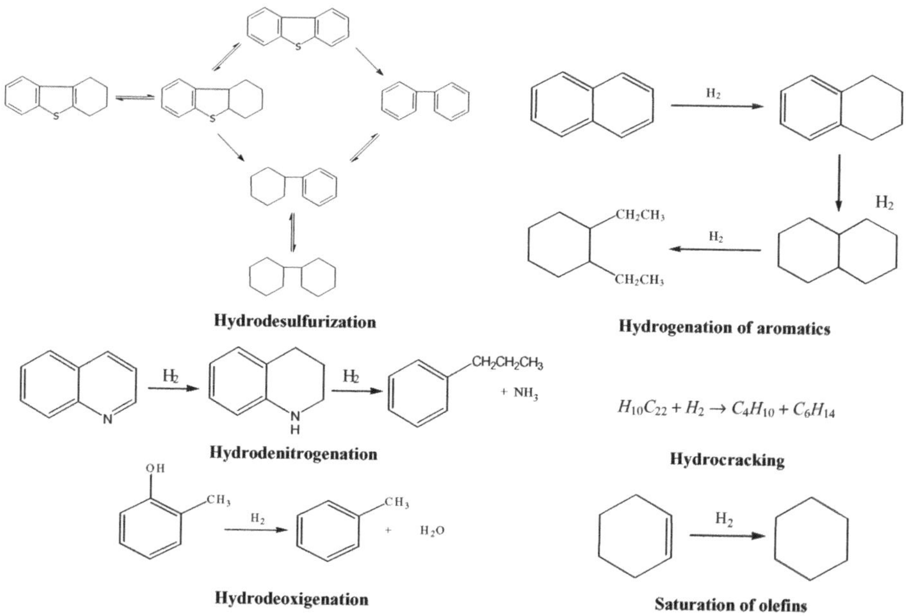

Figure 3.13. Examples of typical hydrotreating reactions.

3.2.2 Thermodynamics

There are fundamental differences in the removal of various impurities, largely because of the structure of the different molecules. HDS and olefin saturation are the most rapid reactions, and HDN and HDA are the most difficult. In contrast to HDS, for HDN the aromatic must first be saturated, and then the nitrogen is removed. Most of the reactions are irreversible with the exception of HDA, which is equilibrium limited at high temperatures, since at these conditions the reverse reaction of naphthene dehydrogenation becomes favored.

All the hydrotreating reactions are exothermic, causing an increase in the reactor temperature as the feed passes through the catalyst bed. The reactor ΔT value depends on the concentration of each heteroatom and the extent of each reaction during hydrotreatment. The heat of the reaction varies significantly among the different reactions and from one compound to the other, as can be seen in Table 3.3 (Ali, 2007). As the number of moles of hydrogen required to remove each organocompound increases, the amount of heat released also increases.

Equilibrium constants of different hydrotreating reactions are also reported in Table 3.3. From these values the following observations can be made:

- The values of K_{eq} of HDS and HDN are positive over a wide range of temperatures (within values commonly reported on a commercial scale), which indicates that these reactions are essentially irreversible

TABLE 3.3. Equilibrium Constants and Standard Enthalpies of Various Hydrotreating Reactions

Reaction	$\log_{10} K_{eq}$ at Temperature (°C):					$\Delta H^{\circ a}$
	25	100	200	300	400	
Hydrodesulfurization						
$C_3H_7-SH+H_2 \Leftrightarrow C_3H_8+H_2S$	10.57	8.57	6.92	5.87	5.15	–57
Thiophene + $3H_2 \Leftrightarrow nC_4H_{10}+H_2S$	30.84	21.68	14.13	9.33	6.04	–262
Benzothiophene + $H_2 \Leftrightarrow$ ethylbenzene + H_2S	29.68	22.56	16.65	12.85	10.20	–203
Dibenzothiophene + $2H_2 \Leftrightarrow$ biphenyl + H_2S	24.70	19.52	15.23	12.50	10.61	–148
Hydrodenitrogenation						
Indole + $3H_2 \Leftrightarrow$ ethylbenzene + NH_3	—	—	—	7.8	5.0	–49
Carbazole + $2H_2 \Leftrightarrow$ biphenyl + NH_3	—	—	—	6.8	5.1	–126
Pyridine + $5H_2 \Leftrightarrow$ *n*-pentane + NH_3	—	—	—	8.9	4.4	–362
Quinoline + $4H_2 \Leftrightarrow$ propylbenzene + NH_3	—	—	—	7.0	3.3	–272
Hydrogenation of aromatics						
Naphthalene + $2H_2 \Leftrightarrow$ tetralin	—	—	1.26	–1.13	–2.80	–140
Tetralin + $3H_2 \Leftrightarrow$ *trans*-decalin	—	—	0.74	–2.95	–5.56	–193
Cyclohexylbenzene + $3H_2 \Leftrightarrow$ cyclohexylhexane	—	—	2.47	–1.86	–4.91	–295
Phenanthrene + $4H_2 \Leftrightarrow$ octahydrophenanthrene	—	—	1.16	–3.64	–7.12	–251

[a]Standard enthalpy of reaction in kJ/mol organic reactant.

and can proceed to completion if hydrogen is present in stoichometric quantity.

- In general, as the temperature increases, the values of K_{eq} decrease, which is in agreement with the exothermicity of the reactions.

Most HDS reactions are straightforward except those of aromatic sulfur species, which must start with ring opening and sulfur removal, followed by saturation of the resulting olefin. In the case of the HDS of dibenzothiophenes, there are two major pathways: direct hydrodesulfurization, in which the sulfur atom is removed from the structure and replaced by hydrogen without hydrogenation of any of the other carbon–carbon double bonds; and the hydrogenation route, which assumes that at least one aromatic ring adjacent to the sulfur-containing ring is first hydrogenated before the removal of sulfur. Also, an aromatic ring may be hydrogenated after sulfur removal. The hydrogenation pathways are subject to thermodynamic equilibrium constraints. Thus, the partially hydrogenated intermediates have lower equilibrium concentrations at higher temperatures, and HDS via the hydrogenation route becomes limited at low pressures and high temperatures.

In the case of the hydrogenation of aromatic ring compounds, it is also an exothermic reaction, and equilibrium yields are favored by low temperatures. The maximum aromatic reduction (i.e., the optimum reaction temperature) is a function of the types and amount of aromatic compounds in the feed, space velocity, hydrogen partial pressure, and catalyst type. Complete hydrogenation of aromatics is not possible, owing to equilibrium limitations under typical hydrotreating conditions.

For the hydrodearomatization reaction to proceed, the polynuclear aromatics are first hydrogenated to three-ring to two-ring to one-ring and to the end products (naphthene rings). Saturation of the final aromatic ring is difficult because of the resonance stabilization of the monoaromatic ring.

Olefin saturation is very rapid and highly exothermic. For example, HDN shows a heat of reaction of 1 Btu/lb of feed for each 100 ft^3 of hydrogen consumed, HDS generates 1 Btu/lb of feed for each 10 ft^3 of hydrogen consumed, and the olefin saturation generates 1 Btu/lb of feed for each 2 ft^3 of hydrogen consumed. Diolefins are readily hydrogenated to olefins at low temperatures (Gary and Handwerk, 2001).

Heats of hydrotreating reactions have been reported in the literature for reactor modeling purposes. For example, Tarhan (1983) used the following values for hydrotreating of straight-run gas oil:

Hydrodesulfurization	–251,000 kJ/kmol
Hydrodeoxygenation	–68,200 kJ/kmol
Hydrodenitrogenation	–64,850 kJ/kmol
Hydrocracking	–41,000 kJ/kmol
Hydrogenation	–125,520 kJ/kmol

Others prefer the use of the overall heat of reaction, as in the case of the HDS of atmospheric residue ($\Delta H_R = -7820\,\text{kJ/kg}$ sulfur $= -250{,}748\,\text{kJ/kmol}$). Overall heats of reaction are mean values derived from heat balances of several similar HDT processes and include the contribution of all the reactions (Shah and Paraskos, 1975; Döhler and Rupp, 1987).

3.2.3 Kinetics

Most of the kinetic studies reported in the literature of the various hydrotreating reactions have been conducted using pure compounds (i.e., model compounds), as well as binary and multicomponent mixtures of them (Girgis and Gates, 1991). The available kinetics data with model compounds are usually represented with pseudo-first-order rate equations or with Langmuir–Hinshelwood rate equations. However, the complexity of the individual reactions occurring in an extremely complex mixture and the interference of the products with those from other components of the mixture is unpredictable. Or the interference of secondary and tertiary products with the course of a reaction, and hence with the formation of primary products, may also be a cause for concern. Hence, caution is advised when applying the data from model compound studies to the behavior of petroleum and its distillates. Kinetic data derived from model compounds cannot be expected to include contributions from the various steric effects that are a consequence of complex molecules containing three-dimensional structures. Indeed, such steric effects can lead to the requirement of additional catalyst and process parameters for the various heteroatoms removed.

For the HDT of real feeds, nth-order kinetics with respect to total concentration of the heteroatom is usually employed, in which the n value depends on several factors, such as type and concentration of the heteroatom, catalyst properties, type of feed, operating conditions, and experimental system, among others.

Hydrodesulfurization The structural differences between the various sulfur-containing molecules make it impractical to have a single rate expression applicable to all reactions in HDS. Each sulfur-containing molecule has its own hydrogenolysis kinetics, which is usually complex.

The complex nature of oil fractions with sulfur compounds exhibiting very different reactivities as well as the presence of others, such as nitrogen (basic and nonbasic), aromatics, and so on, reacting at the same time and competing for the same active sites and also inhibiting the effects of by-products of the same reactions (e.g., hydrogen sulfide) have limited HDS experimental studies to model compounds ranging from easy to desulfurize (e.g., thiophene) to difficult to desulfurize (e.g., 4,6-dimethyldibenzothiophene). These are the main reasons why few works have been reported dealing with experiments with real petroleum feedstocks under industrial conditions, since most of the time it is not simple to extract individual effects and one does not know which one to

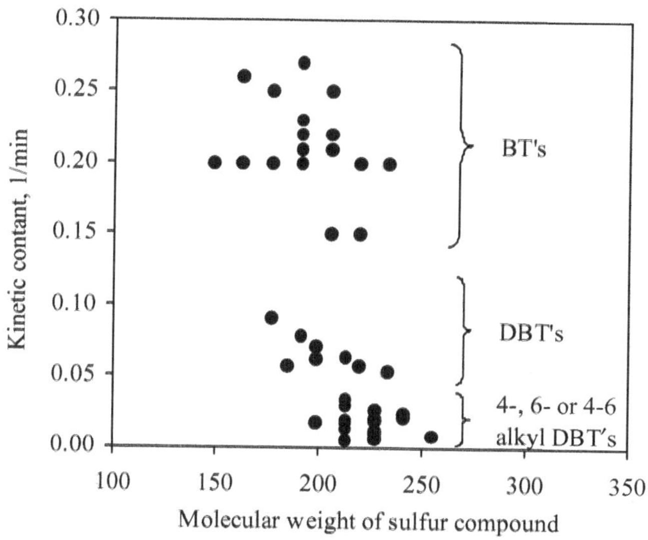

Figure 3.14. First-order kinetic constant values for the HDS of different sulfur compounds in a diesel fraction.

blame. However, when a catalyst formulation is almost ready for commercial application, experiments with real feeds are mandatory. Testing with real feeds not only for exploration of the commercial application of new catalyst formulations but also for process design and optimization studies is a very important step for new technology development. For the latter issue, kinetic data obtained from experiments with real feeds are of great interest, since they are employed for reactor modeling, simulation, and optimization.

When real feeds and their hydrotreated products are characterized in detail, as in the case of sulfur compounds during HDS of straight-run gas oil and other light petroleum distillates, experimental results obtained by gas chromatography with a sulfur chemiluminescence detector have indicated that simple first-order kinetics with respect to the heteroatom is the predominant mechanism by which it is removed from the organic material. However, the various molecules have very different reactivity, as illustrated in Figure 3.14 for several sulfur compounds included in diesel fractions. The differences in HDS reactivity of the sulfur compounds are clearly distinguished. As is well known, DBTs with 4-, 6-, or 4, 6-alkyl positions are the most refractory compounds.

When the analysis is performed as the total content of the heteroatom (e.g., total sulfur content, total nitrogen content), the kinetics is typically represented by *n*th-order kinetics with respect to the total concentration of that heteroatom. The value of *n* for most hydrotreating reactions is in general larger than 1. For example, Table 3.4 reports a compilation of reaction orders and activation energies obtained for the hydrodesulfurization of different real feedstocks. The table has two parts; one shows reaction orders as calculated

TABLE 3.4. Reaction Orders and Activation Energies for Hydrodesulfurization of Different Feedstocks

Feed[a]	Density at 15°C	Sulfur (wt%)	Distillation Range (°C)	n	E_A (kcal/mol)
Reaction Orders Calculated from Experimental Data					
SRGO	0.861	1.31	213–368	1.57	20.3
SRGO	0.843	1.32	188–345	1.53	
HSRGO	0.862	1.33	142–390	1.65	
SRGO-LCO	0.879	1.78	209–369	1.63	16.5
VGO	0.907	2.14	243–514	2.09	33.1
SRGO-LCO	0.909	2.44	199–370	1.78	16.37
Residue oil	0.910	3.45	—	2.0	68.6
Residue oil	0.950	3.72	281–538	2.0	29.0
Residue oil	1.007	5.30	—	2.5	36.1
AR	0.995	5.86	—	2.0	29.0
Assumed Values of Reaction Orders					
Used oil	0.900	0.70	—	1.0	19.6
Residue oil	0.969	1.45	—	1.0	24.0
SRGO	—	1.47	—	1.65	25.0
Residue oil	0.964	2.90	—	1.0	18.3
CGO	0.984	4.27	196–515	1.5	33.0

[a]SRGO, straight-run gas oil; HSRGO, heavy straight-run gas oil; LCO, light cycle oil; VGO, vacuum gas oil; AR, atmospheric residue; CGO, coker gas oil.

with experimental data, and the second presents assumed values of reaction order. For calculated reaction orders, in general an increase in their values in the range 1.5 to 2.5 is observed as the sulfur content is increased. Some data do not follow this trend, which may be due to differences in conditions used during experiments. This tendency with respect to sulfur content in the feed is not observed for activation energies. For different sulfur contents (3.72 and 5.68 wt%), two feeds were reported to have the same reaction order and activation energy ($n_S = 2$ and $E_A = 29$ kcal/mol), or for almost the same sulfur content (3.45 and 3.72 wt%), two other feeds presented very different activation energies (68.6 and 29 kcal/mol) for the same reaction order ($n_S = 2$). Therefore, it is clear that reaction order and activation energy depend on the type and distribution of heteroatom compounds in the oil fraction as well as on the catalyst and reaction conditions employed. Figure 3.15 corroborates the fact that reaction orders can be different for the HDS of middle distillates having the same amount of sulfur but coming from different sources (i.e., crude oils). The development of general kinetic data for the hydrodesulfurization of different feedstocks is complicated by the presence of a large number of sulfur compounds, each of which may react at a different rate because of structural differences as well as differences in molecular weight.

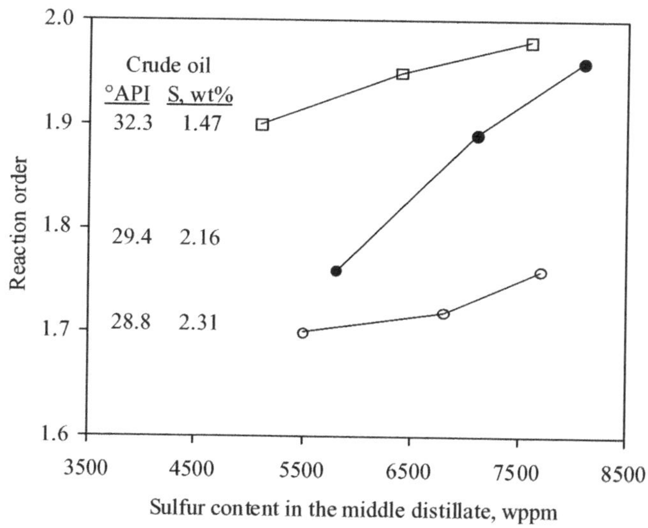

Figure 3.15. Reaction order for the HDS of middles distillates from different crude oils ($CoMo/\gamma\text{-}Al_2O_3$, 340 to 360°C, 1.5 to $2.0\,h^{-1}$ LHSV, $54\,kg/cm^2$, H_2/oil ratio of $2000\,ft^3/bbl$).

One way to represent the HDS reaction is the practical and widely accepted generalized stoichiometric equation, which lumps the HDS reaction of all the sulfur compounds into a single expression:

$$\upsilon_S S_{(liq)} + \upsilon_{H_2} H_{2(gas)} \rightarrow \upsilon_{HC} HC_{(liq)} + \upsilon_{H_2S} H_2S_{(gas)} \tag{3.10}$$

where υ_S, υ_{H_2}, υ_{HC}, and υ_{H_2S} are the stoichiometric coefficients of the organic sulfur compounds, hydrogen, sulfur-free hydrocarbon, and hydrogen sulfide, respectively.

The simplest model that can be used to represent HDS kinetics is the power-law type, which does not take into account the inhibiting effect of H_2S (Cotta et al., 2000):

$$r_{HDS} = k_{HDS} C_S^{ns} p_{H_2}^{m} \tag{3.11}$$

Another simplified kinetic model considers the existence of only two reactive sulfur components by means of the following first-order rate equation (Gates et al., 1979):

$$r_{HDS} = \gamma k_1 C_S + (1-\gamma) k_2 C_S \tag{3.12}$$

where γ, k_1, and k_2 are adjustable parameters. The parameter γ represents the fraction of the easy-to-react sulfur-containing compounds, and $1-\gamma$ is the fraction of the more refractory sulfur-containing compounds. These simplified kinetic models [Eqs. (3.11) and (3.12)] are valid only when partial pressures of hydrogen and hydrogen sulfide are held constant, and these values are incorporated in k_{HDS}, k_1, and k_2.

The most frequently used kinetic expression of HDS is the following Langmuir–Hinshelwood model:

$$r_{HDS} = \frac{k_{HDS} C_S^{ns} C_{H_2}^{m}}{\left(1 + K_{ads}^{H_2S} C_{H_2S}\right)^2} \qquad (3.13)$$

The exponent 2 in the denominator of Eq. (3.13) represents the number of sites in adsorption for hydrogen sulfide.

When HDS is studied together with HDA, the following reaction has been proposed, which is assumed to be irreversible under normal hydrotreating conditions (Chowdhury et al., 2002):

$$A - S + 2H_2 \rightarrow A + H_2S \qquad (3.14)$$

where A represents aromatic compounds. For this HDS reaction, the following form of the Langmuir–Hinshelwood rate equation has been considered:

$$r_{HDS} = \frac{k_{HDS} C_{A-S}^{ns} C_{H_2}^{m}}{1 + K_{ads}^{H_2S} C_{H_2S}} \qquad (3.15)$$

Equations (3.13) and (3.15) include an adsorption-equilibrium constant of hydrogen sulfide (K_{ads}), which is a function of temperature and can be estimated using the van't Hoff equation:

$$K_{ads}(T) = K_0 \exp\left(\frac{\Delta H_{ads}}{RT}\right) \qquad (3.16)$$

The reaction orders for sulfur and hydrogen, n_S and m, respectively, have been reported to range between 1.5 and 2.5 for sulfur, depending on the type of feed as well as the amount and type of sulfur compounds, and between 0.5 and 1.0 for hydrogen. According to the chemical dissociation of the hydrogen molecule on the catalyst surface, the theoretical value of m should be 0.5. However, it approaches 1.0 if the mass transfer rate of hydrogen becomes the limiting step (Cheng et al., 2004).

Apart from H_2S, hydrodesulfurization can also be inhibited by other compounds, so that the term indicating competitive adsorption can be expressed by a sum of terms as follows:

$$r_{HDS} = \frac{k_{HDS} C_S^{ns} C_{H_2}^{m}}{\left(1 + \sum_{i=1}^{N} K_{ads}^{i} C_i\right)^2} \qquad (3.17)$$

For example, when aromatic hydrocarbons and H_2S are to be the inhibiting compounds, Eq. (3.17) becomes

$$r_{\text{HDS}} = \frac{k_{\text{HDS}} C_{\text{S}}^{ns} C_{\text{H}_2}^{m}}{\left(1 + K_{\text{ads}}^{\text{A}} C_{\text{A}} + K_{\text{ads}}^{\text{H}_2\text{S}} C_{\text{H}_2\text{S}}\right)^2} \tag{3.18}$$

A combination of the use of two reactive sulfur components with inhibition by H_2S and aromatics yields the following kinetic model (Avraam and Vasalos, 2003):

$$r_{\text{HDS}} = \frac{\gamma k_1 C_{\text{S}}^{ns} C_{\text{H}_2}^{m}}{\left(1 + K_{\text{ads}_1}^{\text{A}} C_{\text{A}} + K_{\text{ads}_1}^{\text{H}_2\text{S}} C_{\text{H}_2\text{S}}\right)^2} + \frac{(1-\gamma) k_2 C_{\text{S}}^{ns} C_{\text{H}_2}^{m}}{\left(1 + K_{\text{ads}_2}^{\text{A}} C_{\text{A}} + K_{\text{ads}_2}^{\text{H}_2\text{S}} C_{\text{H}_2\text{S}}\right)^2} \tag{3.19}$$

The reaction orders n_S and m in Eqs. (3.18) and (3.19) have been reported to be equal to 1.

It is well accepted that hydrodesulfurization undergoes through two reaction paths: direct HDS (DD) by hydrogenolysis of the reactants, and indirect HDS (ID) by hydrogenation of one aromatic ring followed by C–S bond cleavage of the hydrogenated intermediate products. It is also recognized that deep HDS is inhibited by the nitrogen compounds and aromatics, and they compete with sulfur compounds only on the hydrogenation sites, thus inhibiting the hydrogenation route. Based on these considerations, and assuming that H_2S inhibition is negligible, hydrogen is in excess, and sulfur conversion behaves as a pseudo-first-order reaction, the following kinetic model has been proposed (Liu et al., 2008):

$$r_{\text{HDS}} = k_{\text{HDS}}^{DD} C_{\text{S}} + \frac{k_{\text{HDS}}^{ID} C_{\text{S}}}{1 + K_{\text{ads}}^{\text{A}} C_{\text{A}}^{nA} + K_{\text{ads}}^{\text{N}} C_{\text{N}}^{nN}} \tag{3.20}$$

where k_{HDS}^{DD} and k_{HDS}^{ID} are the rate constants for the direct HDS and indirect HDS routes, respectively.

Hydrodenitrogenation As in the case of HDS, all of the nitrogen compounds present in the feed are sometimes lumped together and the following reaction is assumed for nitrogen removal:

$$\text{R–N} + 2\text{H}_2 \rightarrow \text{R–H} + \text{NH}_3 \tag{3.21}$$

where R–N is the hydrocarbon structure containing nitrogen, and R–H is the nitrogen-free hydrocarbon.

The power-law model and Langmuir–Hinshelwood rate equation have been used for determining nitrogen removal:

$$r_{\text{HDN}} = k_{\text{HDN}} C_{\text{N}}^{nN} p_{\text{H}_2}^{m} \tag{3.22}$$

$$r_{\mathrm{HDN}} = \frac{k_{\mathrm{HDN}} C_{\mathrm{N}}^{n_N} C_{\mathrm{H_2}}^{m}}{\left(1 + K_{\mathrm{ads}}^{\mathrm{NH_3}} C_{\mathrm{NH_3}}\right)^2} \tag{3.23}$$

where n_{N} is the reaction order of the nitrogen compounds. Equation (3.21) is used most frequently when during experiments it is not possible to identify the ammonia content in the gas-phase exit stream.

Not only can ammonia inhibit HDN reactions but other compounds also, such as H_2S and aromatics. To account for the inhibition effect of these other compounds, similar to HDS, the following rate expression can be used:

$$r_{\mathrm{HDN}} = \frac{k_{\mathrm{HDN}} C_{\mathrm{N}}^{n_N} C_{\mathrm{H_2}}^{m}}{\left(1 + \sum_{i=1}^{N} K_{\mathrm{ads}}^{i} C_i\right)^2} \tag{3.24}$$

Since nitrogen is present in petroleum as basic (N_B) and nonbasic (N_{NB}) compounds, the following consecutive reaction scheme in also considered for nitrogen removal, in which nonbasic nitrogen is hydrogenated first to basic nitrogen, which undergoes further reactions to eliminate the nitrogen atom from the molecule:

$$\mathrm{N}_{NB} \xrightarrow{k_{\mathrm{HDN}NB}} \mathrm{N}_B \xrightarrow{k_{\mathrm{HDN}B}} \mathrm{HC} + \mathrm{NH_3} \tag{3.25}$$

The kinetics of HDN has been reported to be represented by the power-law model with a value of $n_{\mathrm{N}} = 1.5$ (Bej et al., 2001). For nonbasic nitrogen,

$$r_{\mathrm{HDN}_{NB}} = k_{\mathrm{HDN}_{NB}} C_{\mathrm{N}_{NB}}^{n_{\mathrm{N}}} \tag{3.26}$$

and for basic nitrogen,

$$r_{\mathrm{HDN}_B} = k_{\mathrm{HDN}_{NB}} C_{\mathrm{N}_{NB}}^{n_{\mathrm{N}}} - k_{\mathrm{HDN}_B} C_{\mathrm{N}_B}^{n_{\mathrm{N}}} \tag{3.27}$$

Hydrodearomatization During hydrotreating, the HDA reaction is controlled kinetically at low temperature but controlled thermodynamically at high temperature. This means that when the reaction temperature increases, the hydrogenation of aromatics increases, passes through the maximum, and then decreases. This behavior can be represented in a general form by the following simple reversible reaction and first-order rate expression:

$$\text{aromatics}\,(\mathrm{A}) + \mathrm{H_2} \underset{k_r}{\overset{k_f}{\rightleftarrows}} \text{naphthenes}\,(\text{naph}) \tag{3.28}$$

$$r_{\mathrm{HDA}} = k_f p_{\mathrm{H_2}}^{m} C_{\mathrm{A}} - k_r C_{\mathrm{naph}} \tag{3.29}$$

where C_{A} and C_{naph} are the concentrations of aromatics and naphthenes, respectively, and k_f and k_r are the forward (hydrogenation) and reverse (dehydrogenation) rate constants. In hydrotreating operations, hydrogen is used in excess so that H_2 partial pressure can be assumed constant as well as the term $k_f p_{H_2}^m$.

Another approach to modeling the HDA reaction is by the separation of the total aromatic content into three groups of aromatics: monoaromatics, containing a single aromatic ring in their structure; diaromatics, with two aromatic rings; and polyaromatics, with three or more aromatic rings. Based on this consideration, the following stoichiometric equations can be established (Chowdhury et al., 2002):

$$\text{polyaromatics (PA)} + H_2 \underset{k_{PA}^r}{\overset{k_{PA}^f}{\rightleftharpoons}} \text{diaromatics (DA)} \tag{3.30}$$

$$\text{diaromatics (DA)} + 2H_2 \underset{k_{DA}^r}{\overset{k_{DA}^f}{\rightleftharpoons}} \text{monoaromatics (MA)} \tag{3.31}$$

$$\text{monoaromatics (MA)} + 3H_2 \underset{k_{MA}^r}{\overset{k_{MA}^f}{\rightleftharpoons}} \text{naphthenes (naph)} \tag{3.32}$$

The naphthenes are the final products of HDA reactions and are composed primarily of alkylmono- to alkylhexacycloparaffins. The HDA reaction rates are expressed as (Cheng et al., 2004)

$$r_{HDA_{PA}} = k_{PA}^f p_{H_2}^{m_1} C_{PA} - k_{PA}^r C_{DA} \tag{3.33}$$

$$r_{HDA_{DA}} = -k_{PA}^f p_{H_2}^{m_1} C_{PA} + k_{PA}^r C_{DA} + k_{DA}^f p_{H_2}^{m_2} C_{DA} - k_{DA}^r C_{MA} \tag{3.34}$$

$$r_{HDA_{MA}} = -k_{DA}^f p_{H_2}^{m_2} C_{DA} + k_{DA}^r C_{MA} + k_{MA}^f p_{H_2}^{m_3} C_{MA} - k_{MA}^r C_{naph} \tag{3.35}$$

$$r_{HDA_{Naph}} = -k_{MA}^f p_{H_2}^{m_3} C_{MA} + k_{MA}^r C_{naph} \tag{3.36}$$

The first reaction order of the aromatics compounds is in general explained by their strong adsorption on the catalyst surface. There is a preference for hydrogenation of the first ring(s) in polyaromatics, since the hydrogenation rate of the first ring in condensed diaromatics (e.g., naphthalene) has been found to be much faster (typically, 20 to 40 times) than that of monoaromatics (including byphenyl, tetralin, and cyclohexylbenzene) (Cooper and Donnis, 1996). In addition, the reaction DA $\leftrightarrow$ MA is more severely limited by equilibrium than is the reaction PA $\leftrightarrow$ DA. Based on this, the previous set of equations can be simplified to (Chowdhury et al., 2002; Bhaskar et al., 2004)

$$r_{HDA_{PA}} = k_{PA}^f p_{H_2}^{m_1} C_{PA} - k_{PA}^r C_{DA} \tag{3.37}$$

$$r_{HDA_{DA}} = k_{DA}^f p_{H_2}^{m_2} C_{DA} - k_{DA}^r C_{MA} \tag{3.38}$$

$$r_{HDA_{MA}} = k_{MA}^f p_{H_2}^{m_3} C_{MA} - k_{MA}^r C_{naph} \tag{3.39}$$

$$r_{HDA_{Naph}} = -r_{HDA_{MA}} = -k_{MA}^f p_{H_2}^{m_3} C_{MA} + k_{MA}^r C_{naph} \tag{3.40}$$

Working at close to constant hydrogen partial pressure allows for lumping rate constants (k_i^f and k_i^r) with $p_{H_2}^{m_i}$ as follows:

$$k_{PA}^{f^*} = k_{PA}^{f} p_{H_2}^{m_1} \tag{3.41}$$

$$k_{DA}^{f^*} = k_{DA}^{f} p_{H_2}^{m_2} \tag{3.42}$$

$$k_{MA}^{f^*} = k_{MA}^{f} p_{H_2}^{m_3} \tag{3.43}$$

Therefore, Eqs. (3.37) to (3.40) reduce to

$$r_{HDA_{PA}} = k_{PA}^{f^*} C_{PA} - k_{PA}^{r} C_{DA} \tag{3.44}$$

$$r_{HDA_{DA}} = k_{DA}^{f^*} C_{DA} - k_{DA}^{r} C_{MA} \tag{3.45}$$

$$r_{HDA_{MA}} = k_{MA}^{f^*} C_{MA} - k_{MA}^{r} C_{naph} \tag{3.46}$$

$$r_{HDA_{Naph}} = -k_{MA}^{f^*} C_{MA} + k_{MA}^{r} C_{naph} \tag{3.47}$$

To determine the forward and reverse reaction rate constants, equilibrium constants are defined as

$$K_{PA} = \frac{k_{PA}^{f^*}}{k_{PA}^{r}} \tag{3.48}$$

$$K_{DA} = \frac{k_{DA}^{f^*}}{k_{DA}^{r}} \tag{3.49}$$

$$K_{MA} = \frac{k_{MA}^{f^*}}{k_{MA}^{r}} \tag{3.50}$$

The equilibrium constants for reversible reactions are determined at different temperatures by using the van't Hoff correlation:

$$K_i = K_i^{\circ}(T_0) \exp\left[\frac{-\Delta H_{HDA_i}^{R}}{R}\left(\frac{1}{T_0} - \frac{1}{T}\right)\right] \tag{3.51}$$

The values of the equilibrium constants decrease when the number of side chains and the number of carbon atoms in each side chain increases until a maximum temperature, where this behavior shifts. For any equilibrium constant, an increase in reaction temperature results in a lower equilibrium constant and a higher equilibrium concentration of aromatics. At low temperatures (<443°C) the order of the equilibrium constant values is $K_{MA} > K_{DA} > K_{PA}$, while at high temperatures (>443°C) the order changes to $K_{PA} > K_{DA} > K_{MA}$ (Chowdhury et al., 2002; Bhaskar et al., 2004).

To simplify the kinetic modeling, HDA is also represented by the following irreversible reactions in series and rate equations:

$$PA \xrightarrow{k_{PA}} DA \xrightarrow{k_{DA}} MA \xrightarrow{k_{MA}} \text{naph} \quad (3.52)$$

$$-r_{PA} = k_{PA}C_{PA} \quad (3.53)$$

$$-r_{DA} = k_{DA}C_{DA} - k_{PA}C_{PA} \quad (3.54)$$

$$-r_{MA} = k_{MA}C_{MA} - k_{DA}C_{DA} \quad (3.55)$$

$$-r_{naph} = k_{MA}C_{MA} \quad (3.56)$$

Sulfur- and nitrogen-containing compounds can reversibly and slowly inhibit aromatic hydrogenation, and the poisoning tendency of various nitrogen compounds seems to be related not only to its basicity, but probably also to its structure.

Hydrogenation of Olefins Olefins react with hydrogen to form saturated hydrocarbons. The following stoichiometric equation is assumed to represent olefin hydrogenation:

$$R{-}CH{=}CH{-}R' + H_2 \xrightarrow{k_{HGO}} R{-}CH_2{-}CH_2{-}R' \quad (3.57)$$

The hydrogenation of olefins (HGO) is represented by a pseudo-first-order reaction with respect to the total concentration of olefins:

$$r_{HGO} = k_{HGO}C_{olef} \quad (3.58)$$

Hydrodeasphaltenization The way in which asphaltenes react is very important during hydrotreating. It is to be remembered that asphaltenes reduce the reaction rate of other reactions during hydrotreating, since asphaltenes are coke precursors that deactivate catalysts by plugging the catalytic sites (Ancheyta et al., 2003a,b). Different approaches have been developed to represent the manner in which asphaltenes react, and kinetic parameters (e.g., reaction order and kinetic constant) can be extracted from such models.

The power-law model is by far the most simplified model to use to represent the kinetics of HDAsp. With this approach, different types of reactivity of asphaltenes are not taken into account. Instead, all asphaltene molecules are considered to react at an average rate expressed in the kinetic constant. For the hydrodeasphaltenization reaction.

$$\text{asphaltenes} + H_2 \xrightarrow{k_{HDasp}} \text{products} \quad (3.59)$$

the power-law kinetics is represented by

$$-r_{\mathrm{HDasp}} = k_{\mathrm{HDasp}} C_{\mathrm{asp}}^{n_{\mathrm{asp}}} C_{\mathrm{H2}}^{m} \tag{3.60}$$

where C_{asp} is the concentration of asphaltene and n_{asp} the reaction order for asphaltenes.

Another approach based on the power-law model is to separate the global HDasp reaction into two reactions in parallel:

$$\mathrm{asp}\begin{cases} \mathrm{asp}_1 = \gamma\mathrm{asp} + \mathrm{H}_2 \xrightarrow{k_{\mathrm{HDasp1}}} \\ \mathrm{asp}_2 = (1-\gamma)\mathrm{asp} + \mathrm{H}_2 \xrightarrow{k_{\mathrm{HDasp2}}} \end{cases}\Big\}\ \mathrm{products} \tag{3.61}$$

where γ is the fraction of hard-to-react asphaltenes (asp_1) and $(1 - \gamma)$ is the fraction of easy-to-react asphaltenes (asp_2). According to the assumptions of Kwak et al. (1992), the reaction order for both types of asphaltenes is 1. The kinetic model is

$$-r_{\mathrm{HDasp}} = \gamma k_{\mathrm{HDasp_1}} C_{\mathrm{asp}}^{n_{\mathrm{asp1}}} C_{\mathrm{H_2}}^{m} + (1-\gamma) k_{\mathrm{HDasp_2}} C_{\mathrm{asp}}^{n_{\mathrm{asp2}}} C_{\mathrm{H_2}}^{m} \tag{3.62}$$

Hydrodemetallization Studies with individual metalloporphyrin model compounds have indicated that the hydrodemetallization mechanisms involve a reversible hydrogenation step of a Ni (or V)-EP to form a Ni (or V)-EPH_2 followed by an irreversible hydrogenolysis, which results in the fragmentation of the porphyrin ring and deposition of the metal on the catalyst, according to the following reaction network:

$$\mathrm{M-EP} + \mathrm{H_2} \rightleftarrows \mathrm{M-EPH_2} + n\mathrm{H_2} \rightarrow \mathrm{deposit} + \mathrm{hydrocarbon} \tag{3.63}$$

where M is Ni or V.

The removal of metals contained mostly in heavy petroleum fractions, such as residues, cannot be represented by the same reaction network, since apart from metalloporphyrin, nonporphyrin metal compounds are also present. The kinetics of HDM is thus commonly expressed by a power-law model:

$$r_{\mathrm{HDNi}} = k_{\mathrm{HDNi}} C_{\mathrm{Ni}}^{n_{\mathrm{Ni}}} \tag{3.64}$$

$$r_{\mathrm{HDV}} = k_{\mathrm{HDV}} C_{\mathrm{V}}^{n_{\mathrm{V}}} \tag{3.65}$$

Hydrocracking The extent of hydrocracking is determined primarily by reaction conditions. During hydrotreating, in general, light and middle distillates exhibit low hydrocracking, whereas for heavy feed it is very high. Kinetic models for HYC, particularly of heavy feeds, have been discussed in detail in Section 2.3, so that in this section, only kinetics of hydrocracking occurring during HDT of light feeds is considered.

The typical support of hydrotreating catalysts is γ-alumina (γ-Al_2O_3), whose acid sites promote mild hydrocracking reactions and produce light hydrocarbons in gas and lighter liquid products. The common hydrocracked products for different feeds are:

Heavy gas oil: diesel and light gases
Light gas oil: naphtha and light gases
Naphtha: C_5–C_6 and light gases

The kinetics of the hydrocracking reaction taking place during hydrotreating can be represented by a three-lump model, as shown in Figure 3.16. From this figure the following pseudo-first-order kinetic equations can be derived. For example, when the feed is light gas oil (LGO), the hydrocracked products are naphtha (NT) and light gases (gas):

$$r_{\mathrm{LGO}} = k_1 C_{\mathrm{LGO}} + k_2 C_{\mathrm{LGO}} \tag{3.66}$$

$$r_{\mathrm{NT}} = -k_2 C_{\mathrm{LGO}} + k_2 C_{\mathrm{NT}} \tag{3.67}$$

$$r_{\mathrm{Gas}} = -k_1 C_{\mathrm{LGO}} - k_3 C_{\mathrm{NT}} \tag{3.68}$$

The concentrations of heteroatoms (sulfur, nitrogen, oxygen, metals) and asphaltenes in a petroleum fraction are usually reported in weight percent. The transformation of mass concentration to molar concentration, which is needed unit for kinetic and reactor modeling purposes, can be done easily using the equation

$$C_i = \frac{(\mathrm{wt\%})_i}{100} \frac{\rho_{\mathrm{oil}}}{\mathrm{PM}_i} \tag{3.69}$$

For olefins, the common analysis to determine the content of unsaturates is the bromide number (Br No.). Therefore, the molar concentration of double bonds can be expressed as follows:

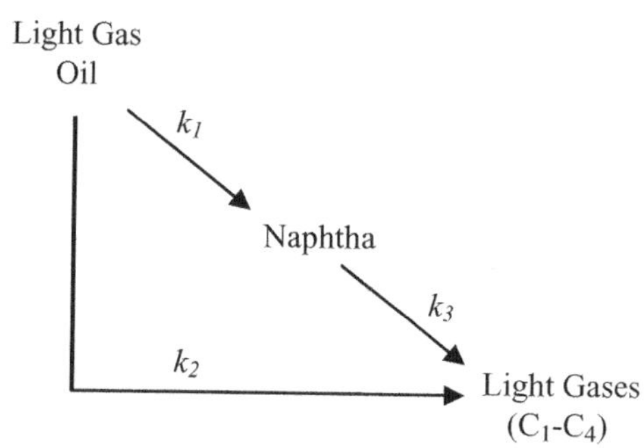

Figure 3.16. Three-lump kinetic model for hydrocracking reaction during catalytic hydrotreating.

$$C_{olef} = \frac{\text{Br No.}}{100} \frac{\rho_{oil}}{\text{PM}_{Br}} \tag{3.70}$$

The content of hydrocrackable compounds (C_{hyc}) can be approximated by the feed concentration as follows:

$$C_{hyc} = \frac{\rho_{oil}}{\text{PM}_{oil}} \tag{3.71}$$

where ρ_{oil} is the density of the hydrocarbon, which varies as feed passes through the catalyst bed, and PM_i, PM_{Br}, and PM_{oil} are the molecular weights of each heteroatom, bromide, and the hydrocarbon, respectively.

3.2.4 Catalysts

Most of the hydrotreating catalysts in commercial use are supported on γ-alumina (γ-Al_2O_3), sometimes with small amounts of silica (SiO_2) or phosphorus (P). Preparation of the support is a very important step during catalyst manufacture to achieve a material with a high surface area and an appropriate pore structure. This high surface area is required to disperse the active metals and promoters uniformly. The typical active metals are molybdenum (Mo) and tungsten (W) sulfides, modified by a promoter: either cobalt (Co) or nickel (Ni) sulfide. The main function of the promoter is to increase the activity of the active metal sulfide substantially. The amount of each component in a commercial catalyst depends on the application desired. In general, the specifications of the feed and the desired product quality will determine which catalyst (or combination of catalysts) will be used.

CoMo and NiMo/γ-Al_2O_3 are the preferred catalysts, for several reasons; they are cheap, highly selective, easy to regenerate, and resistant to poisons. Although being more effective for HDN and HDA, NiW catalysts are seldom used for commercial hydrotreating applications since they are much more expensive than NiMo catalysts. CoMo/γ-Al_2O_3 catalyst is recommended for HDS, and NiMo/γ-Al_2O_3 or NiCoMo/γ-Al_2O_3 for HDN. NiMo catalysts possess higher hydrogenation activity than CoMo catalysts and hence are more suitable for saturation of aromatic rings, although both catalysts will remove both sulfur and nitrogen.

Since the physical and chemical composition of petroleum and its fractions varies considerably depending on their origin, there is not a universal catalyst for hydrotreatment of all the feeds to achieve the desired target in terms of impurities removal and conversion. Thus, the properties of catalysts for hydrotreatment of light and middle distillates are different from those used to hydrotreat heavy oils. Whereas for light distillate hydrotreating the chemical composition of the catalytic surface and the specific surface area are the most important parameters, since metal and coke deposition are not crucial, in the case of heavy feeds, porosity is the determinant as to suitable catalyst activity and life. In both cases the role of the support is crucial. For heavy oil hydrotreatments, support acidity and porosity have to be designed carefully to accomplish the optimum catalyst performance. Acidity must be strictly balanced to perform hydrocracking at a desired reaction extent, but not so much as to produce excessive coking. The acidity of the support is provided primarily by silica, zeolite, and/or phosphorus. In the case of porosity, when hydrotreating light and middle distillates, minimum pore size is required to overcome most diffusional restrictions. However, for hydrotreating heavy oils, pore size needs to be designed properly to handle the complex large molecules (e.g., asphaltenes) contained in such feeds. The ability to adjust pore size to concentrate pores around a particular diameter has a great impact on the hydrotreating catalyst activity either at the beginning of operation (start-of-run) or at the middle or end of operation (middle-of-run or end-of-run).

Usually, hydrotreating catalysts are prepared in the oxide state (e.g., $CoOMoO_3/\gamma\text{-}Al_2O_3$) and they must be activated by converting the metals from oxide form to sulfide form to achieve the maximum activity of the catalyst. This step, also called *presulfiding*, is carried out by four routes (Marroquín et al., 2004):

1. With a nonspiked feedstock, in which sulfiding is conducted with the same sulfur from the normal feedstock
2. With a H_2/H_2S mixture, carried out in the gas phase and most practiced in laboratory experiments
3. With a spiked feedstock, in which sulfiding is done primarily by the sulfur of the spiking agent
4. Ex situ sulfiding, which has been reported to have the same or better activity and stability than in situ sulfiding

In liquid-phase sulfiding, the hydrocarbon carrier aids in wetting and hence in providing a better distribution of sulfur across the bed and sulfiding the catalyst evenly. The hydrocarbon also serves as a sink of heat generated, allowing for better control of the exothermic reaction between sulfur and the metal of the catalyst. This allows for more rapid presulfiding. The sulfiding reaction is highly exothermic, and much care must be taken to prevent excessive temperatures during activation to prevent permanent catalyst deactivation.

It should be remembered that a spiking agent is a sulfur-containing organic compound that releases H_2S at a much lower temperature than do the sulfur compounds present in normal feedstocks. There are various spiking agents reported in the literature and frequently used for activation of HDS catalysts, such as carbon disulfide (CS_2), dimethyl sulfide (DMS), dimethyl disulfide (DMDS), butanethiol, ditertiary nonyl polysulfide (TNPS), ethyl mercaptan (EM), dimethyl sulfoxide (DMSO), and *n*-buthyl mercaptan (NBM). Among them, DMDS has demonstrated better behavior during laboratory and commercial presulfiding.

Before presulfiding, two main steps are recommended to achieve the optimal catalyst activity:

1. Catalyst drying, because due to the hygroscopic nature of the alumina carrier, the catalyst can take up water, and when heating up in wet conditions with oil, the catalyst can be damaged mechanically
2. Catalyst soaking, which is done to wet the catalyst particles properly to prevent the presence of dry areas in the catalyst bed, which eventually lowers the overall activity

The shape and size of hydrotreating catalysts vary depending on the manufacturer. These parameters are important in achieving good catalyst performance and must be matched by the properties of the feed, the process technology, and the type of reactor. As for the size of catalyst, there is a limit to the decrease in particle size [e.g., $\frac{1}{32}$ in. (0.8 mm)], after which particles disintegrate. In addition, such small catalyst particles will cause ΔP problems in fixed-bed reactors. On the other hand, the most common commercial shapes of hydrotreating catalysts are sphere, pellet, cylinder, bilobular, trilobular, and tetralobular. The size and shape of the catalyst particles are usually defined to minimize pore diffusion effects in the catalyst particles and pressure drop across the reactor.

During hydrotreating operation, the performance of the catalyst is measured primarily by the following criteria:

- Initial catalyst activity: measured at the start-of-run condition, and corresponds to the reactor temperature required to achieve the quality of product desired
- Catalyst stability: measured under middle-of-run and end-of-run conditions, and is determined by the rate of temperature increase required to maintain product quality
- Product quality: controlled during the complete operation of the catalyst, and is an indication of the ability of the catalyst to produce products with the specifications desired

During hydrotreating of heavy feeds, the catalyst exhibits a certain degree of deactivation, depending on the nature of the feed, the type of reactor, and reaction conditions. The two main causes of catalyst deactivation are coke deposition and metals deposition. Coke is generally formed by thermal condensation, catalytic dehydrogenation, and polymerization reactions. The main coke precursors are asphaltenes. Coke formation is very rapid during start-of-run, after which it rises to an equilibrium level. During middle-of-run the total amount of coke remains almost constant. In general, the maximum coke laydown is about 20 wt%. Deactivation by coke is temporal since catalyst activity can be restored by regeneration. The recovery of catalyst activity can be about 90% by in situ regeneration and 95 to 97% by ex situ regeneration.

On the contrary, deactivation by metals (mainly Ni and V) is not reversible, and when the catalyst has been deactivated by metals, it needs to be replaced. The deposition of metals takes place at the pore entrances or near the outer surface of the catalyst.

To compensate for catalyst deactivation the reactor temperature needs to be increased continuously to keep the product quality at the level desired. However, exposing the catalytic bed at high temperatures will cause catalyst support sintering, which is another reason for loss of irreversible catalyst.

3.3 REACTOR MODELING

3.3.1 Effect of Catalyst Particle Shape

External Volume and Surface of Catalyst Particles Particle size is defined as the ratio of the total geometric volume to the external area of a catalyst particle ($L_p = V_p/S_p$). Equivalent particle diameter (d_{pe}) is defined as the diameter of a sphere that has the same external surface area (or volume) as the actual catalyst particle shape. These two parameters, L_p and d_{pe}, are very important in calculating others, such as bed void fraction, bed density, surface area of particles per unit volume of the bed, pressure drop, Reynolds number, and effectiveness factor, which are used extensively during reactor and catalyst design.

For regular shapes such as sphere, pellet, or cylinder, the calculation of both V_p and S_p is easy, but for irregular shapes such as polylobes, this calculation involves some considerations regarding the number of lobes and the manner in which they are accommodated. Whereas V_p can be determined through experimentation, S_p needs to be calculated.

The evaluation of external volume and surface of different particle shapes is reported in Table 3.5. More details about how these expressions were derived are given elsewhere (Macías and Ancheyta, 2004). For a pellet it should be remembered that it is a particle that has the same length and diameter ($L = d_p$), and the radius of the cylinder is the particle radius ($r_c = r_p$). For a cylinder its radius is also the particle radius but $L \neq d_p$. The different particle shapes and key geometric parameters are shown clearly in Figure 3.17.

Simulation of an Isothermal HDT Reactor with Different Particle Shapes

Description of the Reactor Model and Experiments The model used to simulate the hydrotreating reactor was described in detail by Macías and Ancheyta (2004) and is based on the following assumptions: operation is isothermal and steady-state in nature, catalyst deactivation is not significant, the reaction is assumed to occur only into a porous solid catalyst uniformly wetted by the liquid, radial and axial concentration gradients are negligible, the kinetics of the HDS reaction is described by a power-law model, the gas and liquid velocities are constant across the reactor, gas–liquid mass transfer was neglected,

TABLE 3.5. Equations for Calculating V_p and S_p of Different Particle Shapes[a]

Shape	V_p	S_p
Sphere	$\frac{4}{3}\pi r^3$	$4\pi r^2$
Pellet	$\pi r_c^2 d_p$	$2\pi r_c^2 + 2\pi r_c d_p$
Cylinder	$\pi r_c^2 L$	$2\pi r_c^2 + 2\pi r_c L$
Lobe-shaped particles	$n_L(\pi r_c^2 L) - A_1 L$	$n_L(2\pi r_c^2 + 2\pi r_c L) \pm 2A_1 - n_L A_2$

	n_L	θ	r_c	A_1	A_2
Two-lobe	2	45°	$\frac{d_p}{3.4142}$	$\frac{d_p^2}{8}\frac{\sin\theta - \sin^2\theta}{(1+\sin\theta)^2} = 8.88348\times10^{-3} d_p^2$	$\frac{\pi}{2} r_c L$
Three-lobe	3	60°	$\frac{d_p}{4}$	$\frac{d_p^2}{8}\frac{(2\sin\theta - 1)^2}{\tan\theta} = 3.86751\times10^{-2} d_p^2$	$\frac{\pi}{3} r_c L$
Four-lobe	4	45°	$\frac{d_p}{4.8284}$	$d_p^2\left[\frac{2\cos\theta - 1}{1+2\cos\theta}\right]^2 = 2.94373\times10^{-2} d_p^2$	$\frac{\pi}{2} r_c L$

[a] n_L, number of lobes; A_1, lateral area of the geometric shape between lobes (two-lobe: rhombus, three-lobe: triangle, four-lobe: frame); A_2, common area between each cylinder and each of the sides of the shape between lobes. For two-lobe particles A_1 takes the sign "–"; for three- and four-lobe particles A_1 takes the sign "+".

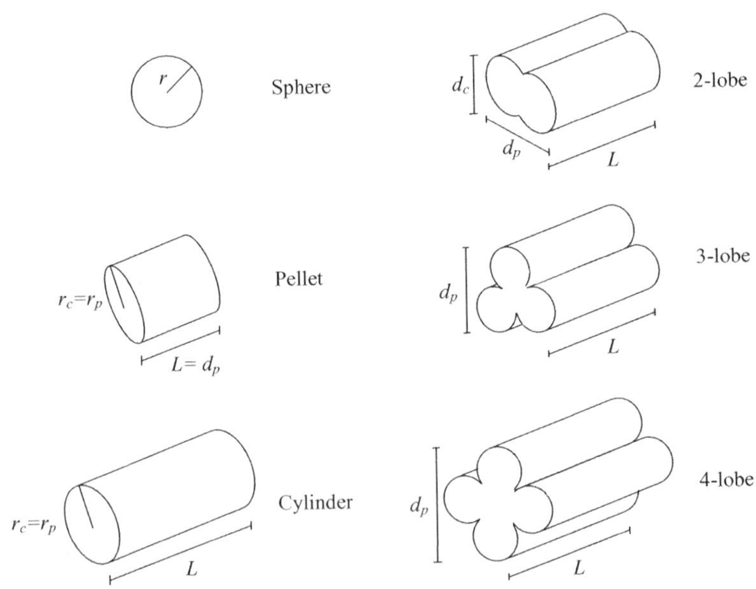

Figure 3.17. Typical particle shapes of commercial HDT catalysts.

and the particle diameter for different shapes was considered to be the equivalent diameter.

To develop the kinetic model and validate the reactor model simulations, experiments at the following conditions were conducted: 340 to 380°C temperature, 1 to 2.5 h^{-1} LHSV, 54 kg/cm^2 pressure, and 2000 ft^3/bbl H_2/oil ratio. The reaction was carried out in an small isothermal reactor (2.54 cm diameter) with a NiMo/γ-Al_2O_3 commercial catalyst sample with trilobular shape (2.4 wt% Ni, 9.5 wt% Mo, 204 m^2/g specific surface area, 0.50 cm^3/g pore volume, and particle density of 1.56 g/cm^3) and a straight-run gas oil (0.8687 specific gravity at 20/4°C, 1.616 wt% total sulfur, 6.83 cSt viscosity at 40°C, 196 to 407°C distillation range).

The heterogeneous isothermal one-dimensional reactor model developed includes correlations to calculate the bed void fraction, bed density, liquid holdup, surface area of the particles per unit volume of the bed, liquid and gas pressure drops, mass balance equations for the liquid and solid phases, effectiveness factor as a function of the generalized Thiele modulus, and physical properties of oil and gas at process conditions.

Characteristics of Particle Shapes To exemplify the simulation of an isothermal hydrotreating small reactor, the following catalytic particle shapes were considered: sphere, pellet, cylinder, bilobular (two-lobe), trilobular (three-lobe), and tetralobular (four-lobe). Except for the sphere, all of the other particle shapes have two external areas: cross-sectional area and lateral area, which are used to determine the total geometric external area of the particle. For cylindrical and lobe-shaped particles, the cylinders are considered symmetrical.

The geometrical characteristics (L_p, d_p, r_c) of all shapes can be calculated by assuming the total geometric volume to be equal to that of the three-lobed particle, which is the commercial sample used as a reference ($V_p = 0.016\,\text{cm}^3$) for the same L_p/d_p ratio of 2.25. For a pellet, $L_p/d_p = 1$. Based on these considerations, the following trends in external area and cylindrical radius can be determined:

S_p: sphere < pellet < cylinder < two-lobe < three-lobe < four-lobe
r_c: four-lobe < three-lobe < two-lobe < cylinder < pellet < sphere

Since lobe-shaped particles exhibit lower d_{pe} values, at similar reactor bed lengths (L_B) they have higher L_B/d_{pe} values, so that the criteria reported in the literature—to avoid deviation from plug-flow behavior due to axial dispersion (e.g., $L_B/d_{pe} > 100$)—are reached much more easily with lobe-shaped catalyst particles.

Total Liquid Holdup The effect of reactor temperature on total liquid holdup for various particle shapes is presented in Figure 3.18. It is observed that the higher the number of lobes in the particle, the higher the value of ε_L, due to

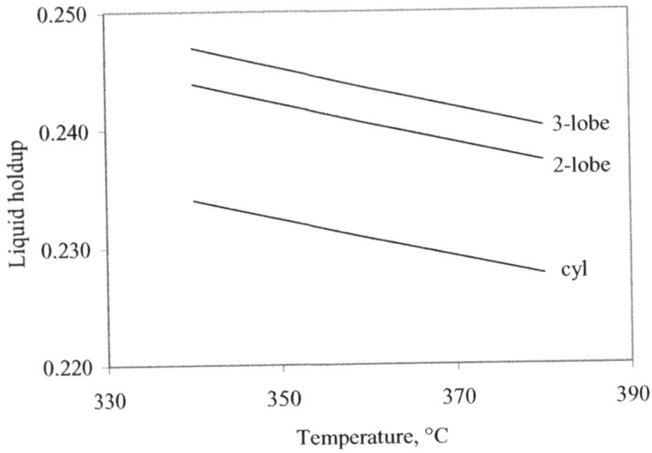

Figure 3.18. Effect of reactor temperature on ε_L for different particle shapes at $1.0\,h^{-1}$ LHSV.

the lower equivalent diameter. ε_L also decreases as the temperature is increased because of reduced liquid density. The increase in liquid holdup is attributed to the decrease in particle size due to higher capillary pressures.

External Concentration Gradients Mass transfer between liquid and solid phases is a function of the liquid flow in contact with the external area of the catalytic particles, which originates at a gradient between the two phases. An increase in this gradient means that the reactant is not being totally transferred to the external area of the particle, which apart from the liquid flow, depends on the shape and size of the particles.

From the simulation results, it was observed that the highest gradients of concentration are found at the beginning of the catalytic bed (Figure 3.19). At the same reaction conditions, the concentration gradients decrease as the external area of the particle is increased, indicating that the liquid mass-transfer to the particle external area is favored by using lobe-shaped particles. The effect of L_p and d_p on external concentration gradients at constant LHSV and temperature is also shown in Figure 3.19 for the three-lobed particle. It is seen that the lower the particle size, the lower the concentration gradients. This behavior implies that when small particles are used, external gradients of concentration are minimal.

Internal Concentration Gradients The values of the Thiele modulus for the three-lobed commercial sample shape as a function of reactor temperature ranged from 1.5 to 7.5. It was assumed that these values are high enough to represent a system with strong internal diffusion limitations; therefore, $\eta \sim 1/\phi$. Diffusion at the interior of the particle depends mainly on its porosity and on

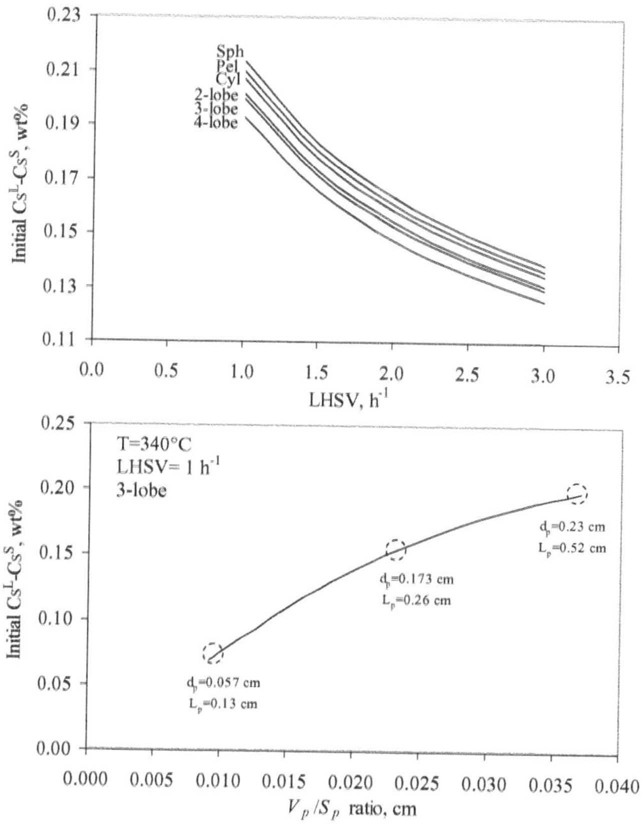

Figure 3.19. Effect of LHSV and particle shape on liquid–solid sulfur concentration gradients.

the size of the molecules being diffused to the pores. To compare the internal gradients with different shapes, effectiveness factors for all shapes are plotted in Figure 3.20 as a function of LHSV at 340°C. These values are within the range of those reported in the literature (0.4 to 0.6). The lobe-shaped particles yield higher effectiveness factors. When the number of lobes is increased, a higher effectiveness factor is also obtained, which means that the external surface of the particle is influencing the internal diffusion, and consequently, a particle exposing a higher external area facilitates internal diffusion, compared with those with less external area for the same particle volume. The lobe-cylinder radius is diminished for the case of particles with a high external area, which reduces the internal path at the interior of the particle pores. With respect to LHSV, an increase in this parameter provokes a marginal reduction in the effectiveness of all particle shapes, indicating that internal diffusion does not depend on the flow rates. The small differences can be attributed to external diffusion, which indeed is a function of flow rates. This implies that for

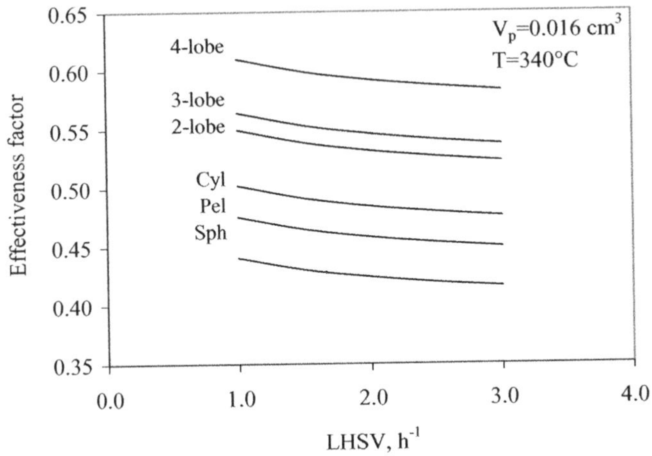

Figure 3.20. Catalyst effectiveness for different particle shapes.

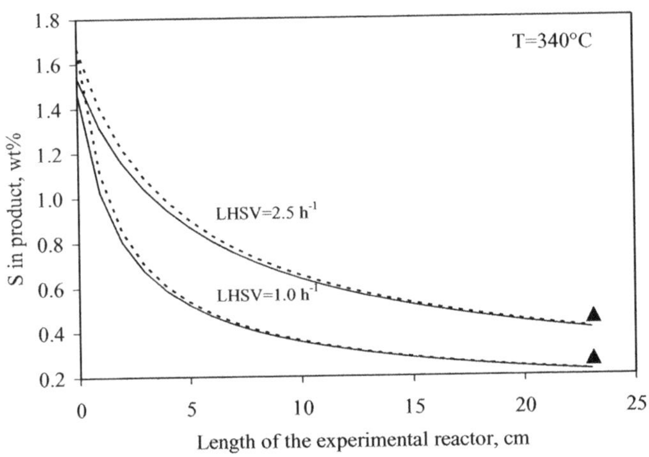

Figure 3.21. Model predictions (lines) versus experimental (▲) sulfur content profiles in liquid (---) and solid (—) phases along the experimental reactor.

maximum catalyst effectiveness the reactor should operate with no interphase mass transfer limitations. These results confirm that a reduction in particle size yields an increase in particle effectiveness, and as the number of lobes is increased, the effectiveness is also incremented.

Effect of the Particle Shape on the Sulfur Content in the Product The average absolute error between experimental and calculated sulfur contents was reported to be lower than 4%. As an example of reactor simulation, the sulfur content profiles predicted at two LHSV values are shown in Figure 3.21. These

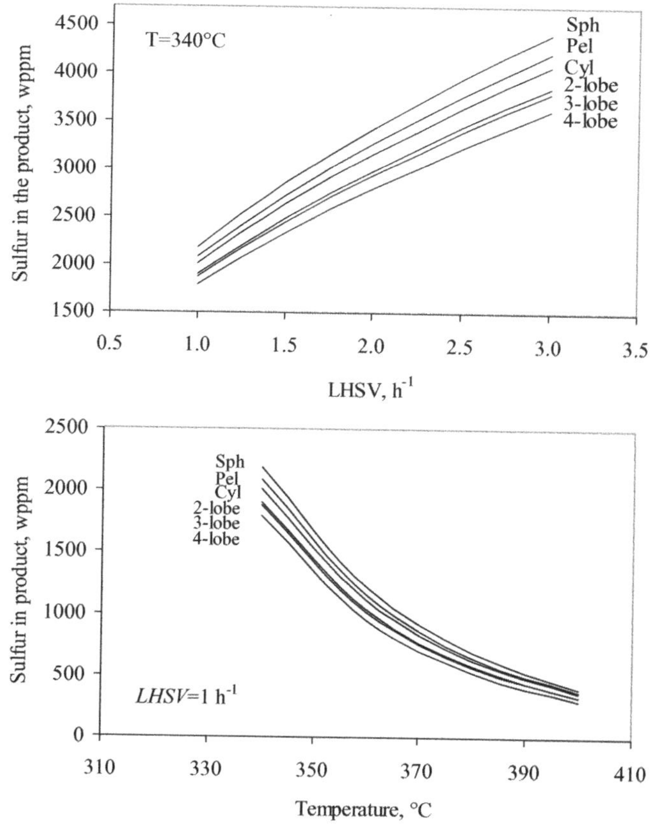

Figure 3.22. Effect of particle shape, LHSV, and temperature on sulfur content in the product.

profiles in both phases (liquid and solid) are not equal along the reactor. The gradients of concentrations between both phases are higher at the initial section of the reactor (0 to 5 cm) and then are reduced further in the bed, and concentrations become more or less equal.

It is known that the best catalyst selected through a previous screening test is formed into a shape, cylindrical or noncylindrical, and then undergoes further experimentation for process development and optimization. This is the point at which the proper particle shape and size selection become very important, since the catalyst needs to be tested in its commercially applied size and shape, to predict its full performance in a commercial operation rather than in a simple preliminary screening. The combined effects of particle shape, temperature, and LHSV indicated that sulfur removal is higher in the four-lobed shape than in other shapes (Figure 3.22). When the reaction temperature is increased, the effect of the particle shape on the sulfur content in the product

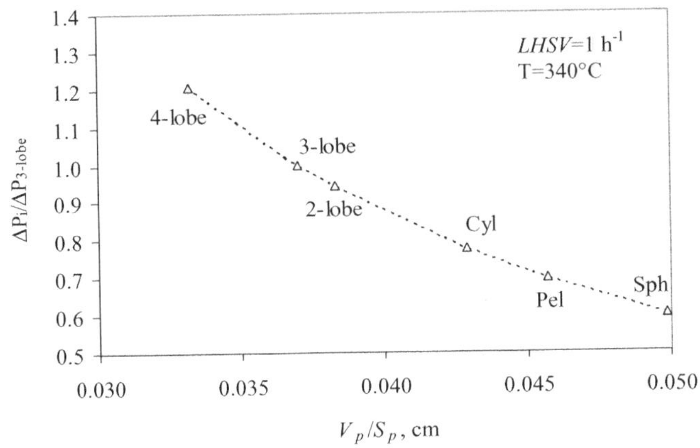

Figure 3.23. Effect of particle size and shape on the relative reactor pressure drop.

is minimal. It was also observed that a reduction in particle size produces higher sulfur conversion, due to the high degree of effectiveness of the small particles. Equivalent particle diameter and the V_p/S_p ratio depend on each particle shape and decrease from sphere to lobe-shaped particles; for this reason, noncylindrical catalysts exhibit a higher void fraction and therefore lower activities than those of cylindrical catalysts with an equal V_p/S_p ratio. A reduction in d_p has a greater effect on sulfur conversion than does a reduction of L_p when they are varied the same percentage, keeping one constant. This behavior is attributed to the reduction in path length at the interior of the particle when d_p is diminished.

Pressure Drop It is well known that the pressure drop in the catalyst bed is related to length, diameter, shape, and bed porosity, and the fact that noncylindrical particles exhibit a lower pressure drop than others, due to their higher bed porosity. To demonstrate this, the effect of particle shapes on pressure drop was studied at constant operating conditions by varying the particle size (the V_p/S_p ratio). The results of relative pressure drop ($\Delta P_i/\Delta P_{3\text{-lobe}}$) are depicted in Figure 3.23. It is confirmed that the smaller the particle size, the greater the pressure drop. These changes in pressure drop are due to the void spaces between particles in the catalytic bed. Increasing LHSV at constant particle volume originates in an increase in pressure drop, which is due to a higher flow velocity. Pressure drop at all LHSV values resultes in higher lobe-shaped particles. During hydrotreating of middle distillates, it is not catalyst deactivation but pressure drop which is the most important factor during commercial operation. This is the reason that catalyst strength requirements are intended to prevent a pressure drop increase due to breakage. In this sense, polylobe particles have an acceptable strength and hence are more highly recommended.

3.3.2 Steady-State Simulation

Description of the Reactor Model To simulate a hydrotreating reactor in steady-state operation, a three-phase reactor model was used, which includes correlations for determining mass transfer coefficients, solubility data, and properties of oils and gases under process conditions using information reported in the literature (Rodriguez and Ancheyta, 2004). The reactor model considers that no reactions occur in the gas phase and that for gaseous components (H_2 and H_2S) the mass balance equations are

$$\frac{u_G}{RT}\frac{dp_i^G}{dz}+k_i^L a_L\left(\frac{p_i^G}{H_i}-C_i^L\right)=0 \tag{3.72}$$

for the gaseous compounds (H_2 and H_2S) in the liquid phase:

$$u_L\frac{dC_i^L}{dz}-k_i^L a_L\left(\frac{p_i^G}{H_i}-C_i^L\right)+k_i^S a_S\left(C_i^L-C_i^S\right)=0 \tag{3.73}$$

for the organic sulfur compounds and the liquid hydrocarbon:

$$u_L\frac{dC_i^L}{dz}+k_i^S a_S\left(C_i^L-C_i^S\right)=0 \tag{3.74}$$

The components transported between the liquid phase and the surface of the catalyst are consumed or produced by a chemical reaction according to the following equation, which is applied for H_2, H_2S, organic sulfur, and hydrocarbons:

$$k_i^S a_S\left(C_i^L-C_i^S\right)=-\rho_B r_j \tag{3.75}$$

For energy balance the following equation was used:

$$\frac{dT}{dz}=\sum[(-\Delta H_{Rj})(r_j)]\frac{\varepsilon_L}{u_G\rho_G c_p^G\varepsilon_G+u_L\rho_L c_p^L\varepsilon_L} \tag{3.76}$$

Simulation of a Small HDT Reactor The use of the reactor model is exemplified with experiments conducted in a small reactor with a NiMo commercial catalyst (175 m^2/g specific surface area, 0.56 cm^3/g pore volume, 127 Å mean pore diameter, 0.8163 g/cm^3 bulk density, 10.7 wt% Mo, 2.9 wt% Ni) and a vacuum gas oil (22°API, 2 wt% sulfur, 1284 wppm total nitrogen, 518 wppm basic nitrogen, 41.9 wt% total aromatics, 441.9 g/mol molecular weight, 267 to 588°C distillation range) as hydrotreating feed. The experimental work was carried out in a bench-scale unit described by Marroquín and Ancheyta (2001). The following reaction conditions were used: pressure of 54 kg/cm^2, H_2/oil ratio of 2000 ft^3/bbl, LHSV of 2 h^{-1}, and a reaction temperature in the range 340 to

TABLE 3.6. Rate Equations, Kinetic Parameters, and Heats of Hydrotreating Reaction

Reaction	Kinetic Model	E_A (J/mol)	k_0[a]	ΔH_R (kJ/mol)
HDS	$r_{HDS} = k_{HDS}\frac{(C_S^S)(C_{H_2}^S)^{0.45}}{(1+K_{H_2S}C_{H_2S}^S)^2}$	131,993	4.266×10^9	–251
HDN				
HDN_{NB}	$r_{HDN_{NB}} = k_{HDN_{NB}}(C_{N_{NB}}^S)^{1.5}$	164,942	3.62×10^6	–64.85
HDN_B	$r_{HDN_B} = k_{HDN_{NB}}(C_{N_{NB}}^S)^{1.5} - k_{HDN_B}(C_{N_B}^S)^{1.5}$	204,341	3.66×10^{11}	
HDA				
Forward	$r_{HDA} = k_f p_{H_2}^G C_A^S - k_r(1-C_A^S)$	121,400	1.041×10^5	–255
Reverse		186,400	8.805×10^9	

[a]Units of k_0: HDS, $cm^3/g{\cdot}s\ (cm^3/mol)^{0.45}$; HDN, $s^{-1}\ (wt\%)^{-0.5}$; HDA, forward, s^{-1}/MPa, HDA, reverse, s^{-1}.

380°C. All the experiments were performed without H_2 recycle. The rate equations used for reactor model simulations and kinetic parameter values determined from experimental data are summarized in Table 3.6.

The model is first used to simulate the isothermal operation of the reactor; that is, only mass balance equations were used. The results of the reactor model were found to predict the concentrations of sulfur, nonbasic nitrogen, basic nitrogen, and aromatics accurately with an average absolute error <2%. The simulated molar concentration profiles along the reactor are shown in Figure 3.24. It is observed that the concentrations of sulfur, nitrogen, aromatics, H_2, and H_2S in both the liquid and solid phases are not equal along the reactor. However, the gradients of concentrations $(C_i^L - C_i^S)$ are more or less constant along the reactor. H_2 and H_2S exhibit the highest difference in concentration gradients in the solid and liquid phases. The balance between reaction rate and mass transfer determines the overall shape of the H_2 and H_2S profiles. The H_2S concentration increases rapidly and H_2 concentration decreases, which is due to the high reaction rate of the catalyst bed in the initial part of the reactor. Since the HDS reaction is known to be inhibited by H_2S, the inhibiting effect of H_2S will slow the reaction down to a point where mass transfer can keep up with it. This mass transfer in the different phases is very important in modeling HDT reactions, which is not taken into account in simpler models (i.e., the pseudohomogeneous plug-flow model).

Simulation of a Commercial HDT Reactor The model was used to predict the expected behavior of a commercial HDT reactor (internal diameter 3.048 m, total length 9.1 m, catalytic bed length 8.534 m, and catalyst density 816 kg/m^3). For this case, the energy balance given by Eq. (3.76) was solved

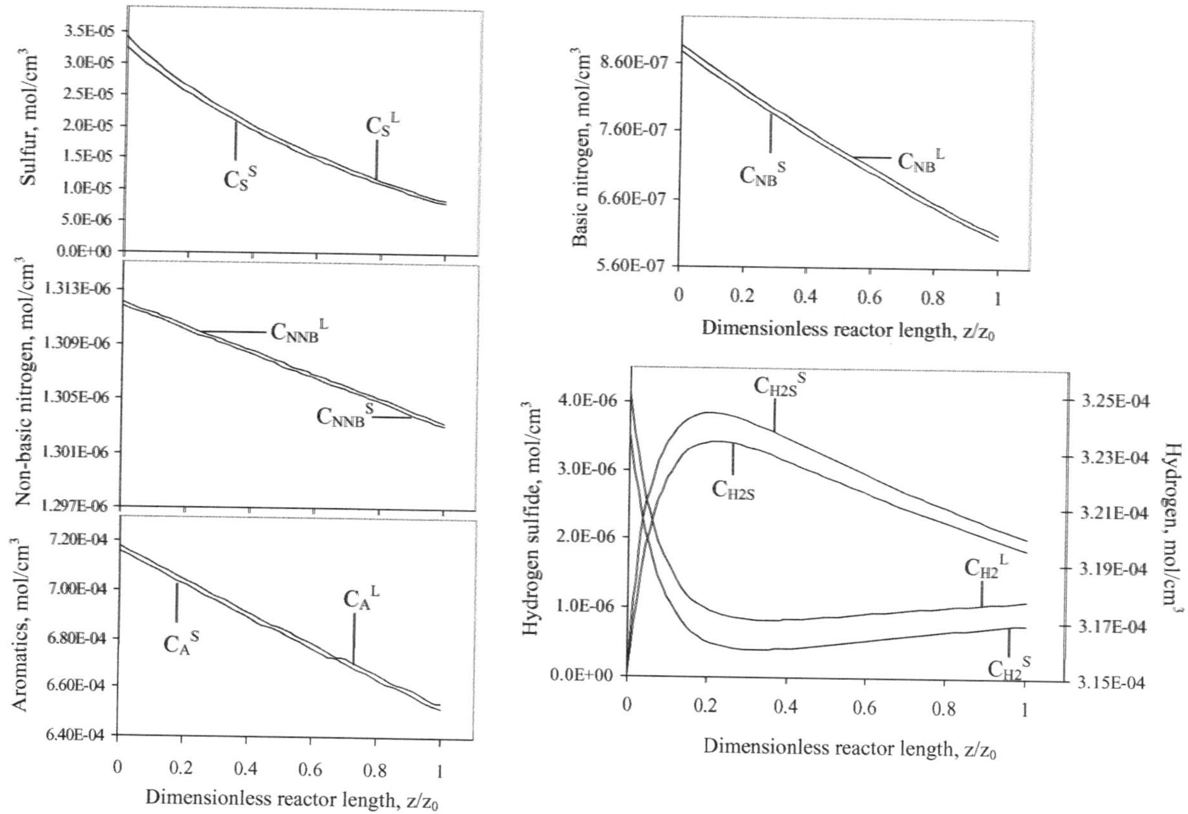

Figure 3.24. Concentrations of impurities of the product, H_2, and H_2S along the reactor in liquid and solid phases.

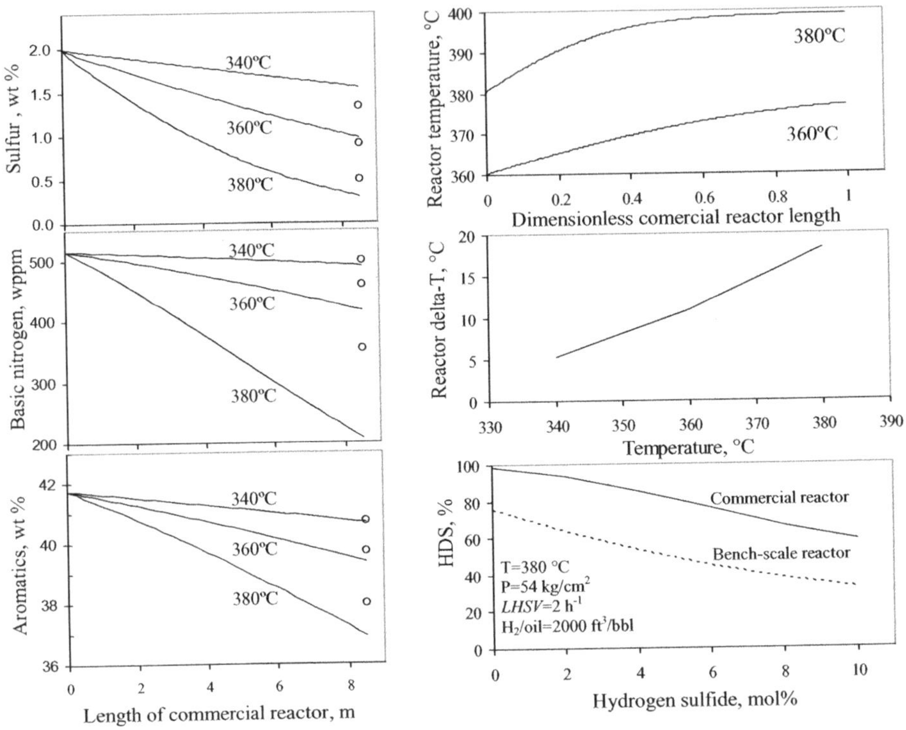

Figure 3.25. Simulation of the commercial HDT reactor (O, experimental pilot-plant data; solid lines, values predicted).

simultaneously using mass balance equations. In contrast to a bench-scale reactor in which a high-purity H_2 stream is used, for the commercial reactor a gas stream with the following molar composition was employed: 81.63% H_2, 3.06% H_2S, and 15.31% lights (methane, ethane, and propane).

Figure 3.25 shows the simulated results for impurities removal as a function of reaction temperature and ΔT along the commercial reactor. The simulation was performed at pressure 54 kg/cm^2, LHSV 2.66 h^{-1}, and H_2/oil ratio 2000 ft^3/bbl. Experimental values from the isothermal bench-scale reactor are included for comparison. The commercial reactor is operated at a higher average temperature due to reaction exothermality, which is why smaller impurities contents are obtained than with an isothermal bench-scale reactor. The axial temperature profiles predicted in the commercial reactor show that the increase in reaction temperature is higher in the initial part of the reactor, which is due mainly to the greater conversion occurring in this zone. Also, as the temperature is increased, reactor ΔT also increases, due to the enhanced reaction rate.

The well-known inhibiting effect of the H_2S produced by the same HDS reaction is shown clearly in Figure 3.25 for both bench-scale and commercial reactors. The literature reports that increasing the content of H_2S in the gas phase entering the reactor up to 10 mol% can reduce the HDS reaction rate to about 50%, which is in good agreement with the simulated results.

Comparison of Bench-Scale and Commercial HDT Reactor Simulations During experiments in isothermal bench-scale reactors, the removal of impurities is generally lower than that observed in commercial reactors under similar reaction conditions. Of course, this depends on the features of the experimental setup. That is the reason that bench-scale experiments are normally compared with adiabatic operations by operating the bench-scale reactor isothermally at a temperature that is equivalent to the average temperature of the commercial unit [i.e., the weighted-average bed temperature (WABT)]. One reason for the differences in bench-scale and commercial results is the incomplete catalyst contact, which is a classical bench-scale problem that can lead to underperformance. During operation of HDT trickle-bed reactors, the liquid hydrocarbon flows over the catalyst particles in films and rivulets from one particle to the next, and the vapor (mostly H_2) flows continuously through the remaining voids. These conditions cause poor catalyst utilization due to incomplete catalyst wetting, axial dispersion, and restricted interphase mass transfer. Because commercial catalyst samples and real feedstocks are commonly used to conduct experiments, the ratio of the bench-scale reactor length to catalyst particle diameter is very low compared with that of commercial reactors, and low liquid velocities are used in small-scale reactors to match the commercial LHSV values. These differences cause several problems in testing catalysts that have sizes and shapes in commercial use.

3.3.3 Simulation of a Commercial HDT Reactor with Quenching

The exothermality of HDT reactions can provoke an increase in reactor temperature beyond design limits, depending on conversion level, reaction conditions, and feed properties—which is why an appropriate temperature control system is required. The reaction temperature is a process variable that affects the conversion degree of the reactants, the selectivity, and the catalyst cycle life. High reactor temperatures enhance HDT reaction rates but may cause excessive hydrocracking of the feed, reducing the desired product yield and increasing gas production. Also, high temperatures enhance coke formation and shorten catalyst cycle life. Since achieving the desired conversion can be limited by temperature increases, the total catalyst volume is frequently divided into several beds separated by interbed zones for introducing quench fluids (Alvarez et al., 2007a,b). Petroleum fractions, particularly highly unsaturated fractions, are known for being H_2-demanding feeds during hydrotreating. Processing of these feeds generates large amounts of heat during commercial HDT, which is reflected in a sharp temperature rise along the reactor. H_2 quenching has been the traditional way to control temperature in most HDT processes; however, the use of liquids has also been reported.

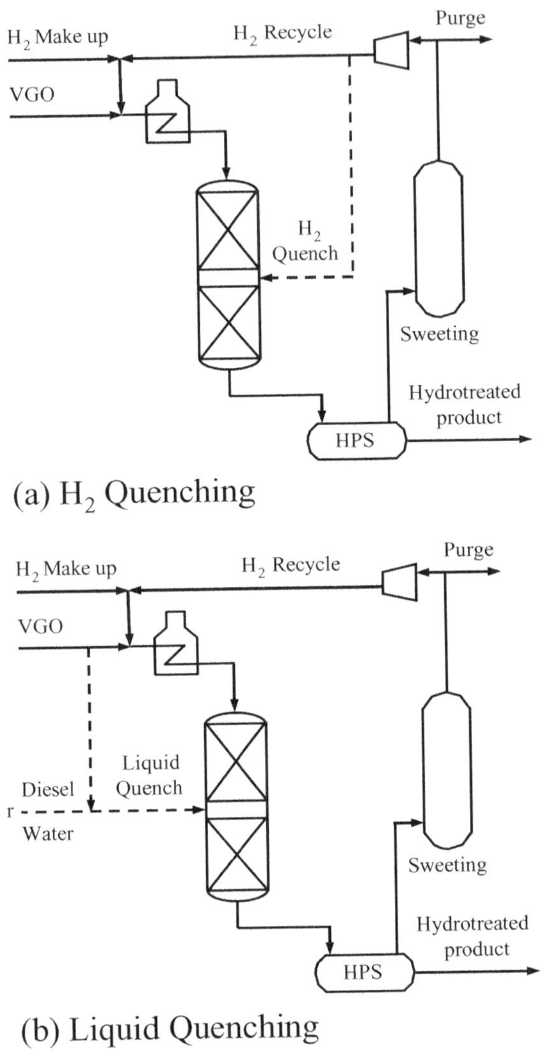

Figure 3.26. H_2 quenching and liquid quenching used for HDT reaction simulations.

Quenching Approaches To simulate the effect of different quenching alternatives, the heterogeneous one-dimensional reactor model described earlier was used. The catalyst, the feed (VGO), and the reaction conditions were also the same as those used previously. Schematic representations of the quenching approaches are shown in Figure 3.26. For one case of liquid quenching, a straight-run gas oils (diesel) was used, which has the following properties: 33°API gravity, 1.1 wt% sulfur, 120 wppm total nitrogen, 37.4 wppm aromatics, 232.6 g/mol molecular weight, and a distillation range of 164 to 370°C.

Hydrogen Quenching This is achieved by introducing part of the H_2 recycle stream into a certain length of reactor, which was previously sweetened. This type of quenching has the advantage of replenishing for some of the H_2 consumed, decreasing H_2S and NH_3 partial pressures and reducing coke formation. However, to recycle H_2 requires more equipment, and a large amount of quench gas, due to the low heat capacity of the gas, and high-pressure drops are experienced due to the elevated gas velocities.

Liquid Quenching Compared with gas quenching the main disadvantage of liquid quenching is that more reactor volume is required to achieve the same conversion. The processes that use liquid quenching can be classified into multiple feed and product recycle processes. *Multiple feed processes* are characterized by previous fractionating of the feed and introducing the heaviest fraction at the top of the reactor while the lighter fractions are used as quench streams/side feed; in this way the quench stream mixed with the main feed is provided with treatment in the following catalytic bed. *Product recycle processes* are based on recycling a previously cooled portion of the reactor effluent for use as a quench stream and providing a second-pass opportunity to unreacted species.

Study Cases To allow for the injection of quench fluids, the total catalyst volume used in the base hydrotreater case was divided into two catalytic beds without considering the increase in the reactor vessel size, in order to keep the same catalyst volume in all cases. For all the quenching alternatives the temperature limit allowed in the first bed was set to 8°C above the inlet temperature (380°C), and once such a limit was reached, an appropriate amount of the quench fluid was injected to reduce the temperature to the inlet value. The study cases are:

- *Base hydrotreater case:* identical to the VGO hydrotreater described earlier, which does not have a quench stream.
- *Hydrogen quenching:* part of the H_2 recycle stream is taken to quench the HDT reactions and improve the composition of the gas phase.
- *VGO feed quenching:* a portion of the main feed is injected between the catalytic beds.
- *Diesel quenching:* a straight-run diesel stream was considered for quenching.
- *Water quenching.*

Modeling the Quench Zone A quench zone was modeled as a stream mixer of the quench stream and the effluent from the first catalytic bed, as shown in Figure 3.27. The expressions provided below describe the quench zone. The global mass balance between the liquid hydrocarbon (l_{out}) and gas leaving the first catalytic bed (g_{out}) and the corresponding streams entering the

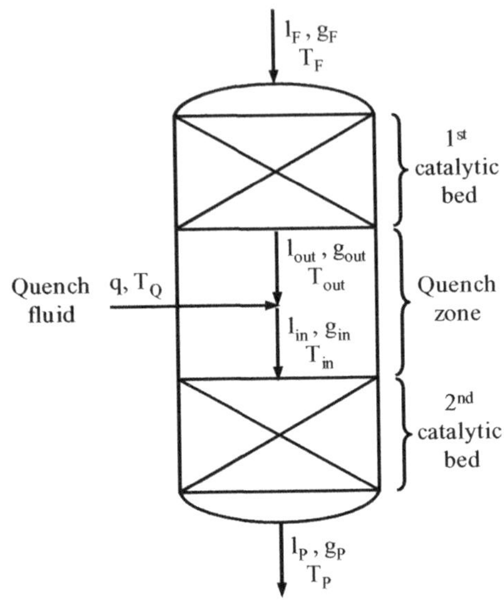

Figure 3.27. Representation of the quench zone model.

following catalytic bed (l_{in} and g_{in}) after mixing with the quench fluid (q) is expressed as

$$q + l_{out} + g_{out} = l_{in} + g_{in} \tag{3.77}$$

The gas mass balance is

$$qv + g_{out} = g_{in} \tag{3.78}$$

The liquid mass balance is

$$q(1-v) + l_{out} = l_{in} \tag{3.79}$$

where v is the vaporized fraction of the quench fluid. The energy balance with temperature-dependent heat capacities is

$$\int_{T_{out}}^{T_{in}} l_{out} c_p^L \, dT + \int_{T_{out}}^{T_{in}} g_{out} c_p^G \, dT + \int_{T_q}^{T_{in}} q c_p^q \, dT = 0 \tag{3.80}$$

Equation (3.80) is used to estimate the cooled mixture temperature (T_{in}) at a fixed quench rate or the quench rate for a desired mix temperature. To solve this equation, the liquid and gas mass flow rates at the exit of the first catalytic bed are both necessary and can be supposed to be equal to the feed flow rates (l_F and g_F) by assuming insignificant hydrocracking.

The gas mixture heat capacity was evaluated using the individual gaseous component heat capacities and the composition at the first bed exit obtained by simulation. The heat capacity of the liquid hydrocarbon was calculated with the following correlation (Perry and Green, 1987):

$$c_p^L = 4.1868\left[\frac{0.415}{\sqrt{\rho_L^{15}}} + 0.0009(T - 288.15)\right] \tag{3.81}$$

where ρ_L^{15} is the liquid hydrocarbon density in g/cm^3 at 15°C. The gas mixture and H_2 quench heat capacity were estimated using the following simple expression:

$$c_p^G = \sum_i c_{pi}^G x_i^G \tag{3.82}$$

where c_{pi}^G and x_i^G are the heat capacity and the mass fraction of compound i in the gas phase, respectively.

For diesel quenching, the value of the liquid quench heat capacity was determined at process conditions using the Hysys process simulator to account for the enthalpy change during vaporization, while for water quenching, the heat capacity term included the liquid-phase sensible heat, heat of vaporization, and gas-phase sensible heat.

Once the quench rate required to have the same temperature as for the entrance of the first catalytic bed has been determined, the initial values of molar concentrations and partial pressures for solving the second catalytic are updated. The main effect with H_2 as quench is the increase in H_2 partial pressure and dilution of H_2S in the gas phase; on the other hand, when using liquid quenching the effect can be on the concentrations of organic compounds (S, N, and aromatics) and partial pressures, depending on the degree of vaporization of the quench liquid. The amount of vaporization of the liquid quench stream at the process conditions can be determined by liquid–vapor equilibrium calculations using the Hysys process simulator. To adjust the concentrations of the reactants to be used as initial values for the following catalytic bed, individual balances for each compound are developed; such balances are derived directly from the quench zone equations. For example, the component balance for an organic compound in the liquid phase is

$$q(1-v)\frac{C_{i,q}^L \mathrm{MW}_i}{\rho_{L,q}} + l_{\mathrm{out}}\frac{C_{i,\mathrm{out}}^L \mathrm{MW}_i}{\rho_{L,\mathrm{out}}} = l_{\mathrm{in}}\frac{C_{i,\mathrm{in}}^L \mathrm{MW}_i}{\rho_{L,\mathrm{in}}} \tag{3.83}$$

This equation is valid only for VGO and diesel quenching schemes, since H_2 and water quenching do not affect the liquid-phase balance. Equivalent expressions can be developed for adjusting the gas-phase concentration for the H_2, water, and diesel quenching cases.

Once the new initial values are obtained, the second catalytic bed is simulated. In diesel quenching, additional mass balance equations for each diesel compound are included in the model to simulate the hydrotreating of the quench stream parallel to the VGO feed. The total conversion is provided by both VGO organic compound and the diesel organic compound removal. This makes it possible to observe the behavior of the system when using liquid quench with more reactive species.

Results of Simulation of Various Quench Approaches Figure 3.28 and Tables 3.7 and 3.8 summarize all the simulation results, from which the following comments can be made.

1. *Reactor* ΔT. The base case presents a reactor $\Delta T = 20°C$, reaching almost 400°C of reactor temperature. Operating at such elevated temperatures will increase the product quality and consequently, the H_2 consumption, compromising the catalyst cycle life. A certain level of hydrocracking may also be present, which would reduce the liquid product yield. In the other cases, when the temperature reaches the established limit, there is a sudden drop to 380°C as a result of quenching. Except for the VGO quenching alternative, in the first catalytic bed all cases have the same temperature rise as that of the base case. The VGO quench stream bypasses the first catalytic bed, causing a slight decrease in the amount of liquid at the reactor entrance, resulting in a decrease in LHSV and, consequently, an increase in reaction severity, which is reflected in a higher reactor ΔT value. In the second bed the diesel quenching alternative has more accelerated temperature rise, resulting in a 9°C reactor ΔT, which is caused by the more reactive organic compounds of the straight-run diesel stream. The H_2 and VGO quenching schemes exhibit $\Delta T = 8.1°C$, while the water scheme presents the lowest ΔT (7.8°C). The small differences in the temperature profiles of each case are explained by the changes in the gas and liquid composition caused by quenching, as for water and diesel, respectively. On the other hand, the effect of changing gas composition is diminished due to the simplicity of the kinetic models employed for the simulation. Only the HDS rate equation takes into account H_2 and H_2S concentration; the HDA model considers H_2 partial pressure. Therefore, such an effect does not have a noticeable impact on the temperature rise of each case.
2. *Profiles of H_2 and H_2S partial pressures.* H_2 partial pressure decreases along the reactor as a result of H_2 consumption and increase in solubility, while H_2S partial pressure has the opposite behavior, due to sulfur removal. After the quench zone, H_2 partial pressures in the H_2 and VGO quenching schemes stand above the base-case line. In the former case a small jump in partial pressure is produced by H_2 quenching. In the latter scheme, H_2 partial pressure is increased due to temperature reduction. Hydrogen solubility is directly proportional to temperature; hence, when

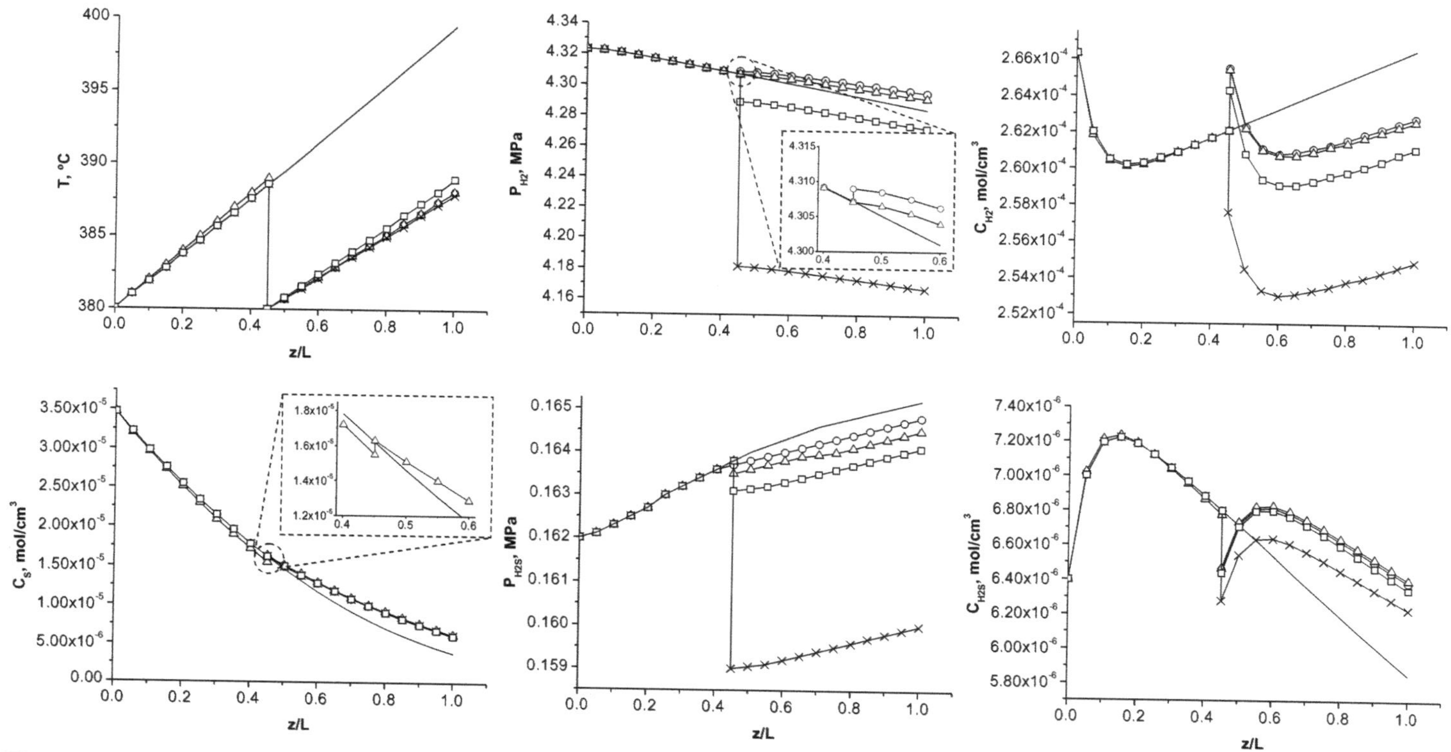

Figure 3.28. Results of the simulation for various quenching alternatives (—, base hydrotreater case; ○, hydrogen quenching; Δ, VGO quenching; □, diesel quenching; ×, water quenching).

TABLE 3.7. Liquid and Hydrogen Balances for the Different Quenching Approaches

Quenching Approach	First Bed			Quench Zone			Second Bed		
	Liquid[a]	H_2/Oil[a]	H_2 Cons.	Quench (std m^3/h)	Liquid/Feed (m^3/m^3)	% Vap.	Liquid[a]	H_2/Oil[a]	H_2 Cons.
Base	1.00	0.95	1.00						
H_2	1.00	0.95	0.42	7.714	—	—	0.98	1.00	0.44
VGO	0.95	1.00	0.44	7.45	0.05	0	0.98	0.88	0.44
Diesel	1.00	0.95	0.42	7.00	0.05	53	1.00	0.86	0.47
Water	1.00	0.95	0.42	1.67	0.01	100	0.98	0.88	0.39

[a] All values are relative to the highest value.

TABLE 3.8. Effect of Quench Position and Temperature for the H_2 Quenching Approach

	L_B							
z	0.1	0.3	0.5	0.7	0.1	0.3	0.5	0.7
Quench temp. (°C)	70	70	70	70	120	120	120	120
Quench rate (std m^3/h)	1622	4896	8198	11,585	1907	5750	9649	13,663
Sulfur in product (wt%)[a]	1.21	1.52	1.61	1.47	1.21	1.52	1.61	1.47
H_2 consumption[a]	0.9	0.83	0.82	0.87	0.9	0.83	0.82	0.87
H_2S in liquid phase[a]	1.03	1.07	1.10	1.12	1.03	1.07	1.10	1.12

[a]All values are relative to those obtained with the base case.

temperature falls, the dissolved H_2 is transferred to the gas phase, increasing its partial pressure. When vaporization of the feed is substantial, quenching causes condensation of the hydrocarbon resulting in an additional increase of H_2 partial pressure. The other alternatives (diesel and water quenching) reduce H_2 partial pressure in different magnitude, especially water, due to vaporization of the fluid at the reaction conditions. As for H_2S partial pressure, at the quenching position in all cases this value is reduced below that of the base case by a dilution effect in the following order: water > diesel > H_2; and by a temperature effect (VGO quenching); in the latter case, the temperature has the opposite effect on H_2S solubility than on that of H_2. Subsequently, in the second catalytic bed the quenching schemes present a slightly smaller H_2S generation rate than that of the base case, due to the lower average reactor temperature. This is reflected in more favorable flatter slopes of H_2S partial pressure profiles compared with the base case. This is particularly important for removing the most refractory sulfur compounds at the exit of the reactor.

3. *Profiles of H_2 and H_2S molar concentrations in the liquid phase.* At the beginning of the catalytic bed, H_2 concentration falls down quickly, while H_2S concentration grows substantially as a result of the elevated reaction rates in this section of the reactor. When using interbed quenching, the shape of H_2 and H_2S molar concentration profiles is repeated in the following catalytic bed. In this section of the reactor, H_2 concentrations are lower while H_2S concentrations are higher than those of the base case. This behavior is attributed primarily to the different temperature profiles of the base and quenching schemes. H_2 solubility increases with temperature, which causes an H_2 release to the gas phase after quenching; on the other hand, H_2S solubility is reduced with temperature, which increases

the amount of dissolved gas in the liquid after quenching. Therefore, since temperature is increased toward the end of a catalytic bed, H_2 tends to concentrate in the liquid phase while H_2S is released to the gas phase. This explains why the base case, where the temperature increases up to 400°C, has higher H_2 and lower H_2S concentrations toward the end of the reactor than do the other quenching schemes. Having low H_2S concentrations toward the last section of the reactor is desirable for eliminating the most refractory sulfur species, whose removal is strongly inhibited by such a compound. Partial pressure also influences the molar concentration profiles; for example, water quenching drastically reduces H_2 and H_2S partial pressures, which results in lower molar concentrations.

4. *Profiles of organic sulfur molar concentration in the liquid phase.* The base case achieves a higher sulfur conversion (~89%) than that of the quenching schemes (~82%). This 7% gain in conversion is attributed primarily to the higher average reactor temperature, which increases the reaction rate. All the quenching schemes present almost the same sulfur removal profiles except for VGO quenching in the first catalytic bed. This difference is explained by the reduced LHSV, which is a characteristic of such an alternative; however, at the quenching position, sulfur concentration and LHSV are increased, which leads to the same sulfur conversion as in the other cases. In the case of diesel, at the quenching point there is also an increase in sulfur concentration and LHSV; however, since the amount of diesel injected is small, the sulfur concentration increases only about 1%, which is unnoticeable in the figure. In the second catalytic bed, all the quenching schemes present a lower sulfur removal rate than that of the base case as a result of a lower reaction temperature. For the diesel case, different sulfur removal rates are expected, due to the presence of more reactive species; however, since diesel contributes only about 1% to the total sulfur amount, the sulfur removal profile is essentially the same as in the other cases. When using water quenching, it could be expected that the substantial H_2 partial pressure reduction would reduce hydrotreating reaction rates; however, such a reduction is relatively small (~3%), which explains the similar sulfur conversion.
5. *Liquid hydrocarbon and H_2 balances (Table 3.7).* In the first catalytic bed, VGO quenching is the only alternative that differs in the amount of liquid feed and H_2 consumption ($-H_2$). Hydrogen quenching replenishes the H_2 consumed in the first catalytic bed and increases the H_2/oil ratio in the following bed. VGO and diesel exhibit similar quench rates, whereas much lower rates of diesel quench would be expected, due to the high percentage of diesel vaporization. However, diesel heat capacity calculations using the Hysys process simulator showed little difference from the values obtained for VGO under the same process conditions, which suggested a small enthalpy change during diesel vaporization, resulting in similar quench rates. On the other hand, when using VGO

or diesel as quench, liquid flow increases, thus increasing LHSV, resulting in decreased severity. Additionally, diesel adds more reactive species, which increases the reaction rates and, consequently, H_2 consumption. In the case of water quenching, a much lower quench rate than those of VGO and diesel quenching is required; this is explained by the total vaporization of water at the process conditions, which absorbs high amounts of heat. Water vaporization modifies drastically the composition of the gas phase, which explains the smallest reactor ΔT and H_2 consumption. Despite the effects of liquid quenching on mass balances, the amount of liquid quench in all cases is relatively small compared with the amount of VGO fed to the unit, which makes valid the assumption that the changes in physical properties due to liquid quenching are negligible. The base case presents the highest H_2 consumption because such a process configuration shows the higher reactor temperatures.

6. *Effect of quench position and reaction temperature for the case of H_2 quenching (Table 3.8).* The quench rate increases with quench temperature, especially with position along the reactor. Quenching near the entrance of the reactor ($z = 0.1$) requires around seven times less quench fluid than that of quenching toward the end of the reactor ($z = 0.7$). However, this modifies substantially the reactor temperature profile, which influences the sulfur content in the product, H_2 consumption, and the amount of H_2S sulfide in the liquid phase. Quenching closer to the entrance of the reactor allows for a high reactor ΔT value in the second catalytic bed, which results in higher sulfur removal and H_2 consumption and reduces the amount of dissolved H_2S in the liquid hydrocarbon compared with other quench positions. If quenching is performed in the middle, the lowest average reactor temperatures are present; thus a higher sulfur content is obtained in a product, and less H_2 is consumed. Nevertheless, quenching toward the end of the reactor augments the quench rate considerably and provokes a high reactor ΔT value in the first bed, which leads to a relatively lower sulfur content in the product. However, lower reactor temperatures in this section of the reactor increase the concentration of H_2S considerably, which is not desirable for the removal of refractory species.

3.3.4 Dynamic Simulation

Description of the Model The steady-state reactor model described in previous sections was taken as a base for developing the dynamic heterogeneous one-dimensional reactor model (Mederos et al., 2006). This model is based on the two-film theory and makes use of correlations to estimate heat and mass transfer coefficients, gas solubilities, and properties of oil and gases at process conditions (Table 3.9).

Model Equations The mass transfer of the compounds in the reactor is described with the following set of partial differential equations (PDEs). The

TABLE 3.9. Correlations for Heat and Mass Transfer Coefficients, Gas Solubilities, and Properties of Oil and Gases

Parameter	Correlation
Oil density	$\rho_L(P,T) = \rho_0 + \Delta\rho_P - \Delta\rho_T$
	$\Delta\rho_P = \left[0.167 + (16.181 \times 10^{-0.0425\rho_0})\right]\dfrac{P}{1000} - 0.01\left[0.299 + (263 \times 10^{-0.0603\rho_0})\right]\left(\dfrac{P}{1000}\right)^2$
	$\Delta\rho_T = \left[0.0133 + 152.4(\rho_0 + \Delta\rho_P)^{-2.45}\right](T - 520)$
	$-\left[8.1 \times 10^{-6} - 0.0622 \times 10^{-0.764(\rho_0 + \Delta\rho_P)}\right](T - 520)^2$
Henry coefficient	$H_i = \dfrac{v_N}{\lambda_i \rho_L}$
Solubility of H_2	$\lambda_{H_2} = -0.559729 - 0.42947 \times 10^{-3}T + 3.07539 \times 10^{-3}\dfrac{T}{\rho_{20}} + 1.94593 \times 10^{-6}T^2 + \dfrac{0.835783}{\rho_{20}^2}$
Solubility of H_2S	$\lambda_{H_2S} = \exp(3.367 - 0.00847T)$
Gas–liquid mass transfer coefficient	$\dfrac{k_i^L a_L}{D_{M,i}^L} = 7\left(\dfrac{G_L}{\mu_L}\right)^{0.4}\left(\dfrac{\mu_L}{\rho_L D_{M,i}^L}\right)^{1/2}$
Dynamic liquid Viscosity	$\mu_L = 3.141 \times 10^{10}(T - 460)^{-3.444}[\log_{10}(\text{API})]^a$
	$a = 10.313[\log_{10}(T - 460)] - 36.447$
Diffusivity	$D_{M,i}^L = 8.93 \times 10^{-8}\dfrac{v_L^{0.267}}{v_i^{0.433}}\dfrac{T}{\mu_L}$
Molar volume	$v_i = 0.285 v_c^{1.048}$
	$v_c^m = 7.5214 \times 10^{-3}(T_{\text{MeABP}}^{0.2896})(d_{15.6}^{-0.7666})$
Liquid–solid mass transfer coefficient	$\dfrac{k_i^S}{D_{M,i}^L a_S} = 1.8\left(\dfrac{G_L}{a_S \mu_L}\right)^{1/2}\left(\dfrac{\mu_L}{\rho_L D_{M,i}^L}\right)^{1/3}$
Specific surface area	$a_S = \dfrac{6}{d_p}(1 - \varepsilon_B)$
Liquid–solid heat transfer coefficient	$j_H = \dfrac{h_{LS}}{c_p^L u_L \rho_L}\left(\dfrac{c_p^L \mu_L}{k_L}\right)^{2/3}$

reactor model considers that there are no reactions in the gas phase, and HDS, HDA, HDN_B, and HDN_{NB} are taking place along the catalytic bed. Based on this, the dynamic mass balance equation in the catalyst bed for the gaseous compounds is

$$\frac{\varepsilon_G}{RT}\frac{\partial p_i^G}{\partial t} = -\frac{u_G}{RT}\frac{\partial p_i^G}{\partial z} - k_i^L a_L\left(\frac{p_i^G}{H_i} - C_i^L\right) \tag{3.84}$$

where $i = H_2, H_2S$, and NH_3. The dynamic mass balance equation in the catalyst bed for the gaseous compounds in the liquid phase is

$$\varepsilon_L\frac{\partial C_i^L}{\partial t} = -u_L\frac{\partial C_i^L}{\partial z} + k_i^L a_L\left(\frac{p_i^G}{H_i} - C_i^L\right) - k_i^S a_S\left(C_i^L - C_i^S\right) \tag{3.85}$$

where $i = H_2, H_2S$, and NH_3.

The model assumes that the organosulfur, organonitrogen, and aromatic compounds, as well as the liquid hydrocarbons, are nonvolatile; therefore, the dynamic mass balance equation for the liquid compounds is

$$\varepsilon_L\frac{\partial C_i^L}{\partial t} = -u_L\frac{\partial C_i^L}{\partial z} - k_i^S a_S\left(C_i^L - C_i^S\right) \tag{3.86}$$

where $i = S, HC, N_B, N_{NB}$, and A. The components transported between liquid and solid phases are consumed or produced by the chemical reaction at the surface of the catalyst, according to the equation

$$\varepsilon_p(1-\varepsilon_B)\frac{\partial C_i^S}{\partial t} = k_i^S a_S\left(C_i^L - C_i^S\right) \pm \rho_B\zeta\eta_j r_{in,j}\left(C_i^S,\ldots,T_S\right) \tag{3.87}$$

where $i = H_2, H_2S, NH_3, S, HC, N_B, N_{NB}, A$, and j = HDS, HDN_{NB}, HDN_B, HDA. The negative sign is for the reactants, and the positive sign is for the products. The reaction rate for ammonia is $r_{NH_3} = -r_{HDN_B} + r_{HDN_{NB}}$.

Because the concentration of hydrocarbons (the main component of the feed) does not change significantly during HDT, Eqs. (3.86) and (3.87) for i = HC are not considered. To model the commercial HDT reactor operating under nonisothermal conditions, the following energy balance equations are used. For the liquid phase,

$$\varepsilon_L\rho_L c_p^L\frac{\partial T_L}{\partial t} = -u_L\rho_L c_p^L\frac{\partial T_L}{\partial z} - h_{LS}a_S(T_L - T_S) \tag{3.88}$$

and for the solid phase,

$$(1-\varepsilon_B)\rho_S c_p^S\frac{\partial T_S}{\partial t} = h_{LS}a_S(T_L - T_S) + \sum_j \rho_B\eta_j r_{in,j}\left(C_i^S,\ldots,T_S\right)(-\Delta H_{Rj}) \tag{3.89}$$

These two equations are usually sufficient for the energy balance since the heat capacity of the gas phase is much lower than that of the liquid phase.

HDT Reaction Kinetics HDS reaction was modeled with the following generalized stoichiometric equation:

$$\upsilon_S S_{(\text{liquid})} + \upsilon_{H_2} H_{2(\text{gas})} \rightarrow \upsilon_{HC} HC_{(\text{liquid})} + \upsilon_{H_2S} H_2S_{(\text{gas})} \tag{3.90}$$

where υ_S, υ_{H_2}, υ_{HC}, and υ_{H_2S} are the stoichiometric coefficients of the organic sulfur compounds, H_2, sulfur-free hydrocarbon, and H_2S, respectively.

The HDA reaction was represented by the following first-order reversible reaction:

$$A \underset{k_r}{\overset{k_f}{\rightleftarrows}} B \tag{3.91}$$

The HDN reaction was modeled by the following consecutive reaction scheme:

$$N_{NB} \xrightarrow{k_{HDN_{NB}}} N_B \xrightarrow{k_{HDN_B}} HC + NH_3 \tag{3.92}$$

Reaction rate expressions of HDS, HDA, HDN_{NB}, and HDN_B reactions are shown in Table 3.6. In the case of the HDS reaction, the following adsorption equilibrium constant of H_2S (K_{H_2S}) is used to account for the influence of the temperature:

$$K_{H_2S}(T) = 41{,}769.8411 \exp\left(\frac{2761}{RT}\right) \tag{3.93}$$

Boundary Conditions Since the reactor model is represented by a system of PDEs with time and spatial coordinates as independent variables, it is necessary to define the following initial and boundary conditions. The initial conditions for $t = 0$ at $z = 0$ are

$$p_{H_2}^G = \left(p_{H_2}^G\right)_0$$

$$p_i^G = 0, i = H_2S, NH_3$$

$$C_i^L = \left(C_i^L\right)_0, i = H_2, S, N_B, N_{NB}, A$$

$$C_i^L = 0, i = H_2S, NH_3$$

$$C_i^S = 0, i = H_2, H_2S, NH_3, S, N_B, N_{NB}, A$$

$$T = T_0$$

and at $z > 0$ are

$$p_i^G = 0, i = \mathrm{H_2, H_2S, NH_3}$$

$$C_i^L = 0, i = \mathrm{H_2, H_2S, NH_3, S, N_{\mathit{B}}, N_{\mathit{NB}}, A}$$

$$C_i^S = 0, i = \mathrm{H_2, H_2S, NH_3, S, N_{\mathit{B}}, N_{\mathit{NB}}, A}$$

$$T = T_0$$

The boundary conditions for $t > 0$ at $z = 0$ are

$$p_{\mathrm{H_2}}^G = \left(p_{\mathrm{H_2}}^G\right)_0$$

$$p_i^G = 0, i = \mathrm{H_2S, NH_3}$$

$$C_i^L = \left(C_i^L\right)_0, i = \mathrm{H_2, S, N_{\mathit{B}}, N_{\mathit{NB}}, A}$$

$$C_i^L = 0, i = \mathrm{H_2S, NH_3}$$

$$C_i^S = 0, i = \mathrm{H_2, H_2S, NH_3, S, N_{\mathit{B}}, N_{\mathit{NB}}, A}$$

$$T = T_0$$

For commercial HDT reactors, values of partial pressures (p_i^G) and liquid molar concentrations (C_i^L) of H_2S and NH_3 at the entrance of the catalytic bed $(z = 0)$ are different from zero.

Integration Method The PDEs describing the heat and mass transfer in the reactor were transformed into a set of first-order ODEs by discretization in the axial direction using the backward finite difference method and leaving the independent variable time without discretize. The ODEs were then solved using a fourth-order Runge–Kutta method.

Dynamic Simulation of an Isothermal HDT Small Reactor Dynamic simulations are carried out to observe the behavior of different product properties and operating conditions with time. Since in this case the reactor is operated isothermally, only the dynamic mass balance equations were considered. Figure 3.29 shows the axial concentration profiles of sulfur, nitrogen (basic and nonbasic), and total aromatic contents in the product at the exit of the reactor as a function of time. All profiles are quite similar. A small amount of hydrotreated product is detected at the exit of the reactor at about 250 s (0.07 h), which corresponds to the mean residence time given by the interstitial velocity of the liquid phase $(u_L' = u_L/\varepsilon_L)$; after that, concentrations increase, and finally, steady state is reached at 2300 s (0.64 h). The concentration of

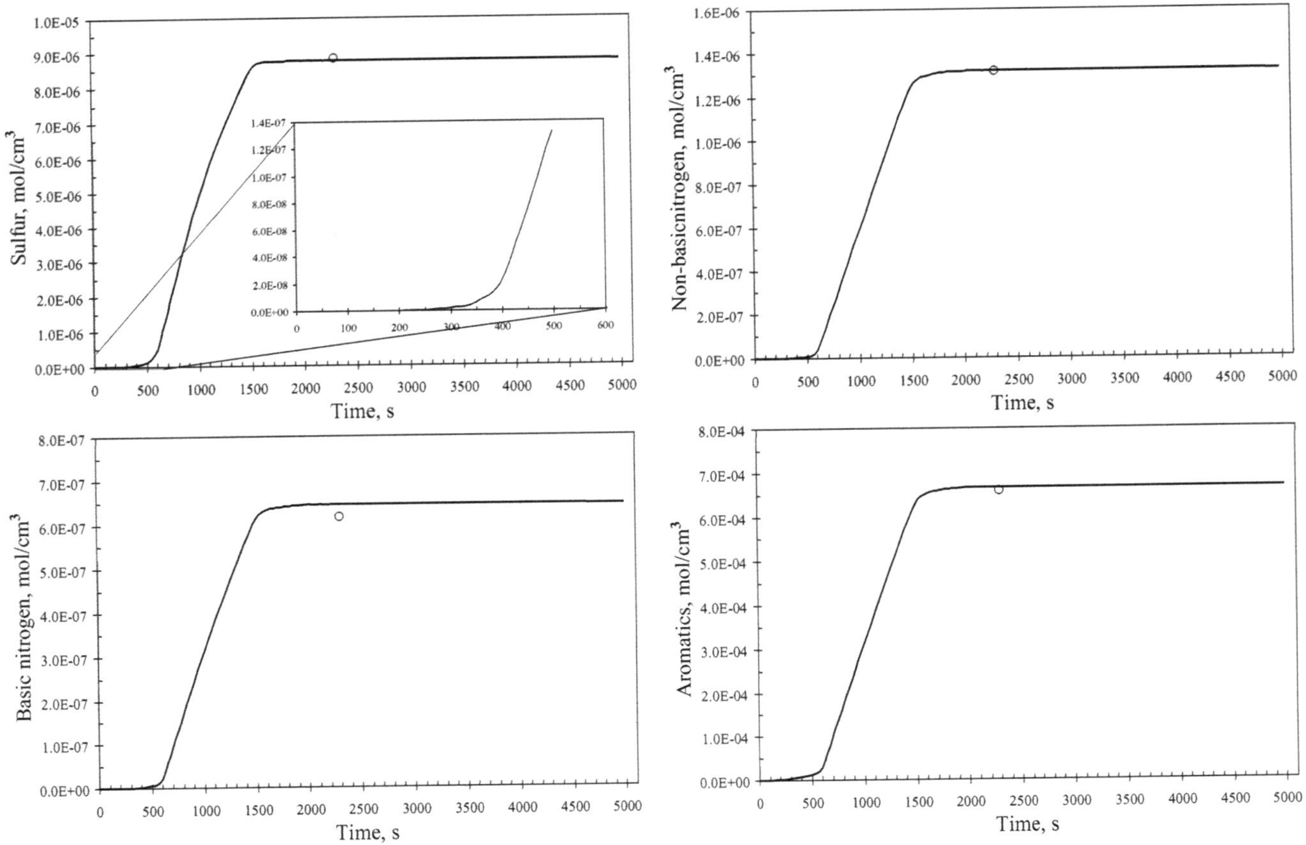

Figure 3.29. Concentration of impurities at the outlet of the bench-scale catalytic bed as a function of time at 380°C and 5.3 MPa (—, simulated; ○, experimental). $z = 31.58$ cm; $u_L = 1.75 \times 10^{-2}$ cm/s; $u_G = 0.28$ cm/s.

aromatics starts increasing earlier, at about 150s. This is because, as opposed to the HDS and HDN kinetic models, the HDA reaction rate takes into account the H_2 partial pressure, and the gas velocity is about 16 times higher than that of the liquid ($u_G/u_L = 16$).

Partial pressure and concentration profiles of H_2 and H_2S along the catalytic bed are shown in Figure 3.30, at times very close to the beginning of the operation (60s), intermediate times (500 and 1000s), and at steady state (2300s). Due to the high gas velocity inside the reactor, the curves of the H_2 partial pressure at 500, 1000, and 2300s overlap. A small amount of H_2S is present only in the first section of the reactor at 60s, and the common H_2S profile is found at 2300s. The overall shapes of molar concentration profiles of H_2 and H_2S are determined by the balance between the reaction rate and mass transfer. The H_2 concentration decreases while the H_2S concentration increases at the beginning of the reactor because of the high reaction rate of the catalyst bed in this zone. This behavior can be seen clearly at different times.

Figure 3.31 presents the dynamic simulated liquid molar concentration profiles of impurities in the product along the reactor at several times. In the case of sulfur, when the model predicts the steady state (at 2300s), the concentrations in both the liquid and solid phases are not equal along the reactor. The simulated results at the steady state obtained with the dynamic model agree reasonably well with experimental concentrations of impurities in the product at the exit of the reactor.

Dynamic Simulation of a Commercial HDT Reactor Because of the nonisothermal operation of the commercial reactor, the energy balance equation is solved simultaneously with the mass balance equations. Figure 3.32 illustrates the results of the dynamic simulation. The values predicted for the isothermal reactor are shown for comparison. It is seen that the steady state in the commercial reactor is reached faster than in the bench-scale reactor. This is not surprising because operating conditions, particularly LHSV and the u_G/u_L ratio, are greater. There are some differences in the sulfur concentration profiles at certain times before the steady state is reached in both reactors, which can be attributed to the operation with different temperature profiles. The experimental concentrations of impurities in the product for the bench-scale reactor are in general higher than those predicted for the commercial reactor because of the higher average catalytic bed temperature observed in the nonisothermal commercial operation.

Figure 3.33 summarizes the simulated dynamic temperature along the catalyst bed as a function of time for the commercial reactor. It is clearly observed that the temperature in the first part of the reactor remains more or less without changes with time, whereas the temperature near the outlet of the reactor changes significantly with time. At short times, the "wrong-way" behavior is observed, and at long times, the reactor temperature stabilizes.

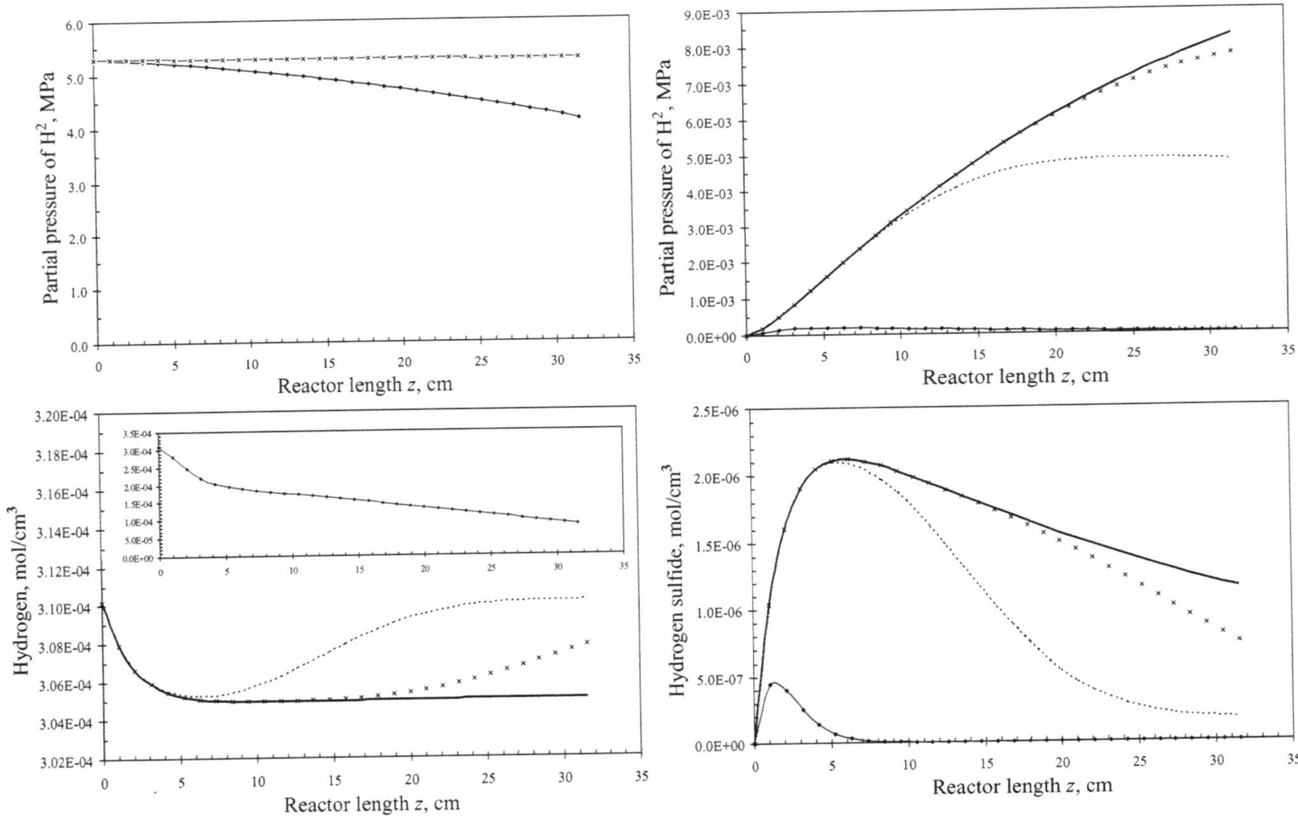

Figure 3.30. Axial H_2 and H_2S partial pressures and concentration profiles in the bench-scale catalytic bed at different times —◆—, 60 s; ---, 500 s; ×, 1000 s; —, 2300 s).

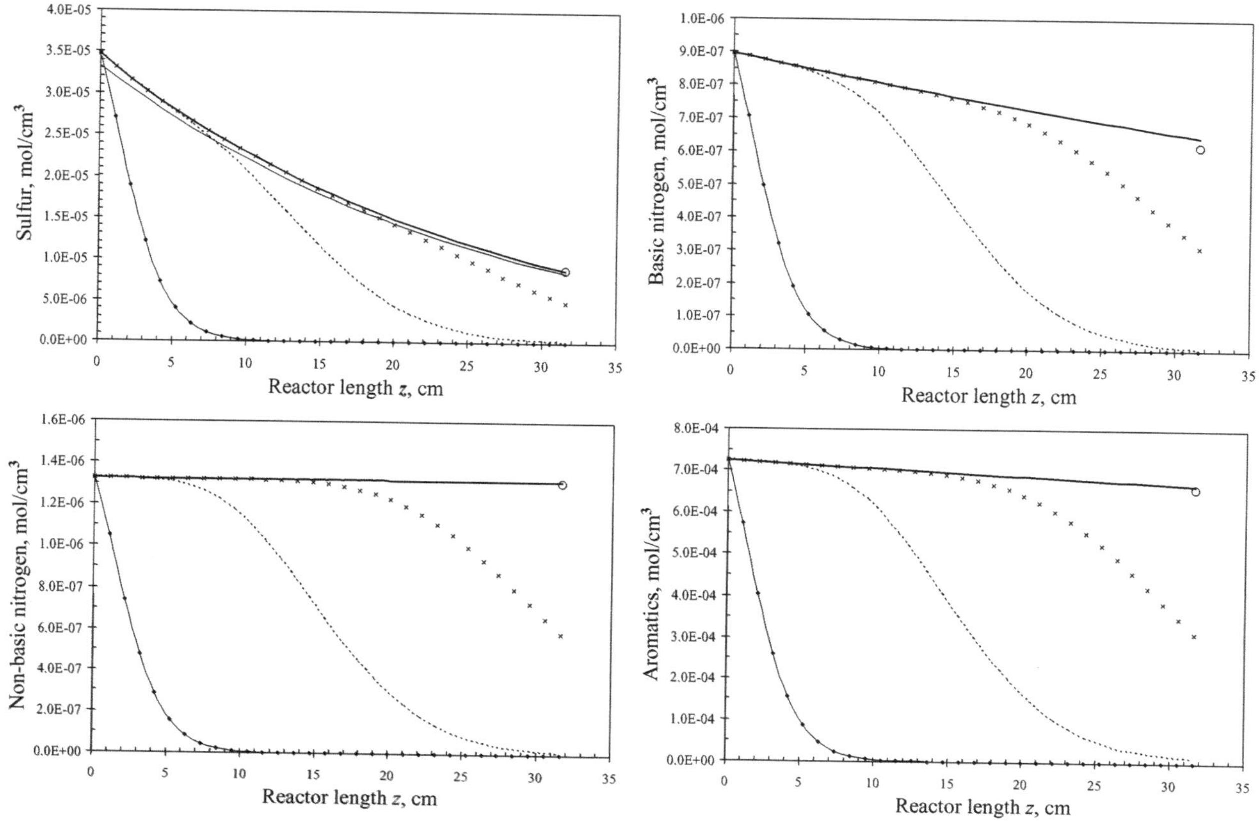

Figure 3.31. Axial profiles of impurities concentrations in the bench-scale catalytic bed at different times (-◆-, 60 s; ---, 500 s; ×, 1000 s; —, 2300 s).

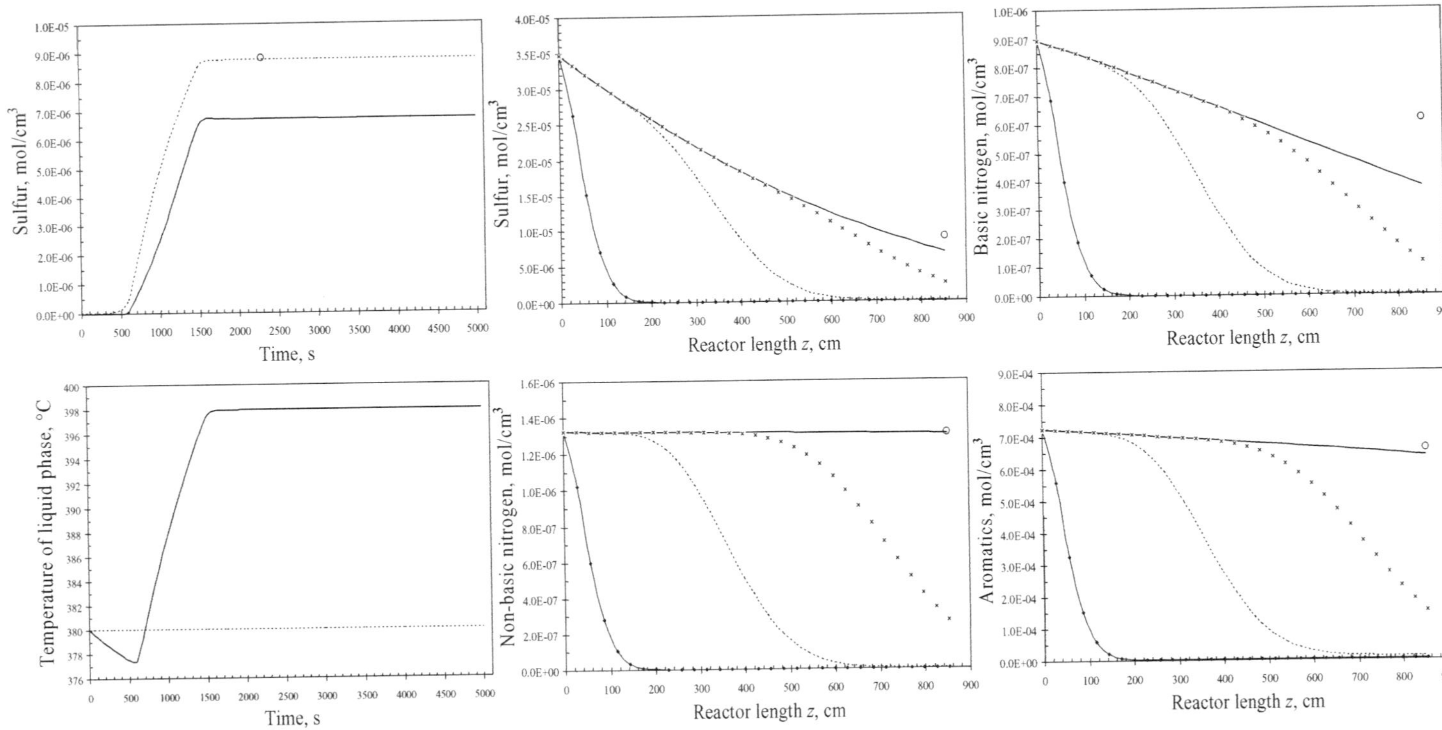

Figure 3.32. Dynamic profiles of concentration of impurities in the product and temperature at the oulet of the catalytic bed as a function of time at 380°C and 5.3 MPa. Commercial reactor: $z = 853.44$ cm, $u_L = 0.63$ cm/s, $u_G = 10.27$ cm/s; bench-scale reactor: $z = 31.58$ cm, $u_L = 1.75 \times 10^{-2}$ cm/s, $u_G = 0.28$ cm/s. (○, Experimental bench-scale reactor; lines, simulated; –◆–, 60 s; ---, 500 s; -×-, 1000 s; —, 2000 s.

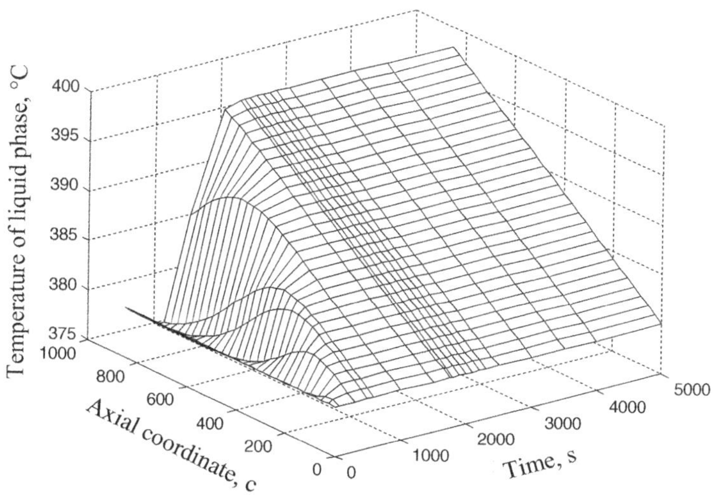

Figure 3.33. Temperature profile in the catalytic bed of the commercial reactor as a function of time and axial position.

3.3.5 Simulation of Countercurrent Operation

Comparison of Modes of Operation The introduction of TBRs with countercurrent flow in a number of refining operations is probably either via redesign of existing reactors or as new technology. The goal of this mode of operation is not an improvement in the reactant mass transfer, which is not rate limiting but enhanced selective removal of by-products that may act as inhibitors (such as H_2S and NH_3) for certain HDT reactions or in situ product separation. The differences between co-current and countercurrent flows become more pronounced for stronger H_2S inhibition, higher liquid rates, and plug-flow conditions. Due to the need for removal of the most refractory sulfur compounds in diesel HDS, in which the H_2S inhibition effect plays an important role, the countercurrent flow operation mode will become more prominent in the future for processes that suffer from by-product catalyst inhibition.

The main problem for countercurrent operation mode in a TBR is the phenomenon of flooding, although it can be undertaken by shaping the catalyst or arranging the packing to create different paths for gas and liquid, reducing momentum transfer between the two phases, thereby shifting the flooding limits to much higher flow rates. In the case of hydrotreating, the major disadvantage of countercurrent reactor for commercial applications is due to hardware limitations. The catalyst loading is 20 to 25 vol% in countercurrent operation (although less catalyst volume is necessary for countercurrent operation to achieve the same conversion), whereas in co-current TBR, the catalyst loading is 60 to 70 vol%. There is, therefore, a need to develop improved hardware configurations that allow countercurrent contacting of gas and liquid in the presence of small catalyst particles and also when the catalyst loading is above 50 vol%.

The countercurrent mode of operation would be much more desirable for deep HDS processes since this reaction is strongly inhibited by the H_2S produced during sulfur removal. The HDS of oil fractions follows a reaction order higher than 1 with respect to sulfur content, due to the presence of a large number of sulfur compounds with different reactivities. The most reactive sulfur compounds are removed in the first part of the reactor, while the less reactive sulfur compounds are eliminated in the final part of the reactor. Therefore, countercurrent is the most favorable mode of operation at those conditions because the larger part of the reactor operates under an H_2S lean regime, H_2S concentration being lowest at the bottom of the reactor, and H_2 partial pressure at the outlet section of the reactor being highest.

The advantages and disadvantages of TBR with countercurrent flow are given below.

Advantages

- Low partial pressure of H_2S and NH_3 in most parts of the catalytic bed
- Improved conversions normally limited by chemical equilibrium
- Favored with respect to a large heat of reaction
- Enables handling more difficult feedstocks to obtain higher conversion
- More favorable flat axial temperature profile
- Large surface area for vapor–liquid mass transfer
- High ratio of active sites to reactor volume
- Easy catalyst handling
- Significant performance for high liquid rates
- Decreased intraparticle resistance by using small particles
- Adapted when larger mean concentration driving forces are needed

Disadvantages

- Excessive pressure drop at high liquid and gas velocities
- Presence of flooding at high liquid throughputs
- Lack of correlations to estimate hydrodynamics and mass and heat transfer parameters
- Extra constraints to catalyst packing, primarily about its size and shape
- Reduced gas–liquid mass transfer
- Less effective temperature control because of gas flow
- Lack of flexibility with respect to the flow rate of the fluid phases
- Low solid/reactor volume ratio
- Low gas–liquid interfacial area
- High axial dispersion effects in the liquid phase

Description of the Countercurrent Reactor Model Both operations, co-current and countercurrent, are simulated with the same dynamic TBR reactor model (one-dimensional heterogeneous) described previously under isothermal and adiabatic conditions (Mederos and Ancheyta, 2007). For countercurrent operation, Eq. (3.84) must be changed to (note that the second term is positive for countercurrent operation and negative for co-current operation)

$$\frac{\varepsilon_G}{RT_G}\frac{\partial p_i^G}{\partial t} = +\frac{u_G}{RT_G}\frac{\partial p_i^G}{\partial z} - k_i^L a_L\left(\frac{p_i^G}{H_i} - C_i^L\right) \tag{3.94}$$

If axial dispersion in the liquid phase is taken into account, Eqs. (3.85) and (3.86) need to be replaced by

$$\varepsilon_L\frac{\partial C_i^L}{\partial t} = -u_L\frac{\partial C_i^L}{\partial z} + \varepsilon_L D_a^L\frac{\partial^2 C_i^L}{\partial z^2} + k_i^L a_L\left(\frac{p_i^G}{H_i} - C_i^L\right) - k_i^S a_S(C_i^L - C_i^S) \tag{3.95}$$

$$\varepsilon_L\frac{\partial C_i^L}{\partial t} = -u_L\frac{\partial C_i^L}{\partial z} + \varepsilon_L D_a^L\frac{\partial^2 C_i^L}{\partial z^2} - k_i^S a_S(C_i^L - C_i^S) \tag{3.96}$$

In these equations, the axial dispersion coefficient of the liquid phase (D_a^L) is needed, which can be determined from the Peclet number:

$$\mathrm{Pe}_{a,m}^L = \frac{d_{pe}u_L}{D_a^L\varepsilon_L} \tag{3.97}$$

The Peclet number can be calculated from different correlations reported in the literature, depending on the mode of operation of the reactor. However, the lack of suitable and reliable correlations to estimate the Peclet number, and hence the axial dispersion coefficient of the liquid phase, makes it difficult to obtain appropriate estimations, since predictions can vary significantly from one correlation to another. Axial dispersion may influence only the results of small-scale reactors, since for commercial reactors it can be neglected. Thus, axial dispersion, if present, affects only bench-scale reactor simulations. Therefore, not having the effect of axial dispersion in both modes of operation is not significant and the conclusions will not change, so that the assumption of plug-flow behavior is justified.

For countercurrent operation it is important to include the gas phase in the energy balance equations given by Eqs. (3.88) and (3.89) to model the heat transfer process accurately in the reactor because upstream heat transfer from the gas phase to the liquid phase will speed up the reaction rate, leading to an even higher concentration of H_2S in the liquid phase at the initial part of the reactor. Therefore, the energy balance equation for gas phase is

$$\varepsilon_G\rho_G c_p^G\frac{\partial T_G}{\partial t} = \pm u_G\rho_G c_p^G\frac{\partial T_G}{\partial z} - h_{GL}a_L(T_G - T_L) \tag{3.98}$$

In the second term, the negative sign is for co-current operation and the positive sign is for countercurrent operation.

Boundary Conditions The reactor model is a system of PDEs and ODEs with time and spatial coordinate as independent variables, and for its solution it is necessary to define the following initial and boundary conditions for the liquid and gas phases and for both modes of operation. The initial conditions for $t = 0$ at $z = 0$ are:

Co-current operation

$$p_i^G = (p_i^G)_0, \qquad i = H_2, H_2S, NH_3$$

$$C_i^L = (C_i^L)_0, \qquad i = H_2, H_2S, NH_3$$

Countercurrent operation

$$p_i^G = 0, \qquad i = H_2, H_2S, NH_3$$

$$C_i^L = 0, \qquad i = H_2, H_2S, NH_3$$

Co-current/countercurrent operation

$$C_i^L = (C_i^L)_0, \qquad i = S, N_B, N_{NB}, A$$

$$C_i^S = 0, \qquad i = H_2, H_2S, NH_3, S, N_B, N_{NB}, A$$

$$T_G = T_L = T_S = T_0$$

at $0 < z < L_B$:

Co-current/countercurrent operation

$$p_i^G = 0, \qquad i = H_2, H_2S, NH_3$$

$$C_i^L = 0, \qquad i = H_2, H_2S, NH_3, S, N_B, N_{NB}, A$$

$$C_i^S = 0, \qquad i = H_2, H_2S, NH_3, S, N_B, N_{NB}, A$$

$$T_G = T_L = T_S = T_0$$

and at $z = L_B$:

Co-current operation

$$p_i^G = 0, \qquad i = H_2, H_2S, NH_3$$

Countercurrent operation

$$p_i^G = (p_{H_2}^G)_0, \qquad i = H_2, H_2S, NH_3$$

Co-current/countercurrent operation

$$C_i^L = 0, \qquad i = H_2, H_2S, NH_3, S, N_B, N_{NB}, A$$

$$C_i^S = 0, \qquad i = H_2, H_2S, NH_3, S, N_B, N_{NB}, A$$

$$T_G = T_L = T_S = T_0$$

The boundary conditions for $t > 0$ at $z = 0$ are:

Co-current operation

$$p_i^G = (p_i^G)_0, \qquad i = H_2, H_2S, NH_3$$

$$C_i^L = (C_i^L)_0, \qquad i = H_2, H_2S, NH_3$$

$$T_G = (T_G)_0$$

Countercurrent operation

$$C_i^L = 0, \qquad i = H_2, H_2S, NH_3$$

Co-current/countercurrent operation

$$C_i^L = (C_i^L)_0, \qquad i = S, N_B, N_{NB}, A$$

$$C_i^S = 0, \qquad i = H_2, H_2S, NH_3, S, N_B, N_{NB}, A$$

$$T_L = (T_L)_0, \qquad T_S = (T_S)_0$$

and at $z = L_B$:

Countercurrent operation

$$p_i^G = (p_i^G)_{LB}, \qquad i = H_2, H_2S, NH_3$$

$$T_G = (T_G)_{LB}$$

Co-current/countercurrent operation with liquid axial dispersion

$$\frac{\partial C_i^L}{\partial z} = 0$$

When a high-purity H_2 stream without gas recycle is used, such as in the case of some laboratory- and bench-scale HDT reactors, or when the gas recycle has been subject to a purification process in commercial units, values of partial pressures (p_i^G) and liquid molar concentrations (C_i^L) of H_2S and NH_3 at the entrance of the catalytic bed ($z = 0$ for co-current operation and $z = L_B$ for countercurrent operation) are equal or very close to zero.

Simulation of a Countercurrent Isothermal HDT Small Reactor The feedstock, catalyst, and reaction conditions are the same as those used in previous sections. There are two ways to simulate the co-current TBR for HDT of oil fractions: (1) by considering that the oil is saturated with H_2 at the entrance of the reactor, and (2) by assuming that the oil is not saturated with H_2; that is, the initial H_2 concentration in the oil is zero $[(C_{H_2}^L)_0 = 0]$. The simulation was done using both approaches.

Figure 3.34 shows the changes in sulfur content in the product at the exit of the reactor as a function of time and along the reactor at steady state for co-current operation with VGO saturated and not saturated with H_2 and for countercurrent operation. The steady state is obtained at about the same time (2300 s) for the three approaches. Before it is reached, the three profiles are almost identical. After 1000 s co-current with unsaturated oil yields higher sulfur contents followed by countercurrent and then by co-current with H_2 saturated oil operations. Co-current with H_2 saturated oil is the approach that better reproduces the experimental value because the kinetic parameters were initially optimized with this condition. Similar to sulfur, nitrogen and aromatics contents follow the same trends for all approaches (i.e., having zero concentration at the exit of the reactor from 0 to ~250 s), since at the beginning these compounds are present only in the hydrocarbon feed which flows down through the catalytic bed, and regardless of the method used to inject H_2, their concentrations diminish from the top to the bottom of the reactor.

For all operational modes, the sulfur content in the liquid phase (as well as other impurities content) decreases through the reactor. The lowest sulfur content in the product is obtained for co-current operation with the oil saturated with H_2 before it enters the reactor. The concentration of sulfur compounds for co-current operation is lower than that of countercurrent operation in about 73% of the reactor length, and then the tendency switches. This behavior is because in countercurrent operation the initial part of the reactor (at the top) has a high H_2S concentration, whereas around the outlet of the reactor (at the bottom) there is a low H_2S concentration, which makes the HDS reaction rate less inhibited by H_2S and thus faster than that of the co-current operation.

Partial pressure and concentration profiles of H_2 and H_2S along the reactor are shown in Figure 3.35 when the steady state is reached for the three approaches. In all cases the overall shape of the molar concentration profiles of H_2 and H_2S is determined by the balance between the reaction rate and mass transfer. The trends for concentration profiles of H_2S in the liquid phase

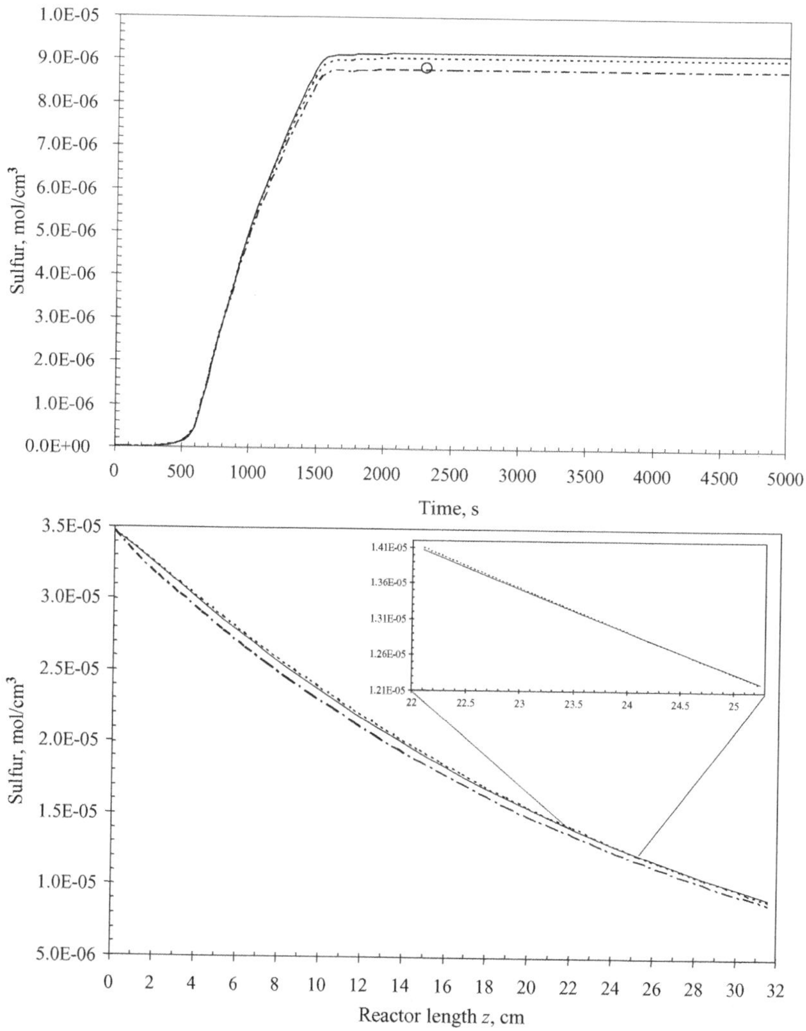

Figure 3.34. Profile of sulfur content in the product at 380°C and 5.3 MPa as a function of time and reactor length at steady state. Bench-scale reactor: $z = 31.58$ cm, $u_L = 1.75 \times 10^{-2}$ cm/s, $u_G = 0.28$ cm/s. (Lines, simulated; ○, experimental; –·–, co-current with H_2 saturated oil; —, co-current with unsaturated oil; ---, countercurrent.)

are similar for the two modes of operation, whereas for partial pressure profiles of H_2 and H_2S, the tendencies are opposite. This opposite pattern is obviously expected since at the top of the reactor ($z = 0$), co-current operation has high H_2 content and low H_2S content, while countercurrent operation has low H_2 content and high H_2S content at the top of the reactor. For co-current operation with H_2 saturated oil, the concentration of H_2 in the liquid phase decreases slightly at the beginning of the reactor until a certain point because

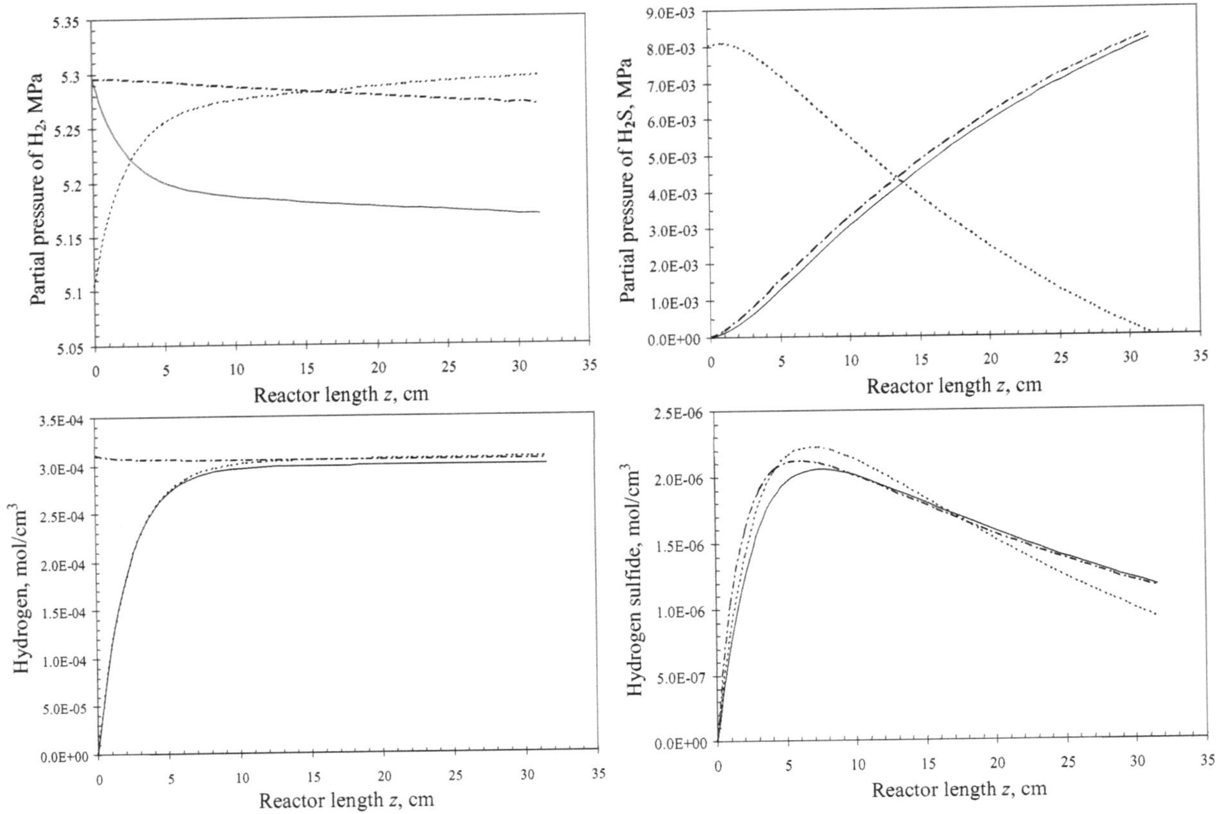

Figure 3.35. Profiles of H_2 and H_2S partial pressures and concentrations at steady state down through the catalytic bed. Bench-scale reactor: $z = 31.58$ cm, $u_L = 1.75 \times 10^{-2}$ cm/s, $u_G = 0.28$ cm/s. (–·–·, Co-current with H_2 saturated oil; —, co-current with unsaturated oil; ---, countercurrent.)

of the high reaction rate in this zone; then it starts to increase by the large mass transfer of H_2 to the liquid phase. For co-current operation with unsaturated oil, the H_2 concentration in the liquid phase increases rapidly due to the high H_2 dissolving rate. For all operation modes the H_2S concentration in the liquid phase increases and then decreases, and for co-current operation, the partial pressure of H_2S always increases through the catalytic bed, and countercurrent operation exhibits a contrary tendency.

Simulation of a Countercurrent Commercial HDT Reactor To simulate the expected behavior of a commercial HDT adiabatic reactor in countercurrent operation, the energy balance equations given by Eqs. (3.88), (3.89), and (3.98) were also solved together with the mass balance equations. To compare co-current and countercurrent modes of operation, the heat and mass transfer coefficients were supposed to be identical independent of the phase flow direction.

Figure 3.36 illustrates the results of dynamic simulation of the commercial reactor. The profiles of sulfur content at the outlet of the catalytic bed are presented for co-current operation with VGO saturated and not saturated with H_2 and for countercurrent operation. For the three modes of operation steady state is reached a little faster than that observed for a bench-scale reactor. Contrary to the situation with a bench-scale reactor, in which the co-current mode with H_2-saturated oil showed the highest impurities removal,

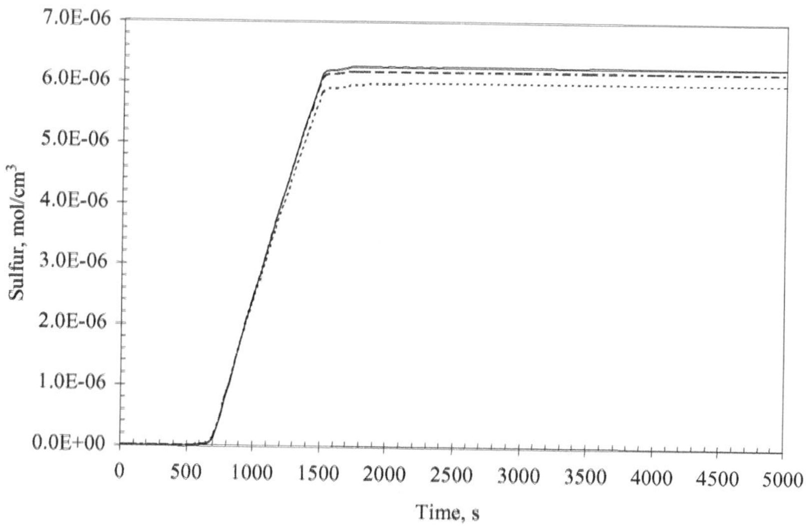

Figure 3.36. Dynamic profiles of concentration of sulfur in the product at 380°C and 5.3 MPa at the outlet of the catalytic bed. Commercial reactor: $z = 853.44$ cm, $u_L = 0.63$ cm/s, $u_G = 10.27$ cm/s. (–·–, Co-current with H_2 saturated oil; —, co-current with unsaturated oil; ---, countercurrent.)

countercurrent operation of a commercial reactor gives the highest conversions. This superior performance of the countercurrent operation mode can be explained as follows: When the reactants move farther down the reactor, mass transfer of H_2S from the liquid to the gas phase prevails and then the liquid H_2S concentration decreases.

The dynamic temperature profiles at the exit of the reactor of the liquid phase, the evolution of temperature of liquid phase along the reactor, and partial pressure and concentration of H_2S in the liquid phase are all depicted in Figure 3.37, for both co-current and countercurrent operations. The temperature profile of the gas phase through the catalytic bed for countercurrent operation is also shown in this figure. Since the countercurrent operation gave the highest sulfur conversion (Figure 3.36), the highest liquid temperature at the exit of the reactor would also be expected; however, this is not the situation. The liquid temperature for countercurrent operation is lower (marginally, but lower) than that predicted by co-current operation with VGO saturated with H_2. This behavior leads to the conclusion that the ascending gas phase cools the liquid phase along the catalytic bed of the reactor. But this happens only at the very end of the reactor. Cooling of the liquid phase by the gas phase is due to the temperature at which the gas is fed to the reactor.

Since sulfur conversion on a commercial scale (adiabatic) is higher than that on a bench scale (isothermal) under similar conditions, the H_2S concentration in the liquid phase is larger in the commercial reactor, and its effect is also greater than that observed on a bench scale. The tendencies of H_2S partial pressure are in general similar to those found on a bench scale. Only co-current operation with VGO not saturated with H_2 presented a decrease in the H_2S partial pressure in the initial part of the reactor because the H_2S is dissolved rapidly in the liquid phase until a point where its saturation is reached, and then the mass transfer is changed from the liquid phase to the gas phase. In comparison with the bench-scale reactor with countercurrent operation, the commercial reactor presents a higher H_2S partial pressure than co-current operation with unsaturated oil beyond the half-part of the reactor due to its improved liquid–gas mass transfer condition, which is called in situ *stripping* of H_2S from the liquid phase.

The high reaction rate in the initial part (10%) of the catalyst bed provokes a rapid increase in the liquid H_2S concentration for all the modes of operation. The accumulation of H_2S in the liquid phase is more pronounced in co-current operation with VGO saturated with H_2, because in the gas entering the reactor there is a certain amount of H_2S as impurity (3.06 mol%), and therefore the VGO is already saturated with H_2S at the entrance to the reactor. The maximum concentration of H_2S in the liquid phase is obtained under countercurrent operation, which is due to the higher mass transfer of H_2S from the gas phase to the liquid phase as a consequence of the high partial pressure of H_2S at the beginning of the reactor.

It is very important to emphasize that in countercurrent operation it is not possible to use the same catalyst particle sizes (1 to 5 mm) as those employed

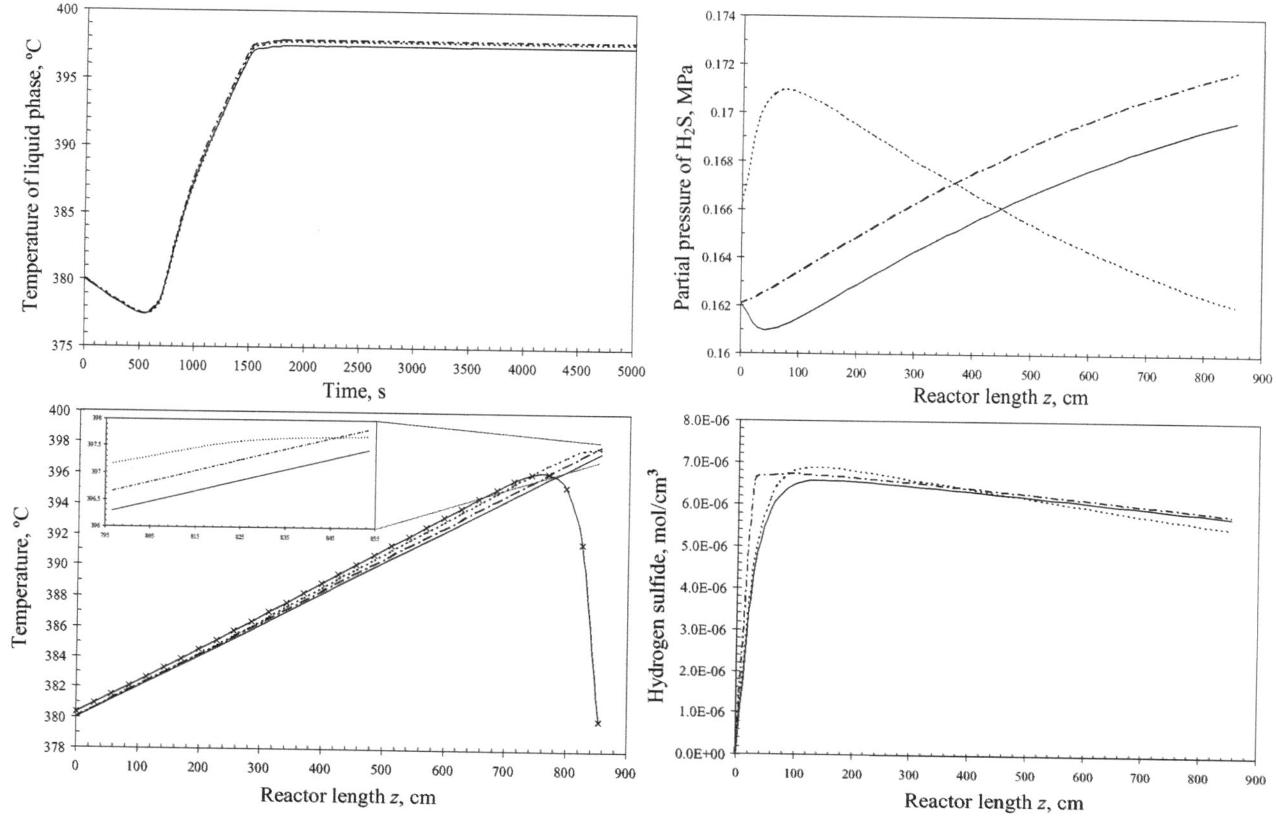

Figure 3.37. Profiles of liquid-phase temperature and of H_2S partial pressure and concentration at the outlet of the catalytic bed as a function of time, and at steady state down through the commercial catalytic bed. Commercial reactor: $z = 853.44\,\text{cm}$, $u_L = 0.63\,\text{cm/s}$, $u_G = 10.27\,\text{cm/s}$. (–·–·, Co-current with H_2 saturated oil; —, co-current with unsaturated oil; ---, countercurrent; —×—, gas-phase countercurrent.

in co-current operation, because of flooding. That is why the use of other types of internals, such as "three levels of porosity" packing, monolithic structures, or random packing (rings, saddles, etc.), has been proposed. In all the simulation results presented here it was assumed that properties of the catalyst used in co-current operation are equal to those proposed to be employed in countercurrent operation.

REFERENCES

Akzo Nobel (2003) Akzo Nobel duplex distributor tray. Presented at the Mexican Institute of Petroleum, Mexico City, Mexico, May 14.

Albermarle (2006) Maximize reactor performance with state-of-the-art PLEX internals. http://www.albermarle.com (accessed Oct. 2006).

Ali, S. A. (2007) Chapter 4: Thermodynamics of hydroprocessing reactions. In: *Hydroprocessing of Heavy Oil and Residua*, Ancheyta, J., Speight, J. G. (eds.). CRC Press, Taylor & Francis, New York.

Altrichter, D. M., Creyghton, E. J., Ouwehand, C., van Veen, J. A. R., Hanna, A. (2004) New catalyst technologies for increased hydrocracker profitability and product quality. AM-04-60. In: *Proceedings of the NPRA Annual Meeting*, San Antonio, TX, Mar. 21–23.

Alvarez, A., Ancheyta, J. (2008) Simulation and analysis of different quenching alternatives for an industrial vacuum gasoil hydrotreater. *Chem. Eng. Sci.* 63: 662–673.

Alvarez, A., Ancheyta, J., Muñoz, J. A. D. (2007a) Comparison of quench systems in commercial fixed-bed hydroprocessing reactors. *Energy Fuels* 21:1133–1144.

Alvarez, A., Ramírez, S., Ancheyta, J., Rodríguez, L. M. (2007b) Key role of reactor internals in hydroprocessing of oil fractions. *Energy Fuels* 21:1731–1740.

Ancheyta, J., Speight, J. G. (2007) *Hydroprocessing of Heavy Oils and Residua. CRC Press*, Taylor & Francis, New York.

Ancheyta, J., Aguilar, E., Salazar, D., Betancourt, G., Leiva, M. (1999a) Hydrotreating of straight run gas oil-light cycle oil blends. *Appl. Catal. A* 180:195–205.

Ancheyta, J., Aguilar, E., Salazar, D., Marroquin, G., Quiroz, G., Leiva, M. (1999b) Effect of hydrogen sulfide on the hydrotreating of middle distillates over Co/MoAl_2O_3 catalyst. *Appl. Catal. A* 183:265–272.

Ancheyta, J., Marroquin, G., Angeles, M. J., Macías, M. J., Pitault, I., Forissier, M., Morales, R. D. (2002a) Some experimental observations of mass transfer limitations in a hydrotreating trickle-bed pilot reactor. *Energy Fuels* 16:1059–1067.

Ancheyta, J., Betancourt, G., Marroquin, G., Centeno, G., Castañeda, L. C., Alonso, F., Muñoz, J. A., Gómez, M. T., Rayo, P. (2002b) Hydroprocessing of Maya heavy crude oil in two reaction stages. *Appl. Catal. A* 233:159–170.

Ancheyta, J., Centeno, G., Trejo, F., Marroquin, G. (2003a) Changes in asphaltene properties during hydrotreating of heavy crudes. *Energy Fuels* 17:1233–1238.

Ancheyta, J., Betancourt, G., Centeno, G., Marroquin, G. (2003b) Catalyst deactivation during hydroprocessing of Maya heavy crude oil (II) effect of reaction temperature during time-on-stream. *Energy Fuels* 17:462–467.

Ancheyta, J., Betancourt, G., Marroquín, G., Centeno, G., Alonso, F., Muñoz, J. A. (2006) Process for the catalytic hydrotreatment of heavy hydrocarbons of petroleum. U.S. patent 20070187294 A1.

Avraam, D. G., Vasalos, I. A. (2003) HdPro: a mathematical model of trickle-bed reactors for the catalytic hydroprocessing of oil feedstocks. *Catal. Today* 79–80(1–4): 275–283.

Babich, I. V., Moulijn, J. A. (2003) Science and technology of novel processes for deep desulfurization of oil refinery streams: a review. *Fuel* 82:607–631.

Ballard, J. H., Hines, J. E. (1965) Vapor–liquid distribution method and apparatus for the conversion of hydrocarbons. U.S. patent 3,128,249.

Ballard, J. H., Hines, J. E. (1970) Apparatus for mixing fluids in concurrent downflow relationship. U.S. patent 3,502,445.

Bej, S. K., Dalai, A. K., Adjaye, J. (2001) Comparison of hydrodenitrogenation of basic and nonbasic nitrogen compounds present in oil sands derived heavy gas oil. *Energy Fuels* 15(2):377–383.

Bhaskar, M., Valavarasu, G., Sairam, B., Balaraman, K. S., Balu, K. (2004) Three-phase reactor model to simulate the performance of pilot-plant and industrial trickle-bed reactors sustaining hydrotreating reactors. *Ind. Eng. Chem. Res.*, 43:6654–6669.

Billon, A., Peries, J. P., Espeillan, M., des Courieres, T. (1991) Hyvahl F versus Hyvahl M, swing reactor or moving bed. Presented at the 1991 NPRA Annual Meeting, San Antonio, TX, Mar. 17–19.

Binghan, F. E., Christensen, P. (2000) Revamping HDS units to meet high quality diesel specifications. Presented at the Asian Pacific Refining Technology Conference, Kuala Lumpur, Malasya, Mar. 8–10.

Bradway, R. A., Tsao, Y. P. (2001) Hydroprocessing reactor and process with gas and liquid quench. U.S. patent 6,299,759.

Cooper, B. H., Donnis, B. B. L. (1996) Aromatic saturation of distillates: an overview. *Appl. Catal. A* 137:203–223.

Cotta, R. M., Wolf-Maciel, M. R., Maciel Filho, R. (2000) A cape of HDT industrial reactor for middle distillates. *Comput. Chem. Eng.* 24(2–7):1731–1735.

Chen, J., Te, M., Yang, H., Ring, Z. (2003) Hydrodesulfurization of dibenzotiophenic compounds in a light cycle oil. *Pet. Sci. Technol.* 21(5–6):911–935.

Cheng, Z.-M., Fang, X.-C., Zeng, R.-H., Han, B.-P., Huang, L., Yuan, W.-K. (2004) Deep removal of sulfur and aromatics from diesel through two-stage concurrently and countercurrently operated fixed-bed reactors. *Chem. Eng. Sci.* 59(22–23): 5465–5472.

Chowdhury, R., Pedernera, E., Reimert, R. (2002) Trickle-bed reactor model for desulfurization and dearomatization of diesel. *AIChE J.* 48(1):126–135.

Daniel, M., Lerman, D. B., Peck, L. B. (1988) Amocos LC-fining residue hydrocracker yield and performance correlations from a commercial unit. Presented at the 1988 NPRA annual meeting.

Davis, T. J. (2002) EMRE Hydroprocessing technologies and low sulfur motor fuels. Presented at the PEMEX Seminar, Mexico City, Mexico, Dec.

Den Hartog, A. P., van Vliet, W. (1997) Multi-bed downflow reactor. U.S. patent 5,635,145.

Döhler, W., Rupp, M. (1987) Comparison of performance of an industrial VGO-treater with reactor model predictions. *Chem. Eng. Technol.* 10:349–352.

Edgar, M. D. (1993) Hydrotreating. Q&A, 1993 NPRA Annual Meeting, San Antonio, TX, Mar.

Euzen, J. P. (1991) Moving-bed process for residue hydrotreating. *Rev. Inst. Fr. Pet.* 46:517–527.

Furimsky, E. (1998) Selection of catalysts and reactors for hydroprocessing. *Appl. Catal. A* 171:177–206.

Gary, J. H., Handwerk, G. E. (2001) *Petroleum Refining: Technology and Economics*, 4th ed. CRC Press, Taylor & Francis, New York.

Gates, B. C., Katzer, J. R., Schuit, G. C. A. (1979) *Chemistry of Catalytic Processes*, 3rd ed. McGraw-Hill, New York.

Girgis, M. J., Gates, B. C. (1991) Reviews: reactivities, reaction networks, and kinetics in high-pressure catalytic hydroprocessing. *Ind. Eng. Chem. Res.* 30:2021–2058.

Gosselink, J. W. (1998) Sulfide catalysts in refineries. *Cattech* 2(2):127–144.

Gruia, A. (2006) Recent advances in hydrocracking. In: *Practical Advances in Petroleum Processing*, Hsu, C. H., Robinson, P. R. (eds.). Springer-Verlag, New York, Chap. 7.

Ho, T. C. (2003) Hydrodesulfurization with RuS_2 at low hydrogen pressures. *Catal. Lett.* 89(1–2):21–25.

Jacobs, G. E., Stupin, W. S., Kuskie, R. W., Logman, R. A. (2000) Reactor distribution apparatus and quench zone mixing apparatus. U.S. patent 6,098,965.

Kam, E. K. T., Al-Mashan, M.-H., Al-Zami, H. (1999) The mixing aspects of NiMo and CoMo hydrotreating catalysts in ebullated-bed reactors. *Catal. Today* 48: 229–236.

Korsten, H., Hoffmann, U. (1996) Three-phase reactor model for hydrotreating in pilot trickle-bed reactors. *AIChE J.* 42(5):1350–1357.

Kwak, S., Longstaff, D. C., Deo, M. D., Hanson, F. V. (1992) Hydrotreating process kinetics for bitumen and bitumen-derived liquids. In: *Proceedings of the Eastern Oil Shale Symposium*, University of Kentucky, Lexington, KY, Nov. 13–15, pp. 208–215.

Langston, J., Allen, L., Davé, D. (1999) Technologies to achieve 2000 diesel specifications. *Pet. Tech. Q.* 2:65.

Litchfield, J. F., Pedersen, M. J., Sampath, V. R. (1996) Optimization of interbed distributors. AM-96-73. In: *Proceedings of the NPRA Annual Meeting*, San Antonio, TX, Mar. 17–19.

Liu, Z., Zheng, Y., Wang, W., Zhang, Q., Jia, L. (2008) Simulation of hydrotreating of light cycle oil with a system dynamics model. *Appl. Catal. A* 339:209–220.

Macías, M. J., Ancheyta, J. (2004) Simulation of an isothermal hydrodesulfurization small reactor with different catalyst particle shapes. *Catal. Today* 98:243–252.

Marroquín, G., Ancheyta, J. (2001) Catalytic hydrotreating of middles distillates blends in a fixed-bed pilot reactor. *Appl. Catal. A* 207:407–420.

Marroquín, G., Ancheyta, J., Díaz, J. A. I. (2004) On the effect of reaction conditions on liquid phase sulfidation of a NiMo HDS catalyst. *Catal. Today* 98:75–81.

McDougald, N. K., Boyd, S. L., Muldowney, G. P. (2006) Multiphase mixing device with baffles. U.S. patent 7,045,103.

Mederos, F. S., Ancheyta, J. (2007) Mathematical modeling and simulation of hydrotreating reactors: cocurrent versus countercurrent operations. *Appl. Catal. A* 332:8–21.

Mederos, F. S., Rodriguez, M. A., Ancheyta, J., Arce, E. (2006) Dynamic modeling and simulation of catalytic hydrotreating reactors. *Energy Fuels* 20:936–945.

Mehra, Y. R., Al-Abdulal, A. H. (2005) Hydrogen purification in hydroprocessing (HPH^{SM} technology). Presented at the 103rd NPRA Annual Meeting, San Francisco, Mar. 13–15.

Minderhoud, J. K., van Veen, J. A. R., Hagan, A. P. (1999) Hydrocracking in the year 2000: a strong interaction between technology development and market requirements. *Stud. Surf. Sci. Catal.* 127:3–20.

Mochida, I., Choi, K. (2004) An overview of hydrodesulfurization and hydrogenation. *J. Jpn. Pet. Inst.* 47(3):145–163.

Mochida, I., Choi, K. (2006) Current progress in catalysts and catalysis for hydrotreating. In: *Practical Advances in Petroleum Processing*, Hsu, C. H., Robinson, P. R. (eds.). Springer-Verlag, New York, Chap. 9.

Mochida, I., Sakanishi, K., Ma, X., Nagao, S., Isoda, T. (1996) Deep hydrodesulfurization of diesel fuel: design of reaction process and catalysts. *Catal. Today* 29: 185–189.

Morel, F., Kressmann, S., Harlé, V., Kasztelan, S. (1997) Processes and catalysts for hydrocracking of heavy oil and residues. In: *Hydrotreatment and Hydrocracking of Oil Fractions*. Studies in Surface Science Catalysis, Froment, G. F., Delmon, B., Grange, P. (eds.). Elsevier Science, Amsterdam.

Moulijn, J. A., van Diepen, A. E., Kapteijn, F. (2001) Catalyst deactivation: Is it predictable? What to do? *Appl. Catal. A* 212:3–16.

Muñoz, J. A. D., Alvarez, A., Ancheyta, J., Rodríguez, M. A., Marroquín, G. (2005) Process heat integration of a heavy crude hydrotreatment plant. *Catal. Today* 109:214–218.

Ouwerkerk, C. E. D., Bratland, E. S., Hagan, A. P., Kikkert, B. L. J. P., Zonnevylle, M. C. (1999) Performance optimisation of fixed bed processes. *Pet. Tech. Q.* 4(1):21–30.

Panariti, N., del Bianco, A., del Piero, G., Marchionna, M. (2000) Petroleum residue upgrading with dispersed catalysts: 1. Catalysts activity and selectivity. *Appl. Catal. A* 204:203–213.

Patel, R. H., Bingham, E., Christensen, P., Müller, M. (1998) Hydroprocessing reactor and process design to optimize catalyst performance. Presented at the First Indian Refining Roundtable, New Delhi, India, Dec. 1–2.

Pedersen, M. J., Sampath, V. R., Litchfield, J. F. (1995) Method and apparatus for mixing and distributing fluids in a reactor. U.S. patent 5,462,719.

Perry, R. H., Green, D. (1987) *Perry's Chemical Engineers' Handbook*, 6th ed. McGraw-Hill, New York.

Peyrot, C. F. (1987) Mixing device for vertical flow fluid–solid contacting. U.S. patent 4,669,890.

Quann, R. J., Ware, R. A., Hung, C., Wei, J. (1988) Catalytic hydrodemetallation of petroleum. *Adv. Chem. Eng.* 14:95–259.

Rana, M. S., Samano, V., Ancheyta, J., Diaz, J. I. A. (2007) A review of recent advances of hydroprocessing of heavy oils and residua. *Fuel* 86(9):1216–1231.

Robinson, P. R., Dolbear, G. E. (2006) Hydrotreating and hydrocracking: fundamentals. In: *Practical Advances in Petroleum Processing*, Hsu, C. H., Robinson, P. R. (eds.). Springer-Verlag, New York, Chap. 7.

Rodriguez, M. A., Ancheyta, J. (2004) Modeling of hydrodesulfurization (HDS), hydrodenitrogenation (HDN), and the hydrogenation of aromatics (HDA) in a vacuum gas oil hydrotreater. *Energy Fuels* 18:789–794.

Scheffer, B., van Koten, M. A., Robschlager, K. W., de Boks, F. C. (1988) The shell residue hydroconversion process: development and achievements. *Catal. Today* 43: 217–224.

Scheuerman, G. L., Johnson, D. R., Reynolds, B. E., Bachtel, R. W., Threlkel, R. S. (1993) Advances in Chevron RDS technology for heavy oil upgrading flexibility. *Fuel Process. Technol.* 35:39–54.

Schulman, B. L., Dickenson, R. L. (1991) Upgrading heavy crudes: a wide range of excellent technologies now available. Presented at the UNITAR 5th International Conference, Caracas, Venezuela, Aug., pp. 105–113.

Seidel, T., Dunbar, M., Johnson, B. G., Moyse, B. (2002) What a difference the tray made. AM-02-52. In: *Proceedings of the NPRA Annual Meeting*, San Antonio, TX, Mar. 17–19.

Shah, Y. T., Paraskos, J. A. (1975) Criteria for axial dispersion effects in adiabatic trickle bed hydroprocessing reactors. *Chem. Eng. Sci.* 30:1169–1176.

Sie, S. T., de Vries, A. F. (1993) Hydrotreating process. European patent application 0553920.

Speight, J. G. (1999) *The Chemistry and Technology of Petroleum*, 3rd ed. Marcel Dekker, New York.

Speight, J. G. (2000) *The Desulfurization of Heavy Oils and Residua*, 2nd ed. Marcel Dekker, New York.

Stefanidis, G. D., Bellos, G. D., Papayannakos, N. G. (2005) An improved weighted average reactor temperature estimation for simulation of adiabatic industrial hydrotreaters. *Fuel Process. Techn.* 86:1761–1775.

Swain, J., Zonnevylle, M. (2000) Are you really getting the most from your hydroprocessing reactors? Presented at the European Technology Conference, Rome, Nov. 15, 2000.

Tarhan, O. M. (1983) *Catalytic Reactor Design*. McGraw-Hill, New York.

Van Ginneken, Van Kessel, M. M., Pronk, K. M. A., Renstrom, G. (1975) Shell process desulfurizes resids. *Oil Gas J.* Apr., 28:59–63.

Van Vliet, W., Den Hartog, A. P., Den Hartog-Snoeij, M. (2006) Mixing device comprising a swirl chamber for mixing liquid. U.S. patent 7,078,002.

Yeary, D. L., Wrisberg, J., Moyse, M. (1997) Revamp for low sulfur diesel: a case study. *Hydrocarbon Eng.*, Sept., pp. 25–29.

NOMENCLATURE

a	Dimensionless number of Glaso's correlation (Table 3.9) or constant of Eq. (3.3)
a_L	Specific gas–liquid interface area

a_S	Liquid–solid interfacial area per unit volume of reactor
A	Aromatic compound
A_1	Lateral area of the geometric shape between lobes according to Table 3.5
A_2	Common area between each cylinder and each side of the shape between lobes, according to Table 3.5
API	API gravity
asp	Asphaltenes
Br No.	Bromide number
c_p	Mass heat capacity
C_i	Molar concentration of compound i
C_i^L	Molar concentration of compound i in liquid bulk phase
C_i^S	Molar liquid phase concentration of i at external surface of solid
d_c	Diameter of cylinder
d_p	Particle diameter
d_{pe}	Equivalent particle diameter
$d_{15.6}$	Specific gravity at 15.6°C
D_a^L	Axial mass dispersion coefficient in the liquid phase
$D_{M,i}^f$	Molecular diffusion coefficient of compound i in phase f
DA	Diaromatics
E_A	Activation energy
g_j	Gas mass flow rate of stream j
G_L	Liquid superficial flow
GHSV	Gas hourly space velocity
h_{GL}	Heat transfer coefficient for gas–liquid interface
h_{LS}	Heat transfer coefficient for liquid film surrounding the catalyst particle
HC	Hydrocarbon
H_i	Henry's law constant for compound i
H_2	Hydrogen
H_2S	Hydrogen sulfide
H_2/oil	Hydrogen-to-oil ratio
j_H	j-factor for heat transfer
k_f	Forward rate constant
k_j	Apparent rate constant of reaction j
k_L	Thermal conductivity of liquid phase
k_r	Reverse rate constant
k_0	Frequency or preexponential factor
k_1, k_2	Adjustable parameters of Eq. (3.12)
k_i^L	Gas–liquid mass transfer coefficient for compound i
k_i^S	Liquid–solid mass transfer coefficient for compound i
K_{ads}^i	Adsorption–equilibrium constant for compound i
K_i	Equilibrium constant

l_j	Liquid mass flow rate of stream *j*
L	Particle length
L_B	Catalyst bed length
L_p	Particle size, particle length
LHSV	Liquid hourly space velocity
m	Reaction order for hydrogen
MA	Monoaromatic
MW	Molecular weight
n_i	Reaction order for compound *i*
n_L	Number of lobes
naph	Naphthenes
N_B	Basic nitrogen compound
N_{NB}	Nonbasic nitrogen compound
NH_3	Ammonia
p_i	Partial pressure of compound *i*
p_i^j	Partial pressure of compound *i* in the *j* phase
P	Reactor total pressure
PA	Polyaromatic
$Pe_{a,m}^L$	Peclet number for axial mass dispersion in liquid phase
q	Quench fluid mass flow rate
r	Radius of sphere
r_c	Radius of cylinder
$r_{in,j}$	Intrinsic rate of reaction *j*
r_j	Rate of reaction *j*
r_p	Radius of particle
R	Universal gas law constant
S_p	Total geometric external area of particle
t	Time
T	Temperature
T_j	Temperature of phase *j*
T_{MeABP}	Mean average boiling point
T_0	Reference temperature or inlet reactor temperature
u_f	Superficial velocity of phase *f* in the reactor
v	Vapor fraction
v_c	Critical specific volume of the gaseous compounds
v_i	Molar volume of solute *i* at its normal boiling temperature
v_L	Molar volume of liquid solvent at its normal boiling temperature
ν_N	Molar gas volume at standard conditions
V_p	Total geometric volume of catalyst
VGO	Vacuum gas oil
W_c	Weight fraction of catalyst in the reactor
WABT	Weight-average bed temperature
WHSV	Weight hourly space velocity

x_i	Mass fraction of compound *i*
z	Axial position along the catalyst bed

Greek Letters

γ	Fraction of the easy-to-react containing compounds
ΔH_{ads}	Adsorption enthalpy of H_2S
ΔH_{Rj}	Heat of reaction *j*
ΔP	Pressure drop
$\Delta\rho_T$	Temperature correction of liquid density
$\Delta\rho_P$	Pressure dependence of liquid density
$\in_B$	Bed void fraction or bed porosity
ε_p	Particle porosity
ε_G	Gas-phase fraction
ε_L	Liquid holdup
ζ	Fractional volume of the catalyst bed diluted by inert particles
η_j	Catalyst effectiveness factor for reaction *j*
λ_i	Solubility coefficient of compound *i*
μ_f	Viscosity of phase *f*
ρ_B	Catalyst bulk density
ρ_{H_2}	Density of hydrogen
ρ_{oil}	Density of oil
ρ_L^{15}, ρ_0	Liquid density at standard conditions (15°C, 1 atm)
ρ_{20}	Liquid density at 20°C
υ_i	Stoichiometric coefficient of compound *i*

Subscripts

A	Aromatics
G	Gas phase
H_2	Hydrogen
H_2S	Hydrogen sulfide
HDA	Hydrodearomatization reaction
HDasp	Hydrodeasphaltenization reaction
HDC	Hydrocracking reaction
HDM	Hydrodemetallization reaction
HDN_B	Hydrodenitrogenation reaction of basic nitrogen
HDN_{NB}	Hydrodenitrogenation reaction of nonbasic nitrogen
HDO	Hydrodeoxygenation reaction
HDS	Hydrodesulfurization reaction
HGO	Hydrogenation of olefin reaction
in	Inlet stream to the following catalytic bed
L	Liquid phase or gas–liquid interface
N_B	Basic nitrogen
N_{NB}	Nonbasic nitrogen
NH_3	Ammonia
olef	Olefins

out	Outlet stream of the previous catalytic bed
q	Quench
S	Organic sulfur compound, solid phase or liquid–solid interface

Superscripts

G	Gas phase
L	Liquid phase or gas–liquid interface
q	Quench fluid
S	Solid phase or liquid–solid interface

4

MODELING OF CATALYTIC REFORMING

4.1 THE CATALYTIC REFORMING PROCESS

4.1.1 Description

Catalytic reforming is a chemical process used to convert petroleum naphtha, particularly low-octane-number straight-run naphtha into high-octane gasoline called *reformate*. In addition to producing reformate, catalytic reforming is also a primary source of aromatics used in the petrochemical industry (BTX: benzene, toluene, and xylenes).

Straight-run naphtha obtained directly from the atmospheric crude oil distillation column is a mixture of paraffins (saturated aliphatic hydrocarbons), naphthenes (saturated cyclic hydrocarbons containing at least one ring structure), and aromatics (hydrocarbons with one or more polyunsaturated rings) in the C_5–C_{12} range with a boiling range between 30 and 200°C, constituting typically 15 to 30 wt% of the crude oil, with some sulfur and small amounts of nitrogen. The typical feed to catalytic reforming is a mixture of straight-run naphthas: 30 to 90°C light naphtha (C_5 and C_6), 90 to 150°C medium-weight naphtha (C_7 and C_9), and 150 to 200°C heavy naphtha (C_9 and C_{12}). These distillation ranges of naphthas differ slightly from those described in Chapter 1 (Table 1.5), but they are more commonly used in catalytic reforming operations rather than those employed for crude oil international assays. The properties of naphthas for various Mexican crude oils are reported in Table 4.1.

Modeling and Simulation of Catalytic Reactors for Petroleum Refining, First Edition.
Jorge Ancheyta.
© 2011 John Wiley & Sons, Inc. Published 2011 by John Wiley & Sons, Inc.

TABLE 4.1. Properties of Naphthas with International Assay Distillation Ranges from Various Crude Oils

Property	Crude Oil				
	10°API	13°API	Maya	Isthmus	Olmeca
	Light Naphtha (TIE–71°C)				
sg, 60°F/60°F	0.6689	0.6659	0.6686	0.6504	0.6503
Total sulfur (wt%)	0.064	0.062	0.050	0.020	0.020
PIONA (vol%)					
n-Paraffins	41.12	39.46	47.75	48.98	47.15
i-Paraffins	44.62	36.20	39.15	40.66	43.63
Oleffins	0.26	0.56	0.00	0.01	0.00
Naphthenes	10.22	18.14	10.89	8.45	8.45
Aromatics	3.69	5.14	2.21	1.90	1.81
Benzene (vol%)	3.28	0.89	2.21	1.90	1.81
	Medium Naphtha (71–177°C)				
sg, 60°F/60°F	0.7550	0.7542	0.7512	0.7448	0.7408
Total sulfur (wt%)	0.432	0.412	0.200	0.030	0.030
PIONA (vol%)					
n-Paraffins	37.59	36.54	30.75	30.79	30.59
i-Paraffins	33.32	31.83	30.69	32.58	31.30
Oleffins	0.89	1.27	0.73	0.27	0.55
Naphthenes	16.65	17.03	19.35	19.74	18.41
Aromatics	10.54	11.18	18.48	16.62	19.15
Benzene (vol%)	0.52	0.55	0.32	0.38	0.46
	Heavy Naphtha (177–204°C)				
sg, 60°F/60°F	0.8012	0.8001	0.7928	0.7920	0.7912
Total sulfur (wt%)	1.589	1.511	0.600	0.100	0.050
Aromatics (vol%)	18.70	15.50	27.08	18.31	22.08

In commercial practice, the most preferred feed for catalytic reforming is naphtha with a boiling range of 85 to 165°C, since the light fraction (85°C–) is not a good feedstock, due to its composition (low-molecular-weight paraffins tending to crack to C_4^- and to be a precursor in benzene formation, which is undesirable because of environmental regulations), and the heavy fraction (180°C+) hydrocracks to excessive carbon laydown on the reformer catalyst.

Prior to catalytic reforming, the naphtha feed needs to be hydrotreated to reduce the impurities content (sulfur, nitrogen, and oxygen compounds) to acceptable levels, which if not removed will poison the reforming catalysts. This pretreatment is mandatory since the catalyst is gradually poisoned, leading to excessive coking and rapid deactivation.

Apart from straight-run naphthas, the following are other streams usually fed to reformer units, which have a boiling range similar to that of typical catalytic reforming feed, and come from a visbreaking unit, coking unit, hydrocracking/HDT unit, or FCC unit. They generally contain high amounts of sulfur, nitrogen, and olefins, which are mostly aromatic and difficult to hydrotreat.

- Visbreaker naphtha, which requires severe hydrotreating in order to prepare a proper reformer feedstock. That is why visbreaker naphtha is usually limited to small percentages of the feed reformer.
- Coker naphtha, whose properties are more or less the same as those of visbreaker naphtha but whose amount available from refineries is higher.
- Hydrocracked and hydrotreated naphtha, which is produced by hydrocracking and hydrotreating of heavier petroleum fractions. This naphtha is a suitable reformer feedstock since it is rich in naphthenes.
- FCC naphtha, which is produced by catalytic cracking of gas oils. Although not being a viable feed to catalytic reforming, some refineries use it, particularly the 75 to 150°C fraction.

The distribution of paraffins, olefins, naphthenes, and aromatics in the feed to catalytic reforming determines the richness of the feedstock, which is normally rated by its naphthenes + aromatics or naphthenes + 2 aromatics value. To convert low-quality naphthas, the catalytic reforming process rearranges (or restructures or reconstructs) the hydrocarbon molecules to form more complex molecular-shaped hydrocarbons with improved octane values. Although a certain degree of cracking occurs, the conversion is done without changing the boiling-point range of the feed. During this transformation, catalytic reforming produces significant amounts of hydrogen, which is used in other processes, such as hydrotreating and hydrocracking, as well as small amounts of methane, ethane, propane, and butanes.

A typical catalytic reforming unit consists of a feed system, several heaters, reactors in series, and a flash drum. Part of the flashed hydrogen is recycled to the feed before it enters the first heater, while the liquid is sent to the fractionation section (stabilizer). The reformate is obtained as a bottoms product from the stabilizer. Off-gas and liquefied petroleum gas (LPG) are recovered from the top of the stabilizer.

Since most of the reforming reactions are endothermic, several heaters are used to maintain the reactor temperature at the desired levels (400 to 500°C). As the feed flows through the catalytic bed in the reactor, the major reaction is the dehydrogenation of naphthenes to aromatics, which is fast and highly endothermic, resulting in a large decrease of temperature within the reactor. The product from the first reactor is reheated and fed to the following reactor. As the feed passes through the reactors in series, the reaction rates decrease and the reactors become larger, the reaction becomes less endothermic, and the temperature differential across them decreases, while the amount of heat required between the reactors also decreases.

4.1.2 Types of Catalytic Reforming Processes

Catalytic reforming processes are commonly classified according to the frequency and mode of catalyst regeneration, into (1) semiregenerative, (2) cyclic regeneration, and (3) continuous regeneration. The main difference among the three types of processes is the need of unit shutdown for catalyst regeneration in the case of a semiregenerative process, the use of an additional swing or spare reactor for catalyst regeneration for the cyclic process, and catalyst replacement during normal operation for the continuous regeneration type. Figure 4.1 illustrates the reaction section of the three types of catalytic reforming processes.

The most used process worldwide is the semiregenerative type, followed by continuous regeneration and by the less common cyclic regeneration. Currently, most catalytic reformers are designed with continuous regeneration, and the former semiregenerative plants are being revamped to operate as continuous regeneration.

Semiregenerative A semiregenerative catalytic reforming process usually has three or four reactors in series with a fixed-bed catalyst system and operates continuously (cycle length) from six months to one year. During this period, the activity of the catalyst diminishes due to coke deposition, provoking a decrease in aromatics yield and in hydrogen gas purity. To minimize the catalyst deactivation rate, the semiregenerative units operate at high pressure (200 to 300 psig). To compensate for catalyst activity decline and to keep conversion more or less constant, the reactor temperatures are increased continuously. When the end-of-cycle reactor temperatures are reached, the unit is shutdown and the catalyst is in situ regenerated. A catalyst cycle ends when the reforming unit is unable to meet its process objectives: octane and yield reformate. Catalyst regeneration is carried out with air as the source of oxygen. A catalyst can be regenerated five to ten times before it is removed and replaced.

Cyclic Regeneration Apart from the catalytic reforming reactors, the cyclic regeneration process has an additional swing reactor, which is used when the fixed-bed catalyst of any of the regular reactors needs regeneration. The reactor with the regenerated catalyst then becomes the spare reactor. By this means, the reforming process maintains continuous operation. Operating at lower pressure (~200 psig) allows the cyclic regeneration process to achieve higher reformate yield and hydrogen production. Compared with the semiregenerative type, in the cyclic regeneration process the overall catalyst activity varies much less with time, so that conversion and hydrogen purity are kept more or less constant during the entire operation. The main disadvantage of

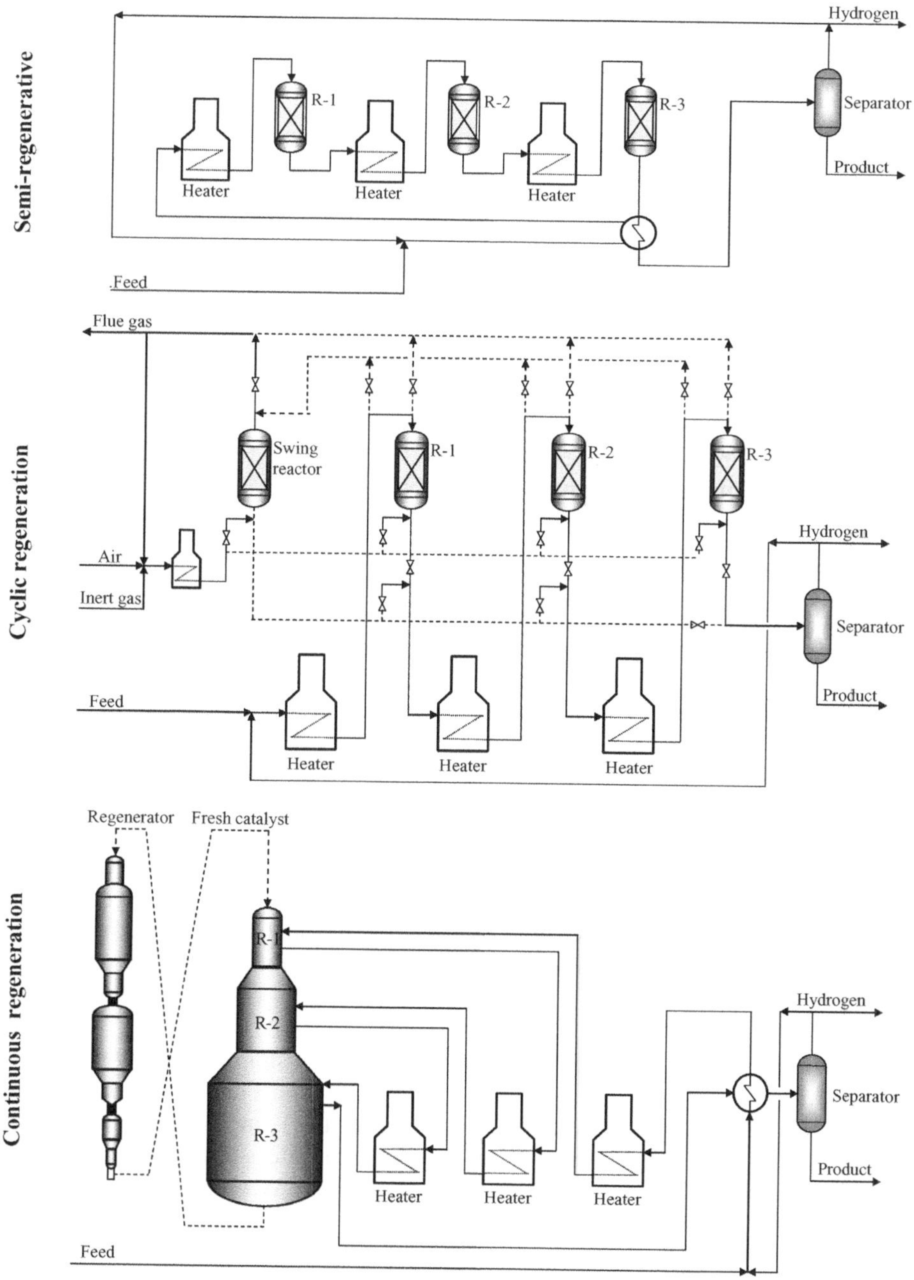

Figure 4.1. Reaction section of the catalytic reforming processes.

this type of catalytic reforming is the complex nature of the reactor switching policy, requiring high safety precautions. Also, to make switches between reactors possible, they need to be of the same maximal size.

Continuous Regeneration The deficiencies in cyclic regeneration reforming are solved by a low-pressure (50 psig) continuous regeneration process, which is characterized by high catalyst activity with reduced catalyst requirements, producing more uniform reformate of higher aromatic content and high hydrogen purity. This type of process uses moving-bed reactor design, in which the reactors are stacked. The catalyst bed moves by gravity flow from top to bottom of the stacked reactors. The spent catalyst is withdrawn from the last reactor and sent to the top of the regenerator to burn off the coke. The transport of catalyst between reactors and regenerator is done by the gas lift method. During normal operation, small quantities of catalyst are withdrawn continuously. Fresh or regenerated catalysts are added to the top of the first reactor to maintain a constant inventory of catalyst.

4.1.3 Process Variables

Similar to the hydrotreating process described in Chapter 3, in the catalytic reforming process there are four principal variables that affect the performance of the unit, either semiregenerative or continuously regenerative: reactor pressure, reactor temperature, space velocity, and H_2/oil molar ratio.

Pressure A reduction in the reactor pressure increases the hydrogen and reformate yield, decreases the required reactor temperature to achieve a constant product quality, and shortens the catalyst cycle by increasing the catalyst coking rate. Due to the pressure drop, the reactor pressure declines across the various reaction stages. The average pressure of the various reactors is generally referred to as the *reactor pressure*. Typical reactor pressures are 200 to 500 psig (semiregenerative and cyclic regeneration) and 60 to 150 psig (continuous regeneration).

Temperature The reaction temperature is the most important variable in catalytic reforming, since the product quality and yields are highly dependent on it. WABT (weighted-average bed temperature) and WAIT (weighted-average inlet temperature) are the two main parameters to express reforming reactor average temperature. The difference between WABT and WAIT is that the former represents the integrated temperature along the catalyst bed, and the latter is calculated with the inlet temperature of each reactor.

WABT is calculated as indicated in Chapter 3 [Eqs. (3.1) and (3.2)], and WAIT is determined as follows:

$$\mathrm{WAIT} = \sum_{i=1}^{N} \mathrm{WAIT}_i Wc_i \tag{4.1}$$

where $WAIT_i$ is the inlet temperature of each reactor, N the number of reactors, and Wc_i is the weight fraction of catalyst in each reactor bed with respect to the total. Semiregenerative units operate at a higher reactor temperature (450 to 525°C) than that of continuous regeneration units (525 to 540°C).

All reaction rates are increased when operating at high temperature. Hydrocracking, which is not desirable in catalytic reforming, occurs to a greater extent at high temperatures. Therefore, to obtain high product quality and yields, it is necessary to control the hydrocracking and aromatization reactions carefully. Reactor temperatures are monitored constantly to observe the extent of each of these reactions.

Space Velocity Both LHSV and WHSV are of typical use in catalytic reforming units to express space velocity. Space velocity and reactor temperature are commonly employed to set the octane of a product. The greater the space velocity, the higher the temperature required to produce a given product octane. The severity of the catalytic reforming unit can be increased either by increasing reactor temperature or by lowering the space velocity. Since the amount of catalysts loaded to the reactors is constant, the reduction of space velocity during operation can be reduced only by decreasing the feed flow rate.

H_2/Oil Ratio In contrast to the catalytic hydrotreating process, in which the H_2/oil ratio is reported in volumetric units [e.g., standard cubic feet of hydrogen per barrel of liquid feed (ft^3/bbl)], in the catalytic reforming process, this ratio is stated on a molar basis [i.e., moles of hydrogen in the recycle gas (a mixture of hydrogen and light gases) per mole of naphtha feed (mol/mol)]. Values of 4 to 6 mol/mol are typical in commercial reforming units. An increase in H_2/oil ratio causes an increase in the hydrogen partial pressure and removes coke precursors from the metal sites. The global effect of this is increased catalyst life. In other words, the rate of coke formation on the catalyst and thus catalyst stability and life is a function of the H_2/oil ratio and hydrogen partial pressure present in the reactor system.

4.2 FUNDAMENTALS OF CATALYTIC REFORMING

4.2.1 Chemistry

A large number of reactions occur in catalytic reforming over bifunctional catalysts, such as dehydrogenation and dehydroisomerization of naphthenes to aromatics, dehydrogenation of paraffins to olefins, dehydrocyclization of paraffins and olefins to aromatics, isomerization or hydroisomerization to isoparaffins, isomerization of alkylcyclopentanes and substituted aromatics, and hydrocracking of paraffins and naphthenes to lower hydrocarbons. All reactions are desirable except hydrocracking, which occurs to a greater extent at high temperature and converts valuable C_5^+ molecules (reformate) into light gases. Some examples of these reactions are shown in Figure 4.2. Most

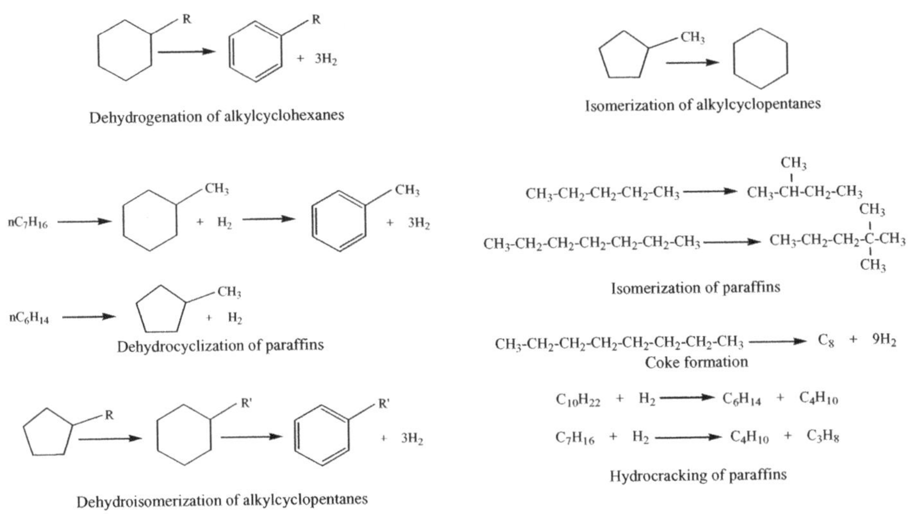

Figure 4.2. Examples of catalytic reforming reactions.

commonly, the reactions occurring during catalytic reforming are classified into the following four types.

1. *Dehydrogenation of naphthenes.* Naphthenes are present in reforming feeds in the form of cyclohexanes and cyclopentanes. Cyclohexanes are dehydrogenated to give aromatics, while cyclopentanes are first hydroisomerized to produce cyclohexanes, which are further dehydrogenated to aromatics. The dehydrogenation of naphthenes to aromatics is probably the most important reaction in catalytic reforming. It is highly endothermic, has the highest reaction rates, and produces hydrogen.
2. *Isomerization of* n-*paraffins.* Paraffins are isomerized to form branched-chain molecules (isoparaffins). Isomerization reactions are so fast that actual concentrations are near equilibrium. Isomerization of *n*-paraffins is a fairly rapid reaction with small heat effects. High H_2/hydrocarbon ratios reduce the hydrocarbon partial pressure and thus favor the formation of isomers. The isomerization of *n*-paraffins does not consume or produce hydrogen.
3. *Dehydrogenation and aromatization of paraffins.* Paraffins undergo dehydrocyclization to produce cycloparaffins. The dehydrocyclization reaction involves dehydrogenation and aromatization steps, and produces hydrogen.
4. *Hydrocracking of* n-*paraffins.* Paraffins are hydrocracked to form smaller molecules. This reaction is the only one that consumes hydrogen and is exothermic. Since it is relatively slow, most of the hydrocracking occurs in the final part of the reaction system.

4.2.2 Thermodynamics

The most rapid reactions (i.e., dehydrogenation of naphthenes) reach thermodynamic equilibrium, while the others are controlled by kinetics. Increasing the reaction temperature and lowering the pressure have both a positive effect on the reaction rate and thermodynamic feasibility as to the dehydrogenation of naphthenes (the most important reaction in catalytic reforming). The effect of these variables on thermodynamic equilibrium for the other reactions is slighter.

Table 4.2 summarizes the thermodynamic effect of the main reforming reactions. Other effects are the following:

- The dehydrogenation of naphthenes and paraffins is rapid and equilibrium concentrations are established in the initial portions of a catalyst bed.
- Olefins are readily hydrogenated, and at equilibrium only small concentrations can exist.
- The isomerization of paraffins is a sufficiently rapid reaction and primarily thermodynamically controlled, which means that actual concentrations are near equilibrium.
- The dehydrocyclization of paraffins is a much slower reaction and kinetically controlled.
- Hydrocracking rates increase with pressure and lower the reformate yield.
- Coking is very slow but increases rapidly at low hydrogen pressure and high temperature.

Thus, it is highly desired to operate reactors at high temperature and low pressure; however, catalyst deactivation due to coke deposition is also favored

TABLE 4.2. General Thermodynamic Comparison of the Major Catalytic Reforming Reactions

	Rate of Reaction	Heat of Reaction	Thermodynamic Equilibrium
Naphthene dehydrogenation	Very fast	Very endothermic	Reached
Naphthene isomerization	Fast	Mildly exothermic	Reached
Paraffin isomerization	Fast	Mildly exothermic	Reached
Paraffin dehydrocyclization	Slow	Very endothermic	Not reached
Paraffin dehydrogenation	Very fast	Endothermic	Not reached
Hydrocracking	Very slow	Exothermic	Not reached

at those conditions. In addition, lowering hydrogen partial pressure results in an increase in the aromatization rate and a decrease in the rate of hydrocracking.

4.2.3 Kinetics

Various kinetic models for catalytic reforming reactions that have been reported in the literature have been the subject of several reviews. The level of sophistication of these models varies from a few lumps to detailed kinetic models and is related to the development of high-speed hardware and large-capacity computers. Since an exhaustive and critical review of the reported kinetic models in outside the scope of this chapter, only a brief mention of the most relevant models will be made. The chronological evolution of the catalytic reforming kinetic modeling is presented in Figure 4.3.

The first attempts to model the kinetics of catalytic reforming reactions were reported more than 50 years ago. The oldest kinetic model, proposed by Smith (1959), divided naphtha feed into three types of hydrocarbons: paraffins, naphthenes, and aromatics. Each of these three hydrocarbon classes is represented by a single compound that has the average properties of that class. No distinction is made on the basis of the number of carbon atoms within each class. A kinetic analysis is developed which describes the reforming operation with satisfactory accuracy. Hydrogen and light gases (ethane, propane, and butane) are also taken into account in the model. This model thus involves five pseudocomponents: paraffins, naphthenes, aromatics, light gases, and

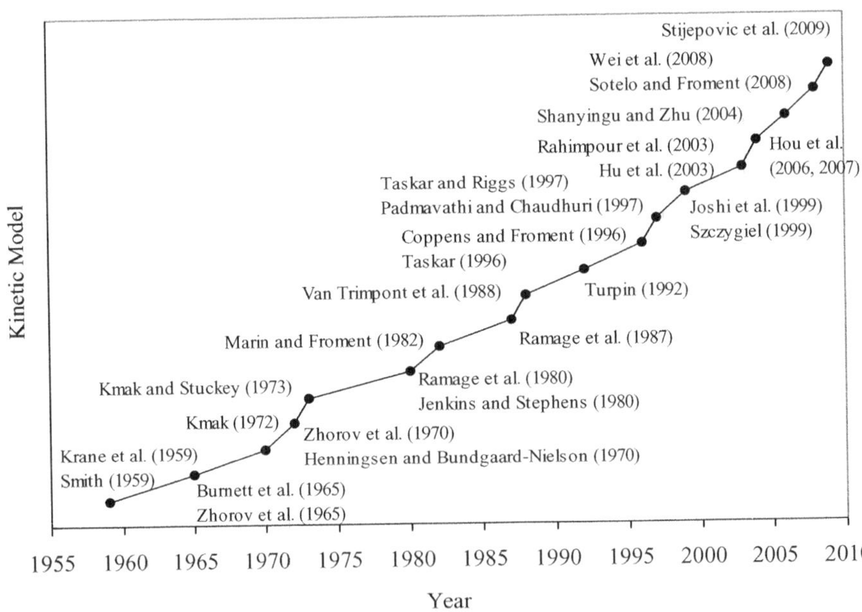

Figure 4.3. Evolution of kinetic modeling for catalytic reforming.

hydrogen. This seems to be the first attempt to "delump" naphtha into various constituents. To simplify the catalytic reforming system, the following four reactions were considered:

Dehydrogenation of naphthenes to aromatics

$$\text{naphthenes } (C_nH_{2n}) \underset{k_{r1}}{\overset{k_{f1}}{\rightleftarrows}} \text{aromatics } (C_nH_{2n-6}) + 3H_2 \tag{4.2}$$

Hydrogenation of naphthenes to paraffins

$$\text{naphthenes } (C_nH_{2n}) + H_2 \underset{k_{r2}}{\overset{k_{f2}}{\rightleftarrows}} \text{paraffins } (C_nH_{2n+2}) \tag{4.3}$$

Hydrocracking of paraffins to lower hydrocarbons

$$\text{paraffins } (C_nH_{2n+2}) + \frac{n-3}{3}H_2 \xrightarrow{k_{f3}} \frac{n}{15}\sum_{i=1}^{5} C_iH_{2i+2} \tag{4.4}$$

Hydrocracking of naphthenes to lower hydrocarbons

$$\text{naphthenes } (C_nH_{2n}) + \frac{n}{3}H_2 \xrightarrow{k_{f4}} \frac{n}{15}\sum_{i=1}^{5} C_iH_{2i+2} \tag{4.5}$$

where

$$\frac{n}{15}\sum_{i=1}^{5} C_iH_{2i+2} = \frac{n}{15}CH_4 + \frac{n}{15}C_2H_6 + \frac{n}{15}C_3H_8 + \frac{n}{15}C_4H_{10} + \frac{n}{15}C_5H_{12} \tag{4.6}$$

Reaction rate equations together with equilibrium and rate constants (k_{fi} for the forward reaction and k_{ri} for the reverse reaction) as a function of temperature (T), partial pressure (p_i), total pressure (P_T), and the inverse of space velocity (1/SV) are summarized in Table 4.3. From this table the following mass balance for paraffins (P), naphthenes (N), aromatics (A), and light ends (C_1–C_5) can be derived:

$$r_P = \frac{dP}{d(1/SV)} = -k_{f2}p_Np_{H_2} + \frac{k_{f2}}{K_{P_2}}p_P - k_{f3}\frac{p_P}{P_T} \tag{4.7}$$

$$r_N = \frac{dN}{d(1/SV)} = -k_{f1}p_N + \frac{k_{f1}}{K_{P_1}}p_Ap_{H_2}^3 - k_{f2}p_Np_{H_2} + \frac{k_{f2}}{K_{P_2}}p_P - k_{f4}\frac{p_N}{P_T} \tag{4.8}$$

$$r_A = \frac{dA}{d(1/SV)} = k_{f1}p_N - \frac{k_{f1}}{K_{P_1}}p_Ap_{H_2}^3 \tag{4.9}$$

$$r_{C_1-C_5} = \frac{d(C_1-C_5)}{d(1/SV)} = k_{f3}\frac{p_P}{P_T} + k_{f4}\frac{p_N}{P_T} \tag{4.10}$$

TABLE 4.3. Kinetic Equations of the Smith (1959) Model[a]

Reaction	Reaction Rate	Equilibrium Constant, K_P		Rate Constant for the Forward Reaction k_f
$N \underset{k_{r1}}{\overset{k_{f1}}{\rightleftarrows}} A + 3H_2$	$-\frac{dN}{d(1/SV)} = k_{f_1} p_N - \frac{k_{f_1}}{K_{P_1}} p_A p_{H_2}^3$	$K_{P_1} = \frac{k_{f_1}}{k_{r_1}} = \frac{p_A p_{H_2}^3}{p_N}$	$K_{P_1} = e^{46.15-46,045/T}$	$k_{f_1} = e^{23.21-34,750/T}$
$N + H_2 \underset{k_{r2}}{\overset{k_{f2}}{\rightleftarrows}} P$	$-\frac{dN}{d(1/SV)} = k_{f_2} p_N p_{H_2} - \frac{k_{f_2}}{K_{P_2}} p_P$	$K_{P_2} = \frac{k_{f_2}}{k_{r_2}} = \frac{p_P}{p_N p_{H_2}}$	$K_{P_2} = e^{-7.12+8000/T}$	$k_{f_2} = e^{35.98-59,600/T}$
$P + H_2 \xrightarrow{k_{f3}} (C_1 - C_5)$	$-\frac{dP}{d(1/SV)} = k_{f_3} \frac{p_P}{P_T}$	—	—	$k_{f_3} = e^{42.97-62,300/T}$
$N + H_2 \xrightarrow{k_{f4}} (C_1 - C_5)$	$-\frac{dN}{d(1/SV)} = k_{f_4} \frac{p_N}{P_T}$	—	—	$k_{f_4} = e^{42.97-62,300/T}$

[a]Units: K_{P_1} atm^3; K_{P_2} atm^{-1}; k_{f_1} mol/h · lb$_{cat}$ · atm^2; k_{f_2} mol/h · lb$_{cat}$ · atm^2; k_{f_3} mol/h · lb$_{cat}$; k_{f_4} mol/h · lb$_{cat}$.

Krane et al. (1959) proposed another more extensive attempt to model catalytic reforming reactions of whole naphtha, which consisted of a reaction network of 20 pseudocomponents with hydrocarbons ranging from 6 to 10 carbon atoms, as well as the difference between paraffins, naphthenes, and aromatics within each carbon number group, which undergo 53 reaction steps. Krane's proposed reaction network can be summarized as follows:

Paraffins

$$P_n \rightarrow N_n \tag{4.11}$$

$$P_n \rightarrow P_{n-i} + P_i \tag{4.12}$$

Naphthenes

$$N_n \rightarrow A_n \tag{4.13}$$

$$N_n \rightarrow N_{n-i} + P_i \tag{4.14}$$

$$N_n \rightarrow P_n \tag{4.15}$$

Aromatics

$$A_n \rightarrow A_{n-i} + P_i \tag{4.16}$$

$$A_n \rightarrow P_n \tag{4.17}$$

$$A_n \rightarrow N_n \tag{4.18}$$

All reactions are represented by a pseudo-first-order rate equation with respect to hydrocarbon concentration. Reaction rate constants were derived from experiments with whole naphtha. More details about the reaction rate equations of the Krane et al. (1959) model will be given in later sections of this chapter, together with various improvements on it.

Diverse modifications and applications of these two pioneer works have been reported in scientific papers. For example, Smith's model was modified by Viñas et al. (1996) to include discrimination between the reaction rates for aromatization of five- and six-ring naphthenes, two types of paraffins with different reactivities, and an overall hydrodealkylation reaction. Bommannan et al. (1989) estimated the values of activation energies from two sets of plant data using Smith's model. Dorozhov (1971) made a distinction between paraffins C_5–C_6 and paraffin C_7 in order to improve the model, which became more complicated, and its predictability capacity was only slightly better. Moharir et al. (1979) incorporated a deactivation function for both acidic and metallic functions of the catalyst into Smith's model, in order to simulate and optimize a naphtha catalytic reforming plant.

Lee et al. (1997) and Lid and Skogestad (2008) modeled a catalytic naphtha reformer with continuous catalyst regeneration using Smith's model with the goal of determining optimal operating conditions. In the work of Lid and Skogestad (2008), the process model is fitted to 21 data sets collected in a two-year period from a commercial naphtha reformer. More recently, Liang et al. (2005) used Smith's model to develop a physical model to simulate a naphtha catalytic reforming radial flow reactor unit with four reactors in series. Kinetics and thermodynamics equations were selected to describe the reforming reactions based on idealizing the complex naphtha mixture by representing the paraffin, naphthene, and aromatic groups by single compounds.

Similarly, Krane's model was refined by Ancheyta et al. (1994, 2000, 2001, 2002) to account for the temperature and pressure effects on the rate constants by an Arrhenius-type equation; to extend the naphtha composition to paraffins, naphthenes, and aromatic hydrocarbons with 11 atoms of carbon; for the inclusion of paraffin isomerization reactions; and for more accurate determination of benzene formation by adding the isomerization reaction of methylcyclopentane to cyclohexane. The modified model includes 71 reactions. The kinetic parameter values were estimated using experimental information obtained in a bench-scale fixed-bed reactor. This model was incorporated in a fixed-bed one-dimensional pseudohomogeneous adiabatic reactor model.

Burnett et al. (1965) proposed a pseudo-first-order kinetic model involving only hydrocarbons with seven atoms of carbon. Zhorov et al. (1965, 1970) incorporated the relationship between the reaction rate constants and the composition of the naphtha feed in a kinetic model consisting of C_5 and C_6 lumps and direct formation of aromatics from paraffins. Henningsen and Bundgaard-Nielson (1970) proposed a different treatment for the C_5 and C_6 ring naphthenes and expressed the reaction rate constants in the form of an Arrhenius-type equation to account for the influence of temperature. Catalyst deactivation was also included in the model. Values for the heat of reaction and activation energies were also provided. Kmak (1972) described a model that incorporated the naphtha catalytic reforming reactions with Hougen–Watson–Langmuir–Hinshelwood kinetics, in which the rate equations account explicitly for the interaction of chemical species with the catalyst. Later, Kmak and Stuckey (1973) used pure components, mixtures, and naphtha feed to develop a detailed catalytic reforming kinetic model over a wide range of reaction conditions. They used this model to simulate the power-forming process within a wide range of operating conditions. The model was capable of determining the concentration profiles of 22 components in four reactors in series.

Ramage et al. (1980, 1987) developed a detailed kinetic model based on extensive studies with pure components and a narrow-boiling fraction of naphthenes in a pilot-plant reactor. The kinetic model involves a reasonable number of lumps and pathways, captures the reactivity differences between particular feeds, and incorporates catalyst deactivation by coke formation. The model considered C_6–C_8 lumps of naphthenes, paraffins, and aromatics and was able to predict interactions between 13 lumps that undergo reactions of hydrocracking, hydrogenation–dehydrogenation, cyclization, and isomerization. Reversible reactions were assumed for hydrocarbons with equal carbon-atom numbers, and irreversible reactions for those between hydrocarbons with different numbers of carbon atoms. Jenkins and Stephens (1980) employed first-order rate equations, including reversible ones, to develop a kinetic model with 78 reactions involving 31 components. The effect of pressure on the reaction rates was simulated by means of a pressure factor with a characteristic exponent for each particular reaction.

Marin and Froment (1982) developed a kinetic model for the catalytic reforming of naphtha by first studying C_6 reforming and then C_7 reforming (Van Trimpont *et al.*, 1988). The model considered 5 to 10 carbon atoms and a reaction network including 23 pseudocomponents and used Hougen–Watson rate equations. Turpin (1992) combined fractionation modeling with kinetic modeling of the reforming processing to determine how best to meet processing requirements associated with the benzene content in reformulated gasoline.

Taskar (1996) and Taskar and Riggs (1997) employed a rigorous kinetic model to optimize the performance of an industrial catalytic reforming plant by studying operating modes and the influence of operational variables. They develop a more detailed kinetic model involving 35 pseudocomponents. Coppens and Froment (1996) improved catalytic reforming models by including diffusional effects in the rate equations. In the model proposed the porous nature of the catalyst support is approximated by a self-similar fractal structure.

Padmavathi and Chaudhuri (1997) developed a simulation model to monitor the performance of a commercial plant, in which details were given as to how the feed and the reacting scheme were lumped, as well as details on parameter estimation and model validation. They proposed a lumped kinetic model with 26 pseudocomponents.

Szczygiel (1999) investigated the kinetics of catalytic reforming by making use of pure components as feed. He reported an algorithm to optimize the porous structure of the reforming catalyst, consisting of three major steps: (1) analysis of kinetic phenomena in the catalyst grain, (2) analysis of diffusion phenomena in the catalyst grain, and (3) construction of a mathematical model to optimize the parameter values for the porous structure of the reforming catalyst grain. For the kinetic analysis, the paraffin reaction paths and the kinetic scheme are determined based on experiments in a flow-through catalytic reactor. It is proposed that the kinetic model be used to optimize the catalyst pore structure.

Joshi et al. (1999) proposed a rigorous pathway-level approach for modeling catalyst reforming consisting of 79 components with 464 reactions. Rahimpour et al. (2003) presented a kinetic and deactivation model for the simulation of an industrial naphtha catalytic reforming unit. Hu et al. (2003) reported a kinetic model for catalytic reforming with 17 lumps and 17 reactions. The

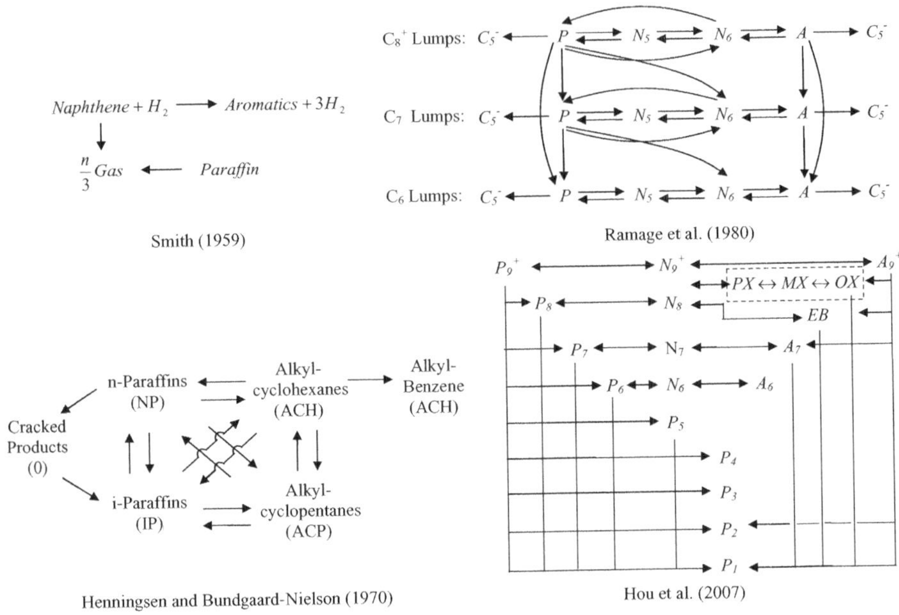

Figure 4.4. Examples of some reaction schemes used to develop catalytic reforming kinetic models.

adsorption and chemical lumps on the catalyst surface and deactivation of catalyst due to coking are also taken into account. Later, Hou et al. (2006, 2007) subdivided the eight-carbon aromatics lump of the Hu et al. (2003) model into their four isomeric compounds: PX (*para*-xylene), MX (*meta*-xylene), OX (*ortho*-xylene), and EB (ethylbenzene). Stijepovic et al. (2009) developed a general framework for modeling the catalytic reforming process. A semiempirical kinetic model was proposed consisting of 18 lumps based on paraffins, *i*-paraffins, naphthenes, and aromatics. Different values of activation energies were considered for each reaction. The model parameters were estimated by benchmarking with industrial data. The model is able to predict the concentration of hydrogen and light gases. Shanyinghu and Zhu (2004) presented a model involving several reactions to illustrate molecular modeling of the naphtha reforming process. Figure 4.4 shows reaction schemes used in the development of some of the kinetic models described above. The rate equations, values of kinetic parameters, properties of the catalyst and feedstock used during experiments, and other details may be found in the respective references.

More recently, Wei et al. (2008) developed an approach to modeling the reaction kinetics of the catalytic reforming by introducing a number of representative pseudocomponents by Monte Carlo simulation. By this means the complexity of the feed was reduced and a reaction network of this synthetic feed was generated by computer using graph theory. Sotelo and Froment

(2009) introduced a fundamental kinetic model for the catalytic reforming process. The model is based on the fundamental chemistry occurring on both the acid and metal sites of a Pt–Sn/Al_2O_3 catalyst. The single–event concept was applied in the development of rate expressions for the elementary steps on the acid sites. The kinetic model was used in pseudohomogeneous and heterogeneous reactor models for the simulation of a commercial adiabatic catalytic reforming unit with three reactors in series with centripetal radial flow.

As a summary of the state of the art in kinetic modeling for naphtha catalytic reforming, it can be observed that on the one hand, most published kinetic models based on the lumping approach report the rate constants to be dependent on feed and catalyst properties. Some models are not capable of predicting the composition of alkylcyclopentanes, the composition of *n*-paraffins and *i*-paraffins, the detailed composition of hydrocracking reaction products, nor the entire range of hydrocarbons present in naphtha composition. The level of sophistication varies from just a few lumps to a very detailed kinetic model. On the other hand, sophisticated models based on fundamental approaches (e.g., a single-event kinetic model), although overcoming some drawbacks of the lumping models still have to be validated under conditions (e.g., other feedstocks) different from those under which they were derived, and provide a more convincing comparison with industrial reality. Also, lumping models involve a reduced number of kinetic parameters and require relatively small amounts of experimental data for their estimation, whereas fundamental detailed models are quite complicated, with a large number of parameters, and frequently need more experiments. Therefore, a kinetic modeler faces a considerable dilemma: One uses either a lumped-kinetic model or a more fundamental approach. The decision is not an easy one to make. However, there are some important points that affect this decision. Most of the time, the model needs to simulate a commercial unit and anticipate the effect on product yield and quality of minor changes in process parameters. If a model with only a few lumps is chosen, the predictive capability surely is not sufficient to represent the desired situation. But if a detailed mechanistic model is selected, it may be too complex to implement, not because of the solution of the model, which with modern computers and algorithms has become a relatively easy task, but due to the cost and amount of experimental information needed to determine the model parameters. Thus, why not use an intermediate approach that maintains the simplicity of the lumping approach and is detailed enough to correctly predict the behavior of a commercial catalytic reforming plant? By an intermediate approach we mean that the number of lumps is such that the composition of the product is predicted with all the components desired. The answer to this question is the reason that lumped kinetic models are still commonly used to characterize reactive groups and to describe the reaction kinetics of complex processes in a tractable manner. It is the general conclusion of all the published scientific papers that the lumping approach is sufficiently reliable in describing the relationships between process variables and reaction rates. Comparisons of simulated results using lumping kinetic models with data obtained at different scales, including those on a commercial scale, support this conclusion.

4.2.4 Catalysts

Catalytic reforming reactions are conducted in the presence of hydrogen over hydrogenation–dehydrogenation catalysts. The dual function of the reforming catalysts is provided by (1) the acid centers of the support (alumina or silica–alumina), and (2) the metallic centers (platinum with other metals dispersed on the support, e.g., rhenium). For maximum catalyst efficiency, a proper balance between the acidic and dehydrogenating functions must be achieved. The activity of platinum is inhibited by sulfur, which adsorbs reversibly on the platinum crystallites. That is why the catalytic reforming feed needs to be hydrotreated to lower its sulfur content to <1 wppm. Water must also be kept at a low content to avoid leaching of chloride and thus loss of acid strength.

In addition to Pt, modern multimetallic catalysts contain highly dispersed rhenium (Re) and in some cases tin (Sn). In fact, two catalyst formulations prevail commercially: Pt–Re/Al_2O_3 and Pt–Sn/Al_2O_3. Pt–Re catalyst is the most stable and is preferred in semiregenerative units, while Pt–Sn has the highest selectivity at low pressure and is the best choice for continuous regeneration reforming units. γ-Alumina is the most common reforming catalyst support. The platinum must be dispersed over the alumina surface such that the maximum number of active sites for dehydrogenation is available. To achieve the appropriate acidic level of the catalyst, chloride is added prior to use.

The noble metals (platinum and rhenium) are considered to be catalytic sites for the dehydrogenation and hydrogenolysis reactions, and chlorinated alumina provides the acid sites needed for isomerization, cyclization, and hydrocracking reactions. Acid-catalyzed reactions together with the Pt-catalyzed dehydrogenation function are largely responsible for hydroisomerization reactions that lead to the formation of aromatics. Paraffins may be isomerized over the acidic function of the catalyst to provide higher-octane branched paraffins. Another acid-catalyzed reaction is paraffin hydrocracking to produce lighter products. Paraffins also undergo cyclization to cyclohexanes, which is believed to proceed through an olefin intermediate, produced by Pt-catalyzed dehydrogenation. The cyclization of the olefin may be catalyzed by the alumina support. During normal operation, the activity of the catalyst is reduced over time by coke deposition. Coking results from secondary reactions of the hydrocarbons, particularly olefins. The catalyst activity can be restored periodically by regeneration (coke burn-off) at high temperature. Coke burning is usually done with air at 400 to 500°C. Normally, the catalyst can be regenerated three or four times. The catalysts used in semiregenerative units must have long catalyst life cycles. In continuous regeneration units, the catalyst flows through the reactors and is regenerated continuously in a separate regeneration vessel that is part of the reactor–regenerator loop.

In semiregenerative units, to compensate for catalyst deactivation, the reactor temperature is increased constantly until the maximum allowable value is reached, or the selectivity to desired products is too much reduced, or the octane of the liquid product is declining. When this happens, the catalytic reforming unit requires a shutdown for regeneration every 6 to 12 months. This relatively long cycle can be obtained by operating under milder conditions of high partial pressure of hydrogen, lower-reactor temperatures, and lower-octane products.

In continuous regeneration units, portions of the catalyst are continually being regenerated outside the process and returned to the reactors so that high selectivity and activity are maintained. Continuous units operate under severe conditions to yield high-octane high-aromatics production, at low hydrogen partial pressures and higher reactor temperatures.

The catalysts are different depending on the type of reforming unit. For semiregenerative units the catalysts contain platinum or platinum modified by rhenium or, to a lesser extent, iridium. The support is most often γ-alumina, although some catalysts are supported on η-alumina. Rhenium or iridium is used to enhance the life of the catalyst. There are two main shapes of catalysts, cylindrical and spherical. The density of the catalysts varies in the range 0.5 to 0.8 g/cm^3. For continuous regeneration units, spherical catalysts are used to transport catalyst from the reactors to the regenerator, and back. The use of spherical catalysts rather than extrudate is to avoid catalyst dusting and breakage. In continuous regeneration units the catalyst is circulated at a rate of about one regeneration per week. Platinum and tin on γ alumina are the typical catalysts used in these units. The tin is used to reduce the hydrogenolysis activity of the platinum and to improve reformate yields. There are reports in the literature of new catalysts used to increase yields of reformate octane and to reduce coke production.

4.3 REACTOR MODELING

4.3.1 Development of the Kinetic Model

The kinetic model used for the simulation of catalytic reforming reactors is an extension of that reported by Krane et al. (1959), which utilizes lumped mathematical representation of the reactions that take place. These representations are written in terms of isomers of the same nature (paraffins, naphthenes, or aromatics). These groups range from 1 to 10 carbon atoms for paraffins, and from 6 to 10 carbon atoms for naphthenes and aromatics. The original model reported by Krane et al. (1959) includes 53 chemical reactions, which are summarized in Table 4.4.

TABLE 4.4. Chemical Reactions Considered in the Original Kinetic Model Reported by Krane et al. (1959), Activation Energies, and Factors for Pressure Effect for Each Reforming Reaction

Reaction[a]	Number of Reactions	E_{Aj} (kcal/mol)
Paraffins		
$P_n \rightarrow N_n$	4	45
$P_n \rightarrow P_{n-i} + P_i$	21	55
Subtotal	25	
Naphthenes		
$N_n \rightarrow A_n$	5	30
$N_n \rightarrow N_{n-i} + P_i$	6	55
$N_n \rightarrow P_n$	5	45
Subtotal	16	
Aromatics		
$A_n \rightarrow A_{n-i} + P_i$	5	40
$A_n \rightarrow P_n$	4	45
$A_n \rightarrow N_n$	1	30
Subtotal	10	

Reaction	α_k
Isomerization	0.370
Dehydrocyclization	–0.700
Hydrocracking	0.433
Hydrodealkylation	0.500
Dehydrogenation	0.000

[a] n, number of atoms of carbon ($1 \leq i \leq 5$).

To account for more reactions and have better naphtha composition predictability, the original model was modified in several ways, described below.

Kinetic Parameters for Hydrocarbons with 11 Atoms of Carbon Typical naphthas used as feed in catalytic reforming include hydrocarbons with up to 11 atoms of carbon, as can be seen in Table 4.5. The original Krane et al. (1959) model considers reactions only for hydrocarbons with 10 atoms of carbon. To maintain the original values of kinetic parameters, it was assumed that the hydrocarbons reported as having 10 atoms of carbon are, in fact, a lump of hydrocarbons with 10 and 11 atoms of carbon: $C_{10}^{+} = C_{10} + C_{11}$. In such a way, the different hydrocarbon species can be delumped as follows:

$$P_{10}^{+} = P_{10} + P_{11} \tag{4.19}$$

$$N_{10}^{+} = N_{10} + N_{11} \tag{4.20}$$

$$A_{10}^{+} = A_{10} + A_{11} \tag{4.21}$$

TABLE 4.5. Typical Compositions of Various Feeds to Catalytic Reforming Process

	Naphthas						Average Value
	1	2	3	4	5	6	
n-Paraffins							
C_4	0	1.568	0	0	0	0	0.261
C_5	1.818	11.368	9.818	10.362	1.983	1.392	6.124
C_6	9.633	8.034	8.356	8.412	9.467	9.477	8.897
C_7	8.116	6.778	7.114	7.148	8.386	8.402	7.657
C_8	6.464	5.326	5.602	5.616	6.640	6.683	6.055
C_9	4.454	3.514	3.858	3.809	4.625	4.68	4.157
C_{10}	1.640	1.403	1.707	1.635	1.948	2.066	1.733
C_{11}	0.297	0.266	0.321	0.292	0.318	0.370	0.311
Subtotal	32.422	38.257	36.776	37.274	33.367	33.07	35.194
i-Paraffins							
C_4	0	0.076	0	0	0	0	0.013
C_5	0.565	6.459	3.191	3.771	0.794	0.453	2.539
C_6	8.868	7.289	7.548	7.495	5.373	5.413	6.998
C_7	6.779	5.676	5.973	5.965	6.943	6.963	6.383
C_8	7.070	5.897	6.310	6.187	7.289	7.344	6.683
C_9	6.241	5.066	5.499	5.311	6.448	6.509	5.846
C_{10}	3.526	2.84	3.384	3.221	3.899	4.402	3.545
C_{11}	0.212	0.203	0.281	0.254	0.289	0.374	0.269
Subtotal	33.261	33.506	32.186	32.204	31.035	31.458	32.275
Naphthenes							
C_5	0.897	0.977	0.973	0.978	0.333	0.286	0.741
C_6	5.069	4.345	4.435	4.434	5.226	5.166	4.779
C_7	6.934	6.038	6.071	6.065	7.179	7.157	6.574
C_8	5.112	4.307	4.593	4.565	5.320	5.461	4.893
C_9	1.842	1.535	1.655	1.578	1.938	1.970	1.753
C_{10}	0.495	0.398	0.558	0.492	0.561	0.641	0.524
C_{11}	0.096	0.085	0.106	0.099	0.105	0.125	0.103
Subtotal	20.445	17.685	18.391	18.211	20.662	20.806	19.367
Aromatics							
C_6	1.393	1.074	1.200	1.199	1.380	1.351	1.266
C_7	3.506	2.676	3.024	3.038	3.634	3.576	3.242
C_8	5.326	4.015	4.529	4.542	5.507	5.428	4.891
C_9	2.908	2.186	2.956	2.671	3.488	3.218	2.905
C_{10}	0.707	0.569	0.903	0.830	0.891	1.056	0.826
C_{11}	0.032	0.032	0.035	0.031	0.036	0.037	0.034
Subtotal	13.872	10.552	12.647	12.311	14.936	14.666	13.164

The reaction rate equation originally reported for each hydrocarbon can then be expressed as

$$\frac{dC_{10}^{+}}{d(1/\mathrm{SV})} = k_{10}^{+} C_{10}^{+} = k_{10}^{+}(C_{10} + C_{11}) \tag{4.22}$$

This equation can also be written as a function of C_{10} and C_{11} and their individual kinetic parameters (k_{10} and k_{11}):

$$\frac{dC_{10}^{+}}{d(1/SV)} = k_{10}C_{10} + k_{11}C_{11} \tag{4.23}$$

By combining Eqs. (4.22) and (4.23), the following relationship can be derived:

$$k_{10} = \frac{k_{10}^{+}(R+1)}{R+K} \tag{4.24}$$

where

$$R = \frac{C_{10}}{C_{11}} \tag{4.25}$$

$$K = \frac{k_{11}}{k_{10}} \tag{4.26}$$

Equations (4.24) to (4.26) can be used to calculate the values of the individual kinetic parameters for hydrocarbons with 10 and 11 atoms of carbon, k_{10} and k_{11}, respectively. For this calculation, the relationships defined by Eqs. (4.25) and (4.26) are needed.

To determine the value of the constant R for each hydrocarbon type [(Eq. (4.25)], typical feed used in a commercial catalytic reforming unit was analyzed during different periods of time. Feed compositions are reported in Table 4.5. From this table, the corresponding values of R for this specific feed are

$$R_P = \frac{P_{10}}{P_{11}} = 9.109 \tag{4.27}$$

$$R_N = \frac{N_{10}}{N_{11}} = 5.106 \tag{4.28}$$

$$R_A = \frac{A_{10}}{A_{11}} = 24.414 \tag{4.29}$$

For calculation of R_P, the total amount of paraffin was used: that is, the sum of n- and i-paraffins. The values of K [Eq. (4.26)] were obtained for each reaction by extrapolation of the relationships calculated with the original kinetic parameters reported by Krane et al. (1959) as a function of the number of atoms of carbon (e.g., k_7/k_6, k_8/k_7, k_9/k_8, and k_{10}/k_9). Figure 4.5 illustrates this procedure for two reactions of hydrocracking of paraffins to paraffins with fewer carbon atoms. Details on the final set of individual kinetic parameters for the reactions of hydrocarbons with 10 and 11 atoms of carbon are provided in Table 4.6.

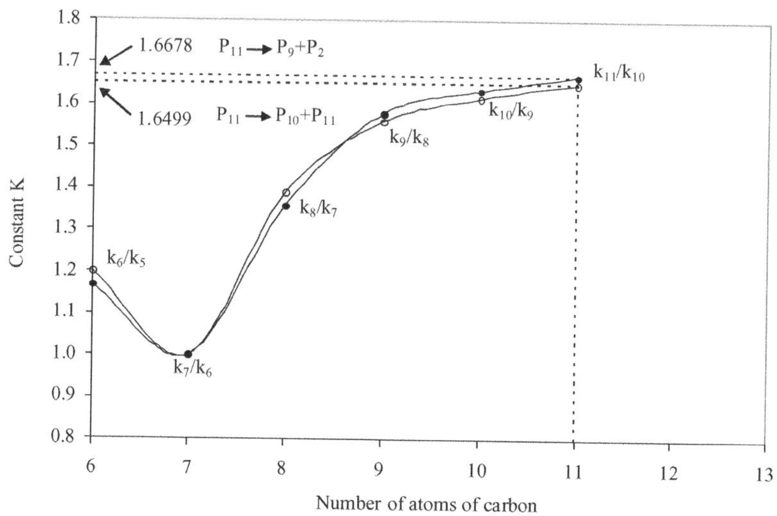

Figure 4.5. Example of the extrapolation procedure to calculate the constant K.

Reactions for the Formation of Benzene The original model proposed by Krane et al. (1959) does not take into account either the formation of cyclohexane (N_6) via methylcyclopentane (MCP) isomerization ($MCP \leftrightarrow N_6$) or the production of MCP from P_6 ($P_6 \leftrightarrow MCP$). The Krane et al. (1959) model considers only the path reaction $P_6 \leftrightarrow N_6 \leftrightarrow A_6$. Due to the importance of benzene content in reformate for accurate prediction, it is necessary to add those reactions in which benzene is taking part. Thus, the reaction network shown in Figure 4.6 and the corresponding contribution to the reaction rate equations were added to the kinetic model. Also, it was assumed that all the benzene is produced via cyclohexane dehydrogenation.

Isomerization of Paraffins The reactions of isomerization of *n*-paraffins to *i*-paraffins are highly desired during catalytic reforming of naphtha, since the *i*-paraffins produced contribute to the increase in octane number of the reformate. Isomerization is a fast reaction catalyzed by acid sites, and it reaches equilibrium at catalytic reforming conditions. Hence, paraffin distribution can be estimated by thermodynamic equilibrium calculation.

For the following general isomerization reaction:

$$nP_i \leftrightarrow iP_i \tag{4.30}$$

the equilibrium constant (K_e) is

$$K_{e_i} = \frac{y_{i-p_i}}{y_{n-p}} = \frac{y_{i-p_i}}{1 - y_{i-p_i}} \tag{4.31}$$

TABLE 4.6. Individual Kinetic Constants for Hydrocarbons with 10 and 11 Atoms of Carbon

Reaction	C_6/C_5	C_7/C_6	C_8/C_7	C_9/C_8	C_{10}/C_9	C_{11}/C_{10}[a]	k_{10}^{+}[b]	k_{10}[c]	k_{11}[c]
$P \rightarrow N$	—	—	2.2931	1.3609	1.4033	1.4645	0.0254	0.0243	0.0356
$P_n \rightarrow P_{n-1} + P_1$	1.1667	1.0000	1.3571	1.5789	1.6333	1.6678	0.0049	0.0046	0.0077
$P_n \rightarrow P_{n-2} + P_2$	1.2000	1.0000	1.3888	1.5600	1.6154	1.6499	0.0063	0.0059	0.0097
$P_n \rightarrow P_{n-3} + P_3$	—	1.1852	1.3438	1.5814	1.6029	1.6170	0.0109	0.0103	0.0166
$P_n \rightarrow P_{n-4} + P_4$	—	—	—	1.5714	1.6182	1.6212	0.0089	0.0084	0.0135
$P_n \rightarrow P_{n-5} + P_5$	—	—	—	—	—	—	0.0124	0.0117	0.0191
$N \rightarrow P$	—	0.1351	2.3500	1.1489	1.0000	1.0000	0.0054	0.0054	0.0054
$N \rightarrow A$	—	2.2587	2.3678	1.1395	1.0000	1.0000	0.2450	0.2450	0.2450
$N_n \rightarrow N_{n-1} + P_1$	—	—	—	14.111	1.0551	1.0000	0.0134	0.0134	0.0134
$N_n \rightarrow N_{n-2} + P_2$	—	—	—	—	1.0551	1.0000	0.0134	0.0134	0.0134
$N_n \rightarrow N_{n-3} + P_3$	—	—	—	—	—	1.0000	0.0080	0.0080	0.0080
$A_n \rightarrow A_{n-1} + P_1$	—	—	—	5.0000	1.2000	1.0000	0.0006	0.0006	0.0006
$A_n \rightarrow A_{n-2} + P_2$	—	—	—	—	1.2000	1.0000	0.0006	0.0006	0.0008
$A \rightarrow P$	—	—	1.0000	1.0000	1.0000	1.0000	0.0016	0.0016	0.0016

[a] Extrapolated values.
[b] Original kinetic parameters reported by Krane et al. (1959).
[c] Calculated values of kinetic parameters.

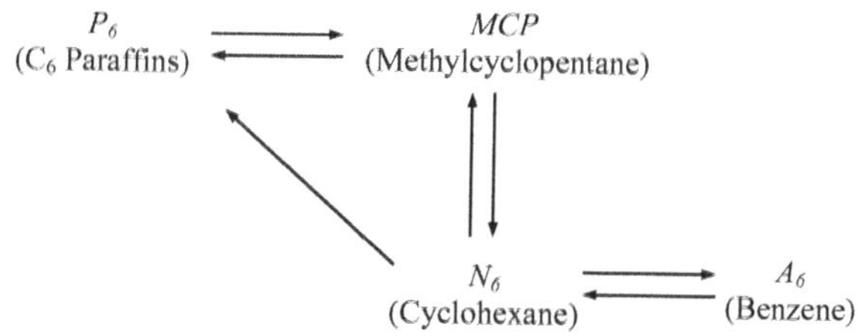

Figure 4.6. Reaction network for benzene formation.

The effect of temperature on equilibrium constant is given by (Smith et al. 1996)

$$\ln(K_{e_i}) = -\frac{\Delta G_i^\circ}{RT} \tag{4.32}$$

where ΔG° is the reaction standard Gibbs energy. ΔG° can be determined as

$$\frac{\Delta G^\circ}{RT} = \frac{\Delta G_0^\circ - \Delta H_0^\circ}{RT_0} + \frac{\Delta H_0^\circ}{RT} + \frac{1}{T}\int_{T_0}^{T} \frac{\Delta Cp}{R} dT - \int_{T_0}^{T} \frac{\Delta Cp}{R}\frac{dT}{T} \tag{4.33}$$

To calculate ΔG° with Eq. (4.33), the following dependence of heat capacity on temperature can be used:

$$C_{p_i} = A_i + B_i T + C_i T^2 + D_i T^3 \tag{4.34}$$

By sustituting Eq. (4.34) in Eq. (4.33), the integrals can be evaluated using the following final forms:

$$\int_{T_0}^{T} \Delta C_p dT = \Delta A(\tau - 1)T_0 + \frac{\Delta B}{2}(\tau^2 - 1)T_0^2 + \frac{\Delta C}{3}(\tau^3 - 1)T_0^3 + \frac{\Delta D}{4}(\tau^4 - 1)T_0^4 \tag{4.35}$$

$$\int_{T_0}^{T} \frac{\Delta C_p}{T} dT = \Delta A \ln(\tau) + \Delta B(\tau - 1)T_0 + \frac{\Delta C}{2}(\tau^2 - 1)T_0^2 + \frac{\Delta D}{3}(\tau^3 - 1)T_0^3 \tag{4.36}$$

where

$$\tau = \frac{T}{T_0} \tag{4.37}$$

To employ the foregoing procedure for equilibrium calculation of the paraffin isomerization reactions that occur in catalytic reforming, some thermodynamic data are required, which are depicted in Table 4.7. For example, for the

TABLE 4.7. Thermodynamic Data of Various Paraffins

Name	*A*	*B*	*C*	*D*	*H°*	*G°*
n-Butane	2.266	7.91E−02	−2.65E−05	−6.74E−10	−30.15	−4.10
i-Butane	−0.332	9.19E−02	−4.41E−05	6.92E−09	−32.15	−4.99
n-Pentane	−0.866	1.16E−01	−6.16E−05	1.27E−08	−35.00	−2.00
2-Methylbutane	−2.275	1.21E−01	−6.52E−05	1.37E−08	−36.92	−3.54
2,2-Dimethylpropane	−3.963	1.33E−01	−7.90E−05	1.82E−08	−39.67	−3.64
n-Hexane	−1.054	1.39E−01	−7.45E−05	1.55E−08	−39.96	−0.06
2-Methylpentane	−2.524	1.48E−01	−8.53E−05	1.93E−08	−41.66	−1.20
3-Methylpentane	−0.570	1.36E−01	−6.85E−05	1.20E−08	−41.02	−0.51
2,2-Dimethylbutane	−3.973	1.50E−01	−8.31E−05	1.64E−08	−44.35	−2.30
2,3-Dimethylbutane	−3.489	1.47E−01	−8.06E−05	1.63E−08	−42.49	−0.98
n-Heptane	−1.229	1.62E−01	−8.72E−05	1.83E−08	−44.88	1.91
2-Methylhexane	−9.408	2.06E−01	−1.50E−04	4.39E−08	−46.59	0.77
3-Methylhexane	−1.683	1.63E−01	−8.92E−05	1.87E−08	−45.96	1.10
2,2-Dimethylpentane	−11.966	2.14E−01	−1.52E−04	4.15E−08	−49.27	0.02
2,3-Dimethylpentane	−1.683	1.63E−01	−8.92E−05	1.87E−08	−47.62	0.16
2,4-Dimethylpentane	−1.683	1.63E−01	−8.92E−05	1.87E−08	−48.28	0.74
3,3-Dimethylpentane	−1.683	1.63E−01	−8.92E−05	1.87E−08	−48.17	0.63
3-Ethylpentane	−1.683	1.63E−01	−8.92E−05	1.87E−08	−45.33	2.63
n-Octane	−1.456	1.84E−01	−1.00E−04	2.12E−08	−49.82	3.92
2-Methylheptane	−21.435	2.97E−01	−2.81E−04	1.10E−07	−51.50	3.05
3-Methylheptane	−2.201	1.88E−01	−1.05E−04	2.32E−08	−50.82	3.28
4-Methylheptane	−2.201	1.88E−01	−1.05E−04	2.32E−08	−50.69	4.00
2,2-Dimethylhexane	−2.201	1.88E−01	−1.05E−04	2.32E−08	−53.71	2.56

2,3-Dimethylhexane	−2.201	1.88E−01	−1.05E−04	2.32E−08	−51.13	4.23
2,4-Dimethylhexane	−2.201	1.88E−01	−1.05E−04	2.32E−08	−52.44	2.80
2,5-Dimethylhexane	−2.201	1.88E−01	−1.05E−04	2.32E−08	−53.21	2.50
3,3-Dimethylhexane	−2.201	1.88E−01	−1.05E−04	2.32E−08	−52.61	3.17
3,4-Dimethylhexane	−2.201	1.88E−01	−1.05E−04	2.32E−08	−50.91	4.14
3-Ethylhexane	−2.201	1.88E−01	−1.05E−04	2.32E−08	−50.40	3.95
2,2,3-Trimethylpentane	−2.201	1.88E−01	−1.05E−04	2.32E−08	−52.61	4.09
2,2,4-Trimethylpentane	−1.782	1.86E−01	−1.02E−04	2.19E−08	−53.57	3.27
2,3,3-Trimethylpentane	−2.201	1.88E−01	−1.05E−04	2.32E−08	−51.73	4.52
2,3,4-Trimethylpentane	−2.201	1.88E−01	−1.05E−04	2.32E−08	−51.97	4.52
2 Methyl-3-ethylpentane	−2.201	1.88E−01	−1.05E−04	2.32E−08	−50.48	5.08
3 Methyl-3-ethylpentane	−2.201	1.88E−01	−1.05E−04	2.32E−08	−51.38	4.76
n-Nonane	0.751	1.62E−01	−4.61E−05	−7.12E−09	−54.74	5.93
2,2,3 Trimethylhexane	−10.899	2.52E−01	−1.71E−04	4.75E−08	−57.65	5.86
2,2,4 Trimethylhexane	−14.405	2.64E−01	−1.84E−04	5.23E−08	−58.13	5.38
2,2,5 Trimethylhexane	−12.923	2.62E−01	−1.85E−04	5.39E−08	−60.71	3.21
3,3 Diethylpentane	−16.067	2.69E−01	−1.91E−04	5.51E−08	−55.44	8.38
2,2,3,3 Tetramethylpentane	−13.037	2.60E−01	−1.81E−04	5.12E−08	−56.70	8.20
2,2,3,4 Tetramethylpentane	−13.037	2.60E−01	−1.81E−04	5.12E−08	−56.64	7.80
2,2,4,4 Tetramethylpentane	−16.099	2.79E−01	−2.06E−04	6.15E−08	−57.83	8.13
2,3,3,4 Tetramethylpentane	−13.117	2.61E−01	−1.82E−04	5.15E−08	−56.46	8.15
n-Decane	−1.890	2.30E−01	−1.26E−04	2.70E−08	−59.67	7.94
3,3,5 Trimethylheptane	−16.808	2.94E−01	−2.07E−04	5.86E−08	−61.80	8.02
2,2,3,3 Tetramethylhexane	−14.052	2.94E−01	−2.11E−04	6.17E−08	−61.66	11.28
2,2,5,5 Tetramethylhexane	−14.890	2.97E−01	−2.14E−04	6.25E−08	−68.32	4.66

Source: Data from Reid (1977).

TABLE 4.8. Example of Calculation of K_e for the Isomerization of *n*-Hexane (T = 400 K)

ΔA	ΔB	ΔC	ΔD	ΔH_0°	ΔG_0°	$\frac{\Delta G^\circ}{RT}$	K_e
			n-*Hexane* ↔ *2-Methylpentane*				
−1.47	0.0087	−1.1E−05	3.8E−09	−1.7	−1.14	−1.200	3.32
			n-*Hexane* ↔ *3-Methylpentane*				
0.484	−0.0031	5.95E−06	−3.5E−09	−1.06	−0.45	−0.303	1.35
			n-*Hexane* ↔ *2,2-Dimethylbutane*				
−2.919	0.0113	−8.6E−06	8.5E−10	−4.39	−2.24	−1.890	6.62
			n-*Hexane* ↔ *2,3-Dimethylbutane*				
−2.435	0.0079	−6.1E−06	7.8E−10	−2.53	−0.92	−0.455	1.58

isomerization of *n*-hexane, four isomers are obtained: 2-methylpentane, 3-methylpentane, 2,2-dimethylbutane, and 2,3-dimethylbutane. Each isomerization reaction needs to be considered separately. The values calculated for all parameters required to evaluate ΔG° are reported in Table 4.8. With ΔG°, K_e is then computed using Eq. (4.32). Once the values of all K_e have been determined, the formula needed to calculate the composition (y_i) of all isomers is deduced from Eq. (4.31):

$$y_i = \frac{K_{e_i}}{1+\sum_{i=1}^{n_{\text{isomers}}} K_{e_i}} \tag{4.38}$$

The values of K_e and the equilibrium composition calculated by using this procedure for isomers of paraffins ranging from 6 to 11 atoms of carbon at different temperatures are summarized in Table 4.9. Splitting the total paraffin lumps as reported by Krane et al. (1959) into *n*- and *i*-paraffins makes it possible to extend the model to consider 48 paraffins plus *i*-paraffin lumps instead of the original seven paraffin lumps. That is, 41 new paraffin lumps are incorporated in the model.

Effects of Pressure and Temperature on Kinetic Parameters The Krane et al. (1959) model does not include the influence of temperature and pressure on the kinetic parameters. To overcome these limitations, an Arrhenius-type variation of the rate constants with activation energy values for each type of reaction, plus a factor that accounts for the pressure effect, can be used. The

TABLE 4.9. Equilibrium Constants and Molar Composition for Hexane Isomerization

	Equilibrium Constants					
Reaction: n-hexane ↔	300 K	400 K	500 K	600 K	700 K	800 K
2-Methylpentane	6.73	3.32	2.20	1.69	1.41	1.24
3-Methylpentane	2.11	1.35	1.04	0.87	0.76	0.69
2,2-Dimethylbutane	41.83	6.62	2.20	1.07	0.65	0.45
2,3-Dimethylbutane	4.60	1.58	0.82	0.54	0.39	0.32
	Equilibrium Molar Composition					
Isomers	300 K	400 K	500 K	600 K	700 K	800 K
2-Methylpentane	11.96	23.94	30.30	32.76	33.51	33.59
3-Methylpentane	3.76	9.76	14.26	16.78	18.09	18.76
2,2-Dimethylbutane	74.33	47.73	30.34	20.73	15.33	12.08
2,3-Dimethylbutane	8.17	11.36	11.34	10.37	9.36	8.52
n-Hexane	1.78	7.21	13.76	19.36	23.71	27.05
Total	100.00	100.00	100.00	100.00	100.00	100.00

equation for the combined effect of temperature and pressure on kinetic parameters can be expressed as

$$k_i = k_i^0 \left[\frac{E_{Aj}}{R} \left(\frac{1}{T_0} - \frac{1}{T} \right) \right] \left(\frac{P}{P_0} \right)^{\alpha_k} \tag{4.39}$$

The values of activation energies (E_{Aj}) for each reaction j and for pressure effect factors (α_k) are given in Table 4.4. Krane et al. (1959) reported values of kinetic parameters (k_i^0) at temperatures ranging from 471 to 515°C, with $P_0 = 300$ psig, an H_2/oil ratio of 2 to 8 mol/mol, and a WHSV of 0.7 to 5.0 h^{-1}. Because for each reaction only one value of the corresponding kinetic parameter for each reaction rate was reported in this range of temperatures, the base temperature was considered to be the average of both values ($T_0 = 493$°C).

The Extended Proposed Kinetic Model Based on all the previous considerations, a kinetic model with 33 lumps plus 41 paraffins isomers was developed. The proposed reaction scheme is shown in Figure 4.7. All reactions are assumed to be pseudo-first order with respect to the hydrocarbon. The reaction rate of each lump is given by the following equations, which comprise the kinetic model. Each rate reaction equation is a function of the kinetic constants (k_i) and the concentration of each pseudocomponent (P_i, N_i, A_i). The values of all rate constants are reported in Table 4.10.

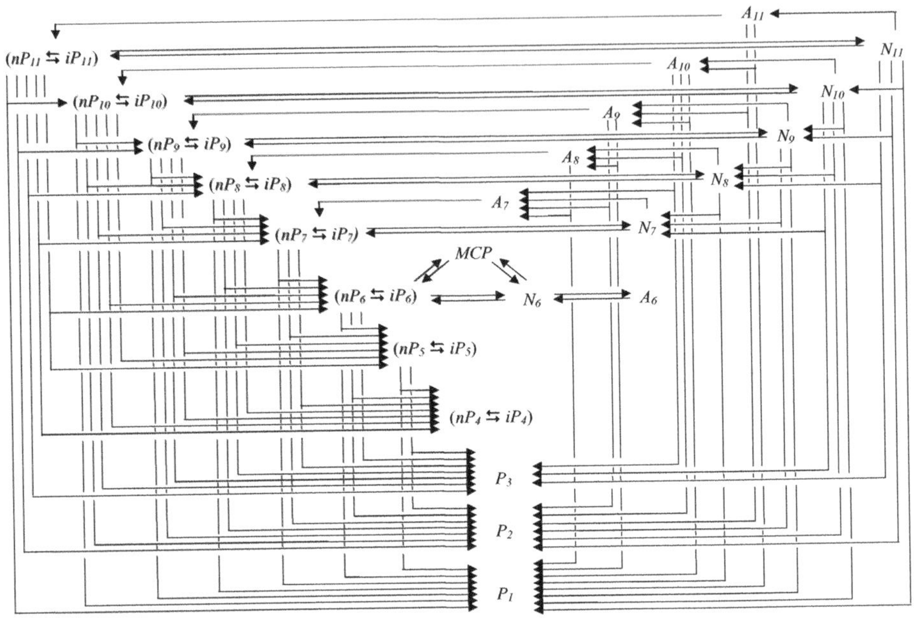

Figure 4.7. Reaction scheme proposed for catalytic reforming of naphtha.

Paraffins

$$\frac{dP_{11}}{d(1/SV)} = k_{34}N_{11} + k_{58}A_{11} - (k_1 + k_2 + k_3 + k_4 + k_5 + k_6)P_{11} \tag{4.40}$$

$$\frac{dP_{10}}{d(1/SV)} = k_2P_{11} + k_{39}N_{10} + k_{61}A_{10} - (k_7 + k_8 + k_9 + k_{10} + k_{11} + k_{12})P_{10} \tag{4.41}$$

$$\frac{dP_9}{d(1/SV)} = k_3P_{11} + k_8P_{10} + k_{44}N_9 + k_{65}A_9 - (k_{13} + k_{14} + k_{15} + k_{16} + k_{17})P_9 \tag{4.42}$$

$$\frac{dP_8}{d(1/SV)} = k_4P_{11} + k_9P_{10} + k_{14}P_9 + k_{48}N_8 + k_{68}A_8 - (k_{18} + k_{19} + k_{20} + k_{21} + k_{22})P_8 \tag{4.43}$$

$$\frac{dP_7}{d(1/SV)} = k_5P_{11} + k_{10}P_{10} + k_{15}P_9 + k_{19}P_8 + k_{51}N_7 + k_{70}A_7 - (k_{23} + k_{24} + k_{25} + k_{26})P_7 \tag{4.44}$$

$$\frac{dP_6}{d(1/SV)} = k_6P_{11} + k_{11}P_{10} + k_{16}P_9 + k_{20}P_8 + k_{24}P_7 + k_{53}N_6 + k_{56}\cdot MCP - (k_{27} + k_{28} + k_{29} + k_{30} + k_{31})P_6 \tag{4.45}$$

TABLE 4.10. Kinetic Parameters of the Proposed Model (T = 490°C)

Reaction Step	k (h^{-1})		Reaction Step	k (h^{-1})		Reaction Step	k (h^{-1})	
$P_{11} \rightarrow N_{11} + H_2$	0.0356	k_1	$P_7 + H_2 \rightarrow P_5 + P_2$	0.0018	k_{25}	$N_8 \rightarrow A_8 + 3H_2$	0.2150	k_{49}
$P_{11} + H_2 \rightarrow P_{10} + P_1$	0.0075	k_2	$P_7 + H_2 \rightarrow P_4 + P_3$	0.0043	k_{26}	$N_8 + H_2 \rightarrow N_7 + P_1$	0.0007	k_{50}
$P_{11} + H_2 \rightarrow P_9 + P_2$	0.0100	k_3	$P_6 \rightarrow N_6 + H_2$	0.0000	k_{27}	$N_7 + H_2 \rightarrow P_7$	0.0019	k_{51}
$P_{11} + H_2 \rightarrow P_8 + P_3$	0.0135	k_4	$P_6 \rightarrow MCP + H_2$	0.0042	k_{28}	$N_7 \rightarrow A_7 + 3H_2$	0.0788	k_{52}
$P_{11} + H_2 \rightarrow P_7 + P_4$	0.0135	k_5	$P_6 + H_2 \rightarrow P_5 + P_1$	0.0018	k_{29}	$N_6 + H_2 \rightarrow P_6$	0.0204	k_{53}
$P_{11} + H_2 \rightarrow P_6 + P_5$	0.0191	k_6	$P_6 + H_2 \rightarrow P_4 + P_2$	0.0016	k_{30}	$N_6 \rightarrow A_6 + 3H_2$	0.1368	k_{54}
$P_{10} \rightarrow N_{10} + H_2$	0.0243	k_7	$P_6 + H_2 \rightarrow 2P_3$	0.0025	k_{31}	$N_6 \rightarrow MCP$	0.0040	k_{55}
$P_{10} + H_2 \rightarrow P_9 + P_1$	0.0015	k_8	$P_5 + H_2 \rightarrow P_4 + P_1$	0.0018	k_{32}	$MCP + H_2 \rightarrow P_6$	0.0008	k_{56}
$P_{10} + H_2 \rightarrow P_8 + P_2$	0.0054	k_9	$P_5 + H_2 \rightarrow P_3 + P_2$	0.0022	k_{33}	$MCP \rightarrow N_6$	0.0238	k_{57}
$P_{10} + H_2 \rightarrow P_7 + P_3$	0.0160	k_{10}	$N_{11} + H_2 \rightarrow P_{11}$	0.0050	k_{34}	$A_{11} + 4H_2 \rightarrow P_{11}$	0.0016	k_{58}
$P_{10} + H_2 \rightarrow P_6 + P_4$	0.0095	k_{11}	$N_{11} \rightarrow A_{11} + 3H_2$	0.6738	k_{35}	$A_{11} + H_2 \rightarrow A_{10} + P_1$	0.0006	k_{59}
$P_{10} + H_2 \rightarrow 2P_5$	0.0095	k_{12}	$N_{11} + H_2 \rightarrow N_{10} + P_1$	0.0134	k_{36}	$A_{11} + H_2 \rightarrow A_9 + P_2$	0.0006	k_{60}
$P_9 \rightarrow N_9 + H_2$	0.0500	k_{13}	$N_{11} + H_2 \rightarrow N_9 + P_2$	0.0134	k_{37}	$A_{10} + 4H_2 \rightarrow P_{10}$	0.0016	k_{61}
$P_9 + H_2 \rightarrow P_8 + P_1$	0.0030	k_{14}	$N_{11} + H_2 \rightarrow N_8 + P_3$	0.0080	k_{38}	$A_{10} + H_2 \rightarrow A_9 + P_1$	0.0006	k_{62}
$P_9 + H_2 \rightarrow P_7 + P_2$	0.0039	k_{15}	$N_{10} + H_2 \rightarrow P_{10}$	0.0054	k_{39}	$A_{10} + H_2 \rightarrow A_8 + P_2$	0.0006	k_{63}
$P_9 + H_2 \rightarrow P_6 + P_3$	0.0068	k_{16}	$N_{10} \rightarrow A_{10} + 3H_2$	0.3198	k_{40}	$A_{10} + H_2 \rightarrow A_7 + P_3$	0.0000	k_{64}
$P_9 + H_2 \rightarrow P_5 + P_4$	0.0058	k_{17}	$N_{10} + H_2 \rightarrow N_9 + P_1$	0.0134	k_{41}	$A_9 + 4H_2 \rightarrow P_9$	0.0016	k_{65}
$P_8 \rightarrow N_8 + H_2$	0.0266	k_{18}	$N_{10} + H_2 \rightarrow N_8 + P_2$	0.0134	k_{42}	$A_9 + H_2 \rightarrow A_8 + P_1$	0.0005	k_{66}
$P_8 + H_2 \rightarrow P_7 + P_1$	0.0019	k_{19}	$N_{10} + H_2 \rightarrow N_7 + P_3$	0.0080	k_{43}	$A_9 + H_2 \rightarrow A_7 + P_2$	0.0005	k_{67}
$P_8 + H_2 \rightarrow P_6 + P_2$	0.0056	k_{20}	$N_9 + H_2 \rightarrow P_9$	0.0054	k_{44}	$A_8 + 4H_2 \rightarrow P_8$	0.0011	k_{68}
$P_8 + H_2 \rightarrow P_5 + P_3$	0.0034	k_{21}	$N_9 \rightarrow A_9 + 3H_2$	0.2205	k_{45}	$A_8 + H_2 \rightarrow A_7 + P_1$	0.0001	k_{69}
$P_8 + H_2 \rightarrow 2P_4$	0.0070	k_{22}	$N_9 + H_2 \rightarrow N_8 + P_1$	0.0127	k_{46}	$A_7 + 4H_2 \rightarrow P_7$	0.0016	k_{70}
$P_7 \rightarrow N_7 + H_2$	0.0076	k_{23}	$N_9 + H_2 \rightarrow N_7 + P_2$	0.0127	k_{47}	$A_6 + 3H_2 \rightarrow N_6$	0.0015	k_{71}
$P_7 + H_2 \rightarrow P_6 + P_1$	0.0027	k_{24}	$N_8 + H_2 \rightarrow P_8$	0.0025	k_{48}			

$$\frac{dP_5}{d(1/SV)} = k_6P_{11} + 2k_{12}P_{10} + k_{17}P_9 + k_{21}P_8 + k_{25}P_7 + k_{29}P_6 - (k_{32} + k_{33})P_5 \tag{4.46}$$

$$\frac{dP_4}{d(1/SV)} = k_5P_{11} + k_{11}P_{10} + k_{17}P_9 + 2k_{22}P_8 + k_{26}P_7 + k_{30}P_6 + k_{32}P_5 \tag{4.47}$$

$$\frac{dP_3}{d(1/SV)} = k_4P_{11} + k_{10}P_{10} + k_{16}P_9 + k_{21}P_8 + k_{26}P_7 + 2k_{31}P_6 + k_{33}P_5 + k_{38}N_{11} + k_{43}N_{10} + k_{64}A_{10} \tag{4.48}$$

$$\frac{dP_2}{d(1/SV)} = k_3P_{11} + k_9P_{10} + k_{15}P_9 + k_{20}P_8 + k_{25}P_7 + k_{30}P_6 + k_{33}P_5 + k_{37}N_{11} + k_{42}N_{10} + k_{47}N_9 + k_{60}A_{11} + k_{63}A_{10} + k_{67}A_9 \tag{4.49}$$

$$\frac{dP_1}{d(1/SV)} = k_2P_{11} + k_8P_{10} + k_{14}P_9 + k_{19}P_8 + k_{24}P_7 + k_{29}P_6 + k_{32}P_5 + k_{36}N_{11} + k_{41}N_{10} + k_{46}N_9 + k_{50}N_8 + k_{59}A_{11} + k_{62}A_{10} + k_{66}A_9 + k_{69}A_8 \tag{4.50}$$

Naphthenes

$$\frac{dN_{11}}{d(1/SV)} = k_1P_{11} - (k_{34} + k_{35} + k_{36} + k_{37} + k_{38})N_{11} \tag{4.51}$$

$$\frac{dN_{10}}{d(1/SV)} = k_7P_{10} + k_{36}N_{11} - (k_{39} + k_{40} + k_{41} + k_{42} + k_{43})N_{10} \tag{4.52}$$

$$\frac{dN_9}{d(1/SV)} = k_{13}P_9 + k_{37}N_{11} + k_{41}N_{10} - (k_{44} + k_{45} + k_{46} + k_{47})N_9 \tag{4.53}$$

$$\frac{dN_8}{d(1/SV)} = k_{18}P_8 + k_{38}N_{11} + k_{42}N_{10} + k_{46}N_9 - (k_{48} + k_{49} + k_{50})N_8 \tag{4.54}$$

$$\frac{dN_7}{d(1/SV)} = k_{23}P_7 + k_{43}N_{10} + k_{47}N_9 + k_{50}N_8 - (k_{51} + k_{52})N_7 \tag{4.55}$$

$$\frac{dN_6}{d(1/SV)} = k_{27}P_6 + k_{57} \cdot MCP + k_{71}A_6 - (k_{53} + k_{54} + k_{55})N_6 \tag{4.56}$$

$$\frac{dMCP}{d(1/SV)} = k_{28}P_6 + k_{55}N_6 - (k_{56} + k_{57}) \cdot MCP \tag{4.57}$$

TABLE 4.11. Coefficients of the Balance Equation of Hydrogen

i	a_i	b_i	c_i
1	$k_1 - (k_2 + k_3 + k_4 + k_5 + k_6)$	$3k_{35} - (k_{34} + k_{36} + k_{37} + k_{38})$	$4k_{58} + k_{59} + k_{60}$
2	$k_7 - (k_8 + k_9 + k_{10} + k_{11} + k_{12})$	$3k_{40} - (k_{39} + k_{41} + k_{42} + k_{43})$	$4k_{61} + k_{62} + k_{63} + k_{64}$
3	$k_{13} - (k_{14} + k_{15} + k_{16} + k_{17})$	$3k_{45} - (k_{44} + k_{46} + k_{47})$	$4k_{65} + k_{66} + k_{67}$
4	$k_{18} - (k_{19} + k_{20} + k_{21} + k_{22})$	$3k_{49} - (k_{48} + k_{50})$	$4k_{68} + k_{69}$
5	$k_{23} - (k_{24} + k_{25} + k_{26})$	$3k_{52} - k_{51}$	$4k_{70}$
6	$k_{27} + k_{28} - (k_{29} + k_{30} + k_{31})$	$3k_{54} - k_{53}$	$3k_{71}$
7	$-(k_{32} + k_{33})$		

Aromatics

$$\frac{dA_{11}}{d(1/SV)} = k_{35}N_{11} - (k_{58} + k_{59} + k_{60})A_{11} \tag{4.58}$$

$$\frac{dA_{10}}{d(1/SV)} = k_{40}N_{10} + k_{59}A_{11} - (k_{61} + k_{62} + k_{63} + k_{64})A_{10} \tag{4.59}$$

$$\frac{dA_9}{d(1/SV)} = k_{45}N_9 + k_{60}A_{11} + k_{62}A_{10} - (k_{65} + k_{66} + k_{67})A_9 \tag{4.60}$$

$$\frac{dA_8}{d(1/SV)} = k_{49}N_8 + k_{63}A_{10} + k_{66}A_9 - (k_{68} + k_{69})A_8 \tag{4.61}$$

$$\frac{dA_7}{d(1/SV)} = k_{52}N_7 + k_{64}A_{10} + k_{67}A_9 + k_{69}A_8 - k_{70}A_7 \tag{4.62}$$

$$\frac{dA_6}{d(1/SV)} = k_{54}N_6 - k_{71}A_6 \tag{4.63}$$

The hydrogen balance can be derived from the stoichiometry of the reactions in which it is involved, either those that consume hydrogen or those that produce hydrogen. The resulting reaction rate for hydrogen is then

$$\frac{dH_2}{d(1/SV)} = \sum_{i=1}^{7} a_i P_{12-i} + \sum_{i=1}^{6} b_i N_{12-i} - \sum_{i=1}^{6} c_i A_{12-i} - k_{56}\mathrm{MCP} \tag{4.64}$$

The values of a_i, b_i, and c_i are reported in Table 4.11.

4.3.2 Validation of the Kinetic Model with Bench-Scale Reactor Experiments

Experiments in an Isothermal Bench-Scale Reactor To validate the modified kinetic model, various experiments were conducted with a hydrodesulfurized

TABLE 4.12. Molar Composition of the Feed for Bench-Scale Reactor and of Different Reformates Obtained at 10.5 kg/cm^2 Pressure, 6.5 H_2/Oil Molar Ratio, and 3.54 h^{-1}WHSV

	Feed	Reaction Temperature		
		490°C	500°C	510°C
n-Paraffins				
nP_{11}	2.20	0.01	0.01	0
nP_{10}	2.97	0.09	0.00	0
nP_9	5.09	0.40	0.28	0.18
nP_8	6.36	1.22	0.91	0.63
nP_7	3.20	2.92	2.45	1.97
nP_6	4.40	5.51	5.22	4.40
nP_5	3.80	5.26	4.97	4.85
Total	28.02	15.41	13.84	12.03
i-Paraffins				
iP_{10}	6.22	0.28	0.17	0.85
iP_9	8.31	1.50	1.24	0.67
iP_8	6.51	3.76	2.75	2.01
iP_7	6.20	8.01	7.29	6.07
iP_6	6.70	9.41	9.50	9.83
iP_5	3.40	6.40	6.19	5.53
Total	37.34	29.36	27.14	24.96
Naphthenes				
N_{11}	0.40	0.00	0.00	0
N_{10}	0.60	0.00	0.00	0
N_9	3.56	0.01	0.02	0.01
N_8	4.71	0.63	0.66	0.38
N_7	5.80	0.33	0.31	0.27
N_6	3.21	0.01	0.01	0.01
MCP	0.42	1.35	1.23	1.15
Total	18.70	2.33	2.23	1.82
Aromatics				
A_{11}	0.30	0.97	1.10	1.25
A_{10}	2.70	5.61	5.76	5.99
A_9	4.21	12.53	13.20	14.17
A_8	4.71	15.66	16.90	18.2
A_7	3.22	12.82	13.91	15.02
A_6	0.80	5.30	5.91	6.56
Total	15.94	52.90	56.78	61.19

straight-run naphtha recovered from a commercial naphtha HDS unit (82 to 168°C distillation range, 0.74 g/mL density at 20°C, and <0.5 wppm sulfur). The detailed composition of this feed for catalytic reforming experiments determined by gas chromatographic analysis is presented in the first column of Table 4.12. The tests were carried out in a fixed-bed bench-scale stainless-steel

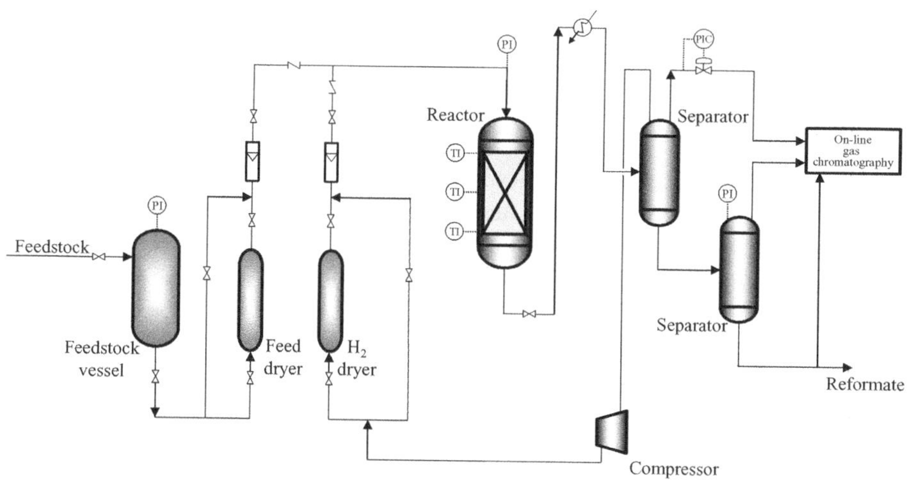

Figure 4.8. Experimental bench-scale unit for catalytic reforming experiments.

reactor (2.5 cm internal diameter and 25 cm length) with hydrogen recycle. The reactor is operated in the isothermal mode by independent temperature control of a three-zone electric furnace. A general diagram of the experimental setup is shown in Figure 4.8. All the experiments were carried out done at a pressure of 10.5 kg/cm^2, an H_2/oil molar ratio of 6.5, and temperatures of 490, 500, and 510°C. To simulate a series of three reforming reactors, the bench-scale reactor was loaded with different amounts of catalyst (i.e., 6, 15, and 30 mL), keeping the same naphtha flow rate at a constant value of 102 mL/h in order to have different space velocities (WHSV): 17.72, 7.09, and 3.54 h^{-1}, respectively. These amounts of catalyst and WHSV were selected in order to have 20% of the total mass of catalyst in the first reactor, 30% in the second reactor, and 50% in the third reactor, which are typical percentages of catalyst loading in commercial catalytic reforming reactors.

The catalyst used in all experiments was a commercially available Pt–Re reforming sample (0.29 wt% Pt, 0.29 wt% Re) with a specific surface area of 221 m^2/g, a pore volume of 0.36 mL/g, and a particle diameter of 1.6 mm. The catalyst beds were diluted with an inert material of uniform particle size in order to achieve a better distribution of heat losses over the reactor, to more easily retain temperature uniformity. The degree of dilution was varied depending on the amount of catalyst loaded in the reactor. The highest dilution was used for experiments with 20% of the total mass of catalyst. The temperature drop, measured with an axial thermocouple, was less than 5°C, so isothermal operation can be assumed. Reformate samples were collected in a high-pressure product receiver. The remaining C_4^- (butane and lighters) cracking products were removed afterward by distillation (stabilization). The stabilized reformate was analyzed on paraffins, *i*-paraffins, naphthenes, and aromatics by gas chromatography.

Results

Reforming Experiments Table 4.12 shows the detailed molar composition of the reformate as a function of reaction temperature. From this table, the following observations can be made:

- The total amount of aromatics increases from 15.94 mol% to 52.90, 56.78, and 61.19 mol% at 490, 500, and 510°C, respectively. The most important increase is observed for lighter aromatics, particularly A_6, A_7, and A_8.
- Naphthenes react relatively easily and are highly selective to aromatic compounds via dehydrogenation. This reaction proceeds essentially to completion. For example, N_6, N_9, N_{10}, and N_{11} disappear completely and the conversions of N_7 and N_8 are higher than 87%. It is also confirmed that naphthene dehydrogenation is favored by high reaction temperature, as they were almost completely converted at temperatures above 490°C (>87% conversion of the total amount of naphthenes). This is the main reason that naphthenes are the most desirable components in reforming feedstocks.
- The paraffin isomerization reaction is very important because naphtha contains a high percentage of *n*-paraffins, which after isomerization yield products with a higher octane number. The naphtha used in the present experiments has a high paraffin content: 28.02 mol% *n*-paraffins and 37.34 mol% *i*-paraffins. This reaction occurs rapidly at commercial operating temperatures and is limited by the thermodynamic equilibrium. The temperature has little influence on it because the heat of reaction is low.
- The most difficult reaction to promote is the dehydrocyclization of paraffins, which consists of molecular arrangements of paraffins to naphthenes. Heavy paraffins (P_9, P_{10}, and P_{11}) have conversions higher than 92%, and lighter paraffins showed lower values. This is because the increase in the probability of ring formation is high as the molecular weight of the paraffin increases. Similar to the naphthene dehydrogenation reaction, paraffin dehydrocyclization is favored at high reaction temperatures.

Validation of the Kinetic Model The kinetic model described previously was incorporated into an isothermal pseudohomogeneous one-dimensional reactor model. The use of diluent and proper particle size of the catalytic bed ensures that the data were collected under a kinetic regime and transport effects can be neglected. To evaluate the product composition as a function of the reactor length, the mass balance equations were solved using the Runge–Kutta method.

Figure 4.9 presents a comparison of the experimental and predicted molar reformate composition for some selected hydrocarbons: nP_5, iP_5, nP_6, nP_7, MCP, N_6, N_7, and A_6) at 510°C as a function of catalyst bed length. The following conclusions can be drawn from this comparison:

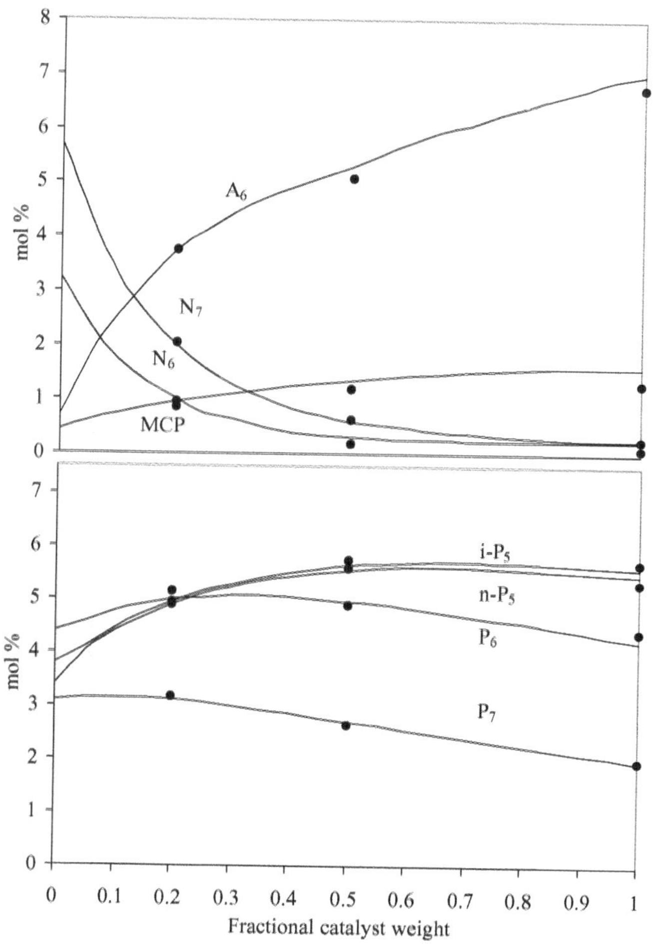

Figure 4.9. Experimental (●) and predicted (—) composition of the reformate obtained at 510°C in the bench-scale reactor.

- The product molar composition predicted agrees very well with experimental data. The average deviation between experimental and predicted values is less than 3%.
- As the naphtha passes through the catalyst bed, the A_6 concentration increases. The same behavior is found for all aromatics compounds.
- The concentrations of N_6 and N_7 and heavy paraffins (P_7 to P_{11}, only P_7 is shown in the figure) decrease as they undergo conversion. A high rate of conversion of naphthenes is found in the first 30% of the catalyst bed. After the point in the 60% catalyst bed, the naphthene concentration approaches a very low steady-state value.

- The relative reaction rates of naphthenes and paraffins are different in the first 20 to 30% of the catalyst bed. Whereas N_6 and N_7 are almost totally converted in this section, MCP and paraffins have a low conversion rate. This means that MCP is much less reactive than N_6 and N_7.
- The A_6 (benzene) composition calculated matches very well with experimental data, with a maximum deviation of 2%, which is the result of a proper description of the benzene reaction mechanism.

Validation of the Effect of Pressure and Temperature on the Kinetic Parameters To evaluate if the correction of the kinetic parameters by temperature and pressure is effective, the model was tested with different options for calculating the parameters of the kinetic model, according to Eq. (4.39): (1) with the effect of pressure only, (2) with the effect of temperature only, and (3) with the effect of both temperature and pressure. When the kinetic parameters were affected only by pressure, the model predicted the same product molar composition for the nations reaction temperatures. This implies that temperature must indeed be taken into account. On the other hand, the original values of kinetic parameters reported by Krane et al. (1959) at 300 psig (21 kg/cm^2) were affected only by the temperature, and the predictions made with them were not satisfactory for the pressure used in the experiments described previously (10.5 kg/cm^2). When both temperature and pressure were considered in calculating the kinetic parameters, the predictions of reformate composition improve substantially.

4.3.3 Simulation of Commercial Semiregenerative Reforming Reactors

The Reactor Model At typical catalytic reforming conditions, the feed (naphtha) enters the reactor in the gas phase and is put in contact with the solid catalyst (i.e., catalytic reforming is conducted as a gas–solid reaction). Radial and axial dispersion effects can also be neglected since the reactor diameter and length are much larger than the diameter of the catalyst particle. In addition, under steady-state conditions the catalyst activity can be assumed to be constant. Thus, commercial semiregenerative reforming reactors can be represented by the one-dimensional pseudohomogeneous adiabatic model. The following ordinary differential equations constitute the reactor model, which are integrated through each reactor bed to describe the reformate composition, temperature, and pressure profiles along the length of the reactors. The Ergun equation was used to predict the total pressure drop of the reactors:

$$-\frac{dy_i}{dz} = \frac{MW_i}{z \cdot WHSV} r_i \tag{4.65}$$

$$\frac{dT}{dz}=\frac{S\sum_{i=1}^{NR} r_i(-\Delta H_{Ri})}{\sum_{i=1}^{NC} F_i C_{Pi}} \tag{4.66}$$

$$-\frac{dP}{dz}=1.75\times10^{-5}\frac{1-\varepsilon}{\varepsilon^3}\frac{G^2}{\rho d_P g_c}+1.5\times10^{-3}\frac{(1-\varepsilon)^2}{\varepsilon^3}\frac{G\mu}{\rho d_P^2 g_c} \tag{4.67}$$

Equations (4.65) to (4.67) are solved simultaneously with the kinetic model rate equation for each component (r_i), described previously using the fourth-order Runge–Kutta method. To solve the energy balance equation, heats of reaction (ΔH_{Ri}) are necessary, which are determined with the following equations:

$$\Delta H_R=\sum v_p H_{fp}-\sum v_r H_{fr} \tag{4.68}$$

$$H_{fi}=H_{fi}^{\circ}+\int_{298\text{K}}^{T} C_p dT \tag{4.69}$$

$$C_p=A+BT+CT^2+DT^3 \tag{4.70}$$

The values of heats of formation H_f° and the constants (A, B, C, and D) for calculating the specific heat (C_{pi}) are reported in Table 4.13.

Reaction Conditions of the Commercial Reformer A commercial semiregenerative reforming unit has four reactors with interstage heaters and operates at the following reaction conditions: 495°C inlet temperature, 10.5 kg/cm^2 reactor pressure, 6.3 mol/mol H_2/oil ratio, and feedstock flow rate of 30,000 bbl/day. The hydrodesulfurized straight-run naphtha used as a feedstock has the following main properties: 0.7406 specific gravity, 60/60°F, 104.8 g/mol molecular weight, 88 to 180°C distillation range, 59.11 mol% total paraffins, 20.01 mol% total naphthenes, and 20.88 mol% total aromatics. The catalyst used in the commercial reforming unit is the same as that employed in the bench-scale experiments described previously.

The main characteristics of each of the four reactors of a semiregenerative reforming unit are detailed in Table 4.14. The inverse of weight hourly space velocity (100/WHSV) is reported in this table since this parameter is used more commonly to indicate the increased reaction severity and reactant reactor positions. As can be seen, the first reactor is always smaller than the other reactors, and the last reactor is the largest. This difference in reactor size is because some of the reactions occurring in the first reactors are very fast, and those taking place in the last reactors are slow.

Results of the Simulation

Composition of the Reformate Figure 4.10 presents the results of simulated reformate molar composition and the commercially reported values at the exit

TABLE 4.13. Properties of Pure Compounds Contained in Reforming Feed and Products

	PM	sg	$T_b(K)$	PVR (bar)	RON	MON	$C_p = A + BT + CT^2 + BT^3$ A	B	C	D	H_f° (J/mol)
Gases											
Hydrogen	2.016	9.0E−5	20.3	—	—	—	2.714E+1	9.274E−3	−1.381E−5	7.645E−9	0.0
Methane	16.043	0.3000	111.6	—	—	—	1.925E+1	5.213E−2	1.197E−5	−1.132E−8	−7.490E+4
Ethane	30.070	0.3564	184.6	—	—	—	5.409E+0	1.781E−1	−6.938E−5	8.713E−9	−8.474E+4
Propane	44.094	0.5077	231.1	13.100	98.90	97.10	−4.224E+0	3.063E−1	−1.586E−4	3.215E−8	−1.039E+5
Butane	58.124	0.5844	272.7	3.558	93.80	89.60	9.487E+0	3.313E−1	−1.108E−4	−2.822E−9	−1.262E+5
n-Paraffins											
Pentane	72.151	0.6310	309.2	1.073	61.70	62.60	−3.626E+0	4.873E−1	−2.580E−4	5.305E−8	−1.465E+5
Hexane	86.178	0.6640	341.9	0.342	24.80	26.00	−4.413E+0	5.820E−1	−3.119E−4	6.494E−8	−1.673E+5
Heptane	100.205	0.6882	371.6	0.112	0.00	0.00	−5.146E+0	6.762E−1	−3.651E−4	7.658E−8	−1.879E+5
Octane	114.232	0.7068	398.8	0.037	−19.00	−15.00	−6.096E+0	7.712E−1	−4.195E−4	8.855E−8	−2.086E+5
Nonane	128.259	0.7217	424.0	0.012	−17.00	−20.00	−8.374E+0	8.729E−1	−4.826E−4	1.031E−7	−2.292E+5
Decane	142.286	0.7342	447.3	0.004	−16.01	−21.10	−7.913E+0	9.609E−1	−5.288E−4	1.131E−7	−2.498E+5
Undecane	156.313	0.7439	469.1	0.001	−14.10	−23.00	−8.395E+0	1.054E+0	−5.799E−4	1.237E−4	−2.705E+5

i-Paraffins											
Isobutane	58.124	0.5631	261.4	4.978	93.00	92.09	−1.390E+0	3.847E−1	−1.846E−4	2.895E−8	−1.346E+5
2-Methylbutane	72.151	0.6247	301.0	1.409	92.30	90.30	−9.525E+0	5.066E−1	−2.729E−4	5.723E−8	−1.546E+5
2,2-Dimethylbutane	86.178	0.6540	322.8	0.679	95.60	93.40	−1.663E+1	6.293E−1	−3.481E−4	6.850E−8	−1.857E+5
2,2-Dimethylpentane	100.205	0.6782	352.4	0.241	92.80	95.60	−5.010E+1	8.956E−1	−6.360E−4	1.736E−7	−2.063E+5
2,2,4-Trimethylpentane	114.232	0.6962	372.4	0.118	100.00	100.00	−7.461E+0	7.779E−1	−4.287E−4	9.173E−8	−2.243E+5
2,2-Dimethylheptane	128.242	0.7146	405.9	0.029	101.10	100.30	−2.089E+1	9.668E−1	−6.120E−4	1.570E−7	−2.470E+5
3,3,5-Trimethylheptane	142.286	0.7469	428.9	0.012	94.20	90.55	−7.037E+1	1.232E+0	−8.646E−4	2.455E−7	−2.587E+5
Naphthenes											
Methylcyclopentane	84.162	0.7536	345.0	0.310	91.3	80.0	−5.011E+1	6.381E−1	−3.642E−4	8.014E−8	−1.068E+5
Cyclohexane	84.162	0.7834	353.8	0.225	83.0	77.2	−5.454E+1	6.113E−1	−2.523E−4	1.321E−8	−1.232E+5
Methylcyclohexane	98.189	0.7740	374.1	0.111	74.8	71.1	−6.192E+1	7.842E−1	−4.538E−4	9.366E−8	−1.549E+5
Ethylcyclohexane	112.216	0.7922	404.9	0.033	45.6	40.8	−6.389E+1	8.893E−1	−5.108E−4	1.103E−7	−1.719E+5
Propylcyclohexane	126.243	0.7977	429.9	0.012	17.8	14.0	−6.252E+1	9.889E−1	−5.795E−4	1.291E−7	−1.934E+5
Butylcyclohexane	140.260	0.8031	454.1	0.004	70.31	68.50	−6.296E+1	1.081E+0	−6.305E−4	1.400E−7	−2.133E+5
Hexylcyclohpentane	154.297	0.8006	476.3	0.001	70.00	68.00	−5.832E+1	1.128E+0	−6.536E−4	1.473E−7	−2.096E+5
Aromatics											
Benzene	78.114	0.8844	353.2	0.222	108.00	98.00	−3.392E+1	4.739E−1	−3.017E−4	7.130E−8	8.298E−4
Toluene	92.141	0.8718	383.8	0.074	120.10	105.00	−2.435E+1	5.125E−1	−2.765E−4	4.911E−8	5.003E−4
Ethylbenzene	106.168	0.8718	409.3	0.025	107.90	97.90	−4.310E+1	7.072E−1	−4.811E−4	1.301E−7	2.981E−4
Propylbenzene	120.195	0.8665	432.4	0.010	101.50	98.70	−3.129E+1	7.486E−1	−4.601E−4	1.081E−7	7.830E−3
Butylbenzene	134.222	0.8646	456.5	0.003	100.40	95.50	−2.299E+1	7.934E−1	−4.396E−4	8.570E−8	−1.382E+4
Pentylbenzene	148.250	0.8624	478.6	0.000	110.00	92.00	−4.218E+1	9.772E−1	−6.262E−4	1.570E−7	−3.380E+4

TABLE 4.14. Main Characteristics of the Commercial Catalytic Reforming Reactors

	Reactor				
	1	2	3	4	Total
Length (m)	4.902	5.410	6.452	8.208	24.972
% of reactor length	19.6	21.7	25.8	32.9	100.0
Accumulated % of reactor length	19.6	41.3	67.1	100.0	
Diameter (m)	2.438	2.819	2.971	3.505	
Catalyst weight (tons)	9.13	13.82	22.82	42.58	88.35
wt% of catalyst bed	10.3	15.7	25.8	48.2	100
Accumulated wt% of catalyst bed	10.3	26.0	51.8	100	
WHSV (h^{-1})	16.0	10.6	6.4	3.4	36.4
Accumulated WHSV (h^{-1})	16.0	26.6	33	36.4	
100/WHSV	6.25	9.43	15.63	29.41	60.72
Accumulated 100/WHSV	6.25	15.68	31.31	60.72	

of the fourth reactor for different hydrocarbons. The following observations can be drawn from this table:

- Good agreement is obtained between simulated and reported commercial values. The content of hydrocarbons with 10 and 11 atoms of carbon and those involved in benzene formation (MCP, A_6, P_6, and N_6), as well as *i*-paraffins, are well calculated. Thus, the reactor and kinetic models predict the reformate composition with good accuracy.
- As the feedstock passes through the reactor in series, the content of all aromatic hydrocarbons increases. The most important change is observed for A_6, A_7, A_8 and A_9, while for A_{10} the increase is low and A_{11} remains essentially unchanged.
- Naphthenes with six and more atoms of carbon react relatively easy and their reactions proceed essentially to completion. Similar behavior is found for heavy paraffins (i.e., P_{10} and P_{11}). P_8 and P_9 also exhibited high conversions, while the content of some light paraffins is increased because they are produced by hydrocracking or hydrogenolysis.
- The contents of N_6 to N_{10} and heavy paraffins (P_8 to P_{11}) decrease as they undergo conversion. High conversion of naphthenes is found in the first and second reactors. After the third reactor, the naphthene content approaches a very low steady-state value. The relative rates of these naphthenes and paraffins conversions are very different in the first two reactors. whereas N_6 and N_7 are almost totally converted in this section, paraffins have a very low conversion.
- There is an insignificant change in *i*-paraffins content in the first and second reactors, and in the last two reactors the change is higher. Similar behavior is found with light paraffins. This means that the increase in

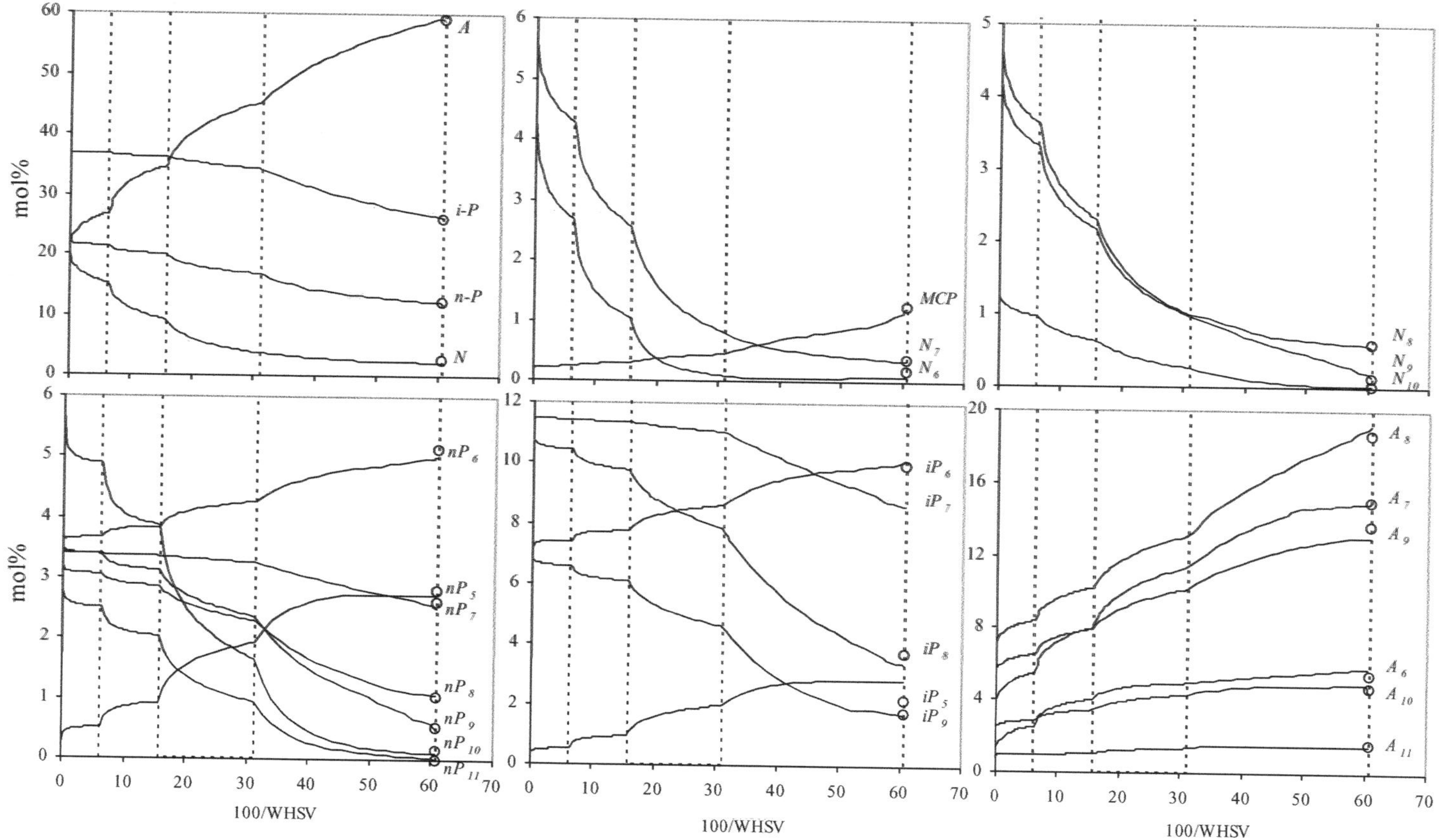

Figure 4.10. Profiles of predicted (—) and commercial (○) molar composition of the reformate.

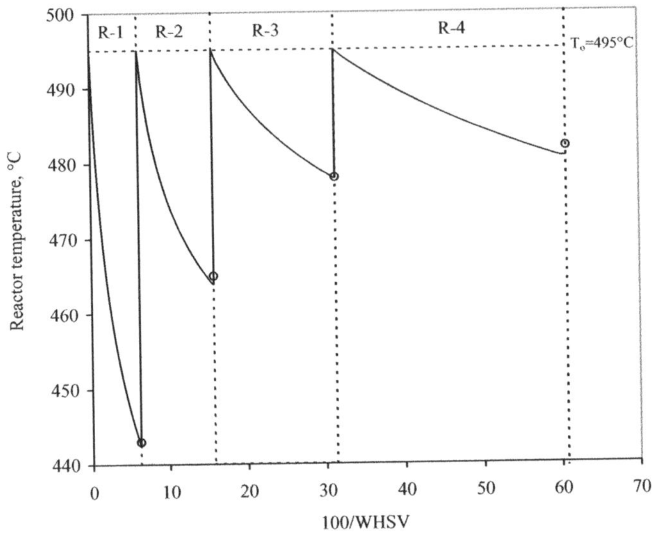

Figure 4.11. Profiles of predicted reactor temperature (—) and commercial value (○).

aromatics content in the third and fourth reactors appears to be solely because of the disappearance of paraffins.

ΔT *of Reactors* Figure 4.11 shows the temperature profiles in the four reactors in series. In the first reactor, the major reactions are endothermic and fast (e.g., dehydrogenation of naphthenes to aromatics), causing a very sharp temperature drop from 495°C to 443°C. This remarkable decrease in temperature ($\Delta T_1 = 52°C$) almost quenches all reactions, and more catalyst in the first reactor would not provide additional conversion.

The outlet stream of the first reactor is reheated to the same temperature as that of the inlet of the first reactor (495°C) and fed to the second reactor. Here, most of the isomerization reaction takes place, the remaining naphthenes are dehydrogenated, and a moderate temperature drop is observed ($\Delta T_2 = 30°C$). From the second reactor the effluent is again reheated to 495°C before entering the third reactor, and finally, the same sequence is followed in the fourth reactor. The temperature drops across the third and fourth reactors are lower (17 and 14°C, respectively), which are due to the exothermic hydrocracking of paraffins reaction. Dehydrogenation and cracking reactions take place in the last two reactors. The average temperature and the amount of catalyst in the last reactors are higher that those of the first reactors. This allows the transformation of paraffins into aromatics by dehydrocyclization and into light paraffins by hydrocracking.

The comparison of simulated and actual temperature drops indicates that the reactor model predictions match very well with the information reported in the commercial reforming unit. The accumulated absolute difference between predicted and actual reactor ΔT values is smaller than 3°C.

4.3.4 Simulation of the Effect of Benzene Precursors in the Feed

Preparation of the Feed The feed used in this case is slightly different from that employed in previous sections. The new feed (feed A) has the following main properties: 0.730 specific gravity 60/60°F, 110 g/mol molecular weight, and 60 to 190°C distillation range. Having an initial boiling point (IBP) of 60°C indicates that the precursors of benzene formation are present in this feed, since the hydrocarbons involved in the benzene formation network have the following normal boiling points: 80.1°C benzene, 68.7°C hexane, 71.8°C methyl cyclopentane, and 80.7°C cyclohexane.

To evaluate the effect of benzene formation precursors, feed A was fractionated to adjust its initial boiling point to 88°C in order to prepare a free-benzene precursor feed. This other feed (feed B) has the following main properties: 0.734 specific gravity 60/60°F, 112 g/mol molecular weight, and 88 to 190°C distillation range.

The compositions of feeds A and B determined by gas chromatography are presented in Table 4.15. It is clearly observed that feed A possesses a certain amount of the precursors of benzene formation and benzene itself (1.12 mol% A_6, 7.69 mol% P_6, 0.64 mol% MCP, and 4.23 mol% N_6), while in feed B their contents are substantially reduced (0.26 mol% A_6, 0.44 mol% P_6, 0 mol% MCP, and 3.45 mol% N_6).

Because of the separation of light compounds, feed B became a little bit heavier than feed A, which is observed as an increase in specific gravity and molecular weight. i-P_5, i-P_6 and MCP were totally removed by adjusting the IBP of feed A, and n-P_5 and n-P_6 were reduced considerably (>94%). Cyclohexane (N_6) exhibited a reduction of 18.4%, while reduction in benzene (A_6) was of 68.3%.

The main benzene formation precursors are N_6 and MCP, and of less importance, n-P_6. By adjusting the IBP of feed A from 60 to 88°C, MCP and n-P_6 are almost totally eliminated, but N_6 is still present in a high amount in feed B. It is expected that at the beginning of the reaction N_6 will not be produced via MCP isomerization, since feed B does not have MCP in its composition. Therefore, benzene will be obtained only by dehydrogenation of the N_6 originally present in feed B. As the reaction proceeds, MCP will be formed from P_6 and N_6, and hence A_6 production will increase.

Results of the Simulations For this case, the simulations were carried out first in the isothermal mode of operation, to compare the reformate composition predicted with the experimental values obtained in the bench-scale reactor, and then in the adiabatic mode to predict the behavior of a commercial reforming unit.

TABLE 4.15. Molar Composition of Feeds with and Without Benzene Formation Precursors

	Feed A	Feed B
n-Paraffins		
nP_{11}	0.77	0.79
nP_{10}	2.72	3.55
nP_9	4.05	5.82
nP_8	5.52	7.88
nP_7	6.77	9.75
nP_6	7.69	0.44
nP_5	6.80	0.09
Total	34.32	28.32
i-Paraffins		
iP_{10}	4.10	5.53
iP_9	4.52	5.20
iP_8	6.50	8.39
iP_7	5.64	6.68
iP_6	6.72	
iP_5	2.46	
Total	29.94	25.80
Naphthenes		
N_{10}	0.87	1.15
N_9	3.56	4.72
N_8	4.04	7.00
N_7	5.95	8.43
N_6	4.23	3.45
MCP	0.64	
Total	19.29	24.75
Aromatics		
A_{11}	0.96	0.90
A_{10}	1.34	1.25
A_9	4.24	5.75
A_8	5.77	8.45
A_7	3.02	4.52
A_6	1.12	0.26
Total	16.45	21.13

Isothermal Model Predictions Versus Experimental Data Figure 4.12 shows a comparison of experimental and predicted reformate molar composition of some selected hydrocarbon types (MCP, N_6, N_7, and A_6) as a function of position in the catalyst bed. It is observed that the calculated compositions agree very well with experimental bench-scale reactor information with an average deviation of less than 3%. Particularly, the calculated A_6 (benzene) composition matches very well with experimental data, with a maximum deviation of 2%.

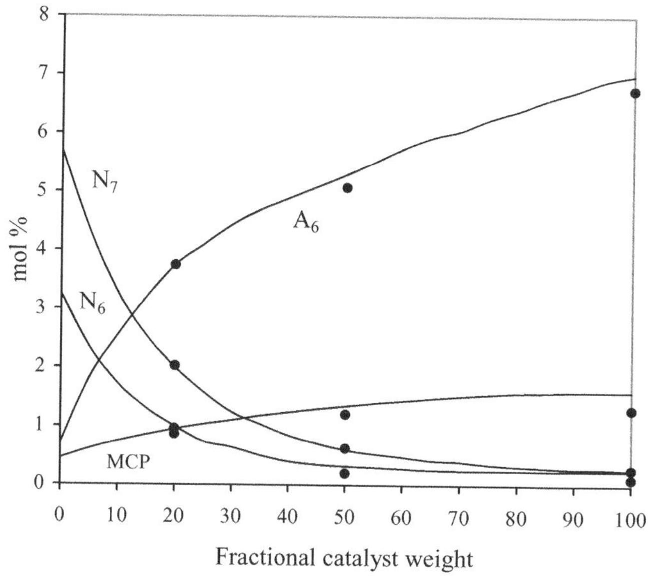

Figure 4.12. Comparison of experimental (●) and predicted (—) molar composition of the reformate obtained at 510°C in the bench-scale reactor.

It is worthy of mention that rate constants of lumped kinetic models usually depend on feedstock and catalyst properties, and it may be inappropriate to use them for simulating reforming reactors for feed conditions different from those from which the parameters were determined. However, if the kinetic model is sufficiently detailed, rate constants can be considered to be independent of initial feedstock composition, and thus can be used to simulate the reactor for other feed conditions. A problem when using more detailed models is that the simplicity of kinetic representations used in models with a small number of lumps is partially lost, since the use of kinetic models with a large number of lumps, where the number of parameters is increased significantly, means that greater amounts of experimental data are also required. Despite this, the results shown in Figure 4.12 clearly indicate that the kinetic model developed is sufficiently detailed to consider the kinetic parameters to be independent of feed composition.

Predictions with the Adiabatic Model Figure 4.13 presents the effects of feed composition on benzene and its formation precursors hydrocarbon as well as the total aromatics content in the reformate and commercial reactor ΔT as a function of the inverse of spacevelocity (100/WHSV) and temperature. The following effects are observed:

- Feed B produces less benzene in the reformate than does feed A (3.4 vs. 6.1 mol% at the exit of the fourth reactor).

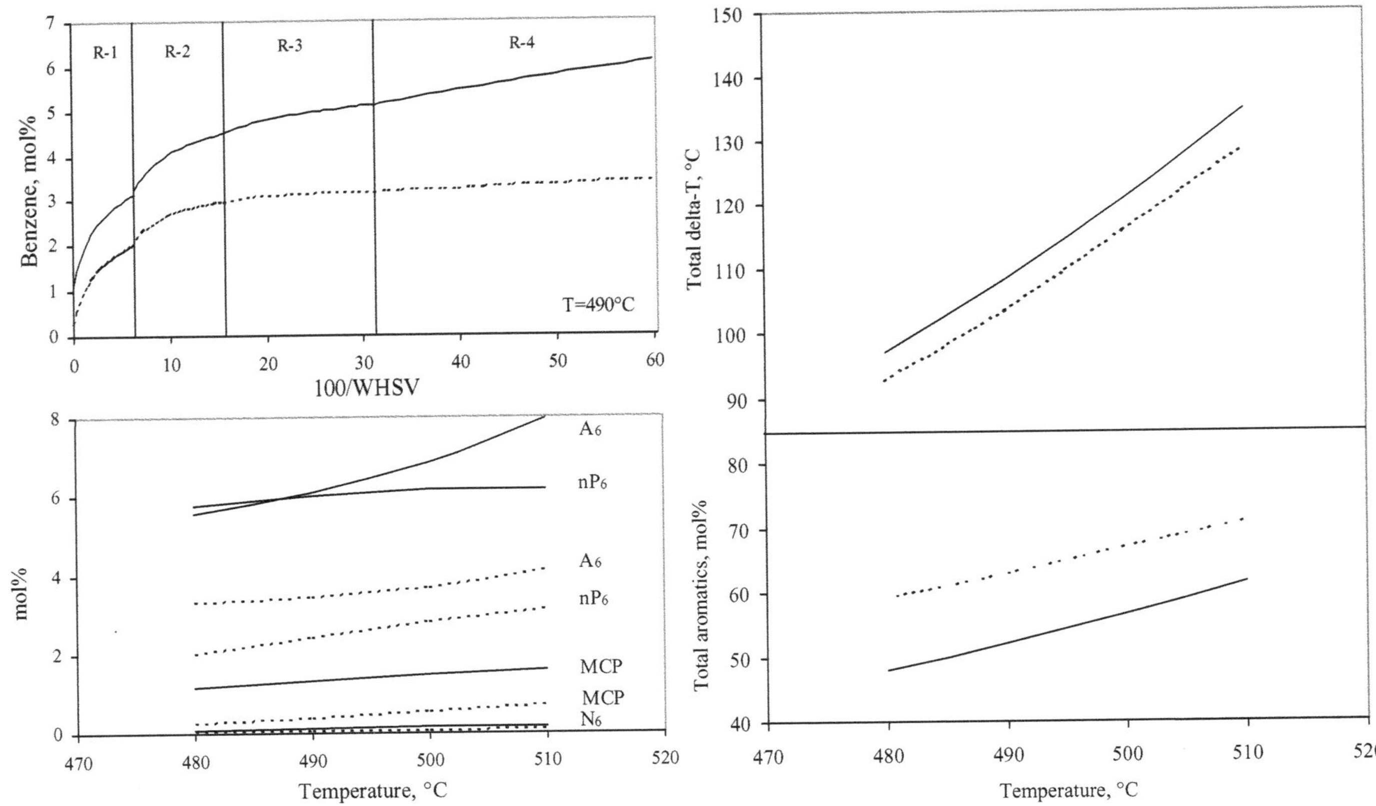

Figure 4.13. Simulation of the operation of commercial reforming reactors with feeds with benzene formation precursors (—, feed A) and without benzene formation precursor (---, feed B).

- The benzene production rate in the first two reactors using feed A is higher than that of feed B. In these two reactors, benzene is produced primarily by N_6 dehydrogenation. After reactor 2 there is no considerable increase in benzene content when feed B is used (from 2.95 mol% at the exit of the second reactor to 3.41 mol% at the exit of fourth reactor), while this increase was indeed higher with feed A (from 4.5 to 6.1 mol% at the exit of the same reactors).
- The complete separation of MCP as in feed B aids in reducing cyclohexane formation, and thus benzene production.
- N_6 conversion is almost 100% at all reaction temperatures for the two feeds. As the reaction temperature is increased, the MCP content in the reformate obtained from both feeds also increases with respect to feed initial values, which is produced mainly from P_6. The increase in MCP contributes to N_6 formation, and hence the benzene content in reformate is also increased.
- The effect of temperature on benzene formation is less in feed A than in feed B. A_6 increases from 1.12 mol% to 8 mol% at 510°C for feed A, while this increase for feed B is from 0.26 mol% to 4 mol%. This indicates that the separation of benzene formation precursors from the reforming feed also helps to decrease A_6 formation as the reaction temperature is increased.
- Reactor ΔT is less when feed B is used, which means that this feed has a lower content of those reacting compounds that contribute more to the reaction exothermality than does feed A.
- Another reason for the differences in reactor ΔT with both feeds is the different heats of reforming reaction, and the place and extent to which they take place. For example, in the first reactor the major reactions are endothermic and very fast, such as dehydrogenation of naphthenes to aromatics; the isomerization takes place primarily in the second reactor, and the remaining naphthenes are dehydrogenated; and exothermic hydrocracking of paraffins and dehydrogenation reactions occur in the third and fourth reactors.
- Feed B produces more aromatics than feed A, which is due to the higher initial contents of aromatics and naphthenes in feed B (21.12 and 24.75 mol%, respectively) compared with feed A (16.45 and 19.29 mol%, respectively).

4.3.5 Use of the Model to Predict Other Process Parameters

Apart from calculating the reformate composition, temperature, and pressure profiles along the reactors system, the model developed can be used to analyze other aspects of the catalytic reforming process. Turpin (1992) proposed a procedure to validate the performance of a catalytic reforming unit. This procedure considers the calculation of the global and hydrogen material balances,

TABLE 4.16. Global Mass and Molar Balances

Stream	Inlet (g/h)	Outlet (g/h)
Naphtha (C_5–C_{12})	119.9916	110.4971
Hydrogen	19.6989	20.8477
Gases (C_1–C_4)	0.0000	8.3458
Total	139.6905	139.6905
	Inlet (mol/h)	Outlet (mol/h)
Naphtha (C_5–C_{12})	1.0857	1.0653
Hydrogen	9.7713	10.3411
Gases (C_1–C_4)	0.0000	0.2064
Total	10.8570	11.6125

a comparison of experimental and calculated feed and product densities, an analysis of the gas composition (calculation of some quotients, isobutene/*n*-butane molar ratio), probability plots of feed and reformate components, and ring balance. All this information and more can be generated with the kinetic and reactor models developed.

As examples, Tables 4.16 and 4.17 show the global and hydrogen mass balances for a study simulated at the following conditions: bench-scale isothermal reactor, 750°C inlet temperature, 10.5 kg/cm^2 reactor pressure, and 6.3 mol/mol H_2/oil ratio. From the data of these tables, the balance errors are calculated with

$$\text{mass balance error} = \frac{\text{feed mass flow rate} - \text{product mass flow rate}}{\text{feed mass flow rate}} \times 100 \tag{4.71}$$

$$H_2 \text{ balance error} = \frac{H_2 \text{ mass flow rate} - H_2 \text{ mass flow rate}}{H_2 \text{ mass flow rate}} \times 100 \tag{4.72}$$

For both balances the error was zero, while the maximum acceptable errors in the global balance and in the hydrogen balance are ±1% and ±0.5%, respectively. Values outside these ranges indicate errors when measuring the flow rates or in analysis of the gas and liquid streams.

Given that it is possible to determine the amount of hydrogen entering the reactor and also that which has been produced or consumed by chemical reactions, the amount of hydrogen leaving the reactor can be calculated with a mass balance as shown in Table 4.18. If the severity of the reforming reactor is modified, determination of the amount of hydrogen at the exit of the reforming reactor becomes important since it gives the potential quantity of hydrogen supply to hydrotreating/hydrocracking units.

TABLE 4.17. Hydrogen Balance

Hydrogen Content in Hydrocarbons	Inlet (g/h)	Outlet (g/h)
n-Paraffins		
nP_{11}	0.0000	0.0000
nP_{10}	2.4727	0.7595
nP_9	2.5915	1.5639
nP_8	2.9667	2.2383
nP_7	2.2658	2.3005
nP_6	1.8799	2.2477
nP_5	0.4216	1.1713
nP_4	0.0000	0.6009
nP_3	0.0000	0.5552
nP_2	0.0000	0.2581
nP_1	0.0000	0.1411
Subtotal	12.5981	11.8365
Naphthenes		
N_{11}	0.0000	0.0000
N_{10}	0.0000	0.0412
N_9	1.2115	0.0615
N_8	0.7757	0.0827
N_7	0.8411	0.1792
N_6	0.4163	0.1305
MCP	0.0000	0.0000
Subtotal	3.2446	0.4951
Aromatics		
A_{11}	0.0000	0.0000
A_{10}	0.3309	0.5416
A_9	0.6146	1.4930
A_8	0.5921	1.2787
A_7	0.1996	0.6850
A_6	0.0420	0.1432
Subtotal	1.7793	4.1416
Total H_2 in hydrocarbons	17.6220	16.4732
Hydrogen gas	19.7003	20.8491
Total	37.2323	37.2323

Once the composition of the reformate product has been calculated with the model, some of its properties can be determined with mixing rules by using properties of the pure components, such as octane numbers (RON, MON), Reid vapor pressure (RVP), density, molecular weight, and average boiling points, among others. Data reported in the literature to calculate some of these properties are summarized in Table 4.13. In general, those properties that are additive (i.e., depend on the mass) are evaluated by linear mixing rules, such as density and molecular weight. Other properties (e.g., RON, MON, RVP)

TABLE 4.18. Amounts of Hydrogen Determined from Chemical Reaction Calculations

	mol/mol	g/g	std ft^3/bbl	m^3/bbl
Inlet H_2	6.3000	0.1157	5600.5	158.6
Produced H_2	1.8094	0.0329	1594.8	45.2
Consumed H_2	−0.7681	−0.0140	−677.0	−19.2
Net H_2 (produced − consumed)	1.0413	0.0189	917.8	26.0
Outlet H_2	7.3413	0.1346	6518.3	184.6

depend nonlinearly on the mixture composition, thus requiring nonlinear mixing rules or other approaches (e.g., structural group contribution).

REFERENCES

Ancheyta, J.; Aguilar, E. (1994) New model accurately predicts reformate composition. *Oil Gas J.*, Jan. 31, pp. 93–95.

Ancheyta, J., Villafuerte, E. (2000) Kinetic modeling of naphtha catalytic reforming process. *Energy Fuels* 14:1032–1037.

Ancheyta, J., Villafuerte, E., García, L., González, E. (2001) Modeling and simulation of four catalytic reactors in series for naphtha reforming. *Energy Fuels* 15: 887–893.

Ancheyta, J., Villafuerte, E., Schacht, P., Aguilar, R., Gonzalez, E. (2002) Simulation of a semiregenerative reforming plant using feedstocks with and without benzene precursors. *Chem. Eng. Technol.* 25:541–546.

Bommannan, D., Srivastava, R. D., Saraf, D. N. (1989) Modeling of catalytic naphtha reformers. *Can. J. Chem. Eng.* 67:405–411.

Burnett, R. L., Steinmetz, H. L., Blue, E. M., Noble, E. M. (1965) An analog computer model of conversion in a catalytic reformer. Presented at the Division of Petroleum Chemistry, American Chemical Society, Detroit meeting, Apr. 17–24.

Coppens, M. O., Froment, G. F. (1996) Fractal aspects in the catalytic reforming of naphtha. *Chem. Eng. Sci.* 51:2283–2292.

Dorozhov, A. P. Moskva. (1971) Ph.D. dissertation.

Henningsen, J., Bundgaard-Nielson, M. (1970) Catalytic reforming. *Bri. Chem. Eng.* 15:1433–1436.

Hou, W., Su, H., Hu, Y., Chu, J. (2006) Modeling, simulation and optimization of a whole industrial catalytic naphtha reforming process on Aspen Plus platform. *Chin. J. Chem. Eng.* 14(5):584–591.

Hou, W., Su, H., Mu, S., Chu, J. (2007) Multiobjective optimization of the industrial naphtha catalytic reforming process. *Chin. J. Chem. Eng.* 15(1):75–80.

Hu, Y., Su, H., Mu, S., Chu, J. (2003) Modeling, simulation and optimization of commercial naphtha reforming process. In: Proceedings of the 42nd *IEEE Conference on Decision and Control*, Maui, HI, Dec., pp. 6206–6211.

Jenkins, J. H., Stephens, T. W. (1980) Kinetics of cat reforming. *Hyd. Proc.* 59:163–167.

Joshi, P. V., Klein, M. T., Huebner, A. L., Leyerle, R. W. (1999) Automated kinetic modeling of catalytic reforming at the reaction pathways level. *Rev. Proc. Chem. Eng.* 2(3):169–193.

Kmak, W. S. (1972) A kinetic simulation model of the powerformimg process. Presented at the AIChE National Meeting, Houston, TX.

Kmak, W. S., Stuckey, T. W. (1973) Powerforming process studies with a kinetic simulation model, Paper 56a. Presented at the AIChE National Meeting, New Orleans, LA.

Krane, H. G., Groh, A. B., Shulman, B. D., Sinfeit, J. H. (1959) Reactions in catalytic reforming of naphthas. In: Proceedings of the Fifth World Petroleum Congress, May 30, Sec. III, pp. 39–51.

Lee, J. W., Ko, K. Y., Jung, Y. K., Lee, K. S. (1997) A modeling and simulation study on a naphtha reforming unit with a catalyst circulation and regeneration system. *Comput. Chem. Eng.* 21:S1105–S1110.

Liang, K., Guo, H., Pan, S. (2005) A study on naphtha catalytic reforming reactor simulation and analysis. *J. Zhejiang Univ. Sci.* 6B(6):590–596.

Lid, T., Skogestad, S. (2008) Data reconciliation and optimal operation of a catalytic naphtha reformer. *J. Process Control* 18:320–331.

Marin, G. B., Froment, G. F. (1982) Reforming of C_6 hydrocarbons on a platinum–alumina catalyst. *Chem. Eng. Sci.* 37(5):759–773.

Moharir, A. S., Agarwal, A. B. L., Saraf, D. N. (1979) Symposium on Science of Catalysis and Its Application in Industry, FPDIL, Sindri, 163–170.

Padmavathi, G., Chaudhuri, K. K. (1997) Modeling and simulation of commercial catalytic naphtha reformers. *Can. J. Chem. Eng.* 75(5):930–937.

Rahimpour, M. R., Esmaili, S., Bagheri, G. N. A. (2003) Kinetic and deactivation model for industrial catalytic naphtha reforming. *Iran. J. Sci. Tech. Trans. B* 27(B2):279–290.

Ramage, M. P., Graziani, K. R., Krambeck, F. J. (1980) Development of Mobil's kinetic reforming model. *Chem. Eng. Sci.* 35:41–48.

Ramage, M. P., Graziani, K. R., Schipper, P. H., Krambeck, F. J., Choi, B. C. (1987) KINPTR (Mobil's kinetic reforming model): a review of Mobil's industrial process modeling philosophy. *Adv. Chem. Eng.* 13:193–266.

Reid, R. C., Prausnitz, J. M., Sherwood, T. K. (1977) *The Properties of Gases and Liquids.* Mc-Graw Hill, 3rd Ed., New York.

Shanyinghu, F., Zhu, X. X. (2004) Molecular modeling and optimization for catalytic reforming. *Chem. Eng. Commun.* 191:500–512.

Smith, R. B. (1959) Kinetic analysis of naphtha reforming with platinum catalyst. *Chem. Eng. Prog.* 55(6):76–80.

Smith, J. M., Van Ness, H. C., Abbott, M. M. (1996) *Introduction to Chemical Engineering Thermodynamics*, Mc-Graw Hill, 5th Ed., New York.

Sotelo, R., Froment, G. F. (2009) Fundamental kinetic modeling of catalytic reforming. *Ind. Eng. Chem. Res.* 48:1107–1119.

Stijepovic, M. Z., Vojvodic-Ostojic, A., Milenkovic, I., Linke, P. (2009) Development of a kinetic model for catalytic reforming of naphtha and parameter estimation using industrial plant data. *Energy Fuels* 23:979–983.

Szczygiel, J. (1999) On the kinetics of catalytic reforming with the use of various raw materials. *Energy Fuels* 13:29–39.

Taskar, U. (1996) Modeling and optimization of a catalytic naphtha reformer, Ph.D. dissertation, Texas Tech University.

Taskar, U., Riggs, J. B. (1997) Modeling and optimization of a semi-regenerative catalytic naphtha reformer. *AIChE J.* 43(3):740–753.

Turpin, L. E. (1992) Cut benzene out of reformate. *Hydrocarbon Process.*, 81–92.

Van Trimpont, P. A., Marin, G. B., Froment, G. (1988) Reforming of C_7 hydrocarbons on a sulfided commercial Pt/Al_2O_3 catalyst. *Ind. Eng. Chem. Res.* 27:51–57.

Viñas, J. M., Gonzalez, M. G., Barreto, G. F. (1996) A kinetic model for simulating naphtha reforming reactors. *Lat. Am. Appl. Res.* 26(1):21–34.

Wei, W., Bennett, C. A., Tanaka, R., Hou, G., Klein, M. T. (2008) Detailed kinetic models for catalytic reforming. *Fuel Process. Tech.* 89:344–349.

Zhorov, Y. M., Panchenkov, G. M., Zel'tser, S. P., Tirakyan, Y. A. (1965) Mathematical description of platforming for optimization of a process (I). *Kineti. Katal.* 6(6):1092–1098.

Zhorov, Y. M., Panchenkov, G. M., Shapiro, I. Y. (1970) Mathematical description of Platforming carried out under severe conditions. *Khim. Technol. Topl. Masel* 15(11):37–40.

NOMENCLATURE

a, b, c	Parameters of the hydrogen reaction rate equation
A	Aromatics
A, B, C, D	Constant for calculating C_p
A_{10}	Aromatics with 10 atoms of carbon
A_{10}^{+}	Aromatics with 10+11 atoms of carbon
A_{11}	Aromatics with 11 atoms of carbon
C_p	Molar specific heat
C_{10}	Hydrocarbons with 10 atoms of carbon
C_{10}^{+}	Hydrocarbons with 10+11 atoms of carbon
C_{11}	Hydrocarbons with 11 atoms of carbon
d_p	Particle diameter
E_A	Activation energy
F	Molar flow
g_c	Force-to-mass conversion factor, 9.8066 $kg_m \cdot m/kg_f \cdot s^2$
G	Superficial mass velocity
ΔG°	Reaction standard Gibbs energy
ΔH	Heat of reaction
k_i	Kinetic constant at T
k_i^0	Kinetic constant at T_0
k_{10}	Kinetic constant for hydrocarbons with 10 atoms of carbon
k_{10}^{+}	Kinetic constant for hydrocarbons with 10 + 11 atoms of carbon
k_{11}	Kinetic constant for hydrocarbons with 11 atoms of carbon
K	Ratio of kinetic constants
K_e	Equilibrium constant
L	Reactor length

LHSV	Liquid hourly space velocity
m	Number of components
MW	Molecular weight
n	Reaction order
N	Number of reactors; naphthenes
N_{10}	Naphthenes with 10 atoms of carbon
N_{10}^{+}	Naphthenes with 10 + 11 atoms of carbon
N_{11}	Naphthenes with 11 atoms of carbon
p_i	Partial pressure of component i
P	Reaction pressure; paraffins
P_0	Base reaction pressure
P_{10}	Paraffins with 10 atoms of carbon
P_{10}^{+}	Paraffins with 10 + 11 atoms of carbon
P_{11}	Paraffins with 11 atoms of carbon
Pe	Peclet number
r_i	Rate of reaction of component i
R	Universal constant of gases; ratio of hydrocarbon compositions
Re_p	Reynolds number based on particle diameter
S	Cross-sectional area
SV	Spacevelocity
T	Reaction temperature
T_0	Base reaction temperature
x	Conversion
WABT	Weighted-average bed temperature
WAIT	Weighted-average inlet temperature
Wc_i	Weight fraction of catalyst in each reactor bed with respect to the total
WHSV	Weight hourly space velocity
y_i	Molar composition of component i
z	Reactor length

Greek Letters

ε	Void fraction of the catalyst bed
μ	Viscosity of the gas mixture
ρ	Density of the gas mixture
ρ_c	Density of the catalyst

5

MODELING AND SIMULATION OF FLUIDIZED-BED CATALYTIC CRACKING CONVERTERS

RAFAEL MAYA-YESCAS
Facultad de Ingeniería Química, Universidad Michoacana de San Nicolás de Hidalgo, Morelia, Michoacán, México

The current fluidized-bed catalytic cracking (FCC) process has evolutioned from thermal hydrocarbon cracking in fixed-bed reactors to the one that uses natural and pillared clays in large fluidized beds. After the 1950s, a revolution was begun with the introduction of *zeolites*, molecular sieves that contain catalytic sites located at specific-sized pores; these catalysts promote selectivity and decrease coke formation by precursor deposition. Zeolites were sealed by harder matrices, some of them containing rare earths to promote thermal stability; thus were born the X and Y types of modern catalysts. Later, type Y zeolites were hydrogenated to produce the first HY type, which served as a basis for the production of an "ultrastable" zeolite, called USY. USY is able to promote reation rates 1000 times faster than those of the original Y type. Following catalyst development, reactor engineering changed from the large fluidized-bed reactors (useful because of their long residence time) to the fast "risers" (upflow transported beds).

Riser reactor engineering is one field that is still under research, because of its complexity. Feedstock arrives at the base of the riser (about 200°C) and is sparged by midpressure steam (about 300°C). Meanwhile, catalyst from the regenerator (about 700°C) is dispersed using the same type of steam, again at the base of the riser. By using this heat, feedstock is evaporated and the catalytic reactions begin. Cracking reactions are moderately endothermal; therefore, they need energy to be performed; they also generate molecules because of the heavy hydrocarbon cracking. Both solid catalyst and vapor mixture of

Modeling and Simulation of Catalytic Reactors for Petroleum Refining, First Edition.
Jorge Ancheyta.
© 2011 John Wiley & Sons, Inc. Published 2011 by John Wiley & Sons, Inc.

feedstock and product travel along the riser for 3 to 5, up to the riser outlet, where solids are separated using a cyclone. Meanwhile, catalyst activity is decaying continuously along the riser, because of deposition of a solid product, coke, on its surface.

Modeling this reactor is complicated, as is obvious, but many simplifications are in common used. For example, both phases in the riser flow follow plug-flow patterns; here the adjustable parameter is the slip velocity, which is the difference between the catalyst and vapor velocities. As will be shown, this is not easy to estimate; however, it does have an effect on the simulation of the entire unit. Finally, riser reactors are always in a transient state; nevertheless, its response time is shorter than that of the regenerator, 3 to 5 s and 4 to 11 min, respectively; therefore, it is usually enough to consider this part of the FCC unit as working in the pseudo-steady state, an assumption that simplifies the dynamics' simulation and control of the unit.

The energy to evaporate the feedstock and sustain the endothermal cracking reactions could be very resource demanding; fortunately, the second-most-important product of the FCC unit (as explained later) is coke, which is deposited on the catalyst surface during cracking reactions. This entity can be burned off at the regenerator, generating the heat necessary to sustain the endothermal reactions and, moreover, to sustain feedstock evaporation and heating up to the reaction temperature. FCC units are adiabatic; therefore, the heat exchange between the riser and the regenerator uses the solid catalyst as a vehicle. This physical situation greatly complicates operation of the unit, as described in the following sections. The regeneartor is a big reactor compared to the riser; therefore, it is necessary to consider this unit as operating in a transient state. Usually, it is enough to consider this unit as a system consisting of three regions (de Lasa et al., 1981; Errazu et al., 1979):

- A two-region system (solid–gas) that is a continuous-stirred-tank reactor (CSTR), consisting of a solid catalyst fluidized by air, called the *dense phase.*
- A top diluted region that consists mainly of the remaining nitrogen and combustion gases, located just ahead of the escape of flue gases from the regenerator, called *freeboard.*

Riser and regenerator are connected by the stripper, which uses midpressure steam to desorb hydrocarbons from the catalyst suface. This is another unit that is under study because of the complexity of the modeling of desorption rates in this system. Stripped catalyst particles go from the stripper to the regenerator; to burn the coke and recover catalyst activity, particles are sent later to the riser by a standpipe. The amount of catalyst sent to the riser controls temperature at its outlet, a parameter that is the control objective during industrial operation.

As a final note in this background section, FCC units are currently used to produce gasoline and its additives (as precursors for MTBE units, for example);

nonetheless, there is a trend toward substituting hydrocarbon fuels in the near future (i.e., 25 to 50 years). Therefore, most refinery processes in use currently will be replaced, or perhaps the entire refinery industry will be altered into something very different from its curent design. It is important here to note that FCC units will continue being part of this new refinery, working to produce lighter products from middle and heavy distillates; these products will go into petrochemical processes to produce new materials. Hence, the future of FCC units is long, and it is important to model and simulate this type of process to be able to effect the changes and challenges that are coming very soon.

5.1 INTRODUCTION

5.1.1 Description of the Process

Because of the large yield to products and their added value, FCC units are today one of the most important processes in oil refineries; a very comprehensive description of this process is given in the classical paper by Venuto and Habib (1978). The FCC process generates more than 40% of the gasoline in the refinery pool; consequently, any small benefit in this process is very profitable. FCC is a very complex process (Salazar-Sotelo et al., 2004) that has at its heart the riser–regenerator couple, known as a converter (Figure 5.1). After preheating, partially evaporated feed enters the riser, where it contacts the regenerated catalyst. The heat absorbed by the catalyst during regeneration provides the energy to evaporate and heat the feed to its desired reaction temperature (about 580°C). Many reactions take place in the vapor phase inside the riser. The products recovered are dry gases (H_2 and C_2), liquid-petroleum gas (LPG, which consists of C_3's and C_4's), gasoline (C_5's, b.p. 221°C), and cyclic oils (considered part of the unreacted feedstock, b.p. > 221°C). Also, there is coke formation; this solid compound deposits on the catalyst surface. The average heat of reaction resulting from feedstock evaporation plus cracking reactions is endothermic.

After reactions occur, catalyst and products are quickly separated in cyclones located at the riser outlet and catalysts fall into a stripper, where steam is used to "strip" the entrained hydrocarbons from catalyst particles; catalyst is transferred to the regenerator, where its activity is restored by burning off the coke with air. One of the most important parts of the FCC is the catalyst regenerator–reactor, because in this vessel the cokized spent catalyst is regenerated in order to recover catalytic activity. Regeneration consists of burning off the coke deposited using atmospheric air in a fluidized-bed reactor, usually a CSTR. The energy generated by the exothermic reactions is employed to vaporize the feedstock and to support the endothermic cracking reactions, which takes place in the riser reactor (Maya-Yescas and Aguilar, 2003). Considering the exothermic nature of the regeneration reactions and the characteristics of the combustion kinetics, which can be described by consecutive reactions, the dynamic behavior of the regenerator is expected to be

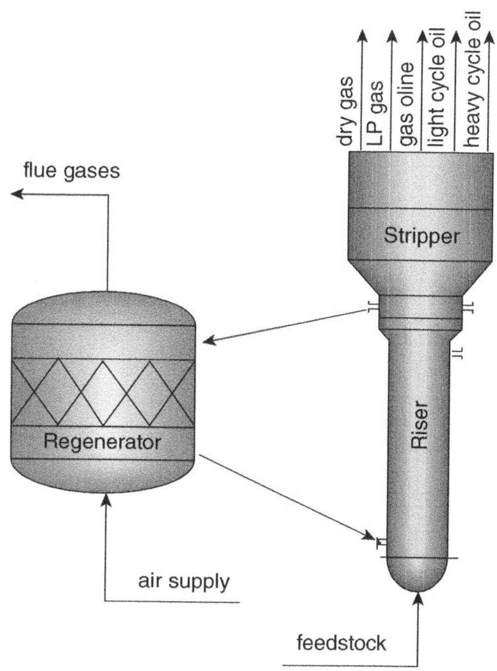

Figure 5.1. Typical FCC converter.

very complex. Phenomena such as steady-state multiplicity, inverse response to control actions, and unstable operating zones may appear. An interesting feature of the system is that a linear approximation of the model exhibits eigenvalues with a positive real part, which is an indication of the instability of the closed-loop internal dynamics (Daoutidis and Kravaris, 1991); this instability will be reflected by control problems around these states (Maya-Yescas and Aguilar, 2003). Regeneration reactions generate the heat that is used to vaporize the feedstock at the riser and to sustain cracking reactions. Finally, hydrocarbon products are fractionated and narrow cuts are separated into commercial-interest products.

On the other hand, FCC units are located at the heart of refineries, which makes them one of the most studied chemical processes. This chapter is devoted to revising some basic concepts of FCC processes and to introduce the reader to modeling techniques, control scheme comparisons and several technological developments that are changing the original sense of this interesting process.

5.1.2 Place of the FCC Unit Inside the Refinery

As stated earlier, because of its physical position, size, and economic impact, FCC units are the heart of a refinery (Figure 5.2). They are preceded by basic

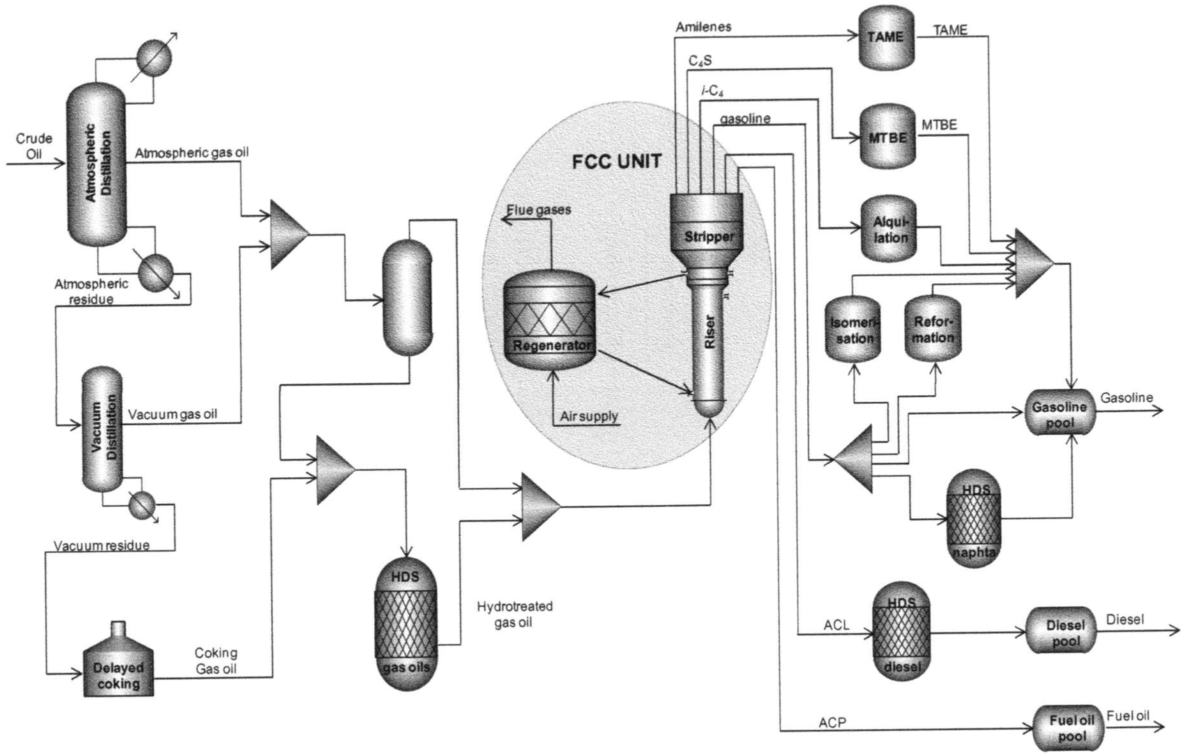

Figure 5.2. Location of the FCC unit in the refinery.

separation processes such as atmospheric and vacuum distillation units; catalytic reacting processes such as hydrotreatment; and minor additional units, pumps, stabilization towers, and so on. Downstream, FCC units supply products mainly to the gasoline pool, but also to other units that require light hydrocarbons. Finally, when there is a high level of sulfur in the gasoline, FCC units supply feedstock for hydrodesulfurisation processes. Therefore, FCC units interact very little with downstream processes and some with upstream processes.

5.1.3 Fractionation of Products and Gas Recovery

The products from an FCC converter are classified as a mixture from dry gas (DG) up to heavy cycle oil (HCO). This blend is separated by the boiling point (atmospheric pressure) in a fractionator followed by a vapor recovery unit. The main fractionator sends heavy products to a nonisothermal decanter, where cycle oils are separated into light (LCO) and heavy (HCO). Top products are compressed and cooled down to separate those that are condensable; these are stabilized (debutanized) in order to obtain gasoline and to separate LPG plus DG. Finally, DG and LPG are separated by condensation, sending DG as a fuel to the rest of the refinery.

5.1.4 Common Yields and Product Quality

The most important product from the FCC unit is gasoline, whose yield ranges between 46 and 51 wt% for standard feedstock and could increase to about 60 wt% for hydrotreated feedstock. Quality parameters for gasoline are mechanical octane number (MON), research octane number (RON), and antiknocking index [AKI = (RON + MON)/2], which are related to the content of paraffins, olefins, naphthenes, and aromatics. When this classification was proposed, it was considered that a perfect gasoline should exhibit AKI = 100, which corresponds to pure 2,2,4-trimethylpentane, usually called *isooctane*. The second commercial product is LPG, especially if the FCC unit supplies it to downstream processes; usually, LPG yield is about 12 to 15 wt%. In LPG, one quality parameter, also related to the quality of gasoline, is the *i*-butane/buthylene ratio.

DG (~5 wt%), LCO (~15 wt%), and HCO (~8 wt%) are not subjected to quality standards, but it is felt that the lower the aromatic content in liquid products, the better. It is possible to incorporate LCO in diesel fuel; however, it requires hydrotreatment to decrease the sulfur and aromatic contents. Finally, about 4 to 6 wt% of the original feedstock converts to coke, which is a solid that deposits on the catalyst surface, blocking the pores and, consequently, decreasing catalyst activity. All these yields might change, depending on the type of feedstock (from heavy to hydrotreated) and the production objectives of the catalyst.

5.2 REACTION MECHANISM OF CATALYTIC CRACKING

One of the most interesting parts of the FCC system from the modeling and operating points of view is the kinetics. This is because there is no simple way to define what *kinetics* means; furthermore, to date it is not known how many reactions take place inside each FCC reactor and how many additional phenomena (i.e., pore blockage, poisoning, sintering, breaking out, etc.) influence the observed (apparent) kinetics. This section is devoted to explaining some of the particular characteristics of the multiple FCC kinetics that have been proposed and currently are under research. We will study the kinetics of the FCC process in terms of cracking reactions, catalyst deactivation, and coke burning during regeneration reactions.

There is no such "reaction mechanism" for catalytic cracking. Even though catalysts for cracking reactions have remained almost unchanged for several years, phenomena involved in catalytic cracking are still under study. The way to approach them has been divided here into transport phenomena, thermodynamics, and kinetics; a lumping approach to understanding; and another complex approach.

5.2.1 Transport Phenomena, Thermodynamic Aspects, and Reaction Patterns

The riser (the transported solid-bed reactor) of the FCC unit is the main reactor in the process. Inside the riser, catalytic cracking of the PNA hydrocarbons takes place. However, prior to the chemical reactions, there are several physical phenomena that are happening on a molecular scale; therefore, they are difficult (or sometimes impossible) to evaluate and require the intuition of the engineer who is trying to model the process.

For the reaction inside the catalytic particle, several sequential and/or parallel steps take place (Figure 5.3):

1. Transport of reactants (A, B, ...) from the fluid bulk to the catalyst surface
2. Transport of reactants inside the catalyst pores
3. Adsorption or reactants to the catalytic site
4. Surface reaction among molecules or atoms adsorbed
5. Desorption of products (R, S, ...)
6. Transport of products from pores to the catalyst surface
7. Transport of products from the catalyst surface to the fluid bulk

Considering the great number of compounds that are present in a feedstock, it is almost impossible to model each step in the catalytic reactions taking place inside the riser. Therefore, global average approaches are commonly used. In addition, during catalytic cracking reactions another family of

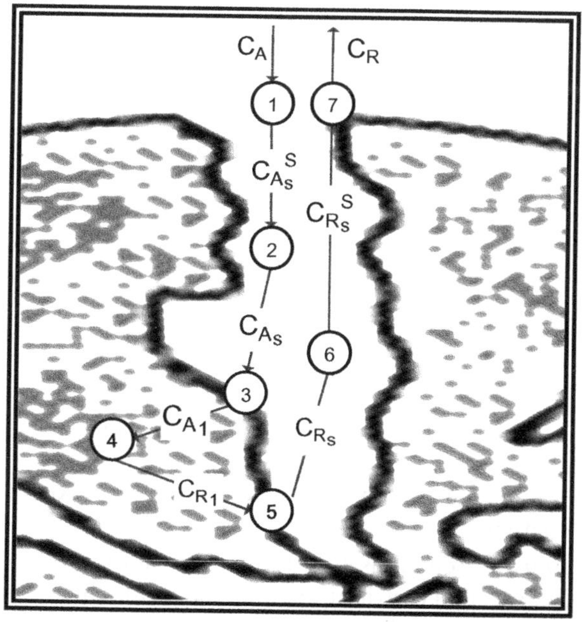

Figure 5.3. Typical steps of a catalytic reaction process (e.g., Froment and Bischoff, 1990).

hydrocarbons, absent in the original feedstock, is produced: olefins. These compounds are characterized by the presence of double bonds between adjacent carbons and exhibit formation energies different from those of their analogous paraffins. A good discussion of this topic may be found in Venuto and Habib (1978). Among the cracking reactions that break C–C bonds, it is possible to mention some of the simpler ones.

Cracking of long paraffins in order to form lighter ones:

$$C_{16}H_{34} \rightarrow C_9H_{18} + C_7H_{16}$$

Cracking of naphthenes to yield olefins:

$$C_{20}H_{40} \rightarrow C_{12}H_{24} + C_8H_{16}$$

Cracking of olefins:

$$C_{12}H_{24} \rightarrow C_7H_{14} + C_5H_{10}$$

Dealkalization of alkali aromatics:

$$C_6H_5{-}C_nH_{2n+1} \rightarrow C_6H_6 + C_nH_{2n}$$

Breaking of alkali side chains of aromatics:

$$C_6H_5\text{–}C_{12}H_{25} \rightarrow C_6H_5\text{–}C_8H_{15} + C_4H_{10}$$

Additionally, there are several second-step reactions, such as hydride transfer (e.g., naphthene + olefin → aromatic + paraffin), isomerization, transfer of alkali groups, condensation reactions, and low olefin disproportion; all of these reactions are important in the final product distribution. Some other reactions, such as paraffin and olefin alkalization, aromatic hydrogenation, and olefin polymerization (except for ethylene polymerization) are not significant. To simplify this reaction scheme, it is common to consider only three main reaction groups:

1. *Primary reaction*: mainly the formation of compounds in gasoline (C_5 to C_{12}), *n*-butane, butenes, and propylenes. It is important to note that gasoline compounds can follow secondary (or overcracking) reactions, due to their reactivity.
2. *Hydride transfer*: reactions that reduce the amount of olefin, influence the molecular weight distribution of products, and increase the selectivity to gasoline; moreover, they increase the AKI. However, they also favor coke formation and, consequently, catalyst deactivation.
3. Coke formation: reactions currently under study that provoke coke deposition on the catalyst surface. Coke is an entity that forms under almost any operating condition and from many precursors, being some of the most important microscopic carbons in feedstocks (León-Becerril and Maya-Yescas, 2007). Also, coke is formed by the polymerization of ethylene, the condensation of aromatics, and so on. Coke is considered to be a carbon formation similar to graphite, whose molecular weight is in the range of 940 to 1010 Da (Wolf and Alfani, 1982).

5.2.2 Lumping of Feedstock and Products

As it is easy to note, consideration of a complete reaction scheme is almost impossible; therefore, some simplifying approaches have been developed. One of the oldest one is *lumping*, the agglomeration of several (or many) chemical compounds into a single compound (called a *lump*), which should exhibit some or several common properties (e.g., boiling point, molecular weight, reactivity). Some fundamentals of this theory may be found in the work of Kuo and Wei, 1969 and Wei and Kuo, 1969.

During the 1960s, the first lumped kinetic scheme for FCC was proposed by Weekman and Nace (1970). This scheme consists of three observable lumps, called feedstock, gasoline, and coke + gases (Figure 5.4). They were chosen because it is possible to measure the first two, and because the coke produced drives the energy dynamics of the entire unit. Additionally, the kinetics depends

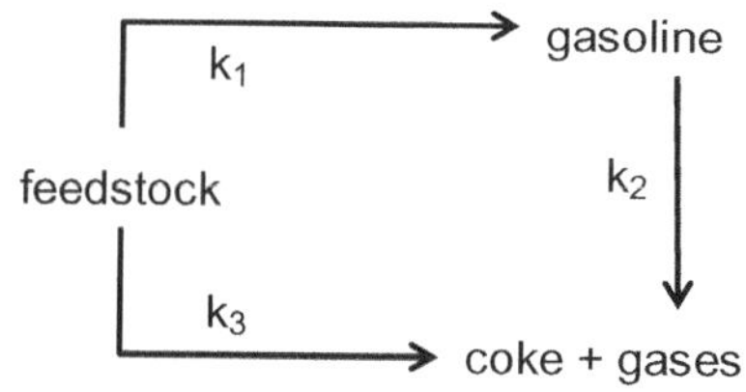

Figure 5.4. Three-lump kinetic scheme as proposed by Weekman and Nace (1970).

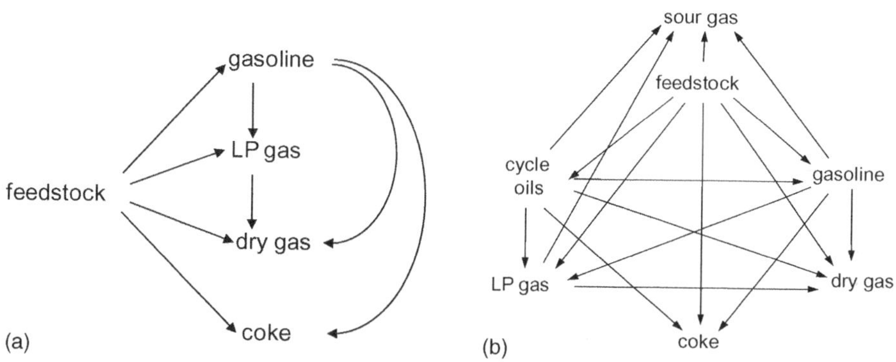

Figure 5.5. Kinetic schemes for an FCC: (a) five-lump (e.g., Ancheyta-Juárez et al., 1997); (b) seven-lump (e.g., Maya-Yescas et al., 2005).

on only three rate parameters ($k_1, k_2,$ and k_3), which are a function of feedstock composition, operating conditions, and so on; however, as there are only three parameters, they are not very difficult to estimate (Ancheyta-Juárez et al., 1997) each time any operating condition changes.

Despite the fact that the three-lump model has been used extensively for more than 30 years, the information that it contains is not sufficient for the prediction of yield to products. Moreover, there are more "observable" products that are not included in the scheme; therefore, the development of kinetic schemes continues. A trend that has been succesful is to expand the three-lump scheme in a systematic way, unfolding the lumps that contain more than one observable product: for example, feedstock = LCO + HCO, coke = coke + DGS, DGS = DG + sour gas, and so on. This strategy improved the use of information about feedstock and products, yielding schemes of five lumps (Ancheyta-Juárez et al., 1997) (Figure 5.5a), seven lumps (Maya-Yescas et al., 2005) (Figure 5.5b), and much more complex schemes (e.g., Sugungun et al.,1998). Another strategy employed to expand the three-lump scheme was to consider the hydrocarbon types in each lump; for example, feedstock is composed of PNA–hydrocarbons. After catalytic cracking, it is possible to find double-bond compounds; therefore, cycle oils, gasoline, and LPG contain olefins (PONA–hydrocarbons). Moreover, it is necessary to account for the alkali side chains present in aromatics and polyaromatics (A_h), which crack in FCC. This strategy yielded a 10-lump scheme (Jacob et al., 1976). Despite the fact that this strategy sounds more informative than the previous lumping methodology, it has not been very succesful. The main problem is that the number of lumped reactions increases very fast when the number of lumps increases. For example, in the 10-lump scheme, the authors consider 20 possible reactions after simplification of pathways that are less probable. Finally, there are many different hydrocarbons in each PONA–A_h group, and as their distribution depends on the feedstock, this elaborate scheme is also dependent on the feedstock composition. Nevertheless, it is possible to find more recent work that continues this approach (e.g., Araujo-Monroy and López-Isunza, 2006).

5.2.3 More Detailed Mechanisms

As noted earlier, lumped schemes solve partially the problem of modeling the kinetics of complex reactions; nevertheless, they exhibit some disadvantages, such as their weak ability to extrapolate results from an adjusted model to be used in case of a change of feedstock. The only advantage of lump schemes (and the main reason they are still in use) is that semiqualitative characterization of the feedstock (PNA analysis, H/C ratio, density, refractive index, boiling-point range) and products (PONA analysis and others) are enough to fit the parameters to the lumped reactions proposed. Therefore, some effort has focused on a fundamental description of the kinetics of all possible chemical reactions using reduced sets of reaction rates and requiring more information about feedstock characterization than is required by lump schemes; for example, they need mass spectrometric data and more detailed hydrocarbon group analysis.

One of the most fundamental approaches to modeling FCC kinetics is the *single step*, developed by Gilbert Froment and co-workers. A discussion of this strategy may be found, for example, in the work of Moustafa and Froment (2003).

5.3 SIMULATION TO ESTIMATE KINETIC PARAMETERS

Due to its very large size, it is impossible to use industrial FCC units to estimate concentration, temperature, and catalyst effects on kinetics. Therefore, it was necessary to find alternative routes to perform this duty. In Section 5.3.1 we discuss laboratory reactors, devices that emulate some operating conditions of industrial units and on the basis of the results obtained, propose some ways to infer the reaction kinetics of industrial units during operation. In Section 5.3.2 we revise some other trends related to the crude estimation of kinetic parameters from operating units.

5.3.1 Data from Laboratory Reactors

One of the most commonly used laboratory reactors is the microactivity test (MAT) (Figure 5.6). Its operating protocol is well established and has been verified several times, most recently in ASTM (2005). The objective of the MAT reactor is to emulate, physically, the catalyst-to-oil ratio (C/O) inside the riser of the industrial unit at "certain operating conditions." During MAT experiments, yield to products in the laboratory reactor is similar to that obtained in the industrial unit; therefore, the main issue of this device is the emulation of yield to products using different types of catalyst and/or feedstock. This unit works under isothermal conditions. To change the C/O ratio, injection times (t_S) are modified for the same amount of catalyst. Liquid products are received in a semibatch accumulator, which mixes them physically. This feature makes it difficult to interpret the results in order to model continuous industrial units, a topic that has been under discussion for awhile (Froment and Bischoff, 1962; Jacob et al., 1976; Kelkar et al., 2003). In contrast to the industrial unit, the MAT reactor is designed and operated in a very different way, as can be seen in Table 5.1.

Therefore, to estimate kinetic parameters it is necessary to modify the operating conditions of a MAT unit. Two of the main changes from the standard protocol are the changed amount of feedstock supplied and the changed operating temperature (e.g., Ancheyta-Juárez et al., 1997; Corella, 2004; Maya-Yescas et al., 2004b). During kinetic parameter estimation, conversion and yield data have to be measured in the m_R stream (see Figure 5.6), which leaves the reactor at its final reaction conditions (instantaneous), instead of data from the semibatch accumulator, whose values are averaged over time. This technique helps to develop Arrhenius plots to determine effective activation energies and to emulate different C/O ratios; it is also useful to estimate activity decay due to coke deposition on the catalyst surface (Maya-Yescas et al., 2004a) by using the proper mathematical model. As an example, the estimation of standard conversions, both instantaneous and averaged, during laboratory experiments at three different C/O ratios and three different reaction temperatures (T_{rx}) were analyzed. One typical industrial feedstock and one

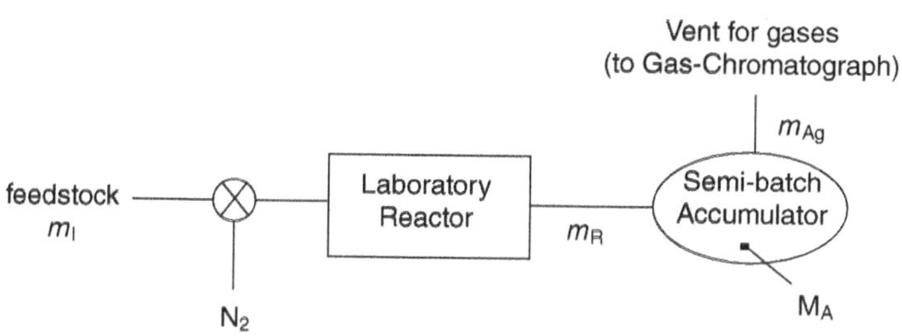

Figure 5.6. Schematic of a laboratory-scale MAT device.

TABLE 5.1. Comparison of Some Operating Aspects of MAT and Industrial Units

Feature	MAT	Industrial FCC
Catalyst	Commercial	Commercial
Catalyst-to-oil ratio	3 to 6	6 to 9
Feedstock	Gas oils diluted with nitrogen	Gas oils aspersed with steam
Type of bed	Fixed	Transported
Heat transfer mode	Isothermal	Adiabatic
Operating temperature, T_{rx}	480–550°C	Ranges from 580–515°C
Catalyst hold up	4 g	~300 metric tons
Feedstock flow rate	1 g/s	~36.8 kg/s
Contact time, t_S	67 s	2–5 s
WHSV	16	~12

TABLE 5.2. Operating Conditions and Results at the MAT Unit

Experiment	T_{rx} (°C)	C/O	y_{fsA}	ω_{CSC}
1	520	3	0.4596	0.01510
2	520	4	0.4050	0.01372
3	520	6	0.2902	0.01322
4	535	3	0.4385	0.01608
5	535	4	0.3784	0.01482
6	535	6	0.2572	0.01355
7	550	3	0.4276	0.01625
8	550	4	0.3616	0.01532
9	550	6	0.2425	0.01397

Source: Adapted from Maya-Yescas et al. (2004a).

commercial catalyst were used. The nine experiments were performed at the same WHSV = 16 (Table 5.2) by triplicate. In all the MAT experiments, $m_I = 0.0177$ g/s was used, as suggested by ASTM method D 5154-05. Averaged data for the remaining nonreacted feedstock (y_{fsA}) and yield to coke specific to catalyst weight (ω_{CSC}) were collected from the standard MAT results.

Now, continuing the experiment, liquid products arriving at the accumulator change the relative mass fraction distribution because of the changing activity and selectivity of the catalyst inside the reactor (Froment and Bischoff, 1962; Jacob et al., 1976; Kelkar et al., 2003; Maya-Yescas et al., 2004a). To extract the instantaneous composition of the liquids coming from the reactor, it is necessary to know the transient mass balance in the accumulator (M_A) as a function of the flow coming from the reactor (m_R) and the flow of gases vented to the gas chromatograph (m_{Ag}):

$$\frac{dM_A}{dt} = m_R - m_{Ag} \qquad \text{initial condition: } M_A(t=0) = 0 \tag{5.1}$$

The mass flow m_R is related to the injection flow rate as $m_R = m_I - m_C$, where m_C is the mass rate of coke generation that can be calculated from the corresponding MAT data. This is possible because coke will remain adsorbed to the catalyst surface.

To calculate the standard conversion ($\chi_R = 1 - y_{fsR}$), a mass balance was performed for the cyclic oils accumulated:

$$\frac{dM_{fs}}{dt} = y_{fsR} m_R \qquad \text{initial condition: } M_{fs}(t=0) = 0 \tag{5.2}$$

where $M_{fs} = M_A y_{fsA}$. It should be noted that the instantaneous mass balance in the accumulator depends on the mass fraction inside this accumulator (y_{fsA}), whereas the conversion reached in the reactor is evaluated using the mass fraction at its outlet (y_{fsR}). The goal is to evaluate y_{fsR} using the data obtained during the laboratory evaluation, y_{fsA}, at different injection times.

Once the data at the reactor outlet have been collected, it is possible to obtain kinetic rate parameters for the feedstock. Following the classic assumption, feedstock cracking exhibits a second-order reaction rate; the mass balance for feedstock inside the fixed-bed reactor is

$$u\frac{dy_{fsR}}{dz} = ky_{fsR}^2\Phi \qquad \text{initial condition: } y_{fsR}(z=0) = 1 \tag{5.3}$$

where z is the axial reactor coordinate, u the gas velocity, and Φ the activity (or deactivation) function. For the used equilibrium catalyst, the initial MAT activity is $\Phi_0 = 0.70$, mass fraction. The residence time inside the reactor is given by the ratio between the gas velocity and the axial coordinate.

This model was used to evaluate the kinetic rate for feedstock conversion in the MAT reactor and the remaining catalyst activity after each of the nine experiments shown in Table 5.2 assuming a hyperbolic deactivation function (Froment and Bischoff, 1962):

$$\Phi = \begin{cases} \Phi_0 & \omega_{\text{CRC}} \le \omega_{\text{CSC}} < \omega_{\text{CSC min}} \\ \dfrac{\Phi_0}{1+\alpha(\omega_{\text{CSC}} - \omega_{\text{CSC min}})} & \omega_{\text{CSC}} \ge \omega_{\text{CSC min}} \end{cases} \tag{5.4}$$

where ω_{CRC} is the mass of coke adsorbed to the equilibrium catalyst surface specific to the mass of catalyst, ω_{CSC} the instantaneous mass of coke adsorbed to the catalyst surface, α the crackability factor, and $\omega_{\text{CSC min}}$ the minimum coke amount that provokes the catalyst to show deactivation.

TABLE 5.3. Instantaneous and Averaged Standard Conversions

Experiment	C/O	y_{fsR}	χ_R (wt%)	χ_A (wt%)
1	3	0.6346	36.54	54.04
2	4	0.6236	37.64	59.50
3	6	0.2902	70.98	70.98
4	3	0.6207	37.93	56.15
5	4	0.6189	38.11	62.16
6	6	0.2572	74.28	74.28
7	3	0.6167	38.33	57.24
8	4	0.5999	40.01	63.84
9	6	0.2425	75.75	75.75

Source: Adapted from Maya-Yescas et al. (2004a).

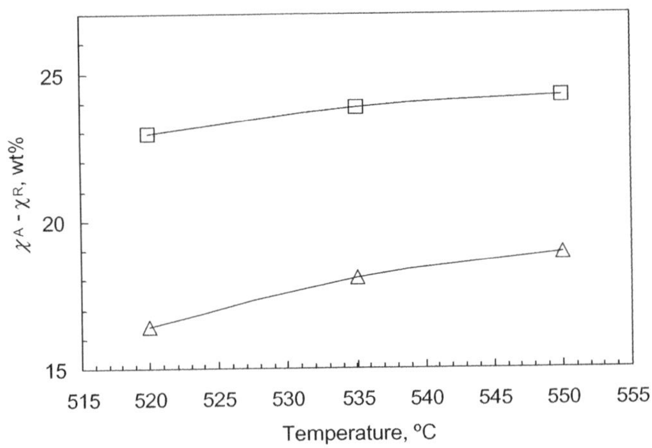

Figure 5.7. Instantaneous minus averaged values for conversion (△, C/O = 3; □, C/O = 4). (Adapted from Maya-Yescas et al., 2004a.)

By integration of the mass balances [Eqs. (5.1) and (5.2)] on discrete intervals for the nine experiences, the instantaneous value for the yield to cyclic oils, y_{fsR} , was calculated. Then, on the basis of y_{fsA} and y_{fsR}, instantaneous (χ_R) and averaged (χ_A) standard conversions were calculated (Table 5.3).

Numerical values for standard conversion differ depending on the data considered; moreover, if averaged data are used for kinetic parameter evaluation, an increasing estimation error is introduced (Figure 5.7). Also, it is possible to note that C/O has a great effect on the difference between instantaneous and averaged values; the differences between values at C/O = 4 exhibit a lower slope than do those measured at C/O = 3. Therefore, to solve the mass balances in the most accurate way, it is necessary to perform more experiments at different times.

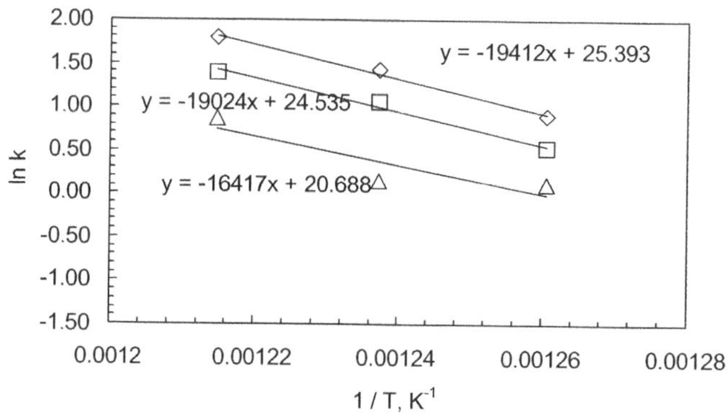

Figure 5.8. Arrhenius plot for feedstock conversion in the fixed bed (△, C/O = 3; □, C/O = 4; ◇, C/O = 6). (Adapted from Maya-Yescas et al., 2004a.)

MAT data reflect averaged trends related primarily to the C/O ratio. However, if the aim is to estimate kinetic parameters and remaining catalyst activity, it is necessary to have instantaneous data from the reactor outlet stream rather than those from the semibatch accumulator, which are averaged. To obtain these instantaneous data, it is necessary to evaluate the mass balances for the accumulated liquids and for the desired yield; the same is true for gaseous products.

Solving the model for the fixed-bed reactor [Eq. (5.3)], it is possible to evaluate the apparent activation energy for feedstock conversion using an Arrhenius plot (Figure 5.8) as $E_A/R_g = (17.153 \pm 1.327)$K. The aparent activation energy is a property of the system; it depends on intrinsic kinetics and on mass transfer resistances by intraparticle diffusion and interphase transport (e.g., Froment and Bischoff, 1990) instead of reactor configuration. Because the Arrhenius plot shows almost parallel lines, it is possible to infer that the activation energy obtained is the effective (apparent) activation energy for observable cracking reactions.

Now, knowing the value of the remaining catalytic activity for these experiments, it is possible to separate out the value of the apparent frequency factor. Then it was extrapolated to the other six experiments using the deactivation function described below. The best value for the frequency factor for the experiments at 37 s is $k_0 = 1.7153 \times 10^{11}$ s^{-1}, and numerical values for the remaining catalytic activity are shown in Table 5.4. Once the Φ values are known, it is possible to fit these data to the deactivation function desired [Eq. (5.4)]; the parameters obtained are $\alpha = 1013.8\ g_{\text{coke}}^{-1} g_{\text{cat}}$ and $\omega_{\text{CSC min}} = 0.01315\ g_{\text{coke}} g_{\text{cat}}^{-1}$. As is clean, Eq. (5.4) fits the experimental data very accurately (Figure 5.9).

Now, the activation energy and the activities obtained by evaluating Eq. (5.4) in terms of coke yield can be used to model a riser reactor. Although these values are not influenced by the semibatch accumulation time in the

TABLE 5.4. Catalytic Activity (Mass Fraction) for Cracking in a MAT Reactor

	T_{rx} (°C)		
C/O	520	535	550
3	0.235	0.176	0.169
4	0.441	0.259	0.216
6	0.649	0.497	0.381

Source: Adapted from Maya-Yescas et al. (2004a).

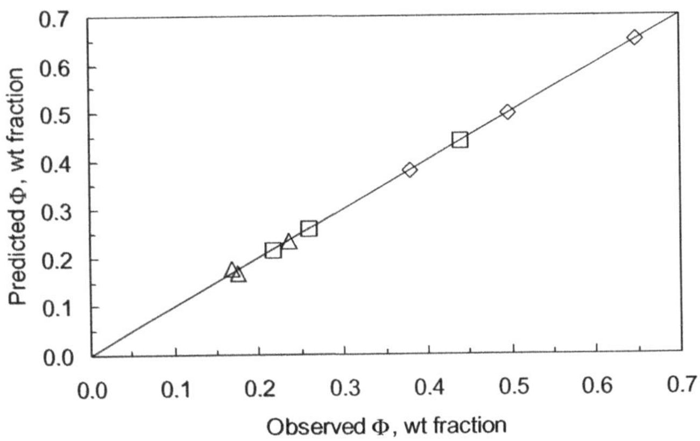

Figure 5.9. Values predicted for deactivation function from Eq. (5-3.4) (△, C/O = 3; □, C/O = 4; ◇, C/O = 6). (Adapted from Maya-Yescas et al., 2004a.)

MAT device, this approach is valid only to calculate the average feedstock cracking reaction rate, which is different from the rates considered for each individual product.

5.3.2 Data from Industrial Operation

Although the industrial unit operates in a very narrow window of temperatures, pressures, and C/O ratios, some people have tried to estimate kinetic parameters from industrial operating data. This can be done, but some drawbacks must be taken into account. First, because the industrial unit is not designed to be used as a research device, it is not easy to dictate operating conditions for certain catalyst and/or feedstock. Control of the unit is mainly regulatory (Aguilar et al., 2002; Aguilar and Maya-Yescas et al., 2006; Taskin et al., 2006), and the objective of control is the riser outlet temperature. This target is controlled by changing the C/O ratio continuously; therefore, it is not possible to know the instantaneous C/O ratio. Second, as described above, industrial units are adiabatic devices; therefore, there is a temperature profile along the riser (about 60 to 80°C) that changes the reaction rates along the riser. And finally, similar to any other catalytic cracking reactor, catalyst is deactivated continuously as a consequence of coke deposition.

Meanwhile, there are some interesting advantages if industrial operating data are used. The most important is that FCC units are tracked in a continuous way at refineries. Hence, it is common to find large collections of operating data under "similar conditions"; here *similar* means "at the same reactor outlet temperature." If changes in the C/O ratio, also recorded, are not very drastic, the data mentioned could be used to fit some kinetic scheme under average operating conditions; moreover, it is possible to calculate the standard deviation of the fitting. This procedure is not very recommendable but could be the only option when laboratory devices are not available or laboratory data sets are not complete for the estimation desired. As an example of this type of fitting of a kinetic scheme, Araujo-Monroy and López-Isunza (2006) presented a six-lump scheme.

5.4 SIMULATION TO FIND CONTROLLING REACTION STEPS DURING CATALYTIC CRACKING

Kinetic factors estimated by the procedure described above are not useful for modeling industrial risers; however, the relative value of reaction rates can be used to infer the behavior of industrial risers (Ancheyta-Juárez et al., 1997; Maya-Yescas et al., 2004a). Moreover, different researchers estimate different values for the same reaction rates (e.g., Moustafa and Froment, 2003; Corella, 2004), but the relative value of the reaction rate constants is preserved (e.g., Jiménez-García et al., 2007). This situation arises because there is a linear proportionality among kinetic rates determined in the MAT reactor and those that are used to model riser reactors. The only way that this could happen is if the apparent activation energy estimated in laboratory reactors is the same as that used to model industrial risers (Vieira et al., 2004). Now it is necessary to check this assumption by analyzing phenomena involved during the cracking of hydrocarbons, when resistances at the interphase and intraparticle are present (Jiménez-García et al., 2010).

Froment and Bischoff (1990) have shown that catalyst geometry influences the numerical values of the effectivity factor. Additionally, because catalyst particles for FCC are small ($D_p \approx 55$ to $70\ \mu m$) and reactions are moderately (not severely) endothermic, reaction conditions inside the particle can be considered isothermal (e.g., Corella, 2004). For a spherical and isothermal particle, the mass balance for a reactive (A) can be expressed as

$$0 = D_{\text{eff}}\left(\frac{d^2C_A}{dr^2} + \frac{2}{r}\frac{dC_A}{dr}\right) - r(C_A) \tag{5.5}$$

with boundary conditions

$$C_A\big|_{r=0} \rightarrow \text{finite} \tag{5.5a}$$

$$k_g(C_{Ab} - C_A\big|_{r=Dp/2}) = D_{\text{eff}}\left(\frac{dC_A}{dz}\right)\Bigg|_{r=Dp/2} \tag{5.5b}$$

Here D_{eff} is the effective difussion coefficient inside the particle, r_A the intrinsic reaction rate (which exhibits Langmuir–Hinshelwood kinetics), k_g the mass transfer coefficient at the fluid–particle interphase, and $D_p/2$ is half the particle radius.

Solution of Eq. (5.5) yields two important parameters:

1. $\phi_S = \dfrac{V_p}{S_x}\sqrt{\dfrac{\rho_p r_A(C_A)}{D_{\text{eff}}C_A}\dfrac{1+K_e}{K_e}} = \dfrac{D_p}{6}\sqrt{\dfrac{\rho_p r_A(C_A)}{D_{\text{eff}}C_A}\dfrac{1+K_e}{K_e}}$
   the Thiele modulus, which accounts for the ratio of characteristic time of diffusion with respect to the characteristic time of reaction inside the catalyst particle.
2. $\text{Bi}'_m = k_g L / D_{\text{eff}}$, the modified Biot number for mass transfer, which accounts for the ratio of characteristic time of diffusion inside the catalyst particle with respect to the characteristic time of mass transfer at the interphase.

To find the average reaction rate observed (apparent), it is necessary to define the global effectiveness factor,

$$\eta_G = \frac{1/V_p \int_{Vp} r_A(C_A)\,dV_p}{r_A(C_{Ab})}$$

which compares this value with the value observed if there were no resistances for mass transfer, either intraparticle or at the interphase (Jiménez-García et al., 2009):

$$\eta_{Gs,j} = \frac{\text{Bi}'_{m,i}}{\phi_{s,j}^2\left[3\phi_{s,j}\coth(3\phi_{s,j}) + (3\text{Bi}'_{m,i} - 1)\right]}\left[3\phi_{s,j}\coth(3\phi_{s,j}) - 1\right] \tag{5.6}$$

Now, for fast reactions (as in the case of FCC reactions over "ultraactive" catalysts), the Thiele modulus tends to be "large" ($\phi_S > 1.5$) (e.g., Jiménez-García et al., 2007). If this happens, Froment and Bischoff (1990) have noted that the combined resistance to mass transfer at the gas–solid interphase and inside the particle exhibits an apparent first-order reaction rate [Eq. (5.7)]. This is the case during FCC reaction rate evaluation for both laboratory and industrial reactors:

$$(r_A)_{\text{obs}} = \eta_{GS,j}(r_{A,j})_{\text{bulk}} = \frac{\text{Bi}'_{m,j}}{\phi_j^2}(r_{A,j})_{\text{bulk}} = \frac{k_{g,i}}{k_{V,j}}k_{V,j}C_{A,ib} = k_{g,i}C_{A,ib} \tag{5.7}$$

Summing up, if linear translation from a MAT (or similar) laboratory reactor to industrial risers is possible, the entire reaction process at the catalyst level is controlled by mass transfer and the apparent reaction coefficients are the mass transfer coefficients. Hence, linear scall-up of reaction rates yields the ratio between mass transfer coefficients at the interphase evaluated for each reactor at the laboratory or inside the riser:

$$\text{factor}_{\text{MAT}\to\text{riser}} = \frac{k_{g,j}\big|_{\text{MAT}}}{k_{g,j}\big|_{\text{riser}}} \tag{5.8}$$

These important factors are used in Section 5.6 to scale-up apparent kinetic parameters from laboratory to industrial reactors.

5.5 SIMULATION OF STEADY OPERATION OF THE RISER REACTOR

A riser is considered to be a heterogeneous, adiabatic, transported bed reactor. Usually, it is modeled as a steady-state system, due to the great difference in residence time between this reactor (3 to 5) and a regenerator (6 to 11 min). The model is developed by the classical application of pressure [Eq. (5.9)], mass [Eq. (5.10)], and energy [Eq. (5.11)] balances. In order to use kinetic factors from laboratory reactors, a pseudohomogeneous model has been proposed by León-Becerril et al. (2004).

$$\frac{dP}{dz} = -\rho_p g(1-\varepsilon) \tag{5.9}$$

with initial condition

$$P(z=0) = P_o \tag{5.9a}$$

$$u_p \frac{dC_{jp}}{dz} = \varepsilon \rho_g \, C/O \, \Phi \Re_j \tag{5.10}$$

with initial condition

$$C_{jp}(z=0) = 0 \tag{5.10a}$$

where j represents feedstock, gasoline, LP gas, dry gas, or coke.

$$u_p \frac{dT_p}{dz} = \frac{\varepsilon \rho_g \, C/O \, \Phi \Re_j}{Cp_p \rho_p} \sum_j (-\Delta H_{r,j}) \tag{5.11}$$

with initial condition

$$T_p(z=0) = T_{po} \tag{5.11a}$$

TABLE 5.5. Industrial FCC Unit Studied

Type	Adiabatic
Technology	Riser, fluid regenerator
Operating mode	Full combustion
Feedstock type	Gas oils blend
Feedstock capacity (bbl/day)	25,000
Feedstock inlet temperature (K)	450.0
Catalyst type	Synthetic, microspherical
Particle diameter (average) (m^2)	7.0×10^{-5}
Riser outlet temperature (average) (K)	794.0

Source: Adapted from León-Becerril et al. (2004).

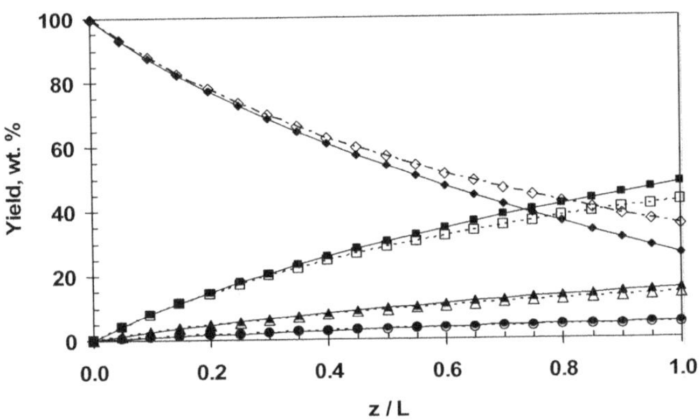

Figure 5.10. Axial profile of mass fractions, (◆, feedstock; ■, gasoline; ▲, LP gas; ●, dry gas). considering pressure profile (solid symbols) and without considering pressure profile (open symbols). (Adapted from León-Becerril et al., 2004.)

By integration of Eqs. (5.9) to (5.11), León-Becerril et al. (2004) developed the following profiles. Operating conditions used to quantify the effect of modeling the pressure gradient are feedstock rate 36.82 kg/s and $C/O = 8$, observed during industrial operation. Comparative profiles for both approaches are shown with respect to dimensionless space velocity in the riser (z/L). Profiles inside the riser are theoretical, because there are no experimental results at these points. Operating conditions are given in Table 5.5.

Axial profiles of mass fractions are shown for both simulation approaches: with the axial pressure profile and without (Figure 5.10). Mass fraction profiles of feedstock show the typical behavior of a reactant, which is not generated inside the riser but is consumed in order to generate products. It is important to note that the conversion predicted is higher when the pressure balance is included in the model; consequently, yields to products predicted are higher. To make more evident the effect of modeling changes in pressure, Figure 5.11

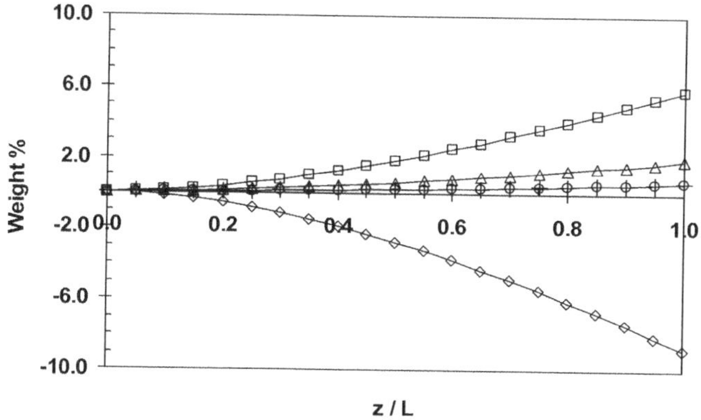

Figure 5.11. Mass fraction differences when modeling the pressure balance (◇, feedstock; □, gasoline; △, LP gas; ○, dry gas). (Adapted from León-Becerril et al., 2004.)

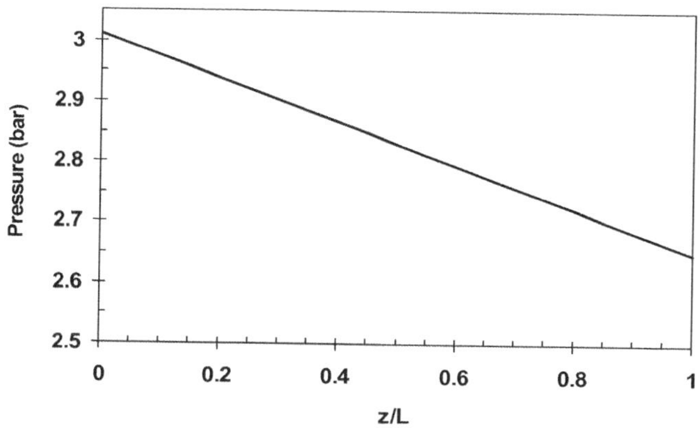

Figure 5.12. Axial pressure gradient. (Adapted from León-Becerril et al., 2004.)

shows the difference between predictions for both cases as a function of the axial nondimensional coordinate of the riser.

Mass fraction predictions at the riser outlet differ by more than 5% for feedstock and gasoline. As the reaction proceeds along the riser, the difference between approaches increases. Moreover, this difference is directly proportional to product yields, which is undesirable for the prediction of gasoline, LP gas, and dry gas. The difference between the two approaches is due to the fact that reaction rates depend on partial pressures, which are a consequence of the total pressure inside the riser. This pressure exhibits a drop of about −0.382 bar (Figure 5.12), in agreement with experimental results (e.g., Theologos and Markatos, 1993); this pressure drop represents more than 20% of the initial pressure; therefore, the results predicted are clearly affected by this change.

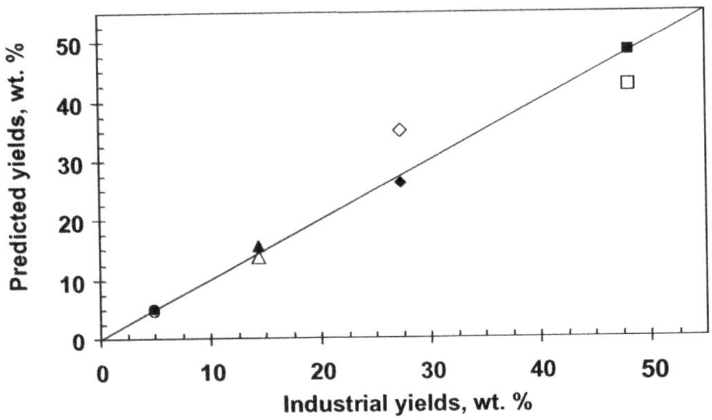

Figure 5.13. Industrial and predicted mass fractions (◆, feedstock; ■, gasoline; ▲, LP gas; ●, dry gas) considering pressure profile (solid symbols) and without considering pressure profile (open symbols). (From León-Becerril et al., 2004.)

Figure 5.13 compares results predicted for conversion (remaining feedstock) and yields to gasoline, liquid petroleum gas, dry gas, and coke against industrial results at the riser outlet. Data were taken under the conditions shown in Table 5.5. Clearly, modeling results are better if the pressure gradient along the riser is taken into account, because of the importance of the pressure drop calculated along the riser.

5.6 SIMULATION TO SCALE UP KINETIC FACTORS

After collection and debugging of laboratory data, the next step is to scale them up to the industrial riser; this procedure also requires mathematical modeling, in this case the riser reactor. As pointed out in the literature (e.g., León-Becerril et al., 2004), it is important to consider mass balance (for hydrocarbons and catalyst), energy balance, and pressure balance in the riser. For example, Jiménez-García et al. (2007) analyzed this particular behavior, finding that the process is controlled by mass transfer resistance at the oil–catalyst interphase; this situation was expected because the relative velocity of catalyst and oil in the riser is small (Villafuerte-Macías et al., 2003), due to the size of the catalyst.

The simple approach suggested here does not imply anything about the intrinsic kinetics of the reaction not being the same in the two reactors, since molecules on the level of chemical reactions do not know in which type of system they exist. To estimate the possible effects of intraparticle mass transfer, Thiele's modulus was calculated for each reaction (Table 5.6) using experimental diffusion data. The conclusion is that although some reactions are limited by intraparticle diffusion instead of chemical kinetics, there are no

TABLE 5.6. Thiele's Modulus for the Reactions Under Study

Reaction	Reaction Temperature (K)		
	793.15	808.15	823.15
Feedstock → gasoline	8.245	8.370	8.447
Feedstock → LPG	2.402	2.424	2.446
Feedstock → dry gas	7.880	7.951	8.025
Feedstock → coke	1.329	1.341	1.353
Gasoline → LPG	4.129	4.130	4.131
Gasoline → dry gas	4.321	4.322	4.323
Gasoline → coke	0.453	0.453	0.453
LPG → dry gas	1.795	1.795	1.796

Source: Adapted from Jiménez-García et al. (2007).

differences in kinetic parameters when scaling them; therefore, intraparticle mass transfer could not be the only controlling step.

The second mass transfer–limiting step could be located at the gas–solid particle interphase. This resistance should be common for the set of reactions, and some later steps are linearly dependent upon that limiting step. However, if pseudohomogeneous models are used, it is not possible to discriminate between the influence of external mass transfer resistances and the intrinsic kinetic rate. The only information known is that those step-limiting phenomena are observed as first-order processes. For a narrow range of temperatures, such as the experimental temperatures used in this work, it is possible by modifying frequency factors, to notice the influence of external mass transfer limitations. As a consequence, the value estimated for the *apparent* activation energy of each reaction at the laboratory reactor can be used directly in a simulation program for the industrial unit; both reactors will exhibit the same apparent activation energy at the same operating conditions. Laboratory and industrial reactors are very different in hydrodynamics; therefore, if frequency factors are modified in order to fit the production observed in industrial reactors, the result is a linear translation of data, a situation that agrees with industrial experience.

Taking advantage of this situation, the reaction scheme proposed by Ancheyta-Juárez et al. (1997) was selected to be scaled up; this kinetic mechanism proposes five lumped compounds and eight possible reactions; the industrial riser reactor was simulated by using the model described in Section 5.5 (León-Becerril et al., 2004). The frequency factors (k_{0j}) proposed by the authors were scaled by dividing each one by the frequency factor ($k_{0f\to g}$) for the reaction feedstock → gasoline:

$$\sigma_j = \frac{k_{0j}}{k_{0f\to g}} \tag{5.12}$$

TABLE 5.7. Observed (Industrial) and Predicted Yields of Products and Error (wt%)

	Feedstock	Gasoline	LPG	Dry Gas	Coke
Yield					
Observed	27.37	48.23	14.35	5.06	4.99
Predicted using values from Ancheyta-Juárez et al. (1997)	0.00	9.81	3.08	41.74	45.37
Error (wt%)	−100.0	−79.66	−78.53	+724.90	+809.22
Scaling predicted $k_{0f \to g}$ Jiménez-García et al. (2007)	26.12	48.23	15.56	5.11	4.98
Error (wt%)	−4.56	0.00	+8.43	+0.98	−0.20

Source: Adapted from Jiménez-García et al. (2007).

During simulation, the yield to gasoline in the industrial riser was fit by modifying the value of $k_{0f \to g}$ for a constant values of the σ_j relationship. To consider catalyst activity decay, a hyperbolic deactivation due to coke deposition was considered.

Table 5.7 shows a comparison of predicted to observed yields when the raw frequency factors from Ancheyta-Juárez et al. (1997) and the scaled frequency factors are used; the industrial riser was operating under typical industrial conditions: 36.82 kg/s and a mass catalyst-to-oil ratio $C/O = 8.0$. The use of crude laboratory estimation does not allow us to infer yields in the industrial riser, which means that mass transfer limitations are different for the two systems. Nevertheless, the fitting of only one of the frequency factors, $k_{0f \to g}$ and the consequent scaling up [as suggested by Eq. (5.12)] of the other seven is enough to fit industrial yields accurately.

This is possible because of the linear relationship between the values of the frequency factors determined in the laboratory and those used to model the industrial riser, which means that the apparent activation energy measured in the MAT reactor is the same as that in the riser (Maya-Yescas et al., 2004a); this also confirms that the entire process is rated by mass transfer (Jiménez-García et al., 2007). It is interesting to note that even when the factor $k_{0f \to g}$ was the only one fitted, prediction of every product is also accurate, as a consequence of the fact that the estimation of apparent activation energies corresponds to that of mass transfer coefficients [Eq. (5.8)].

In addition to the MAT reactor, there are other important laboratory devices featuring different behavior. For example, Kayser Technologies has developed a fluidized-bed laboratory unit, the advanced catalyst evaluations (ACE) reactor (U.S. patent 6,069,012). Its advantage over the MAT unit is that the catalyst is fluidized, improving the homogenety of the reacting media and making it possible to decrease the effective contact time. However, reactor products are managed in a semicontinuous accumulator, similar to the case of

MAT; this design continues mixing products obtained during reaction with partially deactivated catalyst, making difficult the estimation of kinetic parameters. Furthermore, the entire kinetic process continues to be controlled by mass transfer resistances. Another interesting reactor, designed to emulate the catalyst-to-oil ratio, actual contact time, and hydrodynamics of the transported bed industrial reactor, is the CREC riser simulator (Ginsburg et al., 2003). Despite the fact that this laboratory reactor accurately emulates the industrial riser, the final evaluation of whole kinetics is still mixed with mass transfer resistances at the catalyst–fluid interphase. Therefore, the analysis developed in this section continues to be valid for these reactors.

5.7 SIMULATION OF THE REGENERATOR REACTOR

The most important influence of the regenerator on the entire converter is its contribution to the energy balance because of the coke combustion. A typical vacuum gas oil feedstock yields about 4 to 5 wt% (relative to the weight of the feedstock) of coke; this coke is either produced during catalytic cracking or arrives with the feedstock in the form of coke precursors. The coke that arrives as a precursor comes from upstream processes such as vacuum distillation towers; therefore, depending on the boiling cut point of this feedstock, the amount of coke precursor might change slightly; these precursors are evaluated in industry as the amount of Conradson or Ramsbottom carbon (Venuto and Habib, 1978). An increase in Conradson carbon would be reflected by an increase in the effective yield to coke in the riser and the consequent change in the rate of heat production during catalyst regeneration (e.g., Maya-Yescas and Aguilar, 2003), which is able to increase the regenerator temperature. Hence, due to the operating sequence inside the riser–regenerator system, an important variable to control in FCC units is the amount of coke remaining on regenerated catalyst (León-Becerril and Maya-Yescas, 2007). In this section we analyze the management of coke during common operation.

5.7.1 Simulation of the Burning of Nonheterogeneous Coke

The kinetic scheme for coke combustion is unknown because of the amount of heterocompounds contained in this solid hydrocarbon; therefore, it is common to use simplified schemes, such as the one proposed by Errazu et al. (1979) (Figure 5.14). First-order kinetics with respect to gaseous oxygen and constant concentration of coke are assumed for the heterogeneous noncatalytic burning of coke. The coke combustion mechanism considers the activation energies proposed by Errazu et al. It is estimated that $E_\eta/R_g = 6240$ K, and if $CO_2/CO = 1$ between 870 and 920 K (Errazu et al., 1979; Krishna and Parkin, 1985), it is possible to estimate a frequency factor for η with a value between $0.870 < \eta_0 < 1.335$. Simple formulas could be proposed for coke

$$\text{Coke combustion}: CH_{v(s)} + \left[\frac{\eta+2}{2(\eta+1)} + \frac{v}{4}\right] O_{2(g)} \longrightarrow \frac{\eta}{\eta+1} CO_{(g)} + \frac{1}{\eta+1} CO_{2(g)} + \frac{v}{2} H_2O_{(g)}$$

$$\text{Homogeneous CO combustion}: CO_{(g)} + \frac{1}{2} O_{2(g)} \xrightarrow{\text{het}} CO_{2(g)}$$

$$\text{Homogeneous CO combustion}: CO_{(g)} + \frac{1}{2} O_{2(g)} \xrightarrow{\text{hom}} CO_{2(g)}$$

Figure 5.14. Kinetic scheme of coke combustion. (From León-Becerril and Maya-Yescas, 2007.)

(CH_v) and a value for the stoichiometric parameter could be deduced: $v = 4/3$. The kinetic expression is given by

$$R_{\text{coke}} = k_{\text{oxo}} \rho_{\text{coke}} e^{-\gamma_{\text{oxo}}/R_g T} C_{O_2} \tag{5.13}$$

The regenerator is considered to be a two-region system: a *dense region* (dp) considered as a dynamic continuously stirred tank reactor, and a *dilute region* or *freeboard* (fb) considered as a stationary plug-flow reactor (Maya-Yescas et al., 2005). The mathematical model for the regenerator consists of classical mass balances in the dense region [Eqs. (5.14) and (5.15)], mass balances in the dilute region [Eq. (5.16)], and energy balances for both regions [Eqs. (5.17) and (5.18)]. Numerical simulations were performed, solving the coupled models for riser and regenerator simultaneously (León-Becerril and Maya-Yescas, 2007).

Mass balances

Dense region

$$\frac{dN_{j,\text{dp}}}{dt} = G_j^i - G_j + R_j\big|_{dp} \tag{5.14}$$

with initial condition

$$N_{j,\text{dp}}(t=0) = N_{j,\text{dp}}^0 \tag{5.14a}$$

Here, $j = O_2$, CO, and CO_2 and R_{O_2}, R_{CO}, and R_{CO_2} depend on catalyst weight, stoichiometric coefficient, and the rate reaction of coke combustion.

Coke balance in the dense region

$$\frac{dC_{\text{coke}}}{dt} = \frac{m_{\text{coke}}}{W_{\text{cat}}}(\omega_{\text{CSC}} - \omega_{\text{CRC}}) + \frac{r_{\text{coke}}}{MW_{\text{coke}}} \tag{5.15}$$

with initial condition

$$C_{\text{coke}}(t=0) = C_{\text{coke}}^0 \tag{5.15a}$$

Dilute region

$$\left.\frac{dG_j}{dz_{\mathrm{fb}}}\right|_{\mathrm{fb}} = \frac{R_j|_{\mathrm{fb}}}{u_{\mathrm{gch}}} \tag{5.16}$$

with initial condition

$$G_j(z_{\mathrm{fb}} = 0) = G_j^i \tag{5.16a}$$

Here $j = O_2$, CO, and CO_2.

Energy balances

Dense region

$$\begin{aligned}\frac{dT_{\mathrm{dp}}}{dt} &= \frac{(G^i_{O_2} + G^i_{N_2})C^i_{P\mathrm{air}}T^i_{\mathrm{air}}}{W_{Cq}C_{Pp}} \\ &+ \frac{(G_{O_2} + G_{CO} + G_{CO_2} + G_{H_2O} + G^i_{N_2})Cp_{\mathrm{dp}}T_{\mathrm{dp}}}{W_{Cq}Cp_p} \\ &+ \frac{G^i_{\mathrm{vap}}C_{P\mathrm{vap}}(T^i_{\mathrm{vap}} - T_{\mathrm{dp}}) + (m^i_{\mathrm{cat}}T^i_{\mathrm{cat}} - m_{\mathrm{cat}}T_{\mathrm{dp}})C_{Pp} + \sum_{j=1}^{j=5}(-\Delta H_j)\Re_j|_{\mathrm{dp}}}{W_{Cq}C_{Pp}}\end{aligned} \tag{5.17}$$

with initial condition

$$T_{\mathrm{dp}}(t = 0) = T^0_{\mathrm{dp}} \tag{5.17a}$$

Dilute region

$$\frac{dT_{\mathrm{fb}}}{dz_{\mathrm{fb}}} = \frac{(-\Delta H)_{\mathrm{HNC}}\,\Re_{\mathrm{HNC}}|_{\mathrm{fb}}}{(1 - \varepsilon_{\mathrm{fb}})\rho_p C_{Pp} + \varepsilon_{\mathrm{fb}} C_{P\mathrm{gch}}\left(\sum y_j PM_j\right)u_{\mathrm{gch}}} \tag{5.18}$$

with initial condition

$$T_{\mathrm{fb}}(z_{\mathrm{fb}} = 0) = T^i_{\mathrm{fb}} \tag{5.18a}$$

To evaluate the impact of coke deposited on a catalyst surface at the riser outlet (ω_{CSC}), five step increases relative to the amount of coke precursors are simulated: (1) 0.21 wt%, (2) 1.05 wt%, (3) 2.10 wt%, (4) 3.66 wt%, and (5) 3.68 wt% (relative to the weight of the feedstock). The response to these changes of the main operating variables [i.e., regenerator temperature, coke on spent catalyst (ω_{CSC}), and coke on regenerated catalyst (ω_{CRC})] is simulated from the reference operating steady state until the moment the unit reaches a new steady state or runs away. Second, following the same procedure, four step decreases of coke precursors are supplied: (1) –0.21 wt%, (2) –1.05 wt%, (3) –2.10 wt%, and (4) –3.66 wt% (relative to the weight of the feedstock). All simulations are performed in open-loop mode (i.e., without any control action).

A step increase in coke precursors exhibits two opposite effects on regenerator behavior: It tends to increase the operating temperature because of the greater ability of coke to be burned off; however, this response favors coke consumption, the final result being less coke in the regenerated catalyst. Note that after increases in coke precursors, the amount of coke on spent catalyst (ω_{CSC}) always increases (Figure 5.15), and consequently, coke on regenerated catalyst (ω_{CRC}) follows this trend (Figure 5.16). Both amounts of coke

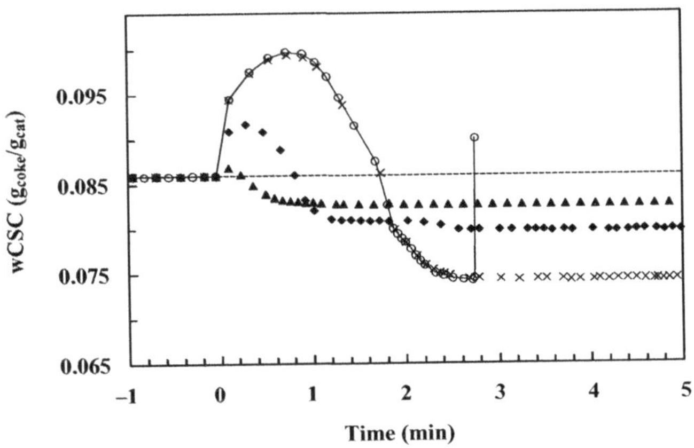

Figure 5.15. Response of coke on spent catalysis for step increases of coke precursors (▲, +0.21 wt%; ◆, +2.10 wt%; ×, +3.66 wt%; -O-, +3.68 wt%; ---, reference). (Adapted from León-Becerril and Maya-Yescas, 2007.)

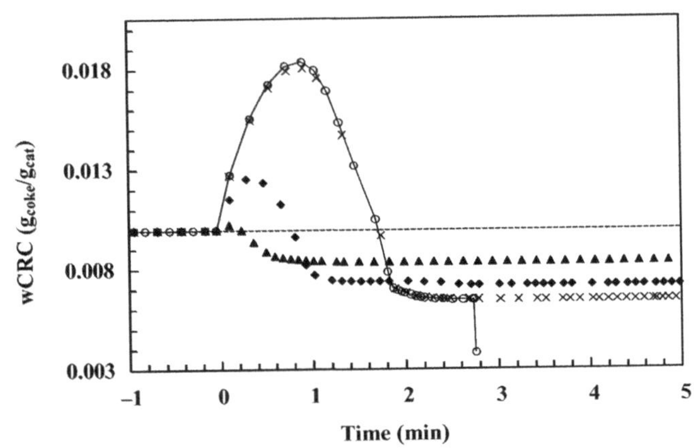

Figure 5.16. Response of coke on regenerated catalysis for step increases of coke precursors (▲, +0.21 wt%; ◆, +2.10 wt%; ×, +3.66 wt%; -O-, +3.68 wt%; ---, reference). (Adapted from León-Becerril and Maya-Yescas, 2007.)

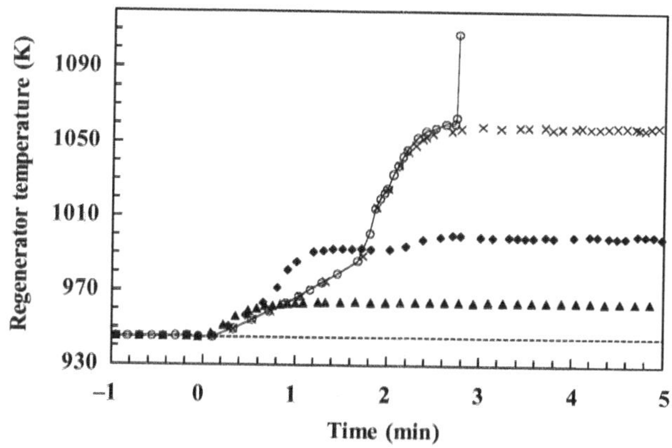

Figure 5.17. Response of regenerator temperature for step increases of coke precursors (▲, +0.21 wt%; ◆, +2.10 wt%; ×, +3.66 wt%; -○-, +3.68 wt%; ---, reference). (Adapted from León-Becerril and Maya-Yescas, 2007.)

reach a new steady state a short time after the step increase arrives, with values lower than the original values. The time necessary to reach the new steady state is directly proportional to the magnitude of the step increase applied, which is difficult to predict a priori during closed-loop operation. Moreover, if the step increase is not large enough (up to about 2.10 wt%), the unit is able to reach a new steady state in a few seconds; however, for a higher step increases, in this case 3.66 wt%, the unit takes longer to reach its new steady state; finally, for an increase of 3.68 wt% the unit is not able to stabilize again.

Any time that coke exhibits higher effective yields, the regenerator receives more raw material to increase heat production; therefore, the regenerator temperature (Figure 5.17) increases proportionately to the magnitude of the step increase applied to the coke precursors. Because the regenerator manages the energy balance, the riser temperature (Figure 5.18) follows the regenerator temperature trend after any disturbance. It is important to note that the unit is not able to stabilize after the last increase provided (the vertical line in Figure 5.17) and the process runs away. In these cases, the yield to coke observed at the riser outlet is affected primarily by the "additive coke" consequence of the increase in coke precursors. Values of the activity function estimated by Eq. (5.13) (Figure 5.19) show that this activity is changing in the opposite way, with coke deposited on the catalyst surface; therefore, the cracking reaction rates are also oscillating with the consequent decrease of selectivity to products. This situation is hidden for empirical activity models, such as Voorhies exponential decay (e.g., Araujo-Monroy and López-Isunza, 2006; León-Becerril and Maya-Yescas, 2007); hence, it is important to notice that deactivation functions should reflect the physical phenomena involved during catalyst deactivation in order to be able to infer the self-stabilization ability of FCC units.

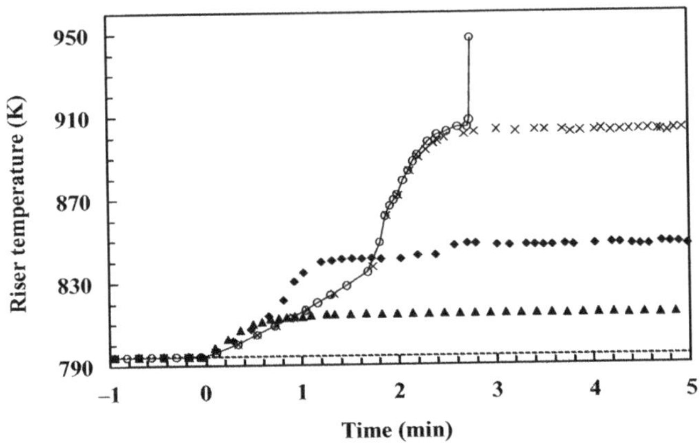

Figure 5.18. Response of riser temperature for step increases of coke precursors (▲, +0.21 wt%; ■, +1.05 wt%; ◆, +2.10 wt%; ×, +3.66 wt%; -○-, +3.68 wt%; ---, reference). (Adapted from León-Becerril and Maya-Yescas, 2007.)

The variables measured during industrial operation are the temperatures of both regenerator (Figure 5.17) and riser (Figure 5.18); therefore, operation ignores the complex response to coke behavior. Additionally, during closed-loop operation, ω_{CRC} is inferred from flue gas composition, and the airflow rate is adjusted based on this value (Álvarez-Ramírez et al., 2004). On the other hand, the riser temperature (Figure 5.18) follows the changes in regenerator temperature because of the difference in thermal holdup (Lee and Kugelman, 1973; Edwards and Kim, 1988; Corella, 2004); however, this change does not reflect which type of coke (either additive or formed by catalytic cracking reactions) is involved in both catalyst activity (Figure 5.19) and energy balance.

To sum up, FCC units exhibit complex responses to increases in coke precursors; nevertheless, inside a certain operating window, the energy balance is able to reach new steady states without severe control actions. Frequently, this period of self-stabilization has not being considered in the control strategies proposed, mainly because of the simple exponential functions, such as the Voorhies decay function, used in the empirical modeling of catalyst deactivation.

A decrease in coke precursors is not as complex as an increase. Both ω_{CSC} (Figure 5.20) and ω_{CRC} (Figure 5.21) reach new state values, with a lower amount of yield to coke with respect to reference. This complex phenomenon is related to the remaining catalyst activity: If the amount of coke deposited on the catalyst surface decreases, the activity increases. Consequently, conversion also increases, yielding larger amounts of each product; for the particular kinetic parameters adjusted to this combination feedstock and catalyst, the yield to coke immediately after the disturbance is followed by a decrease

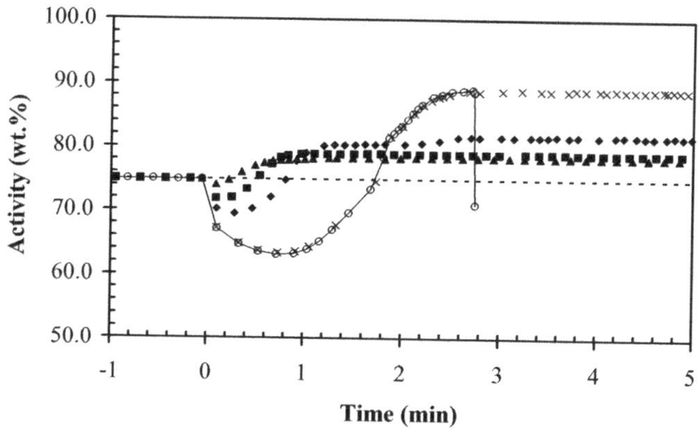

Figure 5.19. Response of catalyst activity for step increases of coke precursors (▲, +0.21 wt%; ■, +1.05 wt%; ◆, +2.10 wt%; ×, +3.66 wt%; -○-, +3.68 wt%; ---, reference). (Adapted from León-Becerril and Maya-Yescas, 2007.)

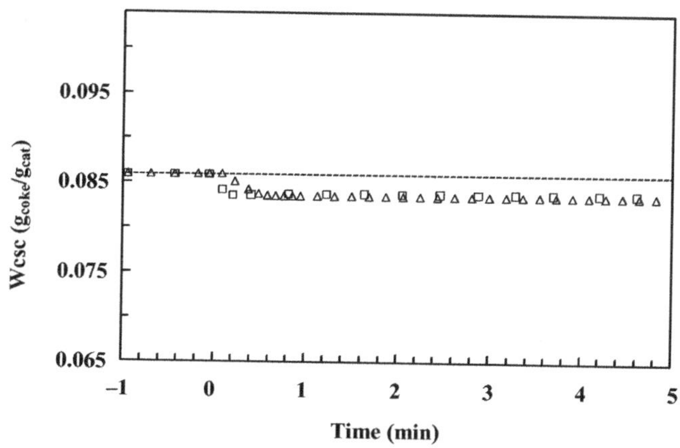

Figure 5.20. Response of coke on spent catalysis for step decreases of coke precursors (Δ, –0.21 wt%; □, –1.05 wt%; ---, reference). (Adapted from León-Becerril and Maya-Yescas, 2007.)

(Figure 5.20). Even after this higher production rate of coke, because there is less additive coke as a consequence of the diminished coke precursors, after 2 min the regenerator has been able to burn off this coke, and the catalyst is returned to the riser with less coke on its surface (ω_{CRC} in Figure 5.21). This situation is observed as a decrease in the effective yield to coke at the riser outlet (Figure 5.20).

After every coke precursors decrease, the regenerator temperature increases up to about 10°C (Figure 5.22). This response seems to be unexpected; however, it is due to the logical response to higher feedstock conversion and,

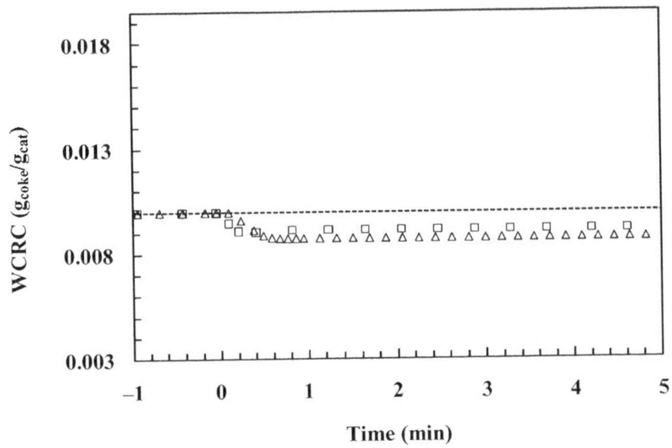

Figure 5.21. Response of coke on regenerated catalysis for step decreases of coke precursors (Δ, –0.21 wt%; □, –1.05 wt%; ---, reference). (Adapted from León-Becerril and Maya-Yescas, 2007.)

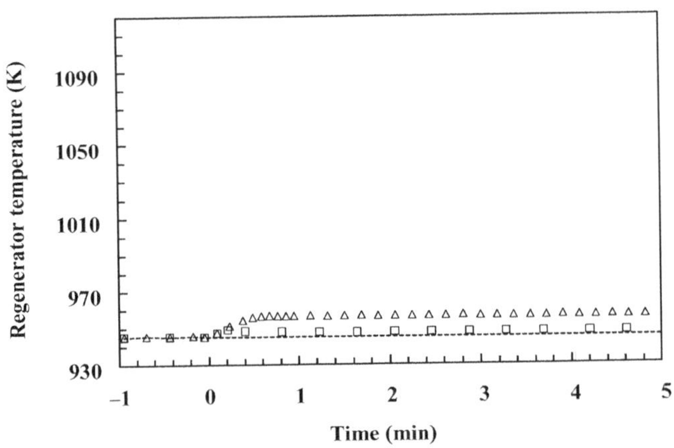

Figure 5.22. Response of regenerator temperature for step decreases of coke precursors (Δ, –0.21 wt%; □, –1.05 wt%; ---, reference). (Adapted from León-Becerril and Maya-Yescas, 2007.)

consequently, higher coke production at the riser (see below). The riser temperature (Figure 5.23) follows the regenerator trend, exhibiting increases of about 17°C. As in the last case, the response of riser temperature, influenced strongly by regenerator temperature, increases conversion.

Due to the decrease in coke precursors, in this case catalyst activity remains higher than that in the reference case (Figure 5.24), a situation that favors the yield to products, including coke. Therefore, the coke that is reflected by ω_{CSC} (Figure 5.20) is mainly "product coke" and not "additive coke." This higher

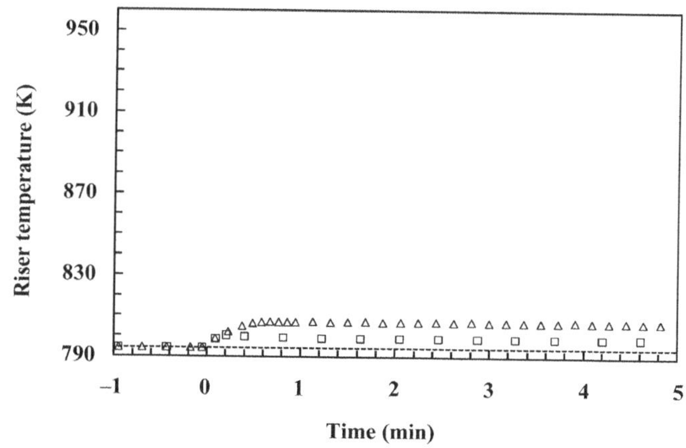

Figure 5.23. Response of riser temperature for step decreases of coke precursors (Δ, −0.21 wt%; □, −1.05 wt%; ---, reference). (Adapted from León-Becerril and Maya-Yescas, 2007.)

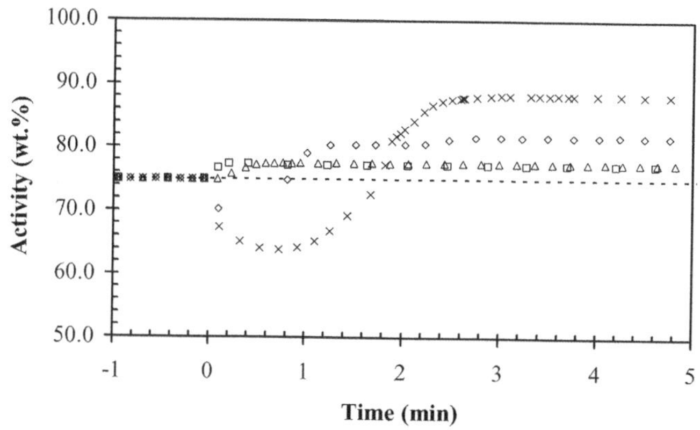

Figure 5.24. Response of catalyst activity for step decreases of coke precursors k (Δ, −0.21 wt%; □, −1.05 wt%; ◇, −2.10 wt%; ×, −3.66 wt%; ---, reference). (From León-Becerril and Maya-Yescas, 2007.)

production of coke relative to the reference case supplies the raw material to heat the system by regeneration reactions, generating an increase in temperature (Figure 5.22) which is followed by the riser temperature (Figure 5.23). As a consequence, the case of small coke decreases does not cause problems during operation (Salazar-Sotelo et al., 2004), due to the smaller amount of coke precursor supplied. Again, the only way to find this response to changes in coke precursor is to model the activity in a more physical way than by using Voorhies exponential decay.

Therefore, it is possible to note that FCC units adapt conveniently to decreases in the amount of coke precursor. The new steady states reached are able to maintain the availability of energy to support the endothermal cracking reactions and do not tend to run away. Therefore, control actions after this disturbance should not be very strong, as has sometimes been pointed out. Moreover, the unit will be able to self-stabilize for a wide change in coke precursor without any control action, a situation that has been under discussion for many years (e.g., Lee and Kugelman, 1973; Edwards and Kim 1988; León-Becerril and Maya-Yescas, 2007). One important aspect the has been pointed out is that it is necessary to model catalyst activity as a function of coke deposited on a catalyst surface (e.g., Jiménez-García et al., 2007), instead of as an exponential decay that does not consider any physical effect on the deactivation phenomenon. In this work a simple hyperbolic function was adjusted to the laboratory data in order to evaluate this functionality point to point during the riser simulations; results show that this approach is better than an empirical approach.

5.7.2 Simulation of Side Reactions During the Burning of Heterogeneous Coke

As was described, during catalytic cracking there is formation of a solid, called *coke*, which deposits on the catalyst surface. The main effect of coke is pore blockage, with apparent loss of active sites. Therefore, it is necessary to burn off this solid in a separate reactor called a regenerator. Coke burning has several important implications that have to be analyzed independently: pollution, energy balance of the entire unit, temperature and temperature changes, catalyst activity and others. Although those issues are complicated, there are others related to environmental constraints, such as the presence of sulfur in feedstock. Part of the sulfur is recovered as sulfhydric acid (desired) and part leaves the FCC unit as sulfided fuels or SO_x in regenerator flue gases. The aim of this section is to analyze how to model and simulate this distribution of feedstock sulfur into all the products of the FCC unit.

Sulfur in Regenerator Flue Gases Villafuerte-Macías et al. (2004) studied the effect of sulfur in the coke generated by catalytic cracking. Coke consists primarily of carbon and hydrogen, and may be polluted by sulfur, nitrogen, or metals from the feedstock. For our goals here, an empirical formula for the coke was modified to take into account the sulfur content after riser reactions: $CH_\nu S_\sigma$. Coke combustion follows the kinetic path.

$$CH_\nu S_\sigma + \left[\frac{\eta+2}{2(\eta+1)} + \frac{\nu}{4} + \frac{2\sigma+3}{2(\sigma+1)}\right] O_2 \rightarrow$$
$$\frac{\eta}{\eta+1} CO + \frac{1}{\eta+1} CO_2 + \frac{\nu}{2} H_2O + \frac{\sigma}{\sigma+1} SO_2 + \frac{1}{\sigma+1} SO_3 \qquad (5.19)$$

Here, ν and σ represent the empirical formulas of the coke, and η is the relationship between CO and CO_2 formation rates evaluated at the catalyst surface (Errazu et al., 1979). As is evident, sulfur in coke is able to form sulfur oxides that are emitted as flue gases.

Numerical simulations can be performed solving the coupled models for riser and regenerator simultaneously. Typical variables (e.g., airflow rate supplied to the regenerator, mass catalyst/oil ratio) were changed to simulate the entire operating zone. Sulfur from feedstock was followed in relation to the sulfur content of fuels, hydrogen sulfide production, and sulfur oxide emissions from the regenerator. It is important to note that catalyst properties and additives are not taken into account explicitly.

A seven-lump kinetic scheme (Figure 5.5b), which specifies the generation of sour gas (H_2S primarily) during catalytic cracking was integrated into the riser mathematical model [Eqs. (5.9) to (5.11)] for an industrial FCC riser. Meanwhile, a comprehensive model for the FCC regenerator [Eqs. (5.14) to (5.18)], including oxidation of sulfur in coke, was coupled to the riser model. Both models were tuned using industrial operating data. Prediction of sour gas formation, sulfur content in final products, and sulfur distribution in regenerator emissions was performed following the entire sulfur balance. This model was a helpful tool for modeling steady-state FCC operation, taking into account valuable clean fuels production and the satisfactory accomplishment of environmental constraints.

Empirical functions related to feedstock conversion were developed to represent sulfur content and sulfur distribution in cycle oils, gasoline, and coke (Venuto and Habib, 1978; McArthur et al., 1981; Cheng et al., 1998; Corma et al., 2001). Following industrial practice, the parameters of these functions were obtained for a particular feedstock and "type of catalyst" and are supported by industrial data. These parameters should be adjusted whenever a different feedstock or another catalyst is used. The operating conditions of the riser do not modify parameter values.

Sulfur content in cracking products (cycle oils, gasoline, and coke) is calculated by applying the function S_{LC} successively:

$$S_{LC} = a + bX + cX^2 \tag{5.20}$$

Here S_{LC} is the sulfur content in cracking products, and X is the feed volume conversion. Values of $a, b,$ and c are fit according to product type (cycle oils, gasoline, and coke). A fourth-order Runge–Kutta method was employed to solve the mathematical model. At each integration interval, temperature, velocity, flux, and density of the reacting mixture; temperature and velocity of the catalyst; mass and heat transfer coefficients; product yields; and sulfur distribution in cracking products are evaluated.

In calibration of the model, 15 sets of parameters selected from operating records of industrial FCC units were utilized; some of them are shown in Table 5.8. Initial numerical values of model parameters were selected from literature

TABLE 5.8. Examples of Operating Conditions, Yields and Properties

	Run 1	Run 2	Run 3
Operating conditions			
Riser outlet temperature (°C)	519.5	514.0	519.0
Feedstock temperature (°C)	213.7	214.3	215.1
Catalyst/oil ratio (kg/kg)	8.77	7.46	8.53
Yields			
Dry gas (m^3/h)	24,600	24,200	23,700
Sour gas (kg/h)	2,104	2,438	2,346
Gasoline (kg/h)	2,313	2,319	2,329
Cyclic oils (kg/h)	758	814	812
Conversion (wt%)	78.6	77.9	77.6
Feedstock supply (kg/h)	4,838	4,849	4,906
Properties			
Feedstock density (kg/m^3)	891	898	901
Sulfur in feedstock (wt%)	2.10	2.17	2.17
Sulfur in gasoline (wt%)	0.212	0.215	0.213
Sulfur in sour gas (wt%)	65.85	67.91	67.43
Sulfur in light cycle oils (wt%)	2.02	2.08	2.11

Source: Adapted from Villafuerte-Macías et al. (2004).

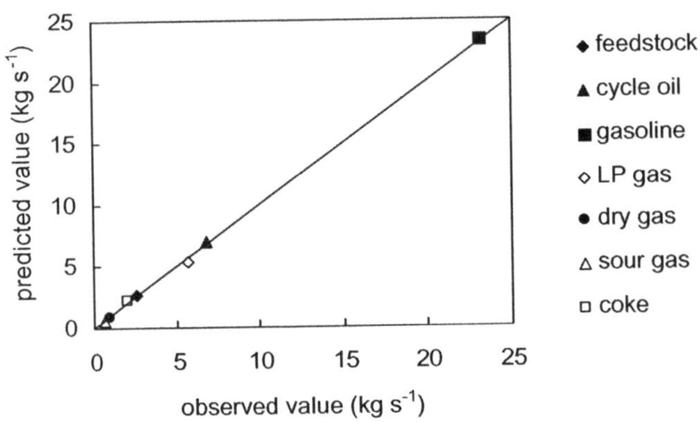

Figure 5.25. Predicted vs. observed values for yield to products. (Adapted from Villafuerte-Macías et al., 2004.)

data (Ancheyta-Juárez et al., 1997; Takatsuka et al., 1987) and fitted to operating data using standard statistical techniques. By using the parameters obtained, a different set of 15 industrial data was simulated. A comparison of predicted yield values of cracking products versus observable values is shown in Figure 5.25. It is important to observe that corresponding points fall in the neighborhood of the 45° line.

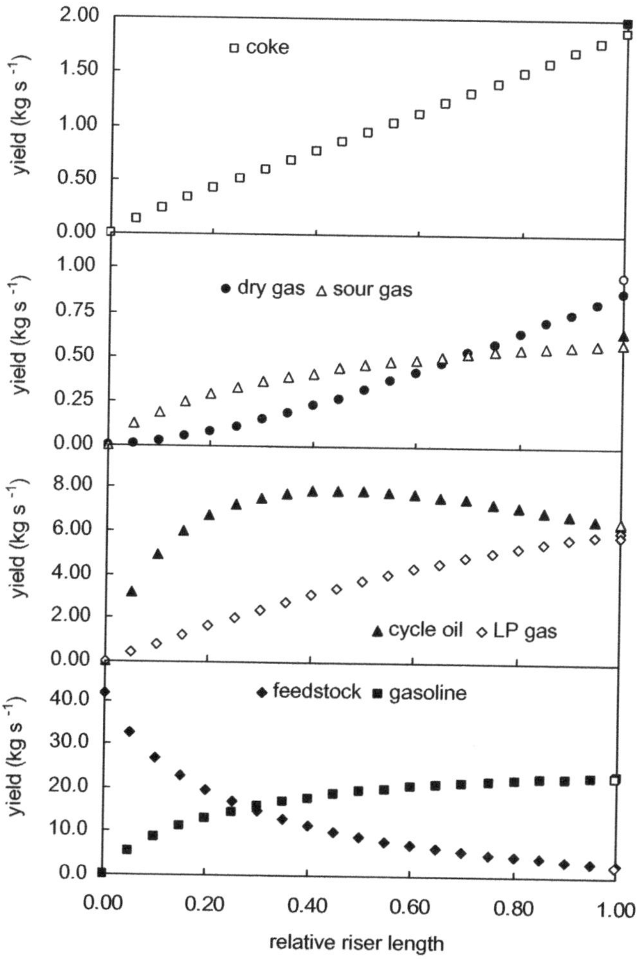

Figure 5.26. Axial profiles of feedstock and products in the riser. (Adapted from Villafuerte-Macías et al., 2004.)

Due to its importance as a process variable, the riser outlet temperature (ROT) is used as a reference variable. In this section, industrial results are taken as a reference. Product yield profiles were modeled along the riser and compared to actual yield values at the riser exit (Figure 5.26). It is possible to observe that the main feedstock cracking occurs before the first third of the riser length. The cyclic oil yield reaches a maximum value before the first half of the riser length, following a predominant soft decreasing yield due to cracking. This last result is in agreement with those intermediate-weight mass products that might be converted to minor molecular weight compounds. Most of the total gasoline final yield (about 90%) is obtained before the first

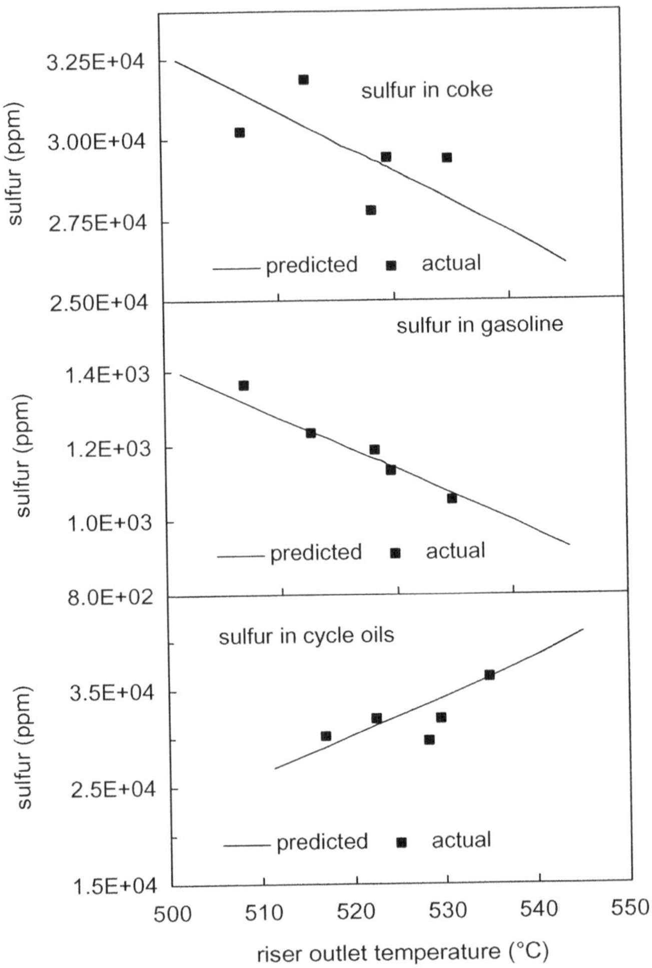

Figure 5.27. Sulfur content in FCC products. (Adapted from Villafuerte-Macías et al., 2004.)

half-length of riser. LPG and dry gas yields are increased continually (Figure 5.26). The greatest sour gas yield is obtained before three-fourths of the riser length, indicating an initial easy link breaking the sulfur hydrocarbons. Coke yield is increased as a result of the condensation of cyclic, heterocyclic, and alkyl compounds on catalyst particles. Industrial results at the riser outlet are close to those predicted by the model. Meanwhile, predicted sulfur content in cyclic oils, gasoline, and coke obtained as a function of ROT is shown in Figure 5.27. It is important to note that the sulfur content in cyclic oils increases in proportion to ROT; meanwhile, the sulfur content of gasoline and coke decreases. Unstable sulfur-linked hydrocarbon compounds crack into sour gas and lighter hydrocarbons; meanwhile, noncracked sulfur compounds

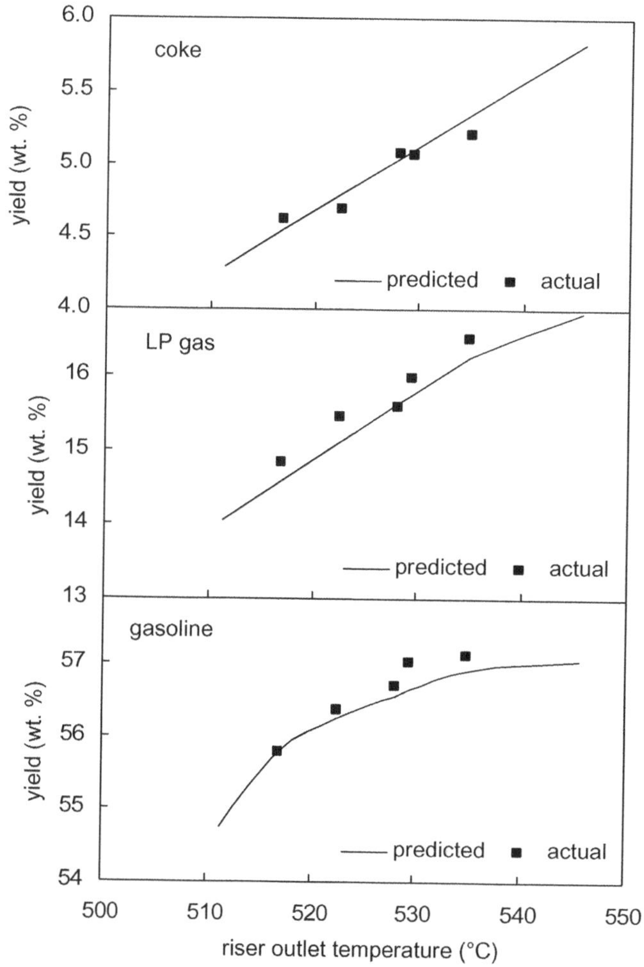

Figure 5.28. Coke, LP gas, and gasoline profiles as a function of ROT. (Adapted from Villafuerte-Macías et al., 2004.)

go to cycle oils and gasoline. The sulfur content predicted in cyclic oils, gasoline, and coke values is in agreement with actual data (see Table 5.8). It should be noted that in order to obtain gasoline with a lower content of sulfur, and a higher coke yield with a lower sulfur content, the unit has to be operated at higher ROT.

Production profiles of gasoline, LPG, and coke as a function of ROT are shown in Figure 5.28, and profiles of cyclic oils, dry gas, and sour gas as a function of ROT are shown in Figure 5.29. Actual data are included. The values predicted comprise a line crossing a neighborhood of actual data. It is observed that the cyclic oil yield predicted decreases as ROT is increased, whereas the

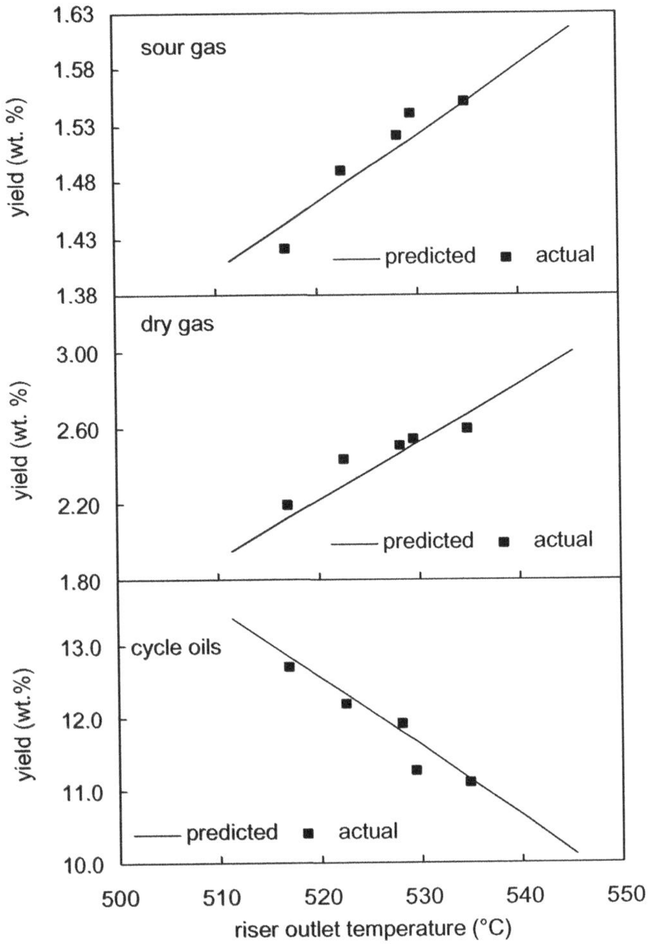

Figure 5.29. Sour gas, dry gas, and cycle oil profiles as a function of ROT. (Adapted from Villafuerte-Macías et al., 2004.)

gasoline, LPG, sour gas, dry gas, and coke yields predicted increase. The decrement in cyclic oil yield is a result of cracking to LPG and dry gas, and some gasoline. At the highest temperature there is only a little or no predicted increase in gasoline yield. It is important to note that an increase in coke yield could be an advantage, because of the relationship between necessary energy in the regenerator and the heat balance of the system. The high sour gas yield predicted results in a lower sulfur content in gasoline and coke, basically an advantage.

As is clear, there is a type of synergy between an increase in gasoline yield and a decrease in sulfur content in this fuel as the ROT increases. Therefore, to preserve the profitability of the operation, FCC units should be operated at the highest possible ROT. At the same time, there is an increase in the sour gas yield, which is also an advantage from an environmental point of view. Both enhancements are made at the cost of a higher sulfur content in cyclic oils. This situation has to be balanced because of the cost of downstream desulfurization. However, the yield to cyclic oils is also decreased, which is also an advantage.

5.7.3 Simulation of the Energy Balance in the Regenerator

As was described at the beginning of this chapter, a regenerator reactor accumulates most of the catalyst during FCC unit operation. In round numbers, a common FCC unit uses about 300 tons of catalyst; about 20 to 30 tons of this total is circulating among the riser and the stand pipes, and of the rest, approximately 270 tons, is in the regenerator following the coke-burning reactions. That is, the energy balance of the entire unit is governed by the regenerator. It is therefore necessary to study this part of the unit, emphasising the role of energy balance.

Because of its physical configuration, the regenerator (the fluidized-bed reactor) cannot be considered to be a homogeneous stirred tank. It is usual to assume the existence of two very different regions, a dense region (at the bottom) and a dilute or freeboard region (the upper part) (see Section 5.7.1). The two regions are different, as is clear from Table 5.9.

The dense and dilute regions interchange mass and energy primarily because of the ascending flow of gases and to a very small extent because of some catalyst particles that move from the dense to the dilute region. However, due to an important difference in the heat capacity, any small production of energy in the dilute region (as a consequence of the CO combustion) provokes a very high increase in temperature. A schematic representation of the FCC regenerator and a common energy balance situation are shown in Figure 5.30.

A great advantage of fluidized-bed reactors is that their dense phase is practically isothermal, exhibiting the same temperature at any point of the phase; this effect is caused by the moving solid particles that agitate the phase.

TABLE 5.9. Characteristics of Dense and Dilute Regions in a Fluidized-Bed Reactor

	Dense Region	Dilute Region (Freeboard)
Location	Lower part	Upper part
Volume fraction of solids	About 20–40%	Less then 1%
Relative density	High	Very low
Relative heat capacity	High	Low
Main reactions	Coke combustion CO combustion	CO combustion

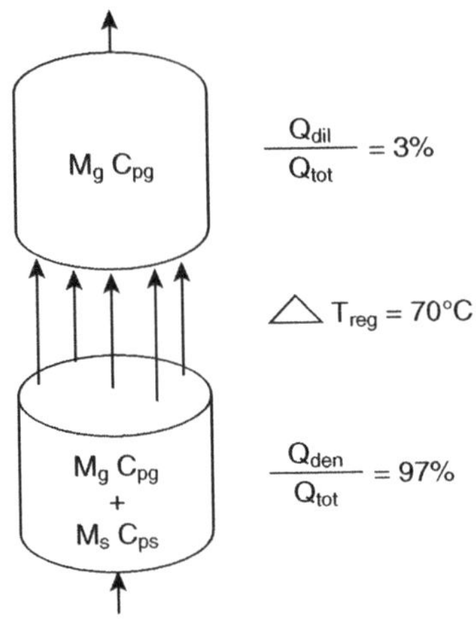

Figure 5.30. Two regions in a fluidized-bed reactor.

In contrast, if the exothermal reactions are not controlled in the dilute phase, the reactor will exhibit severe temperature gradients that cause operating problems and, in extreme situations, destruction of cyclones and other hardware located inside the regenerator. These two physical situations also justify the modeling presented for the regenerator; which considers the dense phase as an isothermal CSTR and the dilute phase as a nonisothermal PFR (Errazu et al., 1979). Also, it will be important to recognize that control of the regenerator reactor, and consequently, of the energy balance of the entire unit, is performed by measuring operating variables in the dense phase—either temperature or oxygen concentration—due to the low variability exhibited by this type of variable in CSTRs.

Another important aspect of the modeling of FCC regenerators is the fact that the computing time used for the solution of equations should "not be too long." If the dense phase of the regenerator is considered as a CSTR, this time is reduced greatly compared to other approaches. Similarly, because the dilute phase exhibits only one entrance and one exit, solution of a PFR model is very efficient. It is accepted that this type of modeling is satisfactory in comercial units.

5.8 MODELING THE CATALYST STRIPPER

Although stripping of hydrocarbons adsorbed to a catalyst surface is a very important step during the catalyst cycle inside the unit, there are not many models available to explain this operation in a faithful way. For example, stripping is assumed to take place in a well-mixed tank. Catalyst arrives at the stripper from the riser and after stripping, moves toward the regenerator; therefore, the dynamic effect of this separator is to create a time delay between the riser and the regenerator. The governing equations of this process are

$$W_{\mathrm{st}} \frac{dC_{\mathrm{st}}}{dt} = F_S (C_{\mathrm{SC}} - C_{\mathrm{st}}) \tag{5.21}$$

$$W_{\mathrm{st}} C_{ps} \frac{dT_{\mathrm{st}}}{dt} = F_S C_{ps} (T_{\mathrm{ro}} - T_{\mathrm{st}}) \tag{5.22}$$

It is clear that this model in not complete, it lacks the kinetics inside the stripper (same as that in the riser), plus the terms of hydrocarbon desoption. However, these types of models are usually chosen to model this part of the converter because the main effect of the stripper from a dynamics point of view is to provoke a time delay. This situation provides a "buffer tank" for the catalyst prior to its arrival at the regenerator, where combustion reactions greatly complicate the dynamics. Parameters for such models are not directly relevant to the subject matter of this section; in fact, they are still under development.

5.9 SIMULATION OF A CONTROLLED FCC UNIT

Although there have been several attempts to develop systematic evaluation of control schemes, the choosing and pairing of control variables in chemical reactors is not an easy task. Intrinsic nonlinearities of these types of systems create dynamic responses that are difficult to predict. In this section, a simple proposition for the evaluation of pairs of controlled and manipulated variables is developed for nonlinear control affine systems. It complements relative gain array (RGA) analysis for nonlinear systems because it is based on the relationship between zero dynamics and control stability. The basic strategy is simple, easy to understand, and easy to employ in the analysis of control schemes; it is also independent of the type of controller used. It is probed in the evaluation of four control options for industrial FCC regenerators (Maya-Yescas and Aguilar, 2003), two of which are employed in industry. The results obtained when evaluating the control strategies are in line with industrial practice and operating experience.

Regulation issues of nonlinear processes are an open problem in the control community, mainly because of the industrial relevance of this type of system. In general, control problems in industrial plants are very complex, so industrial processes are only partially controlled. The complete stabilization and regulation of this type of systems is not assured; therefore, the dynamic behavior of the uncontrolled variables has to be studied in order to predict an approximation of the global performance of the process. Meanwhile, it is common to find control affine schemes for CSTRs (such as the FCC regenerator), which means that manipulated variables appear linearly in the model of the system; this situation provides some advantages, which are studied in this work.

Because of its complexity, the control of nonlinear systems has motivated the development of research in this field. In a nonlinear setting, the stability of the inverse of the system dynamics should itself be checked. Following the ideas of Morari and co-workers (Morari, 1983, García and Morari, 1985: Economou and Morari, 1986a), from a control point of view the perfect control for a process should present dynamic behavior that is given by the inverse of the dynamics of this process. However, due to the existence of time delays, this inverse cannot be realized. Moreover, there are no explicit formulas for the inversion of general MIMO system dynamics. One important attempt is the realization of a minimal order inverse dynamics, which has been defined as zero dynamics (Daoutidis and Kravaris, 1991). In general, the construction of zero dynamics involves complicated algorithmic procedures, which could be unsuccessful; according to Daoutidis and Kravaris (1991), an analysis of the zero dynamics of nonlinear systems yields the same conclusions as those of an analysis of the zeros for linear systems. An interesting idea used for chemical processes was to separate the nonlinear dynamics of a system into invertible and noninvertible parts, the last one containing time-delay terms (Economou and Morari, 1986b).

The papers mentioned present very interesting theoretical frameworks in which some properties of the closed-loop performance of nonlinear systems are highlighted. However, the mathematical tools employed are complex, and it is not yet possible to apply their methodologies to industrial plants. In this section a clear methodology, used as a simple test of controllability, is proposed to analyze control options for nonlinear control affine systems.

5.9.1 Mathematical Background

The procedure of design and control of chemical reacting systems provides a good target for the identification of stable operating steady states, which is a well-known topic (e.g., Isidori, 1999). Nonetheless, the easiness of the regulation of important states will depend on dynamic features related to operating conditions, design, control, and the relationships among them. As we have pointed out, one of the most important characteristics that has to be analyzed when a reacting system is to be controlled is the stability of the zero dynamics.

The mathematical model for the dense phase of the regenerator, which consists of mass balances, energy balances, and equilibrium relationships [Eqs. (5.14) to (5.17)] is the starting point. They can be written as a system of nonlinear equations

$$\dot{x} = f(x) + G(x)u \tag{5.23}$$

where $x \in \Re^n$ is the vector of states, $u \in \Re^q$ the vector control input, $f(x): \Re^n \to \Re^n$ a nonlinear smooth vector field, and $G(x): \Re^{n+q} \to \Re^n$ a matrix that contains the relationship between control and manipulate variables.

Now, consider the following assumptions:

A1. For the control input vector $u(x(t))$ realized ($\|u(x(t))\| \leq u_{\max}$), the nominal closed-loop nonlinear system [Eq. (5.23)] is at least quadratic stable for the controlled states; therefore, there exists a Lyapunov function $V \geq 0$ that satisfies

$$\frac{\partial V}{\partial X}[\Im(X)+\ell(X,U)] \leq -\alpha_1 \|X\|^2 \qquad \left\|\frac{\partial V}{\partial X}\right\| \leq \alpha_2 \|X\|, \quad \alpha_1, \alpha_2 > 0$$

A2. All the trajectories $x(t,x_0), x_0 \in \Re^n$ of the system [Eq. (5.23)] are bounded.

A3. The vector field $G(x)$ is bounded [i.e., for any $x \in \Re^n, \|G(x)\| \leq G^+ < \infty$].

Now, the zero dynamics of a system are defined as the minimal-order dynamics of its inverse. For nonlinear systems, the realization of this inverse could be very complicated or even impossible. However, for control affine systems that are partially controlled, it is possible to assess the stability features of the zero dynamics (Maya-Yescas and Aguilar, 2003) following the dynamics of the uncontrolled (or dynamic) states, $\mathbf{x}_D$, while the system is regulated by the control of a subset of states, $\mathbf{x}_C$:

$$\dot{\mathbf{x}} = \mathbf{f}(\mathbf{x}) + \mathbf{G}(\mathbf{x})\mathbf{u} \rightarrow \begin{cases} \dot{\mathbf{x}}_C = \mathbf{f}_C(\mathbf{x}) + \mathbf{G}_C(\mathbf{x})\mathbf{u} \\ \dot{\mathbf{x}}_D = \mathbf{f}_D(\mathbf{x}) + \mathbf{G}_D(\mathbf{x})\mathbf{u} \end{cases} \tag{5.24}$$

Here

$$\mathbf{x}_C \in \Re^q, \quad \mathbf{f}_C(\mathbf{x}) \equiv \{\mathbf{f}_C : \Re^q \rightarrow \Re^q\}, \quad \mathbf{G}_C(\mathbf{x}) \equiv \{\mathbf{G}_C : \Re^{q \times q}\}$$

and

$$\mathbf{x}_D \in \Re^{n-q}, \quad \mathbf{f}_D(\mathbf{x}) \equiv \{\mathbf{f}_D : \Re^{n-q} \rightarrow \Re^{n-q}\}, \quad \mathbf{G}_D(\mathbf{x}) \equiv \{\mathbf{G}_D : \Re^{n \times q}\}$$

The idea is to find the vector of manipulated inputs $\mathbf{u}$ assuming that the variables regulated will remain steady at the desired set point:

$$\dot{\mathbf{x}}_C^{\text{sp}} = \mathbf{0} \quad \Leftrightarrow \quad \mathbf{u}^{\text{sp}} = -\mathbf{G}_C^{-1}(\mathbf{x})\mathbf{f}_C(\mathbf{x}) \tag{5.25}$$

Then it is possible to substitute the vector of manipulated variables, $\mathbf{u}^{\text{sp}}$, into the balances for dynamic variables:

$$\dot{\mathbf{x}}_D = \mathbf{f}_D(\mathbf{x}) - \mathbf{G}_D(\mathbf{x})\mathbf{G}_C^{-1}(\mathbf{x})\mathbf{f}_C(\mathbf{x}) \tag{5.26}$$

If the evolution of the dynamic behavior of the uncontrolled variables is not stable when operating under this particular set of inputs $\mathbf{u}^{sp}$, it is possible to conclude that zero dynamics are also not stable (Isidori, 1999). Therefore, to ensure complete stability of the zero dynamics and of the control, each balance for the uncontrolled variables should tend to an attractor at the desired set point. The policy proposed is to ask for all the balances $\dot{\mathbf{x}}_D$ to have a negative sign at the operating set point (Maya-Yescas et al., 1998).

Proposition. The controller of the system $\dot{\mathbf{x}}$ will be stable $\forall t > 0$ if and only if $\mathbf{f}_D(\mathbf{x}) + \mathbf{G}_D(\mathbf{x})\mathbf{u}^{sp} < \mathbf{0}$ (i.e., $\dot{x}_D < 0, \forall x_D \in \mathbf{x}_D$).

Proof (Maya-Yescas and Aguilar, 2003): Consider the closed-loop performance of the uncontrolled variables:

$$\dot{\mathbf{x}}_D = \mathbf{f}_D(\mathbf{x}) + \mathbf{G}_D(\mathbf{x})\mathbf{u}^{sp}$$

Now, define the Lyapunov function that follows:

$$V = \varepsilon^{T}\varepsilon = \Big(\underbrace{0 \quad 0 \quad \cdots \quad 0}_{q \text{ times}} \quad \underbrace{\varepsilon_{D_1} \quad \varepsilon_{D_2} \quad \cdots \quad \varepsilon_{D_{n-m}}}_{(n-q) \text{ times}}\Big) \begin{pmatrix} 0 \\ 0 \\ \vdots \\ 0 \\ \varepsilon_{D1} \\ \varepsilon_{D2} \\ \vdots \\ \varepsilon_{Dn-m} \end{pmatrix} \begin{matrix} \\ q \text{ times} \\ \\ (n-q) \text{ times} \\ \\ \end{matrix}$$

$$\Leftrightarrow \quad V = \varepsilon_{D_1}^2 + \varepsilon_{D_2}^2 + \cdots + \varepsilon_{D_{n-q}}^2 = \sum_{i=1}^{n-q} \varepsilon_{D_i}^2$$

Now, applying the stability criterion of Lyapunov, the system $\dot{\mathbf{x}}_D$ will be asymptotically stable if such a function, V, exists and is positive definite and its derivative is negative definite in some domain. Computing the derivative and substituting the expression for ε, the following expression is obtained:

$$\dot{V} = \sum_{i=1}^{n-q} \frac{d\varepsilon_{D_i}^2}{dt} = \sum_{i=1}^{n-q} \frac{d}{dt}\left(x_{D_i} - x_{D_i}^{sp}\right)^2 = 2\sum_{i=1}^{n-q} x_{D_i} \frac{dx_{D_i}}{dt} = 2\sum_{i=1}^{n-q} x_{D_i}\dot{x}_{D_i}$$

Because the values for $\mathbf{x}(t) \geq \mathbf{0}$ and $\boldsymbol{u}^{sp} \geq 0$, the only option for $\dot{V}$ to be negative semidefinite is that the dynamics of variables $\dot{\mathbf{x}}_D(t) < 0$ in a finite number of points. Therefore, the control of the system will be stable, $\forall t > 0$, if and only if

$$\dot{V} \leq 0 \Leftrightarrow \dot{\mathbf{x}}_D \leq \mathbf{0} \Leftrightarrow \mathbf{f}_D(\mathbf{x}) + \mathbf{G}_D(\mathbf{x})\mathbf{u}^{sp} \leq \mathbf{0} \Leftrightarrow \mathbf{f}_D(\mathbf{x}) - \mathbf{G}_D(\mathbf{x})\mathbf{G}_C^{-1}(\mathbf{x})\mathbf{f}_C(\mathbf{x}) \leq \mathbf{0}$$

Remarks: Notice that the full nonlinear model was used without linearization or any other simplification. The vector of manipulated variables was defined as a function of the control policy, in terms of other process variables, and is evaluated at the *desired* set point, whatever it is. Due to these characteristics, the methodology can be used for *any* operating point that would be chosen as a set point. Also, there were no restrictions to the type of controller used. Therefore, this methodology is applicable to *any* system that has control affine structure, in particular to CSTR systems.

It is possible to see that evaluation of the dynamics of a vector of manipulate variables follows a procedure similar to the computing of elements of the relative gain array (RGA) by using partial derivatives of the process model with respect to manipulated variables (Bristol, 1966). The difference is that in this case, the initial assumption is that manipulated variables vector $\mathbf{u}$ is already paired to the corresponding outputs. When used in nonlinear systems, the RGA methodology yields relative gains that depend to some extent on the steady state analyzed and are not constant. Following the methodology proposed in this paper, the dynamic behavior of uncontrolled states, influenced by interactions with controlled states, is obtained without evaluation of the complete RGA. Moreover, in contrast to RGA, this methodology analyses uncontrolled instead of controlled states. Therefore, this methodology could be considered complementary to RGA analysis for nonlinear systems with control affine structure.

5.9.2 Controllability of the Regenerator

Regeneration consists of the burn-off of the deposited coke using atmospheric air, in a fluidized-bed reactor that is considered as a CSTR. The energy generated by the exothermic reactions is employed to vaporize the feedstock and to support the endothermic cracking reactions, which take place in the riser reactor (Arbel et al., 1995). Considering the exothermic nature of the regeneration reactions and the characteristics of the combustion kinetics, which can be described by consecutive reactions, the dynamic behavior of the regenerator is expected to be very complex. Phenomena such as steady-state multiplicity, inverse response to control actions (Figure 5.31), and unstable operating zones might appear. An interesting feature of the system is that linear approximation of the model exhibits eigenvalues with a positive real part (Figure 5.32), which are indicative of instability of the closed-loop internal dynamics (Daoutidis and Kravaris, 1991); this instability will be reflected by control problems around these states (Arbel et al., 1996).

One of the most common problems reflected when controlling adiabatic FCC units is the insufficient understanding of regenerator dynamics. Due to restraints in mechanical design, operating conditions and control actions are limited; therefore, it is necessary to study stability and dynamic resilience, taking these physical limits into account. Since the riser is a plug-flow reactor that carries out only endothermic reactions and feedstock vaporization, it

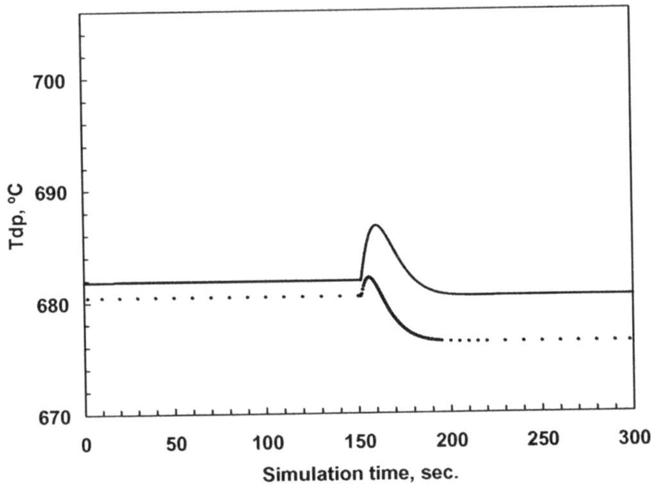

Figure 5.31. Regenerator temperature inverse response at commercial operating conditions: change from $F_{air} = 0.90F_{air}^{base\ case}$ to $F_{air} = F_{air}^{base\ case}$ (solid line) and from $F_{air} = 0.83F_{air}^{base\ case}$ to $F_{air} = 1.05F_{air}^{base\ case}$ (dotted line). (From Maya-Yescas and Aguilar, 2003.)

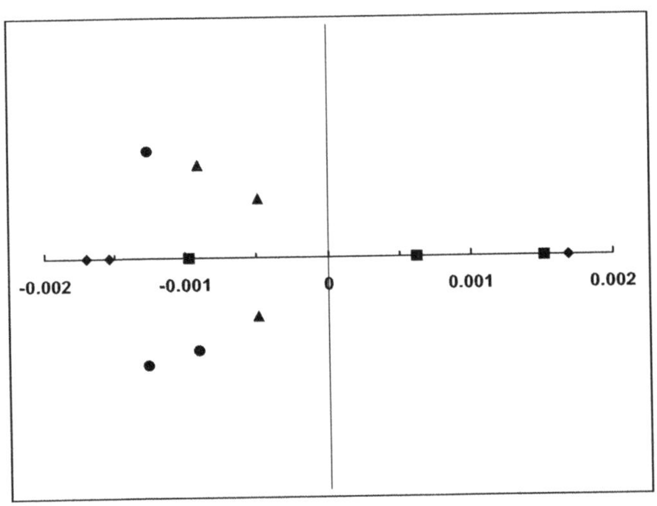

Figure 5.32. Eigenvalues for linearized approximations of the model at commercial operating conditions (◆, $F_{air} = 0.75F_{air}^{base\ case}$; ■, $F_{air} = 0.90F_{air}^{base\ case}$; ▲, $F_{air} = F_{air}^{base\ case}$; ●, $F_{air} = 1.50F_{air}^{base\ case}$). (From Maya-Yescas and Aguilar, 2003.)

presents only one stable steady state for each operating condition. Hence, it is necessary to analyze the dynamics of the regenerator exclusively (Maya-Yescas and Aguilar, 2003).

Additionally, in adiabatic FCC regenerators there are very few manipulated variables, a problem that limits the control design options (Venuto and Habib, 1978). The most common variables to manipulate are the flow of air supplied to a regenerator (F_{air}), the mass flows of catalyst between reactors (m_{cat}), and the preheat temperature of the feedstock (T_{feed}). In more sophisticated units it is also possible to manipulate the oxygen concentration in the air supplied to the regenerator and the regenerator cooling rate; however, these cases are not discussed. Then the problem of control design yields a "pairing game" between a small set of manipulated variables and a large set of control targets on a narrow and constrained operating window (Maya-Yescas et al., 2004a).

As in any reacting system, FCC regenerators are nonlinear. Research related to the dynamic performance of FCC units has pointed out that operating in a partial combustion mode (no CO in regenerator flue gases) is pseudostable (Lee and Kugelman, 1973; Edwards and Kim, 1988). Also, it has been said that a change to a full combustion mode, by increasing the airflow rate, could eliminate steady-state multiplicity at industrial operating conditions (Edwards and Kim, 1988); this phenomenon has been described by a convergence of the intermediate steady state with the ignited state (Maya-Yescas et al., 1998). This fact is also in agreement with the idea that an increase in the operating temperature of the regenerator dense phase is able to stabilize the system in most cases, even when operating in a partial combustion mode (Venuto and Habib, 1978; Maya-Yescas and Aguilar, 2003). Again, it is necessary to follow as precisely as possible the dynamics of the energy balance inside the regenerator. An industrial FCC unit is used as an example; its main characteristics are summarized in Table 5.10.

The states that are followed are oxygen concentration in the dense phase, carbon on regenerated catalyst, CO concentration in the dense phase, and dense phase temperature; all of them are important operation variables, so the vector of states is defined as $\mathbf{x} = (y_{O_2} \quad \omega_{CRC} \quad y_{CO} \quad T_{dp})^T$; here y_{O_2} is the oxygen mole fraction and y_{CO} is the mole fraction of carbon monoxide in flue gases, ω_{CRC} is the mass fraction (relative to mass of catalyst) of coke in regenerated catalyst, and T_{dp} is the temperature of the dense phase.

TABLE 5.10. Main Operating Data of the FCC Unit Studied

Type of unit	Adiabatic; riser/fluid regenerator
Operating mode	Full combustion
Unit feedstock capacity (bbl/day)	25,000
Average coke production (tons/day)	160
Average airflow rate (m^3/h)	75,000

Source: Adapted from Maya-Yescas and Aguilar (2003).

The mathematical model of the regenerator developed in Section 5.7 is used to simulate the dynamics of the gaseous entities in the dense phase; it is written using Eq. (5.23) to apply the methodology developed above. The kinetic scheme and parameters from Section 5.7.1 are used. A comparison of the easiness to control is performed for different operating steady states, when either T_{dp} or y_{O_2} is chosen as a control target.

$$\dot{\mathbf{x}} = \begin{pmatrix} (y^i_{O_2} - y_{O_2})F_{air} + r_{O_2} \\ r_{coke} + m_{cat}\dfrac{\omega_{CSC} - \omega_{CRC}}{W_{rgn}} \\ (y^i_{CO} - y_{CO})F_{air} + r_{CO} \\ \dfrac{Q^i_{steam} + (C^i_{P_g}T^i_g - C_{P_g}T_{dp})F_{air} + \sum_{j=1}^{j=3}(-\Delta H_r)_j r_j + m_{cat}C_{P_p}(T^i_{cat} - T_{dp})}{W_{rgn}C_{P_p}} \end{pmatrix} \tag{5.27}$$

To obtain the operating points desired, a commercial simulator/optimizer (Maya-Yescas and Aguilar, 2003; Maya-Yescas et al., 2004a) was tuned up against some industrial data and used to simulate the operating regions for two different objective functions: maximum C_4-olefin production and maximum gasoline production. Industrial and simulation results are plotted against riser outlet temperature, which is the industrial reference set point in industrial operation (Figures 5.33 and 5.34). Results taken from the simulator are the set points of the riser outlet temperature and some of the dependent operating variables.

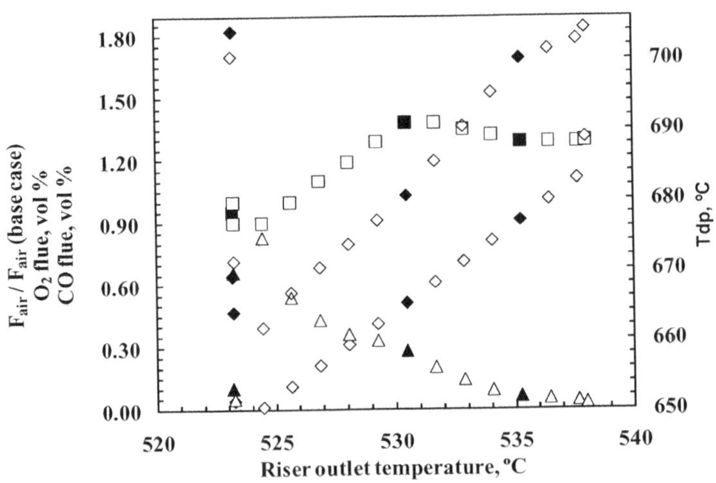

Figure 5.33. Operating (solid symbols) and simulated (open symbols) steady states in the region of maximum production of olefins (■, F_{air}; ◆, T_{dp}; ▲, y_{CO}; ●, y_{O_2}). (Adapted from Maya-Yescas et al., 2004a.)

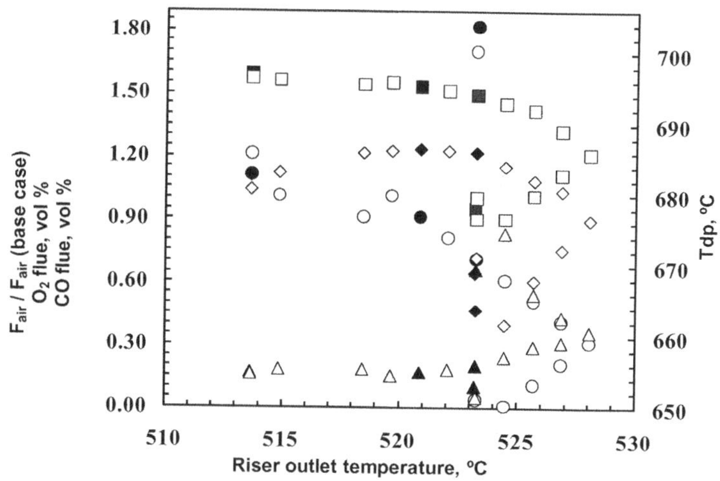

Figure 5.34. Operating (solid symbols) and simulated (open symbols) steady states in the region of maximum production of gasoline (■, F_{air}; ◆, T_{dp}; ▲, y_{CO}; ●, y_{O_2}). (Adapted from Maya-Yescas et al., 2004a.)

After that, the dynamic model described previously is used to determine the sign of the mass balances for y_{O_2}, ω_{CRC}, y_{CO}, and the energy balance for T_{dp}, when applicable. The aim is to regulate the regenerator during normal operating conditions. Hence, the two control policies described (control of either y_{O_2} or T_{dp}) are studied using the methodology developed in Section 5.9.1, in order to characterize the dynamics at different operating steady states. Note that both control policies are analyzed using the same operating data, so the different control problems exhibited by the system are the consequence, of the variable pairing only. In industry, F_{air} is the only variable available to regulate the regenerator; therefore, it is possible to regulate only one control target. The following operation policies analyze the effect of two different elections of this control target.

First Operating Policy

First Control Policy The first case study is the analysis of the dynamic behavior of the FCC regenerator when an arbitrary initial steady state is changed to the maximum C_4-olefin production point. The first *control policy* to be analyzed is that proposed for full combustion regenerators (i.e., y_{O_2} is the control target). Following the methodology proposed [Equation (5.24)]; the vector of states is divided into controlled and uncontrolled variables:

$$\mathbf{x}_C = (y_{O_2})$$
$$\mathbf{x}_D^{\mathrm{T}} = (\omega_{CRC} \quad y_{CO} \quad T_{dp}) \tag{5.28}$$

Now, as proposed by Eq. (5.25), the value of the manipulated variable is then calculated for the desired set point:

$$\mathbf{u}^{\text{sp}} = F_{\text{air}}^{\text{sp}} = -\frac{r_{\text{O}_2}}{y_{\text{O}_2}^{i} - y_{\text{O}_2}^{\text{sp}}} \tag{5.29}$$

Once the manipulated variable is known, it is possible to calculate the dynamics of the uncontrolled variables, $\dot{\mathbf{x}}_D$ [as was done in Eq. (5.26) when the unit is operating following this control policy]:

$$\dot{\mathbf{x}}_D = \begin{pmatrix} r_{\text{coke}} + \dfrac{\omega_{\text{CSC}} - \omega_{\text{CRC}}}{W_{\text{rgn}}} m_{\text{cat}} \\ r_{\text{CO}} - \dfrac{y_{\text{CO}}^{i} - y_{\text{CO}}}{y_{\text{O}_2}^{i} - y_{\text{O}_2}^{\text{sp}}} r_{\text{O}_2} \\ \left[Q_{\text{steam}}^{i} + \sum_{j=1}^{j=3} (-\Delta H_j) r_j + m_{\text{cat}} C_{p_p} (T_{\text{cat}}^{i} - T_{\text{dp}}^{\text{sp}}) - \dfrac{C_{p_g}^{i} T_g^{i} - C_{p_g} T_{\text{dp}}}{(y_{\text{O}_2}^{i} - y_{\text{O}_2}^{\text{sp}})} r_{\text{O}_2} \right] \dfrac{1}{W_{\text{rgn}}} \end{pmatrix} \tag{5.30}$$

The relative values of the balances for $\dot{\mathbf{x}}_D$ for different steady states when maximum olefin production is required are shown in Figure 5.35. As can be seen, at the first three operating states the unit was predicted to be around a difficult operating point. This was due to the fact that the air supplied was not enough. In fact, the CO concentration in flue gases was higher than the accepted limit (>0.05%), which means that there is partial combustion. From the fourth state, the air amount was increased, changing the operating conditions to more favorable ones. As can be noted in Figure 5.34, this operating region presents

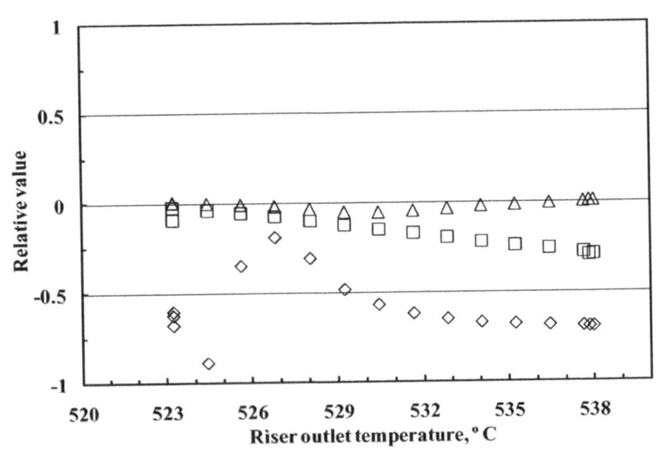

Figure 5.35. Values of the balances for $\dot{\mathbf{x}}_D$ when controlling y_{O_2} in the region of maximum olefin production (◇, T_{dp}; △, y_{CO}; □, ω_{coke}). (Adapted from Maya-Yescas and Aguilar, 2003.)

local problems of the stability of the zero dynamics because of the positiveness of some eigenvalues. This situation is found during operation because at this operating point, temperatures tend to change "too fast" and sometimes in an undesired direction. This is also reflected by the possibility of inverse response in this range of F_{air} values (Figure 5.31). Once the airflow is increased, these problems disappear, in agrement with Edwards and Kim (1988).

This case illustrates the common industrial control policy. Because this unit operates at full combustion, the control policy should be the regulation of y_{O_2}. Once the airflow rate is incremented, this operating problem disappears. Therefore, it is possible to note that between 525 and 535°C the unit works satisfactorily. The change of signs of CO balance at about 536°C was discussed with refinery operators. They said that there is a "kind of limit" in the riser outlet temperature which is reflected by control problems if it is exceeded. Their rule of thumb is to establish, a priori, a maximum temperature and never cross over it. This analysis provides this maximum temperature (~536°C) simply following the time evolution of the signs of mass and energy balances. The actual maximum temperature limit depends on operating conditions and would not easily be estimated a priori; however, using this methodology, this limit is predicted from steady-state simulations.

Second Control Policy This control policy is the one proposed for partial combustion regenerators (i.e., T_{dp} is the control target). Following the methodology proposed, the vector of states is divided into controlled and uncontrolled variables:

$$\begin{aligned} \mathbf{x}_C &= (T_{\text{dp}}) \\ \mathbf{x}_D^{\text{T}} &= (y_{O_2} \quad \omega_{\text{CRC}} \quad y_{\text{CO}}) \end{aligned} \tag{5.31}$$

Next, the value of the manipulated variable is calculated for the desired set point:

$$\mathbf{u}^{\text{sp}} = F_{\text{air}}^{\text{sp}} = -\frac{Q_{\text{steam}}^i + \sum_{j=1}^{j=3}(-\Delta H_j)r_j + m_{\text{cat}}C_{p_p}(T_{\text{cat}}^i - T_{\text{dp}}^{\text{sp}})}{C_{p_g}^i T_g^i - C_{p_g} T_{\text{dp}}^{\text{sp}}} \tag{5.32}$$

Once the manipulated variable is known, it is possible to calculate the dynamics of the uncontrolled variables, when the unit is operating under this control policy:

$$\dot{\mathbf{x}}_D = \begin{pmatrix} r_{O_2} - \dfrac{Q_{\text{steam}}^i + \sum_{j=1}^{j=3}(-\Delta H_j)r_j + m_{\text{cat}}C_{p_p}(T_{\text{cat}}^i - T_{\text{dp}}^{\text{sp}})}{C_{p_g}^i T_g^i - C_{p_g} T_{\text{dp}}^{\text{sp}}}(y_{O_2}^i - y_{O_2}) \\ r_{\text{coke}} + \dfrac{\omega_{\text{CSC}} - \omega_{\text{CRC}}}{W_{\text{rgn}}} m_{\text{cat}} \\ r_{\text{CO}} - \dfrac{Q_{\text{steam}}^i + \sum_{j=1}^{j=3}(-\Delta H_j)r_j + m_{\text{cat}}C_{p_p}(T_{\text{cat}}^i - T_{\text{dp}}^{\text{sp}})}{C_{p_g}^i T_g^i - Cp_g T_{\text{dp}}^{\text{sp}}}(y_{\text{CO}}^i - y_{\text{CO}}) \end{pmatrix} \tag{5.33}$$

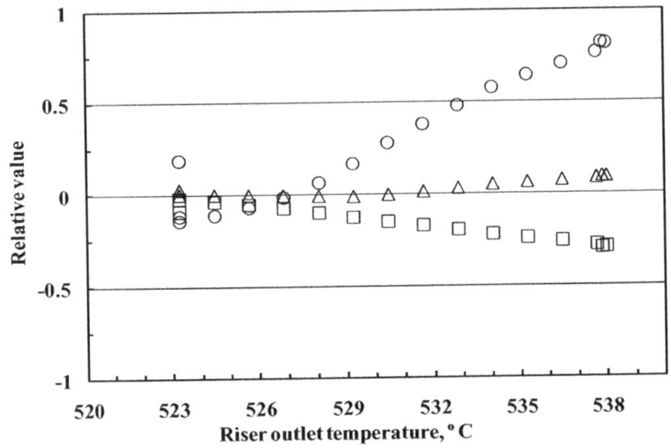

Figure 5.36. Values of the balances for $\dot{\mathbf{x}}_D$ when controlling T_{dp} in the region of maximum olefin production (○, y_{O_2}; △, y_{CO}; □, ω_{coke}). (Adapted from Maya-Yescas and Aguilar, 2003.)

Following this control policy, the relative values of the balances for $\dot{\mathbf{x}}_D$ using the same operating data as those of the first operating policy are as shown in Figure 5.36. It is possible to note that the control of T_{dp} will perform adequately only for the "bad operating points" (i.e., when the unit is operating in partial combustion mode). This is theoretically correct, because for FCC units that operate in partial combustion, T_{dp} is the control target. However, to operate in a full combustion mode, it is necessary to ensure the burn of CO produced; therefore, the temperature of the dense phase should change freely. The problems of inverse response in temperature and the positive eigenvalues illustrated by Figures 5.31 and 5.32, respectively, are closely related to this control instability (Maya-Yescas et al., 1998). The signs of the balances also exhibit this problem. If T_{dp} is regulated, the O_2 and CO balances tend to move away from the desired steady state; hence there will be control problems. Summarizing, if full combustion is achieved, control of T_{dp} always exhibits controllability problems.

Second Operating Policy The second operating policy to be analyzed is the maximum gasoline production. Analysis of the control policies follows the same steps as in the case of the first operating policy. The first control policy analyzed is that proposed for full combustion regenerators (i.e., y_{O_2} is the control target), and the second control policy is the use of T_{dp} as a control target. The sign of the mass and energy balances of both control policies are shown in Figures 5.37 and 5.38.

When optimizing gasoline production, the best operating policy is to take the reaction temperature to values smaller than that of the base case. This

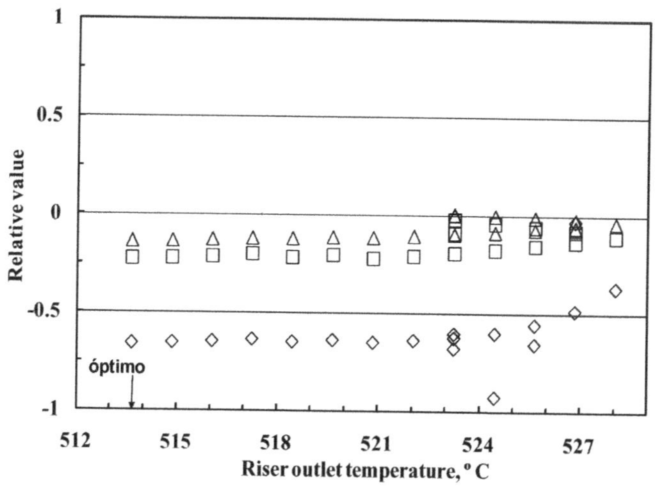

Figure 5.37. Values of the balances for $\dot{\mathbf{x}}_D$ when controlling y_{O_2} in the region of maximum gasoline production (◇, T_{dp} ; △, y_{CO} ; □, ω_{coke}). (Adapted from Maya-Yescas and Aguilar, 2003.)

situation arises because higher temperatures improve conversion to lighter products, such as LPG and dry gases. The first three steady states behave the same as in the other cases analyzed, and the control of y_{O_2} later is always adequate (Figure 5.37). In this case, between 513 and 526°C, the system is working properly in a region that does not present controllability problems.

In contrast to the last analysis, the control of T_{dp} always presents controllability problems (Figure 5.38). This is the situation expected because of the industrial control mode (i.e., full combustion). It is important to notice that the maximum of gasoline production takes place at lower operating temperatures. And even for these "softer" conditions, an incorrect control policy will cause control problems during normal operation.

Again, following the signs of the balances, it is possible to predict the best control policy for different operating conditions. For both situations, results are coherent with industrial experience. The methodology was able to explain the rule of thumb for maximum reaction temperature in this particular unit and, even better, was able to predict the numerical value of *this* maximum temperature. Therefore, this rule of thumb is supported by the analysis of mass and energy balances.

5.9.3 A Technique to Regulate $T_{regenerator}$ in Partial Combustion Mode

Even though FCC units are one of the most important processes in oil refineries, it is necessary to implement advanced process control techniques (Aguilar and Maya-Yescas, 2004). As described above, the FCC process is physically

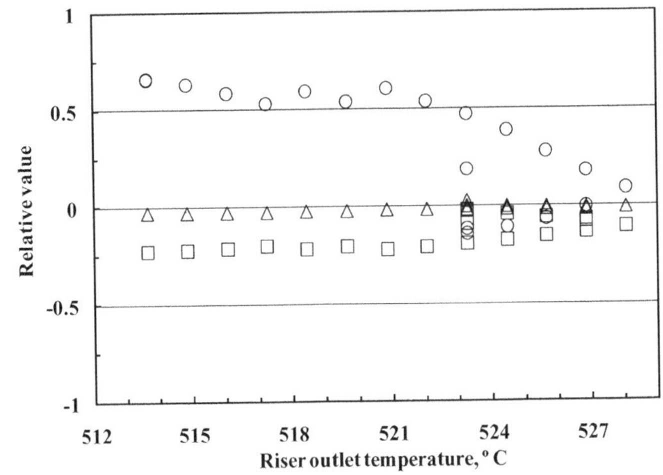

Figure 5.38. Values of the balances for $\dot{\mathbf{x}}_D$ when controlling T_{dp} in the region of maximum gasoline production (○, y_{O_2}; △, y_{CO}; □, ω_{coke}). (Adapted from Maya-Yescas and Aguilar, 2003.)

complex and provides several difficulties in operation and control; for example, the regenerator and reactor are highly interactive, they exhibit complex dynamics, and the knowledge of chemical kinetics is usually poor. Most studies on chemical reactor stabilization are based on linearization of the reactor dynamics by Taylor's series under the assumption that uncertainties belongs to a given conic sector (e.g., Barmish et al., 1983). These approaches have several weaknesses; for example, in the local linear approximation the main properties of the chemical reactors are not exploited, and many uncertainties and disturbances cannot be included in conic sectors. This situation can lead to conservative control law designs and, consequently, not the best closed-loop performance. Therefore, it is necessary to use the entire nonlinear model (as described in Section 5.9) to propose a control able to regulate the set points proposed; moreover, a methodology to prove controllability has been developed.

In order to propose such a control, the main task is to design a control law in the presence of significant system uncertainties, such as modeling errors, unknown disturbances, variations of system parameters, and so on. Recent approaches to estimate online uncertainty terms developed for modeling and control of chemical reactors have employed filtering techniques and calorimetric balances (Aguilar-López and Álvarez-Ramírez, 2002; Álvarez-Ramírez et al., 2004). The advantage exhibited by these methods is their easy computational implementation and the fact that their structure exhibits strong physical meaning; additionally, they take advantage of the nonlinear properties of chemical kinetics.

For the partial combustion mode, a common choice of variables to be regulated is the riser outlet temperature (T_{ro}) and the temperature of the regenerator's dense bed (T_{dp}). Because product distribution at the riser outlet is determined by the reaction temperature inside the riser, there is a good incentive to control both temperatures, T_{ro} and T_{dp}. Additionally, control of T_{dp} is critical to avoid irreversible deactivation of the catalyst during coke burning and/or cyclone damage. A common choice of manipulate variables is the regenerated catalyst flow at the riser inlet (F_S) and the airflow rate at the regenerator inlet (F_a). If the pairings $T_{ro}-F_S$ and $T_{dp}-F_a$ are selected to design a decentralized control strategy, a classical riser–regenerator control structure (Hovd and Skogestad, 1993; Isidori, 1999) is obtained.

Ideal Control Law This discussion is devoted to visualizing the control structure that is obtained when the temperature regulation of the pair ($T_{dp}-T_{ro}$) is performed by input–output linearizing state feedback (Aguilar-López et al., 2002). As a first approach it is assumed that all parameters and reaction rates are known; in addition, it is supposed that all states are available for online measurements. Of course, these are not realistic assumptions; however, they will be used as intermediate assumptions toward the final control law, designed in the section "Actual Control Law Using State Estimation."

Control of T_{dp} The energy balance equation for the regenerator can be expressed as

$$\frac{dT_{dp}}{dt} = L_{rg} + D_{rg}F_a + Q_{rg} + \varpi_{rg} \tag{5.34}$$

where L_{rg} includes the linear terms of Eq. (5.17), D_{rg} is the coefficient of the control input F_a (the airflow rate), and Q_{rg} is the heat generated by the burning of coke in the regenerator's dense bed; $\varpi_{rg} \approx N[0, B_{rg}]$ is the model of the corresponding error of measurement due to white noise, with average zero and covariance B_{rg}.

Now employing a linearizing state feedback law (Aguilar-López et al., 2002) to regulate T_{rg}, the following expression is obtained:

$$F_a = \frac{-L_{rg} - Q_{rg} - g_{rg}(T_{dp} - T_{dp}^*)}{D_{rg}} \tag{5.35}$$

Here $g_{rg} > 0$ is the control gain and T_{dp}^* is the desired regenerator temperature (set point). Note that $D_{rg} \neq 0$ at all operating conditions. This feedback linearizes the dynamics of the temperature of the regenerator's dense bed; hence, the closed-loop dynamics behave as an asymptotically stable linear system:

$$\frac{dT_{dp}}{dt} = -g_{rg}(T_{dp} - T_{dp}^*)$$

Control of T_{ro} The dynamic behavior of the riser reactor is governed by a distributed parameter model (under the plug-flow reactor assumption), and it is then possible to spatially discretize its balance Eq. (5.36) in order to derive a linearizing feedback controller. A first-order discretization of the spatial derivative of Eq. (5.36) at the reactor outlet conditions ($z = 1$) provides an approximation to the dynamics of the temperature at the riser reactor outlet:

$$\frac{dT_{ro}}{dt} = \frac{F_S - F_f}{W_{ri}}\left(\frac{\Delta T_{ro}}{\Delta z} + \varpi_{ro}\right) + Q_{ri} \tag{5.36}$$

Here, the term $\varpi_{ro} \approx N[0, B_{ro}]$ exhibits average zero and covariance B_{ro}; the original spatial derivative has been replaced by a ratio of changes plus white noise:

$$\frac{\partial T_{ro}}{\partial z} = \frac{\Delta T_{ro}}{\Delta z} + \varpi_{ro} \quad \text{and} \quad Q_{ri} = \frac{\Delta H_f K_1(T_{ro})[\text{COR}]\Phi y_f^2}{\rho_{ri} C_{pri}}$$

To calculate the control input F_S, we must take the following into consideration: T_{ro}^* is the set-point temperature at the riser outlet. Therefore, the control objective is that the regulation error $\varepsilon_{ro} = T_{ro}(t) - T_{ro}^*$ should exhibit closed-loop exponentially stable dynamic behavior. Regrouping the dimensionless consumption of heat by the reaction, it is possible at arrive at

$$\breve{Q}_{ri} = Q_{ri} + \frac{F_s - F_f}{W_{ri}}\varpi_{ro} \tag{5.37}$$

given in terms of measurable quantities. Next, the control input is calculated in terms of the same quantities:

$$F_S = -F_f + W_{ri}\left(-\breve{Q}_{ri} - g_{ri}(T_{ro} - T_{ro}^*)\right)\frac{\Delta z}{\Delta T_{ro}} \tag{5.38}$$

Therefore, state feedback, and consequently T_{ro}, should converge asymptotically to the desired temperature T_{ro}^*.

Note that in industrial practice, riser temperature measurements are available only at the reactor outlet and reactor inlet; therefore, for practical controller implementation it is considered that $\Delta z = L$, L being the reactor length, and $\Delta T_{ro} = T_{r\text{-outlet}} - T_{r\text{-inlet}}$.

Actual Control Law Using State Estimation

The control law presented in the preceding section assumed perfect knowledge of reaction rates; however, this is impossible in practice, particularly for the FCC process, where the conversion of feedstock to lighter compounds and the burning of coke during catalyst regeneration follow complex reaction networks (Vieira et al., 2004), which are poorly known. Nonetheless, for temperature regulation, exact knowledge of reaction rate functionalities is not necessary. Instead, a control strategy that has access to the instantaneous temperatures of input and output flows and to instantaneous heat production rates due to the reactions inside reactors is enough. In most cases it is suitable to assume that the convective transport of energy is known; therefore, to obtain good performance by temperature regulation, problems of estimating heat production rates have to be confronted.

Uncertainty Estimation by Kalman Filtering A useful methodology that has been proposed to estimate heat generation rates is by Kalman filtering (Aguilar-López and Maya-Yescas, 2006), considering that uncertainty terms can be expressed as additional state variables (whose structure is unknown, of course) to built a new system of higher dimension ($n+1$). Therefore, the model of the process will include these new dynamic equations.

1. *Regenerator.* The heat balance equation for the regenerator reactor can be expressed as

$$\dot{T}_{\rm dp} = L_{\rm rg} + D_{\rm rg}F_a + Q_{\rm rg} \tag{5.39}$$

Here $Q_{\rm rg}$ is the uncertainty term related to heat generation by chemical reaction, and its dynamic behavior is given by

$$\dot{Q}_{\rm rg} = F(T_{\rm rg}, O_d, C_{\rm rc}) \tag{5.40}$$

By taking regenerator temperature as the system output, the subsystem given by Eqs. (5.39) and (5.40) can be written in observability form (Deza et al., 1992), which satisfies the uniform observability condition; moreover, in practice, $Q_{\rm rg}$ can be reconstructed by means of a state observer. Such an observer would be structured as a copy of the subsystem [Eqs. (5.39) and (5.40)] corrected by an observation error. However, this typical observer structure is not realizable because the functionality $F(\cdot)$ in Eq. (5.40) is unknown. Then to develop an estimate of the uncertain term, the following Kalman filter is proposed (Aguilar-López and Maya-Yescas, 2006):

$$\dot{\hat{T}}_{\rm dp} = \hat{L}_{\rm rg} + \hat{D}_{\rm rg}F_a + \hat{Q}_{\rm rg} + K_{\gamma g1}(T_{\rm dp} - \hat{T}_{\rm dp}) \tag{5.41}$$

$$\dot{\hat{Q}}_{\rm rg} = K_{\rm rg2}(T_{\rm dp} - \hat{T}_{\rm dp}) \tag{5.42}$$

The corresponding expressions for covariance estimation errors are given by the Riccati equations:

$$\dot{P}_{\rm rg1} = 2\left(\frac{-F_S C_{ps} - F_a C_{pa}}{W C_{ps}} P_{\rm rg1} - P_{\rm rg2}\right) + q_{\rm rg1} \tag{5.43}$$

$$\dot{P}_{\mathrm{rg2}} = \frac{-F_S C_{PS} - F_a C_{Pa}}{W C_{PS}} P_{\mathrm{rg2}} - P_{\mathrm{rg3}} - \frac{P_{\mathrm{rg1}} P_{\mathrm{rg2}}}{r_{\mathrm{rg1}}} \tag{5.44}$$

$$\dot{P}_{\mathrm{rg3}} = -q_{\mathrm{rg2}} - \frac{P_{\mathrm{rg2}}^2}{r_{\mathrm{rg1}}} \tag{5.45}$$

Using these expressions, the gains of the corresponding observer in the steady state are

$$K_{\mathrm{rg1}} = \frac{-F_S C_{PS} - F_a C_{Pa}}{W C_{PS}} + \sqrt{\left(\frac{-F_S C_{PS} - F_a C_{Pa}}{W C_{PS}}\right)^2 - \frac{\sqrt{q_{\mathrm{rg2}} r_{\mathrm{rg1}}} - q_{\mathrm{rg1}}}{r_{\mathrm{rg1}}}} \tag{5.46}$$

$$K_{\mathrm{rg2}} = \sqrt{\frac{q_{\mathrm{rg2}}}{r_{\mathrm{rg1}}}} \tag{5.47}$$

Finally, the practical control law that uses the estimate $\hat{Q}_{\mathrm{rg}}$ is expressed by:

$$F_a = \frac{-L_{\mathrm{rg}} - \hat{Q}_{\mathrm{rg}} - g_{\mathrm{rg}}(T_{\mathrm{dp}} - T_{\mathrm{dp}}^*)}{D_{rg}} \tag{5.48}$$

2. *Riser.* By using a similar methodology, the structure of the uncertainty estimator is given by

$$\dot{\hat{T}}_{\mathrm{ro}} = \frac{F_S + F_f}{W_{\mathrm{ri}}} \frac{\Delta z}{\Delta T_{\mathrm{ro}}} + \hat{Q}_{\mathrm{ro}} + K_{\mathrm{ro1}}(T_{\mathrm{ro}} - \hat{T}_{\mathrm{ro}}) \tag{5.49}$$

$$\dot{\hat{Q}}_{\mathrm{ro}} = K_{\mathrm{ro2}}(T_{\mathrm{ro}} - \hat{T}_{\mathrm{ro}}) \tag{5.50}$$

The corresponding expressions for covariance errors for the riser reactor are given by

$$\dot{P}_{\mathrm{ro1}} = -2\zeta P_{\mathrm{ro2}} + 2P_{\mathrm{ro3}} + q_{\mathrm{ro1}} - \frac{P_{\mathrm{ro1}}^2}{r_{\mathrm{ro2}}} \tag{5.51}$$

$$\dot{P}_{\mathrm{ro2}} = -\zeta P_{\mathrm{ro2}} + P_{\mathrm{ro3}} - \frac{P_{\mathrm{ro1}} P_{\mathrm{ro2}}}{r_{\mathrm{ro2}}} \tag{5.52}$$

$$\dot{P}_{\mathrm{ro3}} = q_{\mathrm{ro4}} - \frac{P_{\mathrm{ro2}}^2}{r_{\mathrm{ro2}}} \tag{5.53}$$

Therefore, the observer gains in steady state are

$$K_{\mathrm{ro1}} = -\zeta + \sqrt{\zeta^2 + 2\sqrt{\frac{q_{\mathrm{ro4}}}{r_{\mathrm{ro2}}}} - \frac{q_{\mathrm{ro3}}}{r_{\mathrm{ro2}}}} \tag{5.54}$$

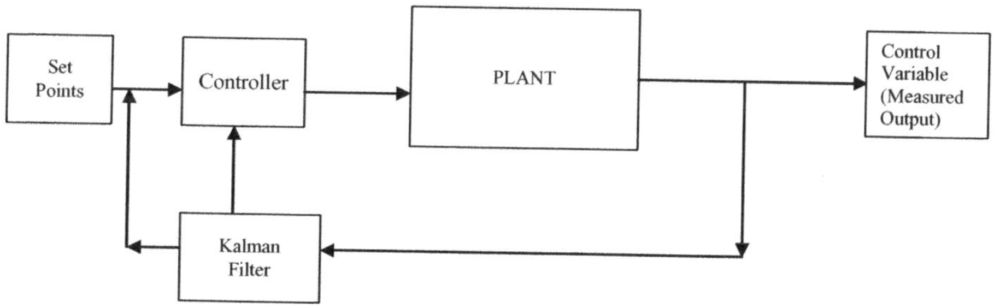

Figure 5.39. Block diagram of the controller. (Adapted from Aguilar-López and Maya-Yescas, 2006.)

$$K_{\mathrm{ro2}} = \sqrt{\frac{q_{\mathrm{ro4}}}{r_{\mathrm{ro2}}}} \tag{5.55}$$

Finally, the practical control law is expressed by

$$F_S = -F_f + \left(-\hat{\eta}_2 - g_{\mathrm{ri}}(T_{\mathrm{ro}} - T_{\mathrm{ro}}^*)\right) W_{\mathrm{ri}} \frac{\Delta z}{\Delta T_{\mathrm{ro}}} \tag{5.56}$$

The control laws for T_{dp} and T_{ro} mentioned above serve as practical stabilizers for both reactors. Practical stability means that controller equations (5.48) and (5.56) are able to drive temperature trajectories as close to the setpoints as desired in a finite period of time. The block diagram for the proposed controller is shown in Figure 5.39, including the position of the filtering process.

In addition to the developments above, we assume that control inputs are subjected to (*not modeled*) actuator dynamics:

$$\tau_r \frac{dFa_r}{dt} + Fa_r = Fa_0 \tag{5.57}$$

Here, Fa_r and Fa_0 are the actual (measured) and the calculated [Eq. (5.25)] control inputs, respectively, at the regenerator.

This control scheme is stable, as proved by Aguilar-López and Maya-Yescas (2006). The following section shows the results obtained when this control is used to regulate the pair $(T_{\mathrm{dp}} - T_{\mathrm{ro}})$ in FCC units that work in partial combustion mode.

Temperature Stabilization Using Extended Kalman-Type Estimators The main features of the dynamic behavior of FCC units, as simulated by the mathematical model, are shown in Figures 5.40 and 5.41. Due to disturbances, if the unit would be operated in open-loop mode, T_{dp} (called T_{rg} in Figure 5.40) and T_{ro} (Figure 5.41) are not maintained at stationary values. Even when these changes are not enough to provoke the unit to run away, they have adverse

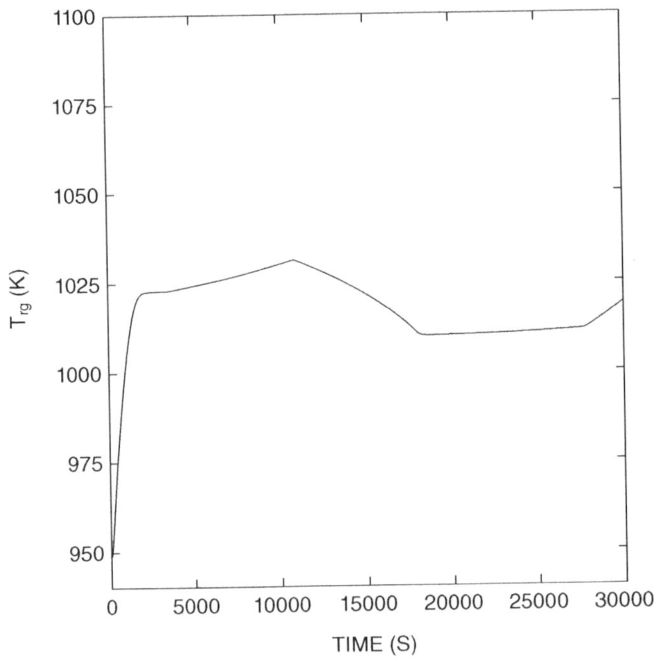

Figure 5.40. Behavior of regenerator temperature during open-loop simulation. (Adapted from Aguilar and Maya-Yescas, 2006.)

effects on gasoline production and catalyst activity; therefore, control actions have to be taken to regulate the temperature of both reactors.

The performance of control and manipulated variables is shown by numerical simulations, comparing open-loop behavior with closed-loop behavior under PI control (tuned following IMC guidelines) and using controllers eqs. (5.48) and (5.56). To realize the linearizing feedback controller equations (5.48) and (5.56), temperatures and flows have to be available for measurement, as is done routinely during operation of FCC units. Additionally, a ±1-K offset in measurements is considered.

During simulations, discussed below, the following sequence of step disturbances entering the regenerator and riser reactors are considered: A 5-K step increase in T_a occurs at $t = 30$ min, a 5-K decrease in T_f occurs at $t = 180$ min, a 2.5% increase in K_C occurs at $t = 270$ min, and finally, a 4 kg/min decrease in F_f occurs at $t = 390$ min. These step disturbances are expected during industrial operation (Hovd and Skogestad, 1993). In addition, several changes in set points occur for both reactors: The original set point for the regenerator temperature (T_{rg}) changes from 970 K to 965 K at $t = 85$ min and from 965 K to 960 K at $t = 250$ min, both in steps. For the riser outlet temperature (T_{ro}), the set point changes from 765 K to 770 K at $t = 165$ min and

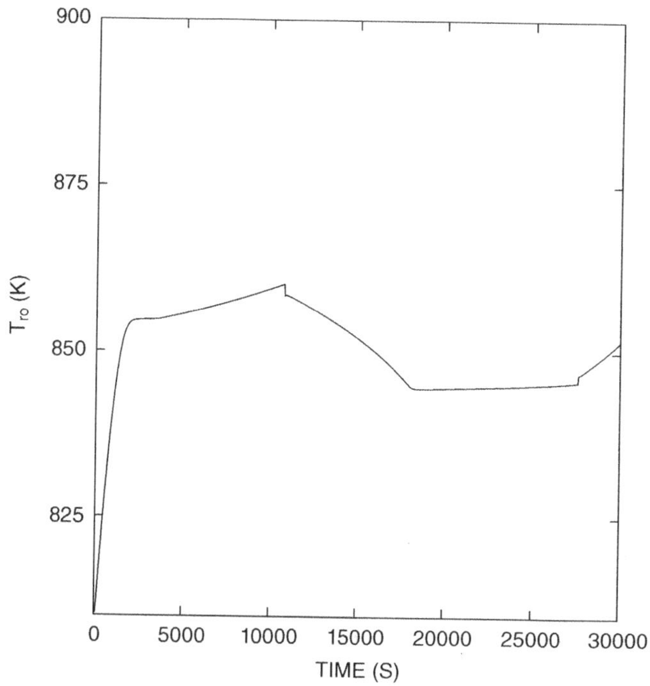

Figure 5.41. Behavior of riser temperature during open-loop simulation. (Adapted from Aguilar and Maya-Yescas, 2006.)

step-changes again from 770 K to 760 K at $t = 333$ min. The effects of the disturbances on T_{rg} and T_{ro} are followed in the open-loop operating mode. For this case, the nominal values of the control inputs (F_a and F_S) were considered.

If standard PI control is used to regulate temperatures, during the first minutes a trajectory varying from the set points (T_{rg} in Figure 5.42 and T_{ro} in Figure 5.43) is noted. Moreover, after the period of stabilization, the control is saturated for the airflow rate (Figure 5.44) and is not able to eliminate the offset that has been produced since the first disturbance arrived at the system. Control of T_{ro} is, even worse, due to the lack of proper response of the variable being manipulated (F_S; Figure 5.45); this difficulty in control arises because of the higher energy inventory in the regenerator, which manipulates the riser temperature in addition to the change in catalyst flow rate. During industrial practice this situation is faced by the elimination of automatic control actions and manual adjustment of the steady state desired, which is not the control strategy considered to be most efficacious.

In contrast, when the controller proposed by Eq. (5.48) is used to regulate the regenerator temperature (Figure 5.46), stabilization is very fast, immediately eliminating any offset in this variable. Following T_{rg} and control actions

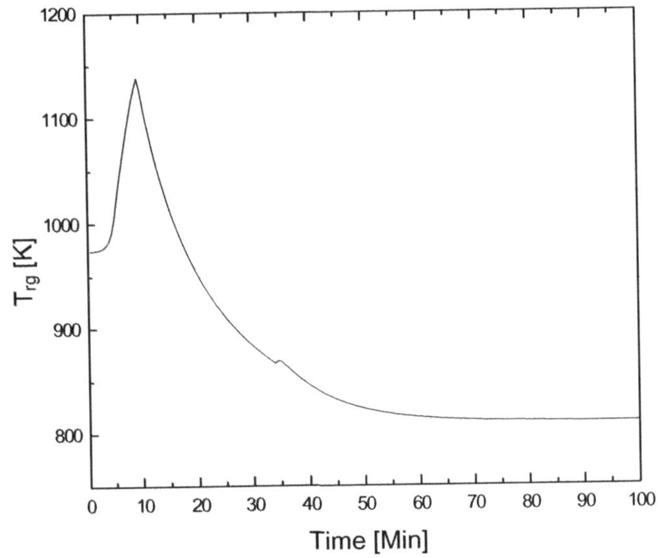

Figure 5.42. Closed-loop performance of regenerator temperature using the PI-IMC. (Adapted from Aguilar and Maya-Yescas, 2006.)

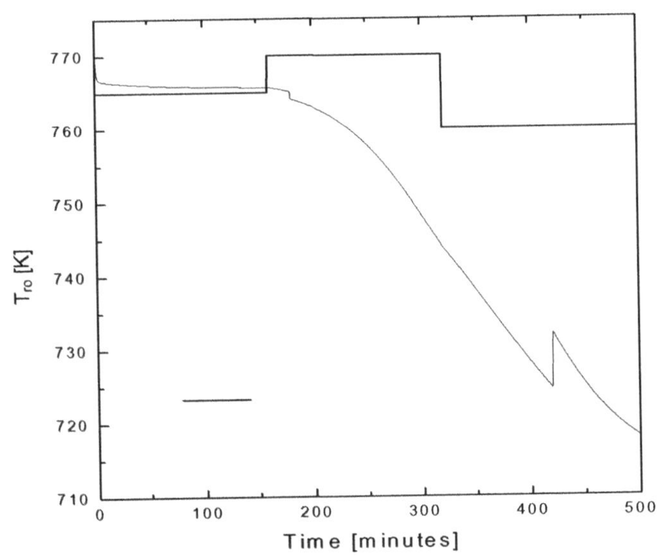

Figure 5.43. Closed-loop performance of riser temperature using the PI-IMC. (Adapted from Aguilar and Maya-Yescas, 2006.)

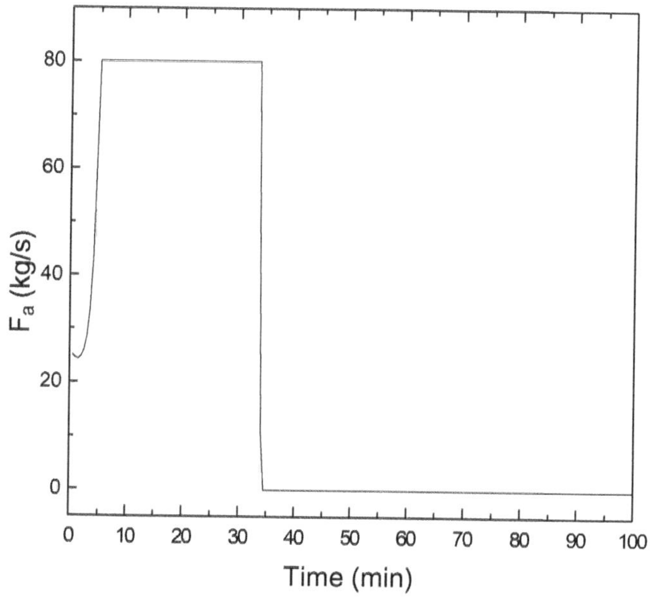

Figure 5.44. Closed-loop performance of regenerator control input using the PI-IMC. (Adapted from Aguilar and Maya-Yescas, 2006.)

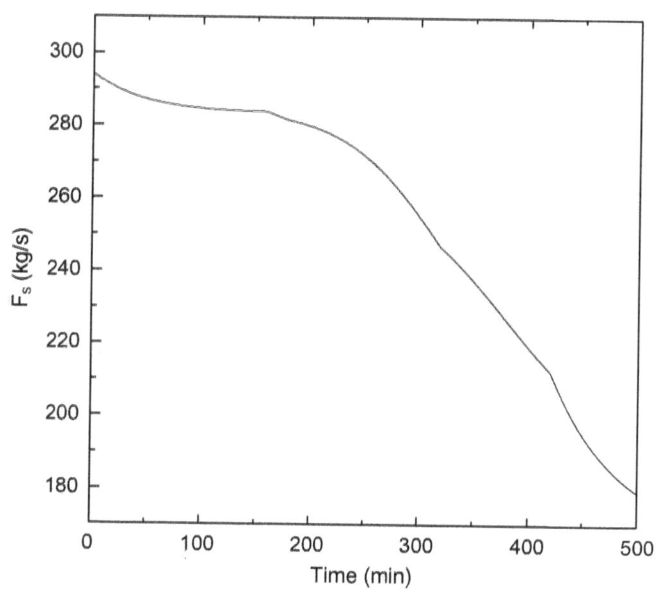

Figure 5.45. Closed-loop performance of riser control input using the PI-IMC. (Adapted from Aguilar and Maya-Yescas, 2006.)

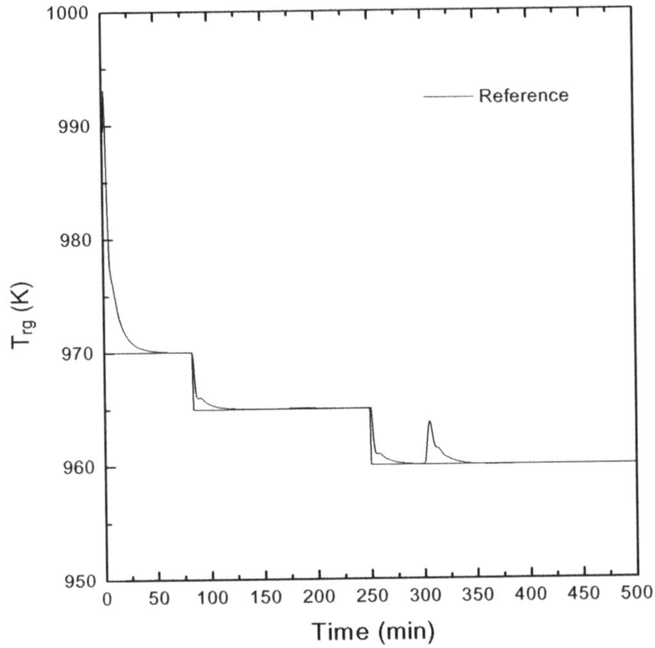

Figure 5.46. Closed-loop performance of regenerator temperature using the controller proposed. (Adapted from Aguilar and Maya-Yescas, 2006.)

from Eq. (5.56), T_{ro} also stabilizes very rapidly and does not exhibit offset after any disturbance or set-point change (Figure 5.47). This situation is the result of the estimation of the actual value of the energy balance instead of the use of the crude measurement of T_{rg}; consequently, the control considers this balance in its actions. Due to the good behavior of the regenerator temperature, and taking advantage of the estimation of the energy balance in the riser, control of T_{ro} (Figure 5.47) is also satisfactory.

Another important advantage of the controller proposed in this work can be noted following the control actions that are taken. It is important to recognize that both controllers use the same measures to change the airflow rate: Figure 5.44 for the PI and Figure 5.48 for Eq. (5.48) or the catalyst flow rate, Figure 5.46 for the PI, and Figure 5.49 for Eq. (5.56), in order to reject these changes and keep the temperatures at their desired values. Nevertheless, control actions are completely different in the two cases. For the PI controller the airflow rate exhibits continuous increases, independent of the disturbance; this situation might be a consequence of the complex behavior of the energy balance, which has been analyzed several times (e.g., Kurihara, 1967; Grosdidier et al., 1993; Hovd and Skogestad, 1993; Taskin et al., 2006).

PI control does not receive information about the energy inventory in the regenerator, which encloses the control inside the operating region in which

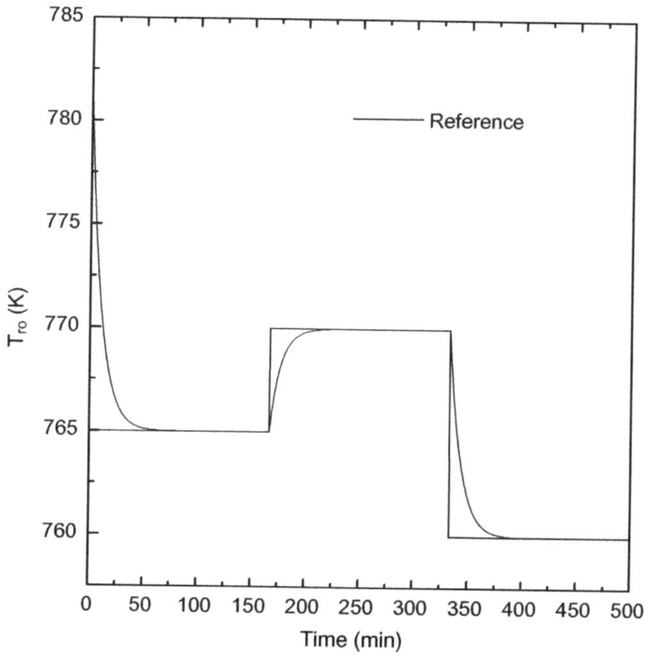

Figure 5.47. Closed-loop performance of riser temperature using the controller proposed. (Adapted from Aguilar and Maya-Yescas, 2006.)

it was tuned. In contrast, for the case of the linearizing control developed in this work, the 5-K step increase in T_a (at $t = 30$ min) is detected by Eqs. (5.41) and (5.42) as an excess of energy entering the regenerator; to counteract the effect of this disturbance, controller equation (5.48) increases the airflow rate (Figure 5.48), so more heat is extracted from the regenerator via output convective flow. Now, when a 2.5% increase in K_C occurs at $t = 270$ min, more coke is produced, increasing the heat production in the regenerator; consequently, gasoline production decreases due to an additional catalyst deactivation. To reject the effect of this perturbation, controller equations (5.48) and (5.56) increase the flow rate of regenerated catalyst (Figure 5.49) and air (Figure 5.48). When a step change in the set point occurs in both reactors, the main effect consists of an excess or diminishing of energy in the process, the changes noted above are detected by the estimation algorithm, and the information generated is used for the controller that counteracts to increase or diminish the flow rate of air and regenerated catalyst in order to keep temperatures at their desired values.

The dynamics of the estimation of uncertainties by the Kalman-like observer, heat generation in the regenerator (Figure 5.50) and in the riser (Figure 5.51) are also shown. In the riser, the effect of the noisy temperature measurements is transmitted to the estimation scheme, which gives a noisy estimate for the

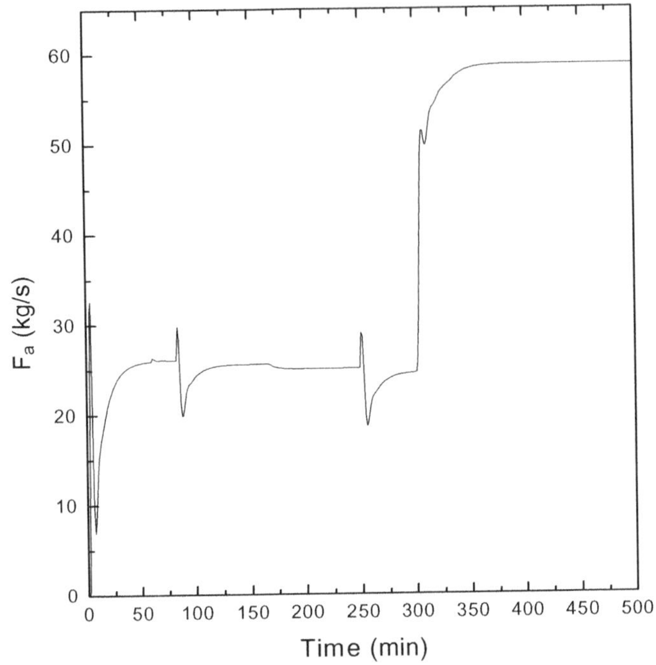

Figure 5.48. Closed-loop performance of regenerator control input using the controller proposed. (Adapted from Aguilar and Maya-Yescas, 2006.)

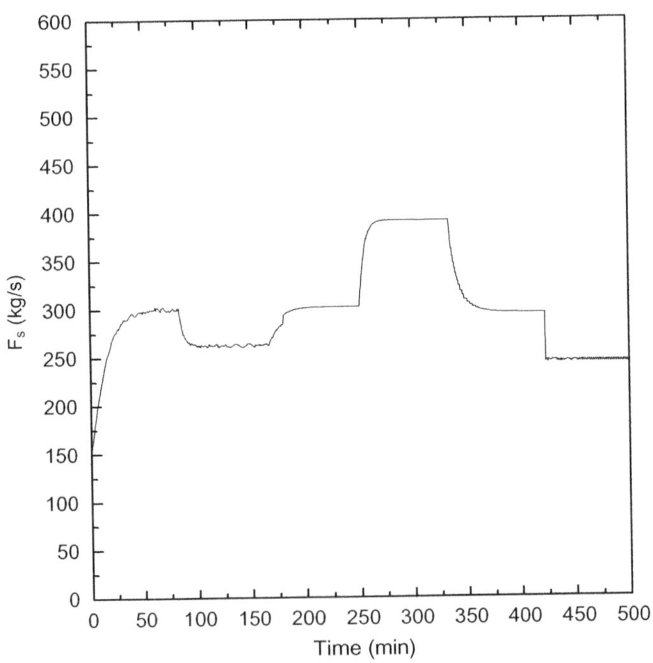

Figure 5.49. Closed-loop performance of riser control input using the controller proposed. (Adapted from Aguilar and Maya-Yescas, 2006.)

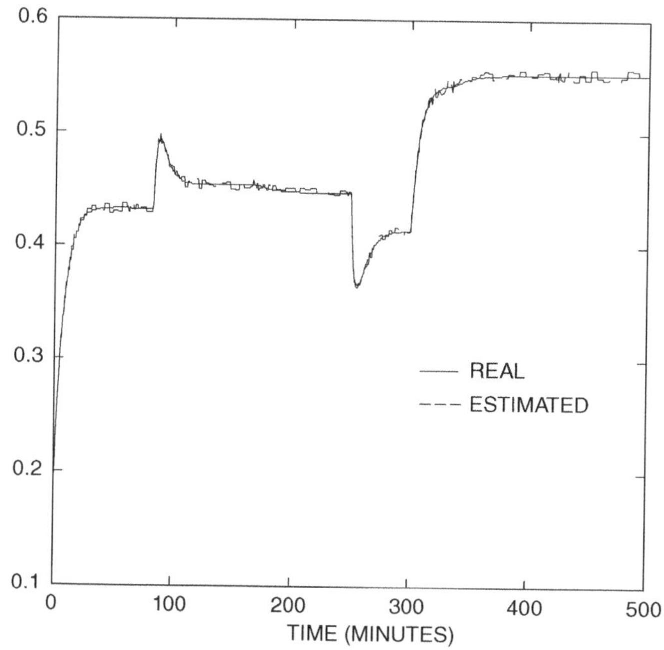

Figure 5.50. Closed-loop estimation of the uncertainty (heat of reaction) in the regenerator. (Adapted from Aguilar and Maya-Yescas, 2006.)

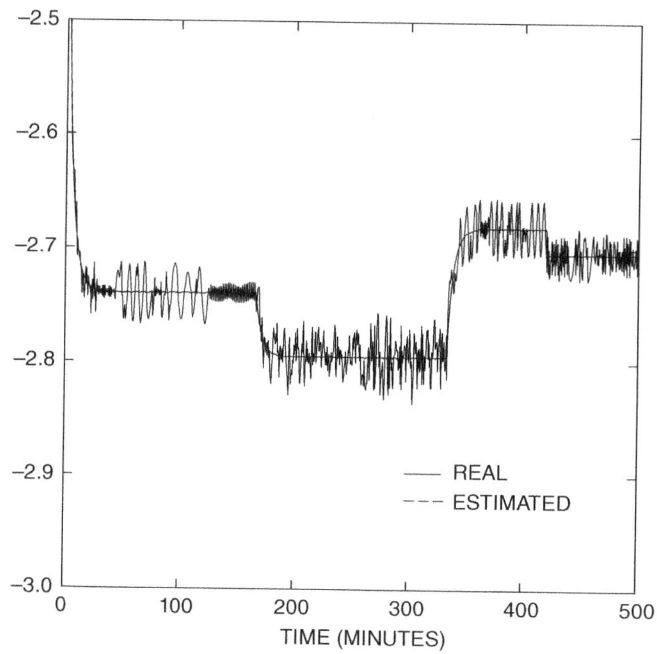

Figure 5.51. Closed-loop estimation of the uncertainty (heat of reaction) in the riser. (Adapted from Aguilar and Maya-Yescas, 2006.)

uncertainty term. However, the control law is able to regulate the temperature despite this situation; moreover, this regulation is reached without great effort. In the regenerator case, the uncertainty is estimated in a smooth way, which supports controller performance.

Finally, when a time delay in the control action of the regenerator is considered, it acts like a filter of the peaking phenomena that occurs every time a disturbance occurs in the process, producing smooth behavior in the airflow rate control action. As noted, a model-based control strategy for temperature regulation in FCC units was considered. The unmeasurable modeling terms related to the kinetics of the process are considered unknown; consequently they are estimated by means of a Kalman-like filter. This estimation procedure makes it possible to realize adaptive input–output linearizing controllers, which are robust against uncertainties and set-point changes. The resulting controllers are similar in form to standard input–output linearizing controllers and can be tuned using standard techniques.

5.10 TECHNOLOGICAL IMPROVEMENTS AND MODIFICATIONS

As noted earlier, the FCC process will continue to be one of the most important parts of a refinery, moving toward higher conversion of middle and heavy distillates to obtain raw products for the new materials industry. Nonetheless, to fulfill economic and environmental requirements, which change continuously, it needs to evolve in several aspects, such as hydrotreatment of feedstock to decrease pollutant content, energy recovery if combustion energy turns out to be "too much", and control of emissions by modification of operating conditions. In this section we analyze these problems and review some options proposed in recent years to ensure the operation of FCC units for the longest possible time.

5.10.1 Effect of Feedstock Pretreatment

Demand for low-sulfur fuels has been increasing during the last 20 years due to environmental concerns about SO_x emissions from processing plants and engines. Due to its high contribution to the gasoline pool, hydrotreating FCC feedstock offers several advantages, such as the increased conversion and yields of gasoline and liquid-phase gas; meanwhile, the sulfur content in fuels is diminished. However, there are more important factors to be considered when hydrotreating FCC feedstock.

In this subsection, two FCC feedstocks, typical (TF) and hydrotreated (HF), were converted in a microactivity test (MAT) reactor, as described by ASTM D-3907-92, at different severities and using two commercial catalysts. Feedstock conversion, product yields, and selectivity to valuable products were compared against industrial-scale results predicted using commercial FCC simulation software (Salazar-Sotelo et al., 2004). An expected increment in conversion and yield to profitable products was observed when hydrotreated feedstock was used; simulation results follow acceptable MAT results.

Often, FCC is the primary conversion process in an integrated refinery, playing a key role in the refinery's profitability. Successful operation of this unit determines whether or not the refiner can remain competitive in the market. An issue that has been gaining importance in the refining industry comes from the fact that the demand for low-sulfur distillates has been increasing during the last 20 years due to environmental concerns about SO_x emissions. FCC feedstock's sulfur content is becoming too high and, consequently, FCC liquid products contain excessive sulfur compounds that need to be treated prior to being used. Worldwide, about 45% of all gasoline comes from FCC and ancillary units; moreover, about 90% of the sulfur in the gasoline pool is supplied by the FCC unit.

An interesting solution to this problem is the hydrotreatment (HDT) of the FCC feedstock. There is some industrial concern about the advantages of feedstock hydrotreatment, mainly because this operation require substantial investment. Nevertheless, in addition to the aggregated environmental and economic value due to the production of cleaner fuels, there are increments in yields to valuable products. This situation is a consequence of several factors, such as the higher catalytic activity, the consequence of the minor amount of metallic pollutants, and the better selectivity to liquid fuels that are obtained when a hydrogen-enriched feedstock is used (Leuenberger et al., 1998, Mariaca-Domínguez et al., 2003, 2004); also, sulfur oxide emission by the regenerator is diminished (Maya-Yescas et al., 2005).

As described, FCC is a very complex process, having at its heart the riser–regenerator couple known as a converter (Figure 5.52). After preheating, partially evaporated feed enters the riser, where it contacts the regenerated catalyst. The heat absorbed by the catalyst during regeneration provides the energy to evaporate and heat the feed to its desired reaction temperature. Many reactions take place in the vapor phase inside the riser. The products recovered are dry gases (H_2–C_2's), liquid-phase gas (LPG, which consists of C_3's and C_4's), gasoline (C_5, b.p. 221°C), and cyclic oils (considered part of unreacted feedstock, b.p. > 221°C). Also, there is coke formation; this solid compound deposits on the catalyst surface. The average heat of reaction, resulting from feedstock evaporation plus cracking reactions, is endothermic.

After reactions occur, catalyst and products are quickly separated in cyclones located at the riser outlet and catalyst falls into a stripper, where steam is used to remove the entrained hydrocarbons between catalyst particles. Catalyst is transferred to the regenerator, where its activity is restored by burning off the coke with air. These regeneration reactions generate the heat that is used to vaporize the feedstock at the riser and to sustain cracking reactions. Finally, hydrocarbon products are fractionated and narrow cuts are separated into commercial-interest products.

Feedstock composition is one of the most important factors affecting the yields and product quality in fluid catalytic cracking (Dahl et al., 1996;

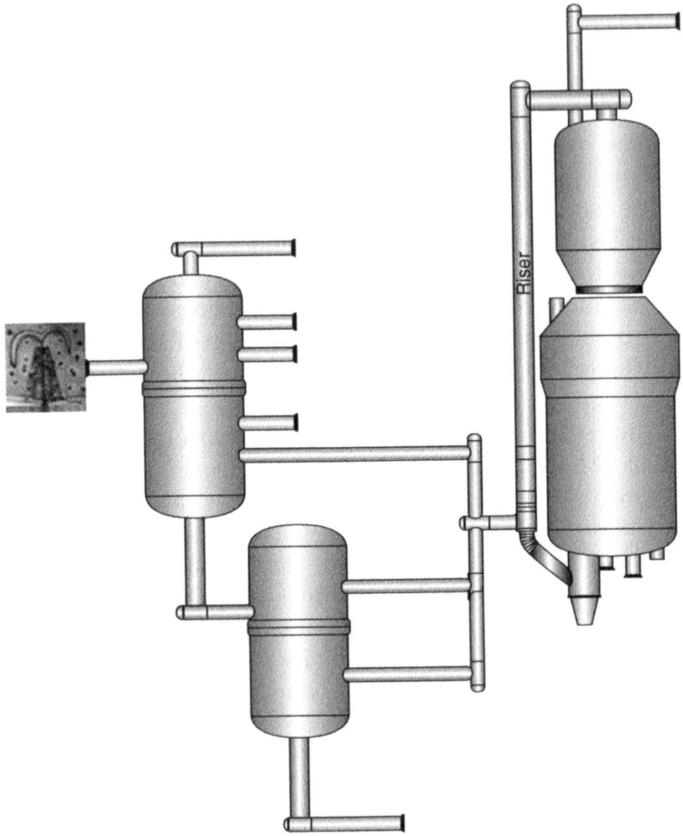

Figure 5.52. Typical FCC unit.

Leuenberger et al., 1998; Mariaca-Domínguez et al., 2003, 2004). A conventional feedstock for FCC units is a blend of gas oils that come from atmospheric and vacuum distillation towers and delayed coker units. This feedstock is usually characterized by high levels of nitrogen, sulfur, metallic pollutants, and microcarbon residue; it also exhibits a high aniline point (Mariaca-Domínguez et al., 2002). These low-value gas oils are converted into valuable products such as gasoline and olefin-rich LPG by using microspherical zeolite catalysts.

The sulfur content of FCC products depends on the catalyst, feedstock, conversion, and operating conditions of the reactor. FCC feedstock contains sulfur linked to organic compounds of high molecular weight; these heterocompounds are concentrated in its heavy end. Cracking of these molecules produces either sour gas or sulfur-containing fuels; sour gas can be recovered and treated downstream to produce solid sulfur or sulfuric acid, whereas sulfur contained by fuels will produce sulfur oxide emissions when burned in internal combustion engines (Maya-Yescas et al., 2005).

Despite the fact that postprocessing of FCC products is apparently the easiest solution, it presents several disadvantages. Postprocessing is more complex since the products have to be treated separately. Some strategies to reduce sulfur in FCC gasoline include naphtha hydrofinishing and lowering the gasoline endpoint. Hydrofinishing lowers the octane of FCC gasoline, which depends, among other factors, on the presence of unsaturated compounds; meanwhile, lowering the gasoline endpoint can diminish gasoline yield significantly.

In contrast, HDT of FCC feedstock is an integral solution and offers several advantages. Hydrotreating FCC feedstock costs more than hydrotreating cracked gasoline but results in economic benefits through increased yields to gasoline and light olefins and longer catalyst life, in addition to compliance with sulfur specifications of products. The benefits vary depending on the feed characteristics, HDT severity, and FCC operating conditions (Lavanya et al., 2002).

The changes in composition that feedstock experiments undergo during hydrotreating enhance the FCC operation for several reasons:

- Sulfur distribution in FCC products changes, so with hydroprocessed feeds, about 5% of lower feed sulfur content ends up in gasoline. For nonhydroprocessed feeds, sulfur in gasoline is typically 10% of feed sulfur.
- Partial elimination of nitrogen helps FCC catalyst to be more active.
- Conversion increases at the same operating conditions for the more severely hydrotreated feedstocks because of saturation of noncrackable aromatic rings to naphthenic rings and reduction in the nitrogen level during hydrotreating.
- Increase in the H/C ratio (Dahl et al., 1996; Mariaca-Domínguez et al., 2003, 2004) enhances gasoline production because of higher feedstock conversion *and* better selectivity.
- Reduction of the metals content in the feedstock (iron, sodium, copper, nickel, and vanadium) increases the FCC conversion and selectivity. Also, catalyst life is incremented, which diminishes operating costs.
- Polyaromatics and Conradson carbon reduction results in less coke production, making the FCC catalyst more efficient and lowering the regenerator temperature and air supply requirements.
- The *i*-butane/butylenes ratio increases, which correlates with better selectivity to gasoline because of a decrease in overcracking (e.g., Leuenberger et al., 1998; Mariaca-Domínguez et al., 2002).

During HDT, difficulty in elimination of the aforementioned pollutants depends on the operational severity according to the sequence metals $<$

TABLE 5.11. Feedstock Properties

Property/Feedstock	TF	HF
Density (g/cm^3, 20°C, ASTM D-1298)	0.9071	0.8887
Viscosity (cSt, 40°C, ASTM D-88)	37.57	58.46
RI (20°C, ASTM D-1218)	1.505	1.496
Conradson carbon (wt%, ASTM D-524)	0.14	0.05
Sulfur (wt%, ASTM D-2622)	1.45	0.14
Basic nitrogen (ppm, ASTM D-4629)	233	67
Composition (P/N/A)	62/20/18	65/21/14
H_2 content (wt%)	13.07	13.83
Distillation data (ASTM D-2887)		
10 vol% (°C)	367	366
50 vol% (°C)	449	463
90 vol% (°C)	528	541

Source: Adapted from Salazar-Sotelo et al. (2004).

sulfur < nitrogen < aromatics. It is important to note that there is an economic optimum limit to HDT severity. As severity is increased, it comes to a point where the introduction of more hydrogen to the feedstock does not improve the yield to FCC gasoline. Hence, as sulfur reduction has become essential for meeting quality specifications of fuel products, finding a point of equilibrium between feedstock hydrotreatment and profitable yields has acquired growing industrial importance.

Process Emulation in a MAT Laboratory Reactor A typical feedstock for an industrial FCC unit was hydrotreated in a hydrodesulfurization unit for gas oils; properties of a typical feedstock (TF) and a hydrotreated feedstock (HF) were measured (Table 5.11). As can be noted, the effects of hydrotreating include decreased density, viscosity, and refraction index; this is a consequence of the change in hydrocarbon distribution, where aromatics are diminished and paraffins and naphthenes increased. Also, microscopic coke precursors (Conradson carbon) are diminished. TF and HF were converted using two equilibrium industrial catalyst samples, C1 and C2. Catalyst properties and measurement methods are given in Table 5.12. It is important to note that the microactivity test (ASTM D-5154-05) conversion, called MAT activity, of both catalysts differs by 11.5 wt%. High active catalysts tend to increase conversion, coke, and LPG, not greatly affecting olefins or light gas yields (Leuenberger et al., 1998); in contrast, low-activity catalysts tend to maintain constant gasoline yields.

TF and HF (Table 5.11), which are obtained from the refining of a mixture of Mexican crude oils, were tested in a standard MAT reactor. For both feedstocks, experiments were carried out at two different temperatures (520 and 550°C), and three different C/O ratios (3, 4, and 6). Each experiment was

TABLE 5.12. Catalyst Properties

Property	C1	C2
Density (g/cm^3)	1.0531	0.9852
Specific area (BET) (m^2/g)	88	128
Average particle size (μm)	65	53
MAT activity (ASTM D-5154-05) (wt%)	55.6	67.1
Metals content (AA, IMP-QA-031)		
Cu (ppm)	18.02	23.23
Fe (wt%)	0.62	0.56
Na (wt%)	0.72	0.74
Ni (ppm)	372.05	473.28
V (wt%)	0.12	0.26

Source: Adapted from Salazar-Sotelo et al. (2004).

carried out with catalysts C1 and C2. The catalyst amount in the reactor was 4 g, and the feedstock injection rate was 1.3 g/min, varied in order to emulate different C/O ratios. Liquid products were quantified by simulated distillation according to (ASTM D-2887), gaseous products were analyzed using an online gas chromatograph, and coke on the catalyst surface was quantified on an elemental carbon analyzer.

Process Simulation Another option to predict the behavior of an industrial unit under different operating conditions is computer simulation. In this work a commercial simulator (Salazar-Sotelo et al., 2004) was used to predict the performance of a commercial FCC unit. The commercial simulator used by Salazar-Sotelo et al. (2004) is a steady-state simulation tool, used widely around the world, based on an engineering framework. It simulates the riser–regenerator system, following heat and mass balances at the operating conditions required. Among its features there are kinetic models for riser and regenerator, as well as a catalyst deactivation function of the Voorhies (1945) type. A carefully chosen operating test run from an industrial unit was used to adjust the theoretical model.

Salazar-Sotelo et al. (2004) used an industrial operating point as test run data for the calibration of the commercial simulator using TF and catalyst C2. Some important characteristics of this industrial unit are given in Table 5.13, along with the operating data used for calibration. Once calibration was complete, predictions were carried out at three riser outlet temperatures (520, 535, and 550°C), two of them used in MAT experiments, using TF and HF as well as C1 and C2.

Comparison of Typical Versus Hydrotreated Feedstock Conversion and product yields from MAT experiments were compared against results obtained by Salazar-Sotelo et al. (2004) using the simulation software for TF and HF.

TABLE 5.13. Characteristics of the FCC Unit and Operative Parameters

Type	Riser Reactor/Adiabatic Regenerator
Feedstock capacity (bbl/day)	25,000
Riser outlet temperature (°C)	519
Preheat temperature (°C)	183
Regenerator operating mode	Complete combustion
Regenerator dense-phase temperature (°C)	663
Regenerator dilute-phase temperature (°C)	689

Source: Adapted from Salazar-Sotelo et al. (2004).

As expected, conversion of feedstock is improved by feedstock hydrotreating; this effect can be noticed with both catalysts. In the case of catalyst C1, there is a conversion improvement of 9 wt% at 520°C a C/O value of 6, and an even higher increment of 16 wt% units at 550°C and C/O = 3 (Figure 5.53). These differences in conversion are also predicted by simulation results that follow almost the same trend for C/O = 6. It is evident that simulated conversion reports a shift that is similar to the laboratory shifts when changing from TF to HF. The simulated temperature effect is also similar to average MAT results. Due to the lack of industrial reference data, the simulated results should be used with care, considering that they depend on the software extrapolation capabilities. Additionally, it is important to take into account that C/O industrial values are not shown here, as they are not comparable to MAT values. In industrial practice, this parameter is defined from the heat balance with values from 6 to 14.

Catalyst C2 is more active and presents higher MAT activity, as shown in Table 5.12. For hydrotreated feedstock, conversions exhibit different improvement ranges but similar behavior, since the highest increase (11 wt%) is observed at a C/O value of 3 and the lowest (2 wt%) at a C/O value of 6 (Figure 5.54). Simulation predicts conversion values and temperature effects similar to MAT results when using TF. However, the industrial conversions predicted are higher than MAT values when HF is used, even if the temperature effect follows closely.

Conversion is not the only variable to look at in FCC—the main point of interest is the production of gasoline. In the case of catalyst C1, differences in gasoline yields are about 6 wt% for any C/O ratio at 520°C and about 13 wt% at 550°C (Figure 5.55), both favorable to hydrotreated feedstock, due to its higher crackability (Mariaca-Domínguez et al., 2004). It is again important to note that a simulator tends to estimate higher HDT benefits than MAT. Another important difference is that MAT data at 550°C exhibit a decrease in gasoline yield for C/O = 6, which is probably due to overcracking (Leuenberger et al., 1998); however, simulation only exhibits a change of slope; it does not reach this phenomenon.

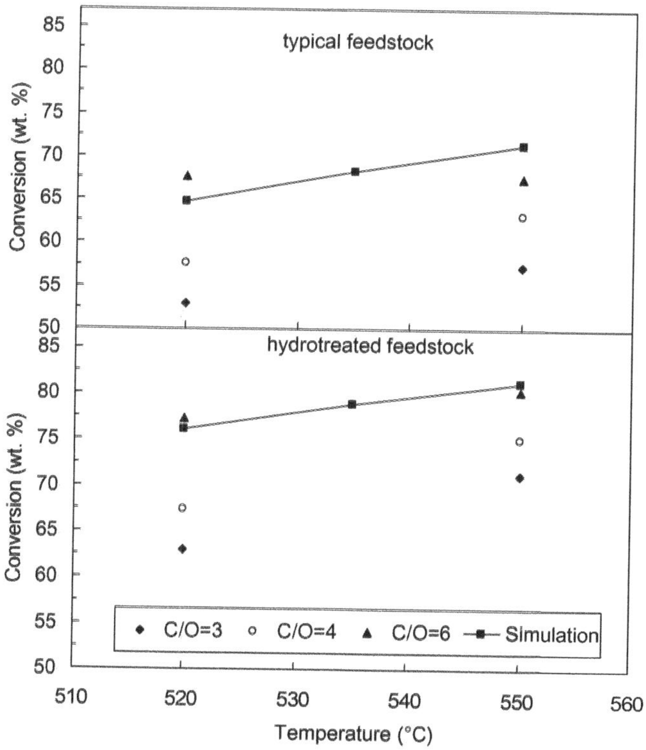

Figure 5.53. Conversion of TF and HF on C1. (Adapted from Salazar-Sotelo et al., 2004.)

In contrast, in the case of catalyst C2, the increase in gasoline yields is greater at 520°C (about 8wt%) than at 550°C (about 3wt%) (Figure 5.56), both favoring hydrotreated feedstock. In this case the more interesting behavior is that simulation predicts overcracking of gasoline for both feeedstocks in the experimental temperature range. This prediction is due to the high initial MAT activity of the equilibrium catalysts, in contrast to C1, which exhibits very low MAT activity (Leuenberger et al., 1998).

Yield to gasoline is closely related to conversion, which is very dependent on catalyst selectivity, at specific reaction conditions. For hydrotreated feedstocks, catalyst selectivity may be more relevant than conversion, again at otherwise specific conditions. Considering uncertainty from experimental errors, gasoline selectivity is considered similar for both catalysts (Figure 5.57).

One key concern during FCC operation is coke formation. Coke attaches to the catalyst surface, blocking the catalytic sites and decreasing catalyst activity. Also, this entity regulates the energy balance of the entire unit, and as the feedstock is heavier, it is possible to form higher amounts of coke that increase regeneration temperatures (León-Becerril and Maya-Yescas, 2007). Higher

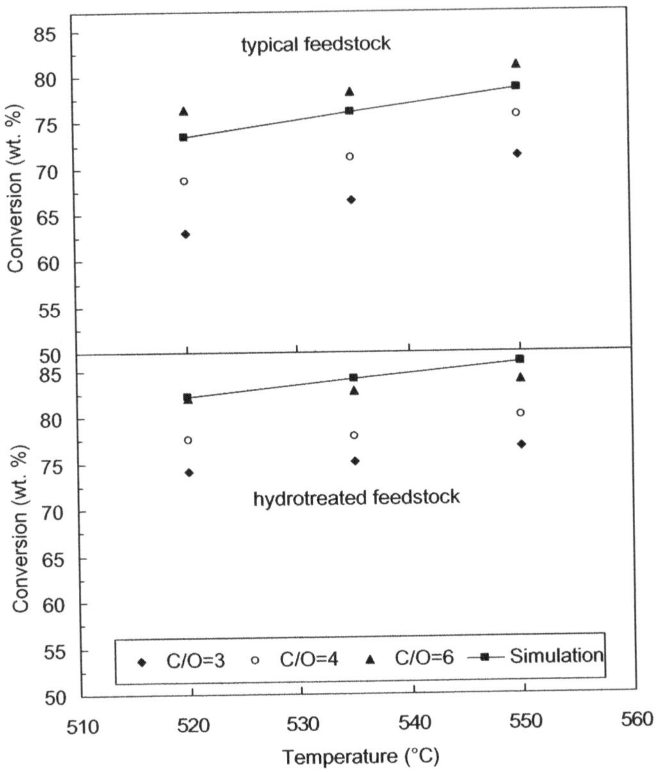

Figure 5.54. Conversion of TF and HF on C2. (Adapted from Salazar-Sotelo et al., 2004.)

temperatures could also mean faster catalyst deactivation. It is not possible to compare MAT yield to coke to industrial simulation results because yield to coke may behave different in laboratory reactors (Kelkar et al., 2003). After feedstock hydrotreating, coke precursors are diminished, as can be noticed when catalyst C1 is used (Table 5.14), and yield to coke decreases as much as 1 wt% at 550°C and C/O = 6. It is important to notice that yield to coke is almost proportional to the C/O ratio for the typical feedstock, whereas its increase is significantly lower for the hydrotreated feedstock; this is a consequence of the decrease of microscopic coke precursors after HDT (León-Becerril and Maya-Yescas, 2007). As we saw earlier, higher C/O ratios mean higher conversion and higher yield to gasoline, at constant temperature, as long as overcracking is not attained.

It is possible to note from these data that when using hydrotreated feedstock, higher C/O ratios are reachable, whereas with typical feedstock this condition produces too high amounts of coke. This is a well-known benefit of HDT in industrial FCC operation, resulting in economic profit that allows higher severity and conversion levels as well as longer catalyst life.

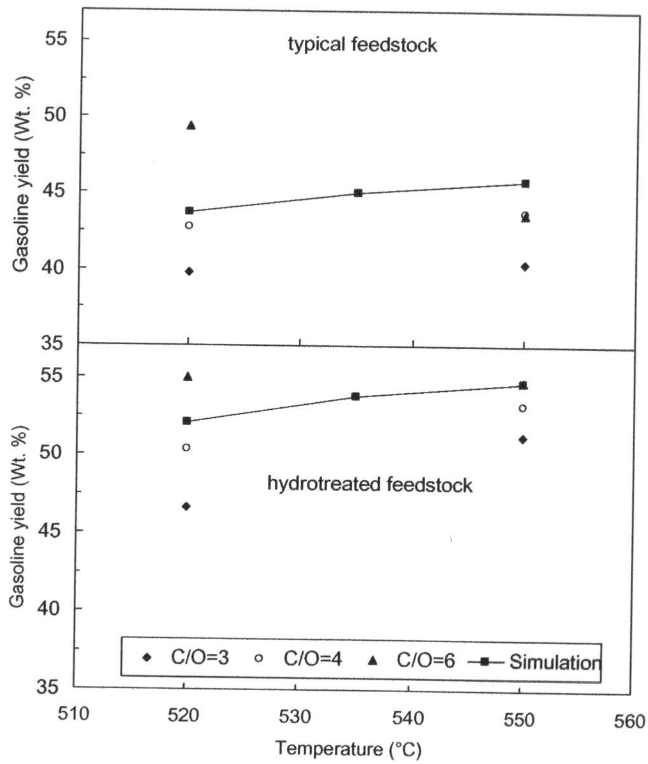

Figure 5.55. Yield to gasoline of TF and HF when using C1. (Adapted from Salazar-Sotelo et al., 2004.)

In the case of catalyst C2 (Table 5.14), which is more active, the difference in yield to coke is more evident. For a typical feedstock, yields reach 7 wt% at 550°C and C/O = 6, which is high for MAT laboratory reactors. This situation, when extrapolated to the industrial unit, could translate into operating problems due to high regeneration temperatures (Leuenberger et al., 1998). For a hydrotreated feedstock, yield to coke remains level, a situation that benefits the operation and increases profitability.

The second more important reason to avoid excessive coke production is that this entity diminishes yield to gasoline, due to the decreasing conversion. For a typical feedstock there is a maximum yield to gasoline of about 54 wt%; in contrast, hydrotreated feedstock exhibits a maximum of about 56 wt% and lower yield to coke (Figure 5.58). Lines are more separated in the case of typical feedstock than in the case of hydrotreated feedstock. This is due to the fact that activity is less decremented when yield to coke changes are smaller. This is an additional advantage of feedstock hydrotreating.

Currently, there are other interesting products from FCC units: LPG, especially C_3 and C_4 olefins. Because of the partial opening of polynuclear

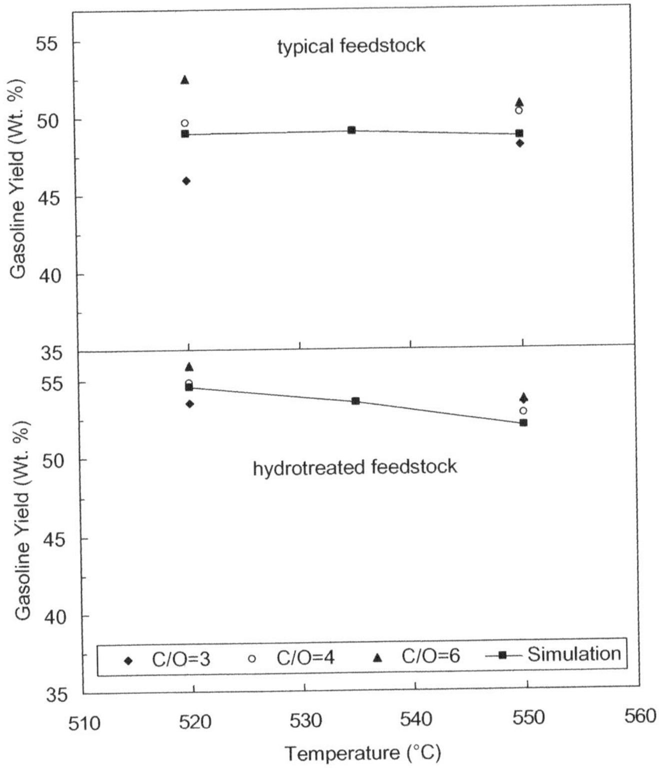

Figure 5.56. Yield to gasoline of TF and HF when using C2. (Adapted from Salazar-Sotelo et al., 2004.)

hydrocarbons, production of these molecules is also favored by hydrotreating. This is an important difference, because propane is used as fuel, propylene is a feedstock for petrochemicals, *i*-butane is a feedstock for downstream processes, and the rest of the C_4 olefins are valuable products. Therefore, a change in LPG yields and distribution of products results in a profit. Dry gas (H_2, C_1, and C_2) also exhibits changes in weight distribution when hydrotreated feedstock is used.

For catalyst C1, a comparison of light products composition is given in Table 5.15 for C/O ratios of 4 and 6 at 520°C. Simulation results are also shown for both TF and HF. The first important point to note is that LPG production is increased in both experimental cases; similarly, simulation also predicts an increase in this product when HF is used.

Dry gas is expected to decrease after hydrotreating of the feedstock (Leuenberger et al., 1998). For C1 catalyst this is the case at low severities; however, at higher severities dry gas increases. Because of the low amounts produced in the MAT reactor, this unexpected behavior could be a conse-

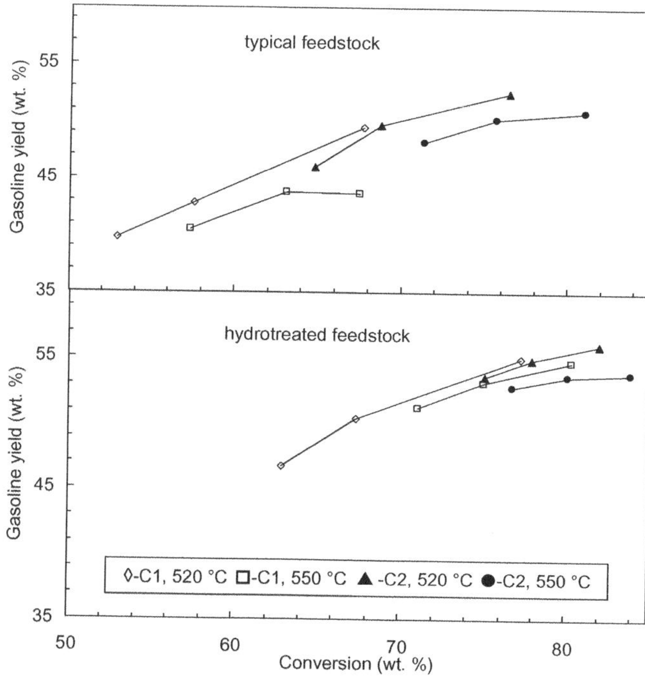

Figure 5.57. Yield to gasoline as function of conversion at different C/O ratios. (Adapted from Salazar-Sotelo et al., 2004.)

TABLE 5.14. Coke Production for Both Catalysts at Different Severities (wt%)

	C/O = 3		C/O = 4		C/O = 6	
	520°C	550°C	520°C	550°C	520°C	550°C
TF + C1	2.27	2.63	2.58	3.38	3.64	4.66
HF + C1	2.17	2.23	2.65	2.66	3.48	3.71
TF + C2	3.83	4.08	4.62	4.91	6.60	7.08
HF + C2	3.33	3.36	3.99	4.01	5.53	5.49

Source: Salazar-Sotelo et al. (2004).

quence of experimental error. Simulation also predicts higher production of this product when using HF.

For catalyst C2, a comparison of light product composition is given in Table 5.16 for C/O ratios of 4 and 6 at 520°C. Simulation results are also shown for both TF and HF. For this catalyst, every compound in the LPG fraction is increased when HF is used, which results in a profit. In contrast to catalyst C1, for both laboratory operating conditions (C/O = 4 and 6) dry gas exhibits

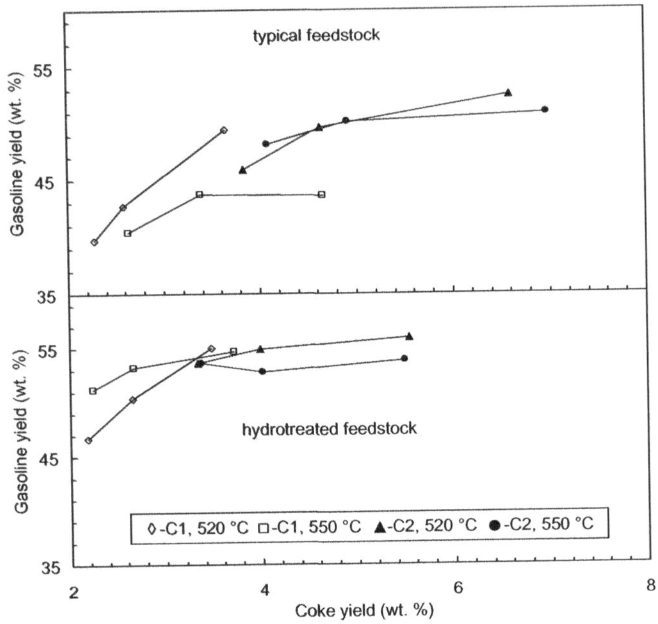

Figure 5.58. Yield to gasoline as function of coke yield at different C/O ratios. (Adapted from Salazar-Sotelo et al., 2004.)

TABLE 5.15. Light Products Yields and Composition at 520°C Using Catalyst C1 (wt%)[a]

		C/O = 4		C/O = 6		Simulation	
Feedstock		TF	HF	TF	HF	TF	HF
Dry gas	Total	0.88	0.80	1.29	1.44	2.89	3.54
	Hydrogen	0.11	0.08	0.14	0.12	0.08	0.09
	Methane	0.30	0.24	0.43	0.48	1.03	1.27
	Ethylene	0.32	0.28	0.43	0.52	0.78	0.95
	Ethane	0.16	0.20	0.29	0.31	1.00	1.22
LPG	Total	11.32	13.31	13.36	17.42	10.95	14.64
	Propane	0.35	0.43	0.47	0.62	1.02	1.41
	Propylene	2.91	3.61	3.80	4.83	2.70	3.72
	i-Butane	2.06	2.76	2.72	4.02	2.42	3.18
	n-Butane	0.36	0.46	0.50	0.73	0.98	1.29
	1-Butene	0.93	1.11	1.18	1.54	0.93	1.22
	i-Butylene	1.16	1.30	1.39	1.62	1.06	1.40
	t-Butylene	1.18	1.41	1.51	1.98	1.00	1.31
	c-Butylene	0.89	1.05	1.14	1.50	0.81	1.06

[a]For details, see Salazar-Sotelo et al. (2004).

TABLE 5.16. Light Products Yields and Composition at 520°C Using Catalyst C2 (wt%)

Feedstock		C/O = 4		C/O = 6		Simulation	
		TF	HF	TF	HF	TF	HF
Dry gas	Total	1.58	1.49	1.98	1.80	3.81	4.47
	Hydrogen	0.22	0.17	0.26	0.19	0.09	0.09
	Methane	0.53	0.49	0.69	0.62	1.37	1.61
	Ethylene	0.50	0.48	0.64	0.62	1.03	1.21
	Ethane	0.33	0.35	0.39	0.38	1.33	1.53
LPG	Total	14.84	17.63	17.16	18.74	14.62	18.69
	Propane	0.66	0.75	0.81	0.87	1.36	1.80
	Propylene	4.16	4.64	4.87	5.18	3.59	4.74
	i-Butane	3.38	4.38	4.16	5.00	3.28	4.13
	n-Butane	0.67	0.83	0.84	0.97	1.30	1.64
	1-Butene	1.24	1.39	1.40	1.49	1.23	1.55
	i-Butylene	1.31	1.28	1.35	1.28	1.41	1.77
	t-Butylene	1.60	1.80	1.81	1.95	1.33	1.67
	c-Butylene	1.20	1.35	1.37	1.46	1.07	1.35

Source: Adapted from Salazar-Sotelo et al. (2004).

TABLE 5.17. *i*-Butane/Butylenes Ratios

	C/O = 4		C/O = 6	
	TF	HF	TF	HF
C1	0.49	0.57	0.52	0.61
C2	0.63	0.75	0.70	0.81

Source: Adapted from Salazar-Sotelo et al. (2004).

minor production when using HF; nevertheless, simulation predictions show an increase in dry gas production when shifting from TF to HF.

In addition to the increase in LPG yield, it is possible to confirm the better yield to gasoline by the *i*-butane/butylene ratio, as proposed by Leuenberger et al. (1998) and Mariaca-Domínguez et al. (2004). This index is proportional to the yield to gasoline, due to a decrease in overcracking. Table 5.17 shows this ratio at 520°C for the yield to gasoline obtained in the MAT reactor (data from Tables 5.15 and 5.16). For catalyst C1 the *i*-butane/butylene ratio increases when the feedstock is changed from TF to HF. The same is true for catalyst C2, which also exhibits large values of the *i*-butane/butylene ratio (Table 5.17). This ratio is proportional to the increase in yield to gasoline, and this improvement is larger for the catalysts that show the highest activity.

All the improvements described (conversion and yields to valuable products) are made at the cost of the production of light (LCO) and heavy (HCO)

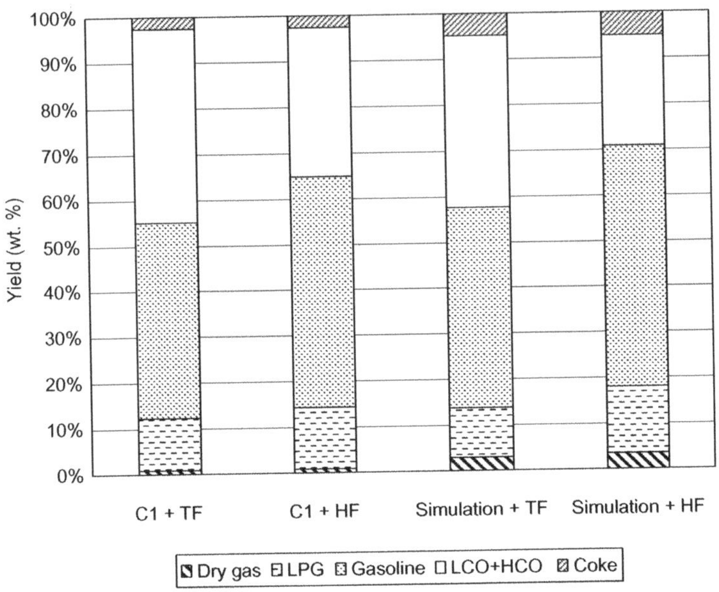

Figure 5.59. Comparison of cumulative yields between MAT at C/O = 4 and simulation at 520°C when using C1 as a catalyst. (Adapted from Salazar-Sotelo et al., 2004.)

cyclic oils. Nevertheless, these products are not desirable because their incorporation in diesel or fuel oil is not as profitable as obtaining more and better gasoline. As the name indicates, these products are rich in polynuclear aromatics; however, if the feedstock is hydrotreated, most of these compounds are partially hydrogenated, which enhances yields to gasoline and LPG (Mariaca-Domínguez et al., 2003).

As expected, for catalyst C1, product distribution improves because of the decrease in cyclic oil yield (Figure 5.59). Simulation also predicts this response when HF is used. It is important to note that simulation of the industrial unit predicts some increase in coke yield referred to MAT results. This situation cannot be validated with laboratory experiments because of the difference in the aforementioned whole heat balance (León-Becerril and Maya-Yescas, 2007). For catalyst C2 the same trend in improvement in valuable products because of reduction of cyclic oils yield is observed (Figure 5.60). In this second case, again, simulation predicts some increase in coke yield, even if changes are minimal.

To sum up, we have compared two commercial catalysts, which exhibit different properties and production objectives, that were used to convert two FCC feedstocks, typical and hydrotreated. Hydrotreating of FCC feedstock decreases heterocompounds that contain sulfur and nitrogen. However, there are other more important effects, such as decreasing yield to dry gases and cyclic oils at constant conversion. The significant increase in the value of the

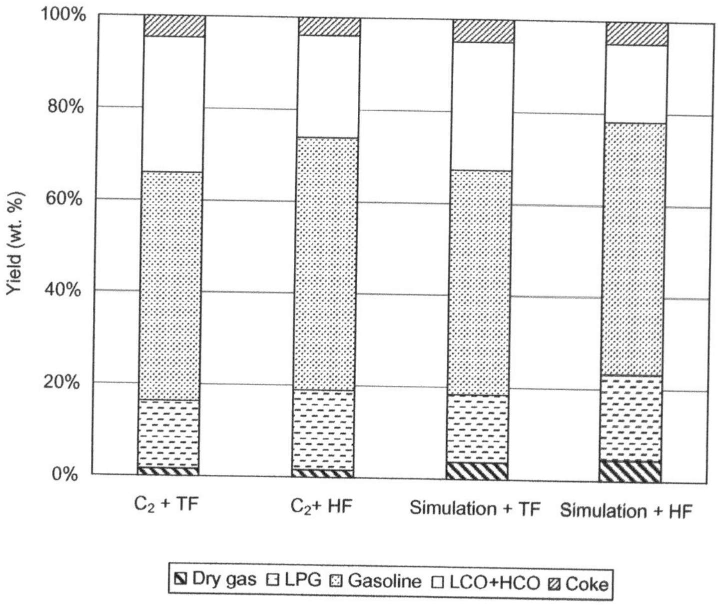

Figure 5.60. Comparison of cumulative yields between MAT at C/O = 4 and simulation at 520°C when using C2 as a catalyst. (Adapted from Salazar-Sotelo et al., 2004.)

product with increasing severity is due primarily to the increase in gasoline production, as a consequence of more efficient catalyst-to-oil interaction, which leads to higher conversion. Coke yield also decreases. Due to the different behavior of MAT and industrial units, simulation results show a slightly different response.

5.10.2 Pilot-Plant Emulation

One of the most desired and difficult issues in the characterization and selection of catalyst, feedstock, and operating conditions is the pilot-plant emulation of industrial units. Due to its high impact on the profitability of refineries, it is often not feasible and/or impossible to try new feedstock or catalysts during FCC industrial operation. On the other hand, laboratory microactivity plants such as MAT and ACE units give quick results that are not directly useful in modeling industrial process performance (Boock and Zhao, 1998; Maya-Yescas et al., 2004a). Due to the need to predict the performance of FCC industrial units accurately after feedstock or catalyst is changed, use of a pilot plant as an emulation device has been increasing. Among them, the recirculation catalyst pilot plant is a small-scale simulator that closely matches the behavior of industrial units. This type of pilot plant consists of a riser, a stripper, a regenerator, and a separation column that perform duties similar to those of industrial units. Additionally, this pilot plant is able to operate using a descendant flow reactor (downer) to emulate several technological trends; the operational mode promises better selectivity. The FCC process involves many variables interconnected in a complex way, making it difficult to predict its performance when operating far from the original design conditions. Nevertheless, using the same catalyst, the same feedstock, and the same main operating conditions of industrial units, it is possible to emulate the performance of industrial units in a pilot plant. Maya-Yescas et al. (2006) have shown the characteristic relationship between a pilot plant and industrial units, with examples of emulation, establishing its importance in the research and support of technical services. Scale-up problems are addressed and solutions that mimic operating data from an industrial plant are found. Conversion results are shown graphically to easily assess industrial potential benefits that can be drawn from pilot-plant emulation.

Pilot-plant scale equipment offers advantages and fewer risks than laboratory equipment to scale-up operating conditions and process performance, quickly generating large enough quantities of products for more complete and detailed analysis (e.g., Leuenberger et al., 1998). Pilot plants are considered, in general, as reduced versions of industrial units (Maya-Yescas et al., 2006); however, scale reduction has implications in the design and operation of pilot plants. Energy balance is the most important difference between industrial and pilot plants, resulting in differences in coke yields and catalyst-to-oil ratio (C/O) when the same conditions are chosen for both plants (Boock and Zhao, 1998).

Pilot-plant size allows confident study and simulation of industrial unit behavior with no risk to production. It is also possible to develop new catalyst formulations for each feedstock in long-term tests and to give recommendations for better operation, aiming to optimize conditions and improve profitability. Pilot-plant information is also essential in developing, adjusting, and validating FCC simulation models. As data sources for simulation, pilot plants provide opportunities for exploring broad ranges of process conditions, useful to process research and development (Dienert et al., 1993). Taking into account the aforementioned arguments and considering the high economic impact of FCC units in refineries, it is highly advisable to use pilot-scale information to provide permanent technical services, to optimize industrial operation and research and development efforts.

Pilot-Plant Description The typical arrangement of a pilot-plant scheme is shown in Figure 5.61. Feedstock is taken from several storage vessels connected to a control system. These storage vessels make possible normalizing operation with a reference feedstock and then the introduction of a new feedstock. A dosage pump with precision control is used to send the feed through the heater and nozzle. There is a system to control and register the flow. A nitrogen or vapor stream can be used as a dispersant, feeding it through an independent heater. Feed vaporizes as soon as it gets in touch with catalyst coming from the regenerator and goes through the riser, as in the industrial

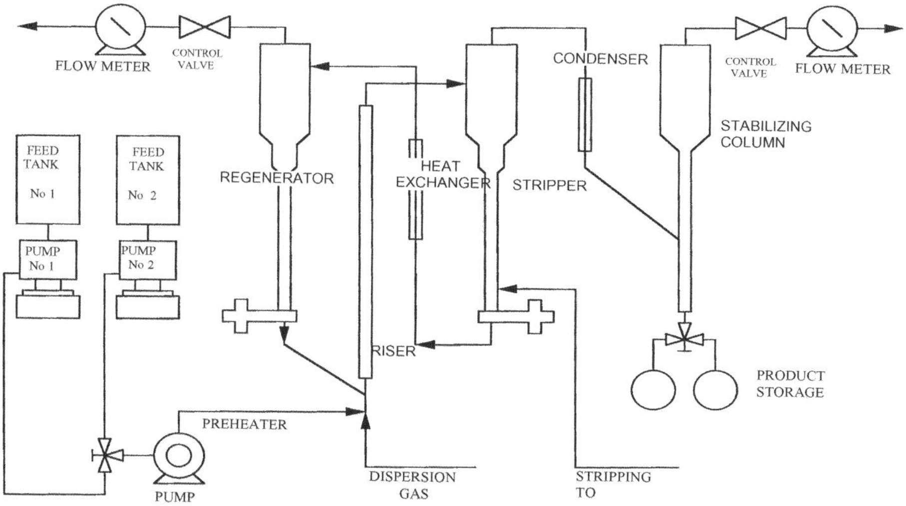

Figure 5.61. FCC pilot plant basic equipment. (From Maya-Yescas et al., 2006.)

unit. At the top of the riser, catalyst and products are separated using cyclones. Spent catalyst falls to a vertical column within a dense-phase fluidized bed, in order to strip hydrocarbons from catalyst with vapor or nitrogen flowing countercurrent. A slide valve controls the flow of stripped catalyst to the regenerator. Stripping temperature, bed height, and vapor/nitrogen flows are controlled to adjust stripper efficiency. Gaseous products go to a stabilizer column to separate components heavier than C_5 and C_6; this liquid product is fractionated to obtain gasoline, LCO, and HCO. Gas products at NTP conditions are analyzed using an online chromatograph. The spent catalyst is sent to the regenerator with a nitrogen or vapor line transfer, part of which is a double-tube heat exchanger using air. Exchanger heat balance provides a reliable way to obtain different C/O ratios. During regeneration reactions, coke deposited on a catalyst surface burns inside a fluidized bed using air and generating flue gas. At the exit of this flue gas, a control valve is used to maintain regenerator and stripper pressure. Regenerated catalyst goes through a slide valve to a return line with independent heating that fixes the inlet temperature to the riser.

The second most important reactor, the regenerator vessel, can be heated to different temperatures. Flue gas and excess air are measured and analyzed continuously (O_2, CO, CO_2, SO_x, and NO_x). The control system, based on the flue gas composition, is used to adjust the air quantity and keep the regeneration level required. Additionally, this pilot plant can be operated using a descendant flow reactor (a downer) in order to study other technological trends. It has been presumed that this operation mode promises greater selectivities (Ikeda and Ino, 1999). Heat balance can be adjusted following two different operating modes: the adiabatic mode and the heat balance mode.

1. *Adiabatic mode.* In this operating mode, riser outlet temperature is fixed and controlled with hot catalyst flow coming from the regenerator. This mode allows industrial emulation of conversion and product yields by using the same temperatures for the outlet riser, preheating, and regeneration. It is only necessary to adjust the C/O ratio, fine-tuning with nitrogen or vapor flows, and using the desired temperature profile inside the riser.
2. *Heat balance mode.* Typical FCC pilot-plant operation cannot follow industrial operation in the control of an important parameter—the regenerator temperature. In industrial units, the regenerator temperature is determined from the thermal balance existing in steady-state operation and is not controlled. The minor size in a pilot plant means that heat losses are comparatively more important than in the industrial case. Consequently, it is necessary to add heat to a pilot-plant regenerator to obtain industrial temperatures. In the heat balance mode it is possible to fix the regenerator temperature, taking the coke yield as a basis. A correlation is used to calculate this parameter and use it in the control system to adjust the temperature of the catalyst at the riser inlet, forcing the pilot plant to react to the industrial unit in a similar way. Using this mode, calculations are made automatically by the control system of the pilot plant.

Methodology Typical feedstock and equilibrium catalyst samples were taken from an industrial FCC unit (Maya-Yescas et al., 2006); standard characterization variables were measured for feedstock (Table 5.18) and catalyst (Table 5.19), and some special tests were performed. To maintain the catalyst properties, the equilibrium catalyst was changed before each of the 50+ experiments carried out. The outlet riser temperature was maintained at the industrial

TABLE 5.18. Typical Feedstock Properties

Density (kg/m^3)	920.0
Conradson carbon (% p)	0.20
ASTM D-1160 distillation (°C)	
10 vol%	346
50 vol%	422
90 vol%	501
Metal content	
Nickel (ppm)	380
Vanadium (ppm)	440
Iron (ppm)	750
Sodium (ppm)	2.5

TABLE 5.19. Typical Equilibrium Catalyst Properties

Activity MAT (% p)	70.0
Specific area (m^2/g)	158
Pore specific area (cm^3/g)	0.172
Coke on regenerated catalyst (g_{coke}/g_{cat})	0.066

TABLE 5.20. Base Parameters in FCC Pilot-Plant Operation

Variable	Range	Independent	Dependent
Outlet riser temperature (°C)	525	×	
Preheating temperature (°C)	100–350	×	
C/O ratio	6.8–25.5		×
Conversion (wt%)	61–77		×
Yield (wt%)	—		×

Source: Adapted from Maya-Yescas et al. (2006).

operation value. Feed preheating temperature values were chosen from a wide range (see Table 5.20), and operations were performed in the adiabatic mode. For each operating point, standard conversion was calculated by analyzing gas products (gas chromatography) and liquid products (simulated distillation); coke was estimated from the heat balance. Regenerator temperature was kept high in all experiments, to assure good regeneration and to keeping residual coke in the catalyst (ω_{CRC}) below 0.05 wt%.

Industrial Plant Emulation Since standard conversion is the main dependent variable in FCC units, it was considered a response variable. Emulation experiments were designed to investigate the conversion trend around the industrial value (73 wt%). It was decided to move the preheating temperature from 100°C to 350°C, maintaining a constant outlet riser temperature (525°C) and keeping the regenerator temperature above 690°C to see the way the pilot plant behaves. The results are shown in Figure 5.62. Results show the expected tendency, as the FCC standard conversion is proportional to the C/O ratio, assuming asymptotic behavior. The upper limit is a function of the feed and catalyst characteristics within the operation conditions range. The data dispersion shows the normal experimental error uncertainty. There is a general similar trend for each series, despite the fact that other variables could have additional effects. Among the results obtained, experimental points that showed conversion similar to the industrial value were selected. Gasoline, LPG, and dry gas yields were compared to industrial yields as shown in Figure 5.63. It is evident that just small deviations were found, the biggest being in dry gas yields, making the emulation results acceptable.

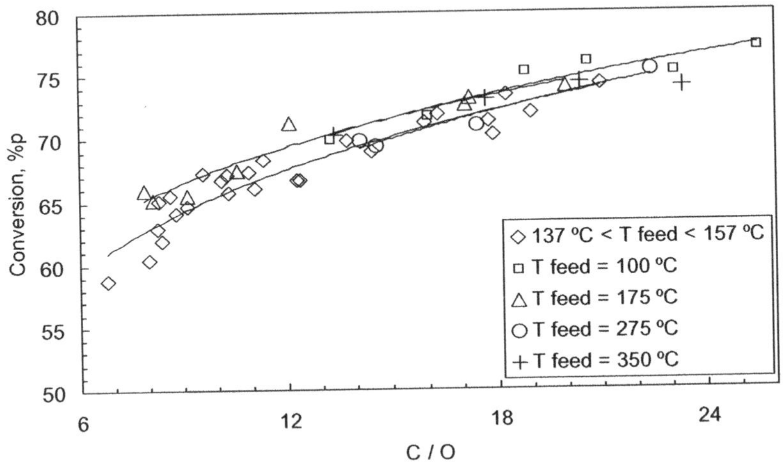

Figure 5.62. Operating region for 525°C outlet riser temperature. (From Maya-Yescas et al., 2006.)

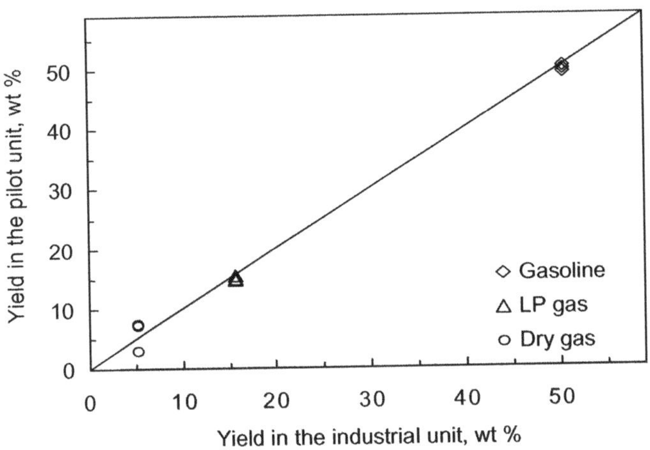

Figure 5.63. Pilot-plant results versus industrial data. (From Maya-Yescas et al., 2006.)

Scaling problems seem to explain the differences in yields. The emulation of industrial operation can be fine-tuned by moving some other variables around. For specific emulations, it is possible to adjust the temperature profile along the riser, moving regenerator temperatures and/or make runs with different dispersion flows in such a way to minimize deviations.

Yield and gas composition were obtained and compared with industrial data; a sample is shown in Table 5.21. The composition is not quite similar; these data need to be studied under the frame kinetic models in order to use

TABLE 5.21. Gas Composition Comparison (wt%)

Product		Pilot Plant	Industrial Unit
Dry gas	Total	2.70	3.73
	Hydrogen	0.11	0.09
	Methane	0.98	1.34
	Ethylene	0.83	1.01
	Ethane	0.77	1.30
LPG	Total	15.78	14.09
	Propane	1.65	1.31
	Propylene	6.38	3.45
	i-Butane	3.51	3.18
	n-Butane	0.79	1.26
	Propadiene	0.03	N.A.
	1-Butene	0.85	1.19
	i-Butylene	0.89	1.36
	t-Butene	0.98	1.29
	c-Butene	0.68	1.04

them for complete emulation. As mentioned earlier, pilot-plant parameters can be tested beyond standard industrial conditions. In this particular case it was possible to obtain values for the C/O ratio in a broad range, with values as high as 25. These values are not exhibited by industrial units, but they could be useful to evaluate behavior during heavy feeds cracking or to explore the maximum conversion attainable for a catalyst–feedstock system.

To sum up, during emulation of the behavior of an industrial FCC unit explored using a pilot plant, standard conversion was the target for different C/O ratios and different feed temperatures at constant temperature at the riser outlet. In this region, unit operation could be oriented to maximize the yield of either gasoline or olefins. The operating region emulated is specific for the catalyst and feedstock used, as well as operating conditions. It is interesting to note that it is possible to obtain parameters from common operation regions by manipulating variables. In this case, C/O ratios greater than 25 were studied, which are not seen in commercial FCC units but are useful when evaluating different combinations of feedstock and catalyst. The yields and composition of the primary products, gasoline and LPG, are well emulated, even if there is still room for fine-tuning by manipulating other variables. Finally, it should be noted that despite the similarity between pilot and commercial units, there are still some scale-up problems to be solved by mathematical modeling and simulation based on fundamental principles of chemical reactor engineering.

5.10.3 The Sulfur Balance

One aspect that has become very important is the heterocompound presence in fuels. Heterocompounds are molecules that contain sulfur, nitrogen, oxygen, or metallic porfirins. The combustion of these heterocompounds produces pollutants that contain sulfur oxides (SO_x, $x = 2, 3$), nitrogen oxides (NO_x, $x = \frac{1}{2}, 1, 2, 3$), and so on. Currently, the most regulated of them are the SO_x; therefore, it is necessary to consider their generation during cracking reactions. As described, at the refinery the FCC process produces about 40% of the total gasoline; however, it contributes more than the 90% of the sulfur content of commercial gasolines. The sulfur content of FCC products depends on catalyst, feedstock, and conversion as well as the reactors operating conditions. The FCC feedstock contains sulfur linked to organic compounds of high molecular weight; these heterocompounds are concentrated at the heavy end. Cracking these molecules produces either sour gas (hydrogen sulfide, which is desirable) or sulfided fuels (undesirable). Sour gas can be recovered and treated downstream in order to produce solid sulfur or sulfuric acid. In contrast, sulfur contained by fuels will produce SO_x emissions at internal combustion engines.

In the case of sulfur in coke, stack gas emissions from an FCC regenerator, including SO_x, NO_x, and catalyst particulates, constitute a major environmental pollution concern. Some other strategies can be implemented after the stack of the regenerator, such as gas desulfurisation and scrubbing; however, these solutions are noneconomic. For fuels produced by FCC units, strategies to reduce sulfur in FCC gasoline include naphtha hydrofinishing and lowering the gasoline endpoint. Hydrofinishing significantly lowers the octane of FCC gasoline, which depends on the presence of unsaturated compounds; meanwhile, lowering the gasoline endpoint can significantly diminish yield to gasoline. Therefore, it is very important to separate sulfur in catalytic cracking processes as sour gas at the riser outlet.

Kinetic schemes and mathematical models presented previously do not consider sour gas generation and sulfur distribution into cracking products; however, there have been some attempts at explicit prediction of sulfur content and distribution into catalytic cracking products. For example, Villafuerte-Macías et al. (2003, 2004) proposed a seven-lump kinetic scheme that considers the individual formation of H_2S. The seven lumps are selected as follows: for liquid products, according their boiling points: feedstock (343 to 560°C), cyclic oils (HCO + LCO, 223 to 342°C), and gasoline (C_5+, 37 to 222°C); and for light products, according to environmental and trade requirements: LPG (C_3–C_4), dry gas (C_1–C_2, H_2), sour gas (H_2S), and finally, solid coke (C). Each product is able to be cracked into lighter products. The feedstock cracking reaction is of second order, and all other cracking reactions are of first order, as has generally been assumed. Rate constants for cracking reactions follow the Arrhenius dependence on temperature. Initial numerical values of kinetic parameters were selected from literature data and were adjusted as well as validated, utilizing a number of sets of industrial refinery product results. Figure 5.64 illustrates the kinetic scheme proposed, and Table 5.22 provides the kinetic parameters used.

Authors also developed empirical functions related to feedstock conversion as well as reactor temperature, to represent not only the sulfur content and

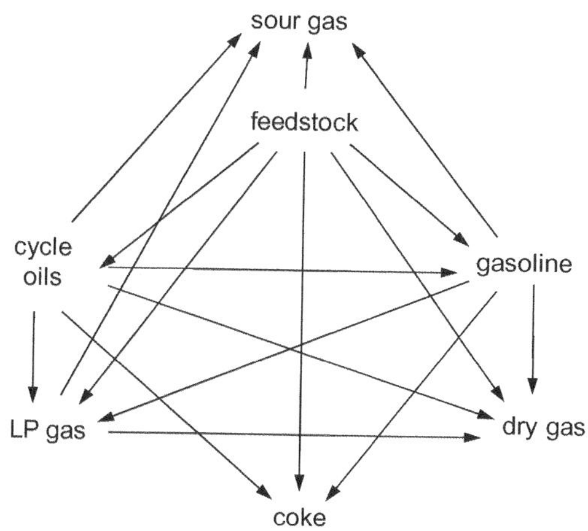

Figure 5.64. Seven-lump kinetic scheme. (Adapted from Villafuerte-Macías et al., 2004.)

TABLE 5.22. Kinetic Parameters Used in the Model

Cracking Reaction	$k_0{}^a$	E (kJ/mol)
Feedstock → cycle oils	240.0	70.0
Feedstock → gasoline	380.0	70.0
Feedstock → LPG	70.5	70.0
Feedstock → dry gas	217.5	80.0
Feedstock → sour gas	2400.0	70.0
Feedstock → coke	0.40	50.0
Cycle oils → gasoline	24.0	60.0
Cycle oils → LPG	30.0	60.0
Cycle oils → dry gas	217.5	60.0
Cycle oils → sour gas	600.0	70.0
Cycle oils → coke	0.60	50.0
Gasoline → LPG	1.0	50.0
Gasoline → dry gas	145.0	70.0
Gasoline → sour gas	300.0	70.0
Gasoline → coke	0.50	50.0
LPG → dry gas	261.0	40.0
LPG → coke	0.40	40.0
Dry gas → coke	1.30	40.0

Source: Villafuerte-Macías et al. (2004).

[a]For feedstock cracking the k_0 units are m^3/kmol·s; for other lump cracking the k_0 units are s^{-1}.

TABLE 5.23. Characteristics of the FCC Unit

Type	Riser reactor/adiabatic regenerator
Operating mode	Complete combustion
Feedstock capacity (bbl/day)	25,000
Average coke production (tons/day)	160
Average airflow rate (m^3/h)	75,000

Source: Villafuerte-Macías et al. (2004).

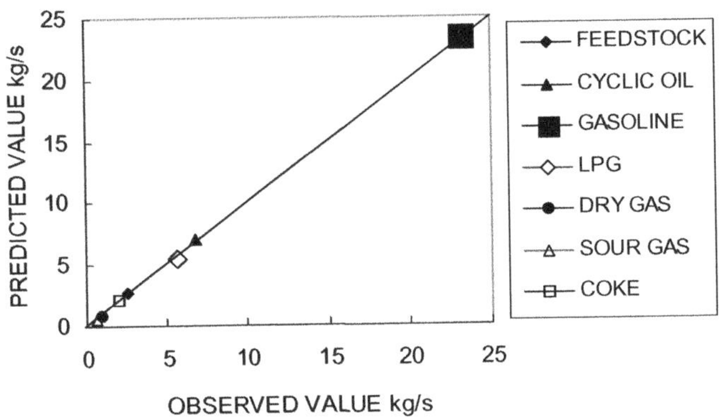

Figure 5.65. Predicted vs. observed values for product yields. (From Villafuerte-Macías et al., 2004.)

sulfur distribution in CO, gasoline, and coke, but also sulfur as sour gas independent of light lumps. Following industrial practice (MAT laboratory evaluation scaled to industrial data), the parameters of these functions were obtained for a particular feedstock and a "type of catalyst" and are supported with industrial (actual) data. These parameters should be fitted whenever different feedstock or "another type of catalyst" is used. The operating conditions of a riser–regenerator system do not modify parameter values.

A complete combustion regenerator will be analyzed. Its main characteristics are listed in Table 5.23. The vector of state variables consists of oxygen and sulfur concentrations, coke on regenerated catalyst, CO concentration, and dense bed temperature. Predicted yield values of cracking products and actual values are compared in Figure 5.65. It is important to observe that prediction points fall in the neighborhood of the 45° line; therefore, the values predicted are close enough to the values observed.

Product yield profiles and actual yield values have been modeled at the riser exit (Figure 5.66). It is possible to observe that the main feedstock cracking occurs before the first third of the riser length. The cyclic oil yield reaches a maximum value before the first half of the riser length, following a predominant soft decreasing yield due to cracking. This last result is in agreement with

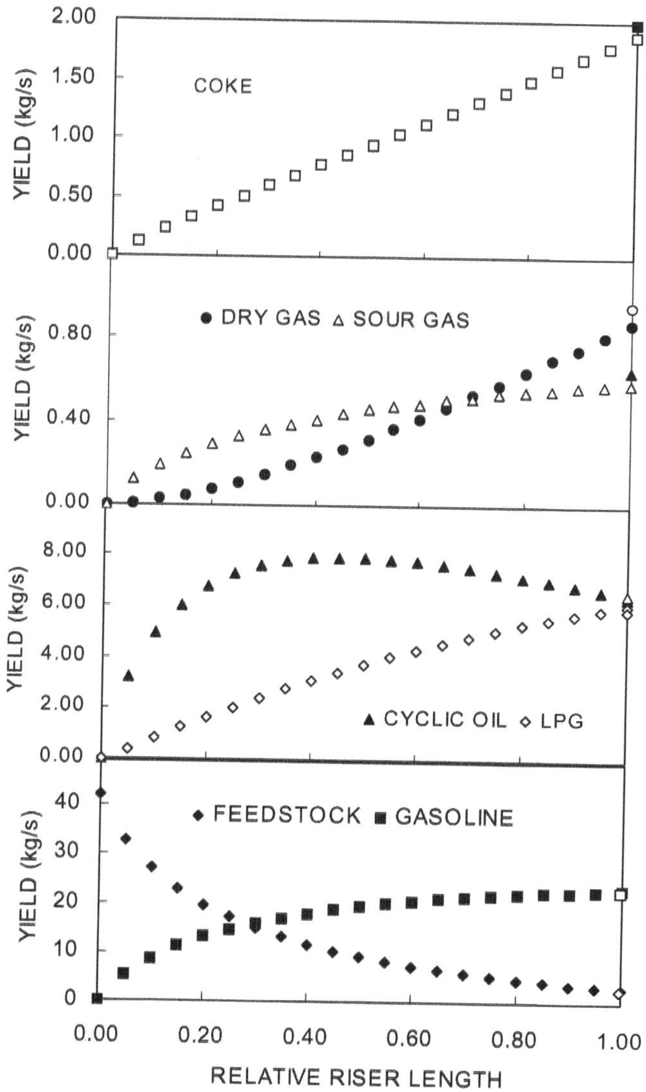

Figure 5.66. Axial profiles of feedstock and products in the riser. (From Villafuerte-Macías et al., 2004.)

those intermediate-weight mass products that might be converted to minor molecular weight compounds. Most (about 90%) of the total gasoline final yield is obtained before the first half-length of riser. Under industrial conditions, it has been observed that once gasoline is produced, it is not easily cracked and also that LPG and dry gas yields are increased continually (Figure 5.67). The greatest sour gas yield is obtained before three-fourths of riser

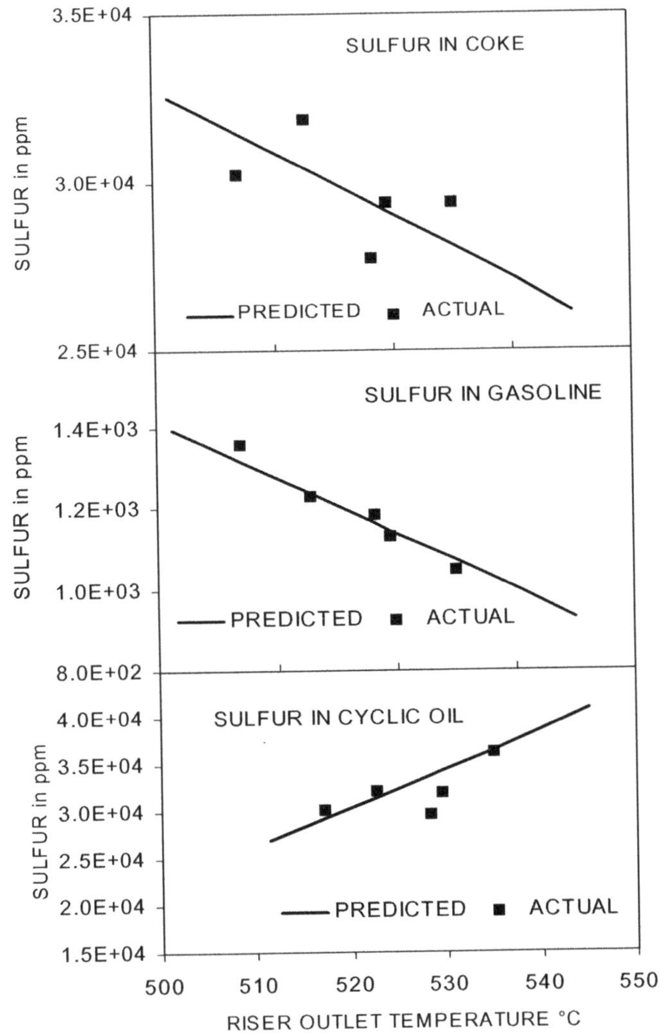

Figure 5.67. Sulfur content of FCC products. (From Villafuerte-Macías et al., 2004.)

length, indicating an initial easy link breaking sulfur hydrocarbons. Coke yield is increased as a result of the condensation of cyclic, heterocyclic, and alkyl compounds on catalyst particles.

The predicted sulfur content of cyclic oil, gasoline, and coke obtained as a function of ROT is shown in Figure 5.67. It is important to note that sulfur content in cyclic oils increases as ROT is increased; meanwhile, the sulfur content of gasoline, as well as of coke, decreases. Unstable sulfur-linked hydrocarbon compounds are cracked into sour gas and more light hydrocarbons; meanwhile, noncracked sulfur compounds go into cycle oils and only a little

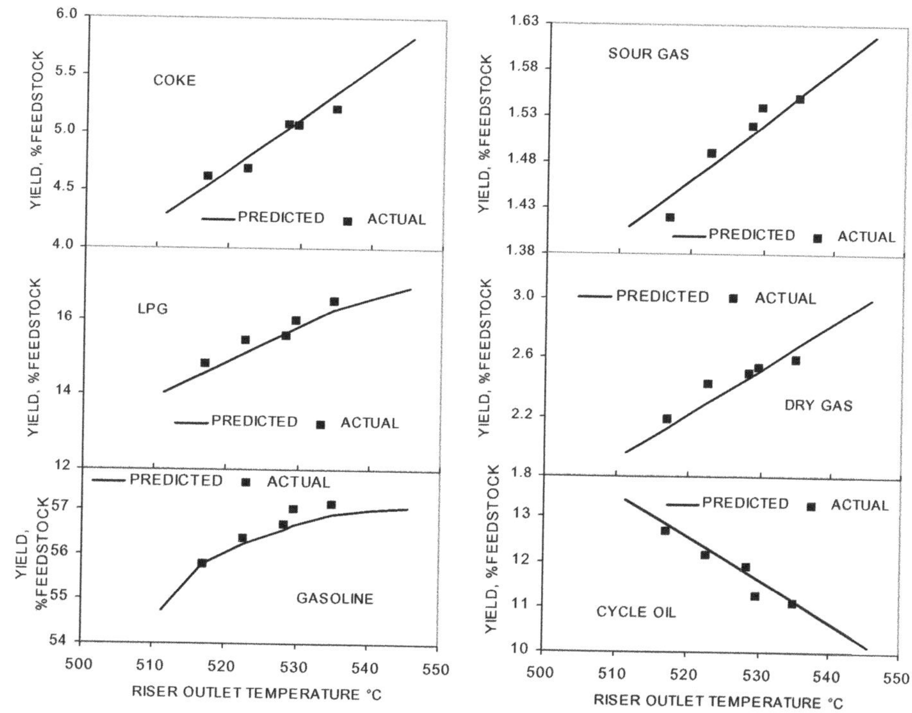

Figure 5.68. Product profiles as a function of ROT. (From Villafuerte-Macías et al., 2004.)

into gasoline. The sulfur content predicted for cyclic oils, gasoline, and coke is in agreement with actual data. It should be noted that to obtain gasoline with a lower sulfur content and a higher coke yield with a lower sulfur content, the unit must be operated at higher ROTs.

Profiles predicted for cyclic oil, gasoline, LPG, dry gas, sour gas, and coke yields obtained when simulating the operation between 510 and 550°C of the ROT are shown in Figure 5.68. Actual data are also included. The values predicted are depicted by a line crossing a neighborhood of actual data. It is observed that the cyclic oil yield predicted decreases as the ROT is increased, whereas the gasoline, LPG, sour gas, dry gas, and coke yields predicted increase. The decrement in cyclic oil yield is a result of cracking to LPG, dry gas, and some gasoline. At the highest temperature there is only a little or no increase predicted for gasoline yield. The industrial practice suggests gasoline cracking at an ROT higher than 550°C. It is important to note that an increase in coke yield is a possible advantage because of the relationship between necessary energy in the regenerator and the heat balance of the riser–regenerator–stripper system. The high sour gas yield predicted induces a lower sulfhur content in gasoline and coke, basically an advantage.

There is a type of synergy between an increase in gasoline yield and a decrease in the sulfur content in this fuel as the ROT increases. Therefore, to preserve the profitability of the operation, FCC units should be operated at the highest possible ROT. At the same time, there is an increase in sour gas yield, which is also an advantage from an environmental point of view. Both enhancements are made at the cost of a higher sulfur content in cyclic oils. This situation has to be balanced because of the cost of desulfurization downstream. However, the yield of cyclic oils is also decreased, which could also be an advantage.

By using a seven-lump kinetic scheme, which clearly specifies H_2S generation, it is possible to account for contributions of H_2S formation, sulfur content in cracking final products, and sulfur distribution in cracking products. In addition, but in common with other models, it helps to predict cracking product distribution. The results predicted are referred to the feedstock volume conversion and the riser outlet temperature range in which FCC units are commonly operated. This information helps to manage the sulfur content of fuels during fuel production. This model is, furthermore, a helpful tool for modeling steady-state FCC operation, taking into account valuable cleaner fuel production and satisfactory environmental control.

5.11 CONCLUSIONS

Fluidized-bed catalytic cracking (FCC) is one of the main processes in petroleum refining. The heart of this process is the converter, which consists of a riser (transported-bed reactor, where the principal reactions take place), a stripper (a fluidized-bed reactor used to desorb gaseous hydrocarbons from the catalyst surface), and a regenerator (a fluidized-bed reactor used to burn off coke produced, recovering catalyst activity and energy to sustain the converter). This converter is a very complex system, due to the variety of compounds used as feedstock and the highly interacting nature of the system as a consequence of the energy balance.

During the analysis and design of the riser, it is necessary to evaluate (properly) the cracking reaction kinetics; nevertheless, this kinetics involves too many compounds and requires intensive feedstock analysis in order to be characterized in an acceptable way. On the other hand, due to the catalyst and reacting fluid process, the entire cracking kinetic process is rated by mass transfer, either at the fluid–particle interface and/or as intraparticle diffusion. As a consequence, evaluation of effective (apparent) kinetic parameters in laboratory devices (such as MAT, ACE, CREC–riser–simulator, and pilot plants) are linearly scalable to industrial riser parameters. This result has been used, empirically, for many years to evaluate the performance of catalyst–feedstock couples at the laboratory scale, prior to use at industrial FCC units.

The regenerator of the FCC manages the energy balance. This reactor is a very complex system because it is able to exhibit a variety of dynamic responses to the many disturbances that commonly occur. There are two principal types of regenerator modes: full combustion (no CO in the flue gases) and partial combustion (CO in the flue gases, followed by a CO combustor). Regenerator study is not finished; therefore, there are several recent papers regarding the dynamics and control of FCC regenerators.

Finally, FCC is able to utilize many different types of feedstock; additionally, it is possible to alter the conversion to products by managing the riser outlet temperature. Hence, these units could help to transform solid pollutants (such as plastics) into gasoline. Meanwhile, the use of gas oils containing heterocompounds that contain sulfur or nitrogen causes the oxides of those heteroatoms to be emitted. This problem is under research in order to improve process operation, to pretreat feedstocks, and to find options that are more environmentally friendly.

Some Perspective on Present and Future Opportunities Everything (i.e. scientific reasearch) can be studied taking as a target the FCC unit. The question is: Why is it that way?

FCC units were developed during World War II; gasoline for airplanes was critical and there was no unit to produce it. Therefore, in a very intrepid and rapid development, some inspired researchers designed the first TCC unit, the precursor of the current FCC. This new unit was very complex, exhibiting very innovative features, including moving-bed reactors and the chance to spend and regenerate the catalyst inside the converter during current operation. However, even though the unit eventually worked successfully, the development brought with it a lot of empirism. Even when the economists running refineries want "explicitly sure results," this is one of the best examples of engineering design.

As market demands had changed, feedstock had to adapt to many other streams inside the refinery, there is much new environmental legislation, and so on. However, the FCC unit is still the heart of the refinery and continues to evolve. Furthermore, whatever the future outlook for petroleum, FCC units will continue working in the production of raw materials for specialty plastics. Therefore, it is necessary to continue to develop better kinetic models. Lumping models have sufficed to date, but they lack many beneficial feedstock and product properties. Use of single-event models could be a good future strategy, as they require detailed analyses of feedstock, products, and byproducts such as coke (the second-most-important FCC product). Catalyst development is one targets of focus around the world, but descriptions of the loss of activity (*deactivation*) are still very empirical. The most advanced part of this study explains the deactivation phenomenon in terms of the specific rate of formation of coke. This is not enough to characterize the phenomenon; there are no explanations regarding coke deposition on a catalyst surface causing a decrease in the pore-mouth diameter (e.g., Jiménez-García et al., 2009). If this phenomenon were not important, we would not see very dark spent catalyst.

Finally, with regard to dynamics modeling and control, do we understand the FCC unit? Are we able to build the next generation of FCC units? Well—work for the future!

REFERENCES

Aguilar, R., Maya-Yescas, R. (2004) Feedback regulation of temperature in FCC regenerator reactors. *Pet. Sci. Technol.* 22:201–218.

Aguilar, R., Poznyak, A., Martínez-Guerra, R., Maya-Yescas R. (2002) Temperature control in catalytic cracking reactors via a robust PID controller. *J Process Control.* 12:695–705.

Aguilar-López, R., Álvarez-Ramírez, J. J. (2002) Sliding-mode control scheme for a class of continuous chemical reactors. *IEE Proc. Control Theory Appl.* 149: 263–268.

Aguilar-López, R., Maya-Yescas, R. (2006) Robust temperature stabilization for fluid catalytic cracking units using extended Kalman-type estimators. *Chem. Prod. Process Model.* 1, Art. 3:1–20.

Álvarez-Ramírez, J., Valencia, J., Puebla, H. (2004) Multivariable control configurations for composition regulation in a fluid catalytic cracking unit. *Chem. Eng. J.* 99: 187–201.

Ancheyta-Juárez, J., López-Isunza, F., Aguilar-Rodríguez, E., Moreno-Mayorga, J. C. (1997) A strategy for kinetic parameters estimation in the fluid catalytic cracking process. *Ind. Eng. Chem. Res.* 36:5170–5174.

Araujo-Monroy, C., López-Isunza, F. (2006) Modeling and simulation of an industrial fluid catalytic cracking riser reactor using a lump-kinetic model for a distinct feedstock. *Ind. Eng. Chem. Res.* 45:120–128.

Arbel, A., Huang, Z., Rinard, I. H., Shinnar, R., Sapre, A. V. (1995) Dynamic and control of fluidized catalytic crackers: 1. Modeling the current generation of FCCs. *Ind. Eng. Chem. Res.* 34:1228–1243.

Arbel, A., Rinnard, I. H., Shinnar, R. (1996) Dynamics and control of fluidized catalytic crackers: 3. Designing the control system: choice of manipulated and measured variables for partial control. *Ind. Eng. Chem. Res.* 35:2215–2233.

ASTM (2005) *Standard Test Method for Determining Activity and Selectivity of Fluid Catalytic Cracking (FCC) Catalysts by Microactivity Test.* ASTM D-5154-05. Philadelphia.

Barmish, B. R., Corless, M., Leitman, G. A. (1983) A new class of stabilizing controllers for uncertain dynamical systems. *SIAM J. Control Optim.* 21:246–255.

Boock, L.T., Zhao, X. (1998) Recent advances in FCC catalyst evaluations: MAT vs. DCR pilot plant results. *Chem. Ind.* 74:131–141.

Bristol, E. H. (1966) On a new mesure of interaction for multivariable process control. *IEEE Trans. Autom. Control*, Jan., pp. 133–134.

Cheng, W. C., Kim, G., Peters, A.W., Xhao, X., Rajagopalan, K., Ziebarth, M. S., Pereira, C. J. (1998) Environmental fluid catalytic cracking technology. *Catal. Rev. Sci. Eng.* 40:39–79.

Corella, J. (2004) On the modelling of the kinetics of the selective deactivation of catalysts: application to the fluidized catalytic cracking process. *Ind. Eng. Chem. Res.* 43:4080–4086.

Corma, A., Martínez, C., Ketley, G., Blair, G. (2001) On the mechanism of sulphur removal during catalyst cracking. *Appl. Catal. A.* 208:135–152.

Dahl, I. M., Tangstad, E., Mostad, H. B., Andersen, K. (1996) Effect of hydrotreating on catalytic cracking of a VGO. *Energy Fuels.* 10:85–90.

Daoutidis, P., Kravaris, C. (1991) Inversion and zero dynamics in nonlinear multivariable control. *AIChE J.* 37:527–538.

Deza, F., Busvelle, E., Gauthier, J. P., Rakotopara, D. (1992) High gain estimation for nonlinear systems. *Sys. Control Lett.* 18:295–299.

Dienert, J. M., Renfree, M., Holden, I. A. (1993) On-line simulation of commercial adiabatic FCCU operation with a FCC pilot plant. *AIChE Symp. Ser.* 88:54–58.

Economou, C. E., Morari, M. (1986a) Internal model control: 5. Extension to nonlinear systems. *Ind. Eng. Chem. Process Des. Dev.* 25:403–411.

Economou, C. E., Morari, M. (1986b) Internal model control: 6. Multiloop design. *Ind. Eng. Chem. Process Des. Dev.* 25:411–419.

Edwards, W. M., Kim, H. N. (1988) Multiple steady states in FCC unit operations. *Chem. Eng. Sci.* 43:1825–1830.

Errazu, A. F., de Lasa, H. I., Sarti, F. (1979) A fluidized bed catalytic cracking regenerator model: grid effects. *Can. J. Chem. Eng.* 37:191–197.

Freeman, R. A., Kokotovic, P. V. (1996) *Robust Nonlinear Control Design: State Space and Lyapunov Techniques*. Birkhauser, Berlin, Chap. 8.

Froment, G. F., Bischoff, K. B. (1962) Kinetic data and product distributions from fixed bed catalytic reactors subject to catalyst fouling. *Chem. Eng. Sci.* 17:105–114.

Froment, G. F., Bischoff, K. B. (1990) *Chemical Reactor Analysis and Design*, 2nd ed. Wiley, Singapore.

García, C. E., Morari, M. (1985) Internal model control: 2. Procedure for multivariable systems. *Ind. Eng. Chem. Process Des. Dev.* 24:472–484.

Ginsburg, J., Pekediz, A., de Lasa, H. I. (2003) The CREC fluidized riser simulator: characterization of mixing patterns. *Int. J. Chem. Reaction Eng.* 1, A52:1–14.

Grosdidier, P., Mason, A., Aitolahti, A., Vanhumaki, V. (1993) FCC unit regenerator-reactor control. *Comput. Chem. Eng.* 17:165–179.

Hovd, M., Skogestad, S. (1993) Procedure for regulatory control structure selection with application to the FCC process. *AIChE J.* 39:1938–1953.

Ikeda, S., Ino, T. (1999) Down-flow FCC pilot plant for light olefin production. In: *Proceedings of the Acs Symposium on Advances in Fluid Catalytic Cracking*, New Orleans, LA, Aug. 22–26.

Isidori, A.(1999) *Nonlinear Control Systems*. 3rd ed. Springer-Verlag, London.

Jacob, D., Gross, B., Voltz, S., Weekman, V. W. (1976) A lumping and reaction scheme for catalytic cracking. *AICHE J.* 22:701–713.

Jiménez-García, G., Aguilar-López, R., León-Becerril, E., Maya-Yescas, R. (2007) Scaling-up of instantaneous data of complex kinetics. *Fuel.* 86:1278–1281.

Jiménez-García, G., Aguilar-López, R., León-Becerril, E., Maya-Yescas, R. (2009) Tracking catalyst activity during fluidized-bed catalytic cracking. *Ind. Eng. Chem. Res.* 48:1220–1227.

Jiménez-García, G., Quintana-Solórzano, R., Aguilar-López, R., Maya-Yescas, R. (2010) Modelling catalyst deactivation by external coke deposition during fluid catalytic cracking. *Int. J. Chem. Reactor Eng.* 8:A2.

Kelkar, C. P., Xu, M., Madon, R. J. (2003) Laboratory evaluation of cracking catalysts in a fluid bed: effects of bed dynamics and catalyst deactivation. *Ind. Eng. Chem. Res.* 42:426–433.

Krishna, A. S., Parkin, E. S. (1985) Modeling the regenerator in commercial FCC units. *Chem. Eng. Prog.* 81:57–62.

Kuo, J., Wei, J. (1969) A lumping analysis in monomolecular reaction systems: analysis of approximately lumpeable systems. *Ind. Eng. Chem. Fundam.* 8: 124–133.

Kurihara, H. (1967) Optimal control of fluid catalytic cracking process. Ph.D. dissertation, M.I.T., Cambridge, MA.

Lavanya, M., Sriram, V., Sairam, B., Bhaskar, M., Meenakshisundaram, A. (2002) Studies on fluid catalytic cracking of hydroprocessed feedstocks. *Pet. Sci. Technol.* 20:713.

Lee, W., Kugelman, A. M. (1973) Number of steady-state operating points and local stability of open-loop fluid catalytic cracker. *Ind. Eng. Chem. Process Des. Dev.* 12:197–204.

León-Becerril, E., Maya-Yescas, R. (2007) Open loop response to changes of coke-precursors during fluidised-bed catalytic cracking. *Fuel.* 86:1282–1289.

León-Becerril, E., Maya-Yescas, R., Salazar-Sotelo, D. (2004) Effect of modelling pressure gradient in the simulation of industrial FCC risers. *Chem. Eng. J.* 100: 181–186.

Leuenberger, E. L., Moorehead, E. L., Newell D. F. (1998) Effects of feedstock type on FCC overcracking. Paper AM-88-51. In: *Proceedings of the 1998 NPRA Annual Meeting*, San Antonio, TX, Mar. 20–22.

Mariaca-Domínguez, M. E., Rodríguez-Salomón, S., Maya-Yescas, R., González-Ortíz, A. (2002) Effect of FCC feedstock quality on product distribution. In: *Proceedings of the 15th International Congress of Chemical Process Engineering*, Prague, Czech Republic, pp. 25–29.

Mariaca-Dominguez, E., Rodríguez-Salomón, S., Maya-Yescas, R. (2003) Reactive hydrogen content: a tool to predict FCC yields. *Int. J. Chem. Reactor Eng.* 1:A46.

Mariaca-Domínguez, E., Maya-Yescas, R., Rodríguez-Salomón, S., González-Ortíz, A., Martínez-Tapia, G. E., Álvarez-Ramirez, R., López-Franco, C. (2004) Reactivity of fluid catalytic cracking feedstocks as a function of reactive hydrogen content. *Pet. Sci. Technol.* 22:13–29.

Maya-Yescas, R., Aguilar, R. (2003) Controllability assessment approach for chemical reactors: nonlinear control affine systems. *Chem. Eng. J.* 92:69–79.

Maya-Yescas, R., Bogle, D., López-Isunza, F. (1998) Approach to the analysis of the dynamics of industrial FCC units. *J. Process Control* 8:89–100.

Maya-Yescas, R., Aguilar-López, R., González-Ortíz, A., Mariaca-Domínguez, E., Rodríguez-Salomón, S., Salazar-Sotelo, D. (2004a) Impact of production objectives on adiabatic FCC regenerator control. *Pet. Sci. Technol.* 22:31–43.

Maya-Yescas, R., León-Becerril, E., Salazar-Sotelo, D. (2004b) Translation of MAT kinetic data for modeling industrial catalytic cracking units. *Chem. Eng. Technol.* 27:777–780.

Maya-Yescas, R., Villafuerte-Macías, E. F., Aguilar, R., Salazar-Sotelo, D. (2005) Sulphur oxides emission during fluidised-bed catalytic cracking. *Chem. Eng. J.* 106:145–152.

Maya-Yescas, R., Salazar-Sotelo, D., Mariaca-Domínguez, E., Rodríguez-Salomón, S., García-Moreno, L. M. (2006) Fluidized-bed catalytic cracking units emulation in pilot plant. *Rev. Mex. Ing. Quími.* 5:97–103.

Morari, M., (1983) Design of resilient processing plants: III. A general framework for the assessment of dynamic resilience. *Chem. Eng. Sci.* 38:1881–1891.

Moustafa, T. M., Froment, G. F. (2003) Kinetic modelling of coke formation and deactivation in the catalytic cracking of vacuum gas oil. *Ind. Eng. Chem. Res.* 42:14–25.

Salazar-Sotelo, D., Maya-Yescas, R., Mariaca-Domínguez, E., Rodríguez-Salomón, S., Aguilera-López, M. (2004) Effect of hydrotreating FCC feedstock on product distribution. *Catal. Today.* 98:273–280.

Sugungun, M. M., Kolesnikov, I. M., Vinogradov, V. M., Kolesnikov, S. I. (1998) Kinetic modelling of FCC process. *Catal. Today.* 43:315–325.

Takatsuka, T., Sato, S., Morimoto, Y., Hashimoto, H. (1987) A reaction model for fluidized-bed catalytic cracking of residual oil. *Intern. Chem. Eng.* 27: 107–116.

Taskin, H., Kubat, C., Uygun, Ö., Arslankaya, S. (2006) FUZZY-FCC: fuzzy logic control of a fluid catalytic cracking unit to improve dynamic performance. *Comput. Chem. Eng.* 30:850–863.

Theologos, K. N., Markatos, N. C. (1993) Advanced modeling of fluid catalytic cracking riser-type reactors. *AIChE J.* 39:1007–1017.

Venuto, P. B., Habib, E. T. (1978) Catalyst–feedstock–engineering interactions in fluid catalytic cracking. *Catal. Rev. Sci. Eng.* 18:1–150.

Vieira, R. C., Pinto, J. C., Biscaia, E. C., Jr. Baptista, C. M. L. A., Cerqueira, H. S. (2004) Simulation of catalytic cracking in a fixed-fluidized-bed unit. *Ind. Eng. Chem. Res.* 43:6027–6034.

Villafuerte-Macías, E. F., Aguilar López, R., Maya Yescas, R. (2003) Cambio de los coeficientes de transferencia en reactores de desintegración catalítica. *Rev. Mex. Ing. Quim.* 2:63–68.

Villafuerte-Macías, E. F., Aguilar, R., Maya-Yescas, R. (2004) Towards modelling production of clean fuels: sour gas formation in catalytic cracking. *J. Chem. Technol. Biotechnol.* 79:1113–1118.

Voorhies, A. (1945) Carbon formation in catalytic cracking. *Ind. Eng. Chem.* 37: 318–322.

Weekman, V. W., Jr. Nace, D. M. (1970) Kinetics of catalytic cracking selectivity in fixed, moving and fluid bed reactors. *AICHE J.* 16:397–404.

Wei, J., Kuo, J. (1969) A lumping analysis in monomolecular reaction systems: analysis of exactly lumpeable systems. *Ind. Eng. Chem. Fundam.* 8:114–123.

Wolf, E. E., Alfani, F. (1982) Catalyst deactivation by coking. *Catal. Rev. Sci. Eng.* 24:329–371.

NOMENCLATURE

C_p	Heat capacity at constant pressure, kJ/kmol·K
$\mathbf{f}$	Vector of terms that are independent of the manipulated variable(s), consistent
F	Volumetric flow, m^3/s
$\mathbf{G}$	Matrix of terms independent of manipulated variable(s), consistent
H	Specific enthalpy, kJ/kmol
k_V	First-order kinetic rate factor, $1\,s^{-1}$
m	Mass flow, kg/s
P	Pressure, bar
Q	Heat flow, kW
$\mathbf{r}$	Reaction rate, $kmol/s{\cdot}kg_{cat}$, $kmol/s \cdot m^3_{gas}$
R_g	Universal constant of ideal gases, $bar{\cdot}m^3/kmol{\cdot}K$
T	Temperature, K
$\mathbf{u}$	Vector of manipulable variables, consistent
V	Lyapunov function, consistent
W	Catalyst mass holdup, kg
$\mathbf{x}$	Vector of states, consistent
y	Mole fraction, dimensionless

Greek Letters

ρ	Mass density, kg/m^3
ω	Mass fraction, kg_{coke}/kg_{cat}

Subscripts

C	Control variable
cat	Referred to catalyst
CRC	Coke on regenerated catalyst
CSC	Coke on spent catalyst
D	Uncontrolled or dynamic variable
dp	Referred to the regenerator dense phase
flue	Referred to regenerator stack gases
rgn	Referred to the whole regenerator

Superscripts

i	Inlet
sp	Set point
T	Transpose
–1	Matrix inverse
•	Time derivative

INDEX

Modeling and Simulation of Catalytic Reactors for Petroleum Refining, First Edition.
Jorge Ancheyta.
© 2011 John Wiley & Sons, Inc. Published 2011 by John Wiley & Sons, Inc.

Printed and bound by CPI Group (UK) Ltd, Croydon, CR0 4YY

16/09/2026

14963460-0005